Lecture Notes in Computer Science 10627

Commenced Publication in 1973
Founding and Former Series Editors:
Gerhard Goos, Juris Hartmanis, and Jan van Leeuwen

More information about this series at http://www.springer.com/series/7407

Xiaofeng Gao · Hongwei Du
Meng Han (Eds.)

Combinatorial Optimization and Applications

11th International Conference, COCOA 2017
Shanghai, China, December 16–18, 2017
Proceedings, Part I

Springer

Editors
Xiaofeng Gao (iD)
Shanghai Jiao Tong University
Shanghai
China

Meng Han (iD)
Kennesaw State University
Kennesaw, GA
USA

Hongwei Du (iD)
Harbin Institute of Technology
Shenzhen
China

ISSN 0302-9743   ISSN 1611-3349 (electronic)
Lecture Notes in Computer Science
ISBN 978-3-319-71149-2   ISBN 978-3-319-71150-8 (eBook)
https://doi.org/10.1007/978-3-319-71150-8

Library of Congress Control Number: 2017959595

LNCS Sublibrary: SL1 – Theoretical Computer Science and General Issues

Printed on acid-free paper

This Springer imprint is published by Springer Nature
The registered company is Springer International Publishing AG
The registered company address is: Gewerbestrasse 11, 6330 Cham, Switzerland

Preface

The 11th Annual International Conference on Combinatorial Optimization and Applications (COCOA 2017) was held during December 16–18, 2017, in Shanghai, P.R. China. COCOA 2017 provided a forum for researchers working in the area of theoretical computer science and combinatorics.

The technical program of the conference included 59 regular papers selected by the Program Committee from 145 full submissions received in response to the call for papers. Each submission was peer-reviewed by at least three, and on average 3.8, Program Committee members or external reviewers. The topics cover most aspects of theoretical computer science and combinatorics related to computing, including classic combinatorial optimization, geometric optimization, complexity and data structures, graph theory, etc. We also selected 19 short papers to demonstrate various applications in the related areas. Some of the papers were selected for publication in special issues of *Algorithmica, Theoretical Computer Science,* and *Journal of Combinatorial Optimization*. It is expected that the journal version of the papers will appear in a more complete form.

We thank everyone who made this meeting possible: the authors for submitting papers, the Program Committee members, and external reviewers for volunteering their time to review conference papers. Our sponsors include the Advanced Network Laboratory (ANL) from Shanghai Jiao Tong University, the GPS Laboratory from Nanjing University, the Research Institute for Interdisciplinary Sciences (RIIS) from Shanghai University of Finance and Economics, and the Cardinal Operations (shanshu. ai) company, China. We would also like to extend special thanks to the chairs and conference Organizing Committee for their work in making COCOA 2017 a successful event.

October 2017
Xiaofeng Gao
Meng Han
Zhipeng Cai
Hongwei Du

Preface

The 12th International Conference on Combinatorial Optimization and Applications (COCOA 2018) was held during December 16–18, 2018, in Shanghai, China. COCOA 2018 provided a forum for researchers working in the area of theoretical computer science and combinatorics.

Organization

General Chairs

Guihai Chen   Nanjing University, China
Minyi Guo   Shanghai Jiao Tong University, China

Vice General Chair

Zhipeng Cai   Georgia State University, USA

Program Co-chairs

Xiaofeng Gao   Shanghai Jiao Tong University, China
Hongwei Du   Harbin Institute of Technology, Shenzhen, China

Publicity Co-chairs

Dongdong Ge   Shanghai Jiao Tong University, China
Chenchen Wu   Tianjin University of Technology, China

Publication Chair

Meng Han   Kennesaw State University, USA

Financial Chair

Fay Zhong   California State University, USA

Local Organization Chair

Sherman Hung   Shanghai Jiao Tong University, China

Web Chair

Shilei Tian   Shanghai Jiao Tong University, China

Program Committee

Xiaohui Bei   Nanyang Technological University, Singapore
Wolfgang Bein   University of Nevada, Las Vegas, USA
Zhipeng Cai   Georgia State University, USA
Gruia Calinescu   Illinois Institute of Technology, USA

T.-H. Hubert Chan	The University of Hong Kong, SAR China
Kun-Mao Chao	National Taiwan University, Taiwan
Vincent Chau	City University of Hong Kong, SAR China
Jing Chen	Stony Brook University, USA
Xujin Chen	Institute of Applied Mathematics, Chinese Academy of Sciences, China
Rajesh Chitnis	Weizmann Institute, Israel
Ovidiu Daescu	University of Texas at Dallas, USA
Haipeng Dai	Nanjing University, China
Thang Dinh	Virginia Commonwealth University, USA
Hongwei Du	Harbin Institute of Technology Shenzhen Graduate School, China
Zhenhua Duan	Xidian University, China
Thomas Erlebach	University of Leicester, UK
Neng Fan	University of Arizona, USA
Bin Fu	University of Texas, Rio Grande Valley, USA
Stanley Fung	University of Leicester, UK
Xiaofeng Gao	Shanghai Jiao Tong University, China
Dongdong Ge	Shanghai University of Finance and Economics, China
Qianping Gu	Simon Fraser University, Canada
Meng Han	Kennesaw State University, USA
Pinar Heggernes	University of Bergen, Norway
Juraj Hromkovic	ETH Zurich, Switzerland
Sun-Yuan Hsieh	National Cheng Kung University, Taiwan
Jie Hu	Wuhan University, China
Hejiao Huang	Harbin Institute of Technology Shenzhen Graduate School, China
Kazuo Iwama	Kyoto University, Japan
Naoki Katoh	Kyoto University, Japan
Donghyun Kim	Kennesaw State University, USA
Minming Li	City University of Hong Kong, SAR China
Xianyue Li	Lanzhou University, China
Guohui Lin	University of Alberta, Canada
Xianmin Liu	Harbin Institute of Technology, China
Xiaowen Liu	Indiana University-Purdue University Indianapolis, USA
Bin Ma	University of Waterloo, Canada
Mitsunori Ogihara	University of Miami, USA
Sheung-Hung Poon	Brunei Technological University, Brunei
Erfang Shan	Shanghai University, China
Gerhard Woeginger	RWTH Aachen University, Germany
Chenchen Wu	Tianjin University of Technology, China
Xiaowei Wu	University of Hong Kong, SAR China
Boting Yang	University of Regina, Canada
Hsu-Chun Yen	National Taiwan University, Taiwan
Huacheng Yu	Harvard University, USA
Chihao Zhang	Shanghai Jiao Tong University, China

Zhao Zhang   Zhejiang Normal University, China
Jiaofei Zhong   California State University, East Bay, USA
Yuqing Zhu   California State University, Los Angeles, USA

Additional Reviewers

Aloupis, Greg
Andro-Vasko, James
Armaselu, Bogdan
Bein, Doina
Boeckenhauer, Hans-Joachim
Boyanapalli, Uday Bhaskar
Burjons Pujol, Elisabet
Cao, Zhigang
Chang, Nai-Wen
Chang, Yi-Jun
Chen, Chi-Yeh
Chen, Ho-Lin
Chen, Li-Hsuan
Chen, Yu-Fang
Chiu, Man Kwun
Dao, Minh-Son
Deineko, Vladimir
Dobrev, Stefan
Doerr, Carola
Fan, Chenglin
Frei, Fabian
Fukagawa, Daiji
Guo, Longkun
Han, Xin
He, Hongjin
He, Simai
Higashikawa, Yuya
Hung, Ling-Ju
Jakoby, Andreas
Jansson, Jesper
Jiang, Bo
Kim, Yeojin
Ko, Euiseong
Kobayashi, Yuki
Komm, Dennis
Larmore, Lawrence
Lee, Chia-Wei

Letsios, Dimitrios
Li, Bo
Li, Yingkai
Liao, Chao
Lin, Bingkai
Lin, Chun-Cheng
Lu, Yue
Malik, Hemant
Mount, David
Möhring, Rolf H.
Nakano, Shin-Ichi
Nguyen, Kim Thang
Nishimura, Naomi
Nistor, Marian Sorin
Nyknahad, Dara
Oda, Yoshiaki
Peng, Sheng-Lung
Polak, Ido
Raichel, Benjamin
Rutter, Ignaz
Saitoh, Toshiki
Shi, Yongtang
Sukegawa, Noriyoshi
Suzuki, Akira
Takizawa, Atsushi
Tan, Zhiyi
Tang, Zhihao Gavin
Teruyama, Junichi
Wang, Hui
Wang, Hung-Lung
Wang, Meng
Wang, Wensheng
Wang, Yinling
Wehner, David
Wei, Chia-Chen
Williams, Derek
Wong, Prudence W.H.

Xiao, Mingyu
Xiao, Tao
Xu, Chunming
Yang, Kai
Ye, Deshi
Ye, Junjie
Yu, Bin

Yu, Tian-Li
Zhang, An
Zhang, Peng
Zhang, Yihan
Zhang, Yong
Zhao, Chenxia

Contents – Part I

Combinatorial Optimization

Contents – Part II

Application

Network

Filtering Undesirable Flows in Networks

Gleb Polevoy[1](✉), Stojan Trajanovski[1,2], Paola Grosso[1], and Cees de Laat[1]

[1] University of Amsterdam, Amsterdam, the Netherlands
G.Polevoy@uva.nl
[2] Philips Research, Eindhoven, the Netherlands

Abstract. We study the problem of fully mitigating the effects of denial of service by filtering the minimum necessary set of the undesirable flows. First, we model this problem and then we concentrate on a subproblem where every good flow has a bottleneck. We prove that unless P = NP, this subproblem is inapproximable within factor $2^{\log^{1-1/\log\log^c(n)}(n)}$, for $n = |E| + |GF|$ and any $c < 0.5$. We provide a $b(k+1)$-factor polynomial approximation, where k bounds the number of the desirable flows that a desirable flow intersects, and b bounds the number of the undesirable flows that can intersect a desirable one at a given edge. Our algorithm uses the local ratio technique.

Keywords: Flow · Filter · MMSA · Set cover · Approximation · Local ratio algorithm

1 Introduction

Denial of Service (DoS) and Distributed DoS [18] are widespread network attacks. These attacks negatively impact functionality, especially when the system needs to be quick (soft real time, for example) [17]. Consequently, fighting the problem is highly important [22]. Filtering the attacking flows [16] is one of the main ways to fight the problem. Filtering is also preferred among practitioners and network operators, rather than, for example, the more complicated and expensive link addition or removal. If we properly select a flow we want to filter, filtering always succeeds, but the required efforts depend on the filtered flow. For example, defining in the firewall which flows to filter is sometimes simple (say, filter all the UDP), but sometimes contrived (e.g., no simple pattern of what to filter exists) [11]. Unlike admission control, here we do not decide whether to allow a connection, but rather how to handle an existing one.

A similar problem is having less important but not malicious flows in the network. We then remove the less important flows to allow the more important ones to optimally utilize the network, and we want to incur the least possible cost from removing the less important flows. This pertains to both computer

---

S. Trajanovski—The research was started while S.T. was with the University of Amsterdam. He is now with Philips Research.

X. Gao et al. (Eds.): COCOA 2017, Part I, LNCS 10627, pp. 3–17, 2017.
https://doi.org/10.1007/978-3-319-71150-8_1

networks and transportation networks. In computer networks, streaming video to prioritized customers may be contractually binding, forcing flows to the other customers to give space to the prioritizes ones. In transportation, for example, a less important freight connection may be removed in favor of the more crucial ones [19].

We define a flow as a single path from the source to the sink and consider a system with some desirable (call them *good*) and undesirable (name them *bad*) flows. Undesirable flows can either model malicious flows or, alternatively, legitimate but dispensable flows. In particular, we model DoS as a set of bad flows that take up the available bandwidth. We study filtering as a coping method, possible within traffic engineering [1,2]. If we filter some bad flows, we can allocate the good flows more *value* because of the freed capacity. We aim to maximally increase the good flows, while spending the minimum necessary filtering effort, or losing the least from filtering the less important flows. Indeed, in the context of DoS, minimizing the filtering effort is practically important: Koning et al. [16] show that the filtering cost can have a significant effect on the overall effectiveness of the response. Therefore, we should not simply filter everything: Example 1 demonstrates that filtering all the bad flows can take arbitrarily more effort than the minimum effort necessary to maximally increase the good flows. In order to autonomously decide which flows to filter, as suggested in [16], we need an algorithm to find which bad flows to filter. In order to cope with large instances in real time, the algorithm has to be polynomial.

Example 1. In Fig. 1, assume the capacity of edge (V_1, V_2) is $2c$ and the capacity of (V_2, V_3) is c. Let the original flows have the value of c each: $v(b) = v(g) = c$, and let $w(b)$ be positive. Because of the saturated edge (V_2, V_3), filtering b would not allow increasing g. Therefore, the optimal set to filter is $\emptyset$ and it costs zero, infinitely smaller than filtering anything.

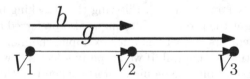

Fig. 1. Consider the network represented by the path graph with the 3 vertices V_1, V_2, V_3. We have a bad flow, b, and a good one, denoted by g.

We assume we know which flows are good and which are bad, either because we know all the flows, they are all good, and we decide which are dispensable and which are not, or, when malicious flows exist, we can identify them by frequent access trials from the same IP group.

Take a look at the following example of using an algorithm that decides which bad flows to filter.

Example 2. In a system where the flows belong to the same organization and carry equally important traffic, assume that our intrusion detection system discovers a DoS attack, and determines which flows are attacking. We need to respond quickly and efficiently. Having determined which flows are desirable, which are attacking and how large the flows are, we first estimate how hard filtering each attacking flow would be. Now, we run our algorithm to obtain an (approximately) easiest set of attacking flows to filter, such that the desirable flows will be able to fully utilize the system.

To pose the problem, we first model it in Sect. 2 and define k as the largest number of good flows that any good flow intersects in a network. We will also need b, defined as the largest number of bad flows that flow through an edge where a good flow also flows. The measures k and b will be used later on. Section 2.1 provides a short primer to the *local ratio* technique, which we employ to approximate the subproblem where every good flow has a bottleneck. At the outset, we prove in Sect. 3 that for any $k \geq 0$, the problem is NP-hard, using a reduction from Set Cover. Even when no bad edges intersect one another, we prove by reduction from Minimum-Monotone-Satisfying-Assignment that the problem is not even approximable within $2^{\log^{1-1/\log\log^c(n)}(n)}$, for $n = |\text{edges in the network}| + |\text{desirable flows}|$ and any $c < 0.5$, unless $P \neq NP$. In Sect. 4 we provide a polynomial approximation of the problem, with the tight approximation ratio of $b(k+1)$. The algorithm uses the *local ratio* technique [3–5]. We conclude and suggest further research directions in Sect. 5.

Our approximation can facilitate remedying the distributed DoS and similar congestion scenarios.

1.1 Related Work

We are not aware of any theoretical flow filtering approximations, but there is literature studying related flow problems. First, we aim to maximize the desirable flows, each flow being on a given path, while the famous max-flow – min-cut problem [8, Chap. 26] aims to maximize the total flow from source to sink, without predefined paths. There exist many famous generalizations of max-flow, such as maximum circulation [15, Chap. 7] and multi-commodity flow [10]. Regarding the allowed actions, we study filtering, thereby completing the studies of network design: edge addition [14], edge deletion [13,21], etc. In particular, deleting edges that can disconnect all the flows from a source to a sink is a famous problem, and Menger's theorem [7, Chap. 3.2] characterizes the minimum number of edges one has to remove in order to disconnect the source from the sink. Of course, finding a minimum cut mentioned above and disconnecting it is an optimal algorithm for this problem.

2 Model

We model the flow network as a directed graph $G = (N, E)$ with (edge) capacities $c \colon E \to \mathbb{R}_+$. A flow f from node a to node z in this network is a path from source a to sink z, each of which edges carries the value of the flow. Formally, $f = (v(f), P(f))$, where $v(f) \in \mathbb{R}_+$ is the value of the flow and $P(f)$ is the set of the edges of the path that the flow takes from a to z. Flow in this paper are not splittable, which meanings that a flow takes a single path. This can also model a splitting flow as separate flows with partially overlapping paths. All the flows together fulfill the capacity constraint, meaning that for every edge $e \in E$, all the passing flows together are bounded in their values by the capacity of the edge, i.e.

$$\sum_{f:e \in P(f)} v(f) \leq c(e).$$

Let us define the basic problem we are considering.

Definition 1. *The* Bad Flow Filtering *problem (BFF) receives the input* $(G = (N, E), c \colon E \to \mathbb{R}_+, F, GF, BF, w \colon BF \to \mathbb{R}_+)$. *Here,* $G = (N, E)$ *is a capacitated network with capacities* c *and flows* $F = \{f_i\}$, *where some flows, denoted* $GF = \{g_i\} \subseteq F$, *are marked as good (desirable), and the rest, denoted* $BF = \{b_i\} \triangleq F \setminus GF$, *are bad (undesirable). The values of the good flows are not given in the input. Every bad flow* f *is endowed with a weight* $w(f)$, *designating how hard filtering that flow would be, or how important the bad flow is, if bad flows model dispensable but legitimate flows.*

A solution *S is a subset of bad flows to filter.*

A feasible solution *is a solution such that the good flows can be allocated values such that the total value of the good flows is the maximum possible (i.e. equal to the total value that can be allocated if all the bad flows are removed).*

We aim to find a feasible solution with the minimum total weight. Intuitively, we aim to optimize the total good flow while investing the minimum filtering effort, or while losing the minimum of the less important flows, depending on what bad flows model.

BFF is monotonic with respect to inclusion, in the sense that filtering more bad flows after having filtered a feasible solution preserves feasibility.

We now define a constrained version of BFF, such that the algorithm can always provably approximate the solution. We need to avoid a situation when decreasing the value of a good flow can allow multiple good flows increase. Intuitively, we achieve this by always having a bottleneck that connects all the good flows that intersect one another, so that decreasing one of them will never be used multiple times to increase others. Formally,

Definition 2. *For any good flow* $g \in GF$, *define a* bottleneck *of* g *be a set of edges* $S(g) \subseteq P(G)$ *such that every other good flow* g' *that intersects* g *contains all these edges, and for every edge* i *where* g *intersects another good flow and for every solution* $BF' \subseteq BF$, *there exists an edge* $e \in S(g)$ *such that* $c(e) - \sum_{b \in BF \setminus BF' : e \in b} v(b) \leq c(i) - \sum_{b' \in BF \setminus BF' : i \in b'} v(b')$.
A BFF *problem where every good flow has a bottleneck is called a* Bottleneck-BFF (BBFF).

We prove BBFF, and therefore, BFF, is hard and approximate BBFF. Let us now present two cases that fall under BBFF.

Common Narrow Link. If every good flow g and all the flows that it intersects pass through an edge of a much smaller capacity than the other edges on the path of this flow, then this edge constitutes a bottleneck of g. This happens in practice when the flows pass through a physically common link.

Uniform Intersection. Intuitively, we require that a set of intersecting good flows all intersect each other at the same edges.

> **Definition 3.** *The* Uniform Intersection Bad Flow Filtering *problem* (UIBFF) *is a restriction of BFF where every $g \in GF$ has a set of edges on its path, $E(g) \subseteq P(g)$, such that every other good flow g' that intersects g shares with g exactly the edges of $E(g)$, i.e. $P(g) \cap P(g') = E(g)$.*

Since the defined $E(g)$ is a bottleneck of g, UIBFF is a subproblem of BBFF. This uniformity can happen, for instance, if the intersecting flows share a source or a destination, and intersect only near those nodes.

We now define the parameters which we will use to express the approximation ratio of our algorithm.

Definition 4. *Given an instance of BFF, let k be the largest possible number of good flows that a given good flow intersects. Formally,*

$$k \triangleq \max \left\{ |\{g' \in GF \setminus \{g\} : P(g') \cap P(g) \neq \emptyset\}| : g \in G \right\}.$$

Definition 5. *For a BFF instance, let b be the largest number of bad flows that intersect a good flow at any given edge. Formally,*

$$b \triangleq \max \left\{ |\{f \in BF : e \in P(f)\}| : g \in G, e \in P(g) \right\}.$$

2.1 Local Ratio Approximation

A typical local ratio r-approximation algorithm for minimization [4,5] is easier to formulate recursively, though practical implementations are usually iterative. It works by manipulating the weights as follows.

1. If a trivial solution (often, the empty set) is feasible, return it.
2. Otherwise, if zero weight elements exist, we remove them, solve the problem without them and add them back afterwards.
3. Otherwise, decompose the weight function $w = w_1 + w_2$ such that every feasible solution would be an r-approximation with respect to w_1. We call such a w_1 weight function r-*effective* and finding it is the main challenge. Then, recursively invoke the algorithm with w_2. The returned feasible solution is an r-approximation w.r.t. w_2, by induction. Since it also is an r-approximation w.r.t. w_1, by the way we decomposed w, the following theorem implies that this solution is also an r-approximation w.r.t. w, as required.

Theorem 1 (Local Ratio Theorem [4]**).** *Let us have a feasible set $D \subseteq \mathbb{R}^n$.[1] Assume we have weight vectors $w = w_1 + w_2$ and that a feasible solution $x \in D$ is r-approximate w.r.t. w_1 and w.r.t. w_2. Then, x is r-approximate w.r.t. w as well.*

We also require from w_1 that at least one element will have the zero weight in $w - w_1$, so that the instance will shrink at the next invocation. Finding a suitable r-effective w_1 for a small r is the crux of the method, requiring an insight about all the feasible solutions of the problem.

3   Hardness

We prove that the decision version of BBFF is NP-complete, and even NP-hard to be approximated within $2^{\log^{1-1/\log\log^c(n)}(n)}$, for $n = |E| + |GF|$, thereby motivating the need to seek an approximation instead of an exact solution.

We first prove that the problem is NP-hard not merely to optimize exactly, but even to approximate.

Theorem 2. *UIBFF (and, therefore, BBFF) is not approximable within $2^{\log^{1-1/\log\log^c(n)}(n)}$, for $n = |E| + |GF|$ and any $c < 0.5$, unless $P \neq NP$. This holds even if no bad edges intersect one another.*

Proof. We prove the hardness of approximation by reducing the Minimum-Monotone-Satisfying-Assignment of depth 3 (MMSA$_3$) problem to UIBFF. Let us remind the definition of MMSA$_3$ [9].

Definition 6. *The input of the MMSA$_3$ problem is a monotone (with no negative literals) Boolean formula, which is a conjunction (AND) of disjunctions (OR), every such disjunction being a disjunction of conjunctions. The goal is finding a satisfying assignment that minimizes the number of variables that are assigned 1.*

An example of an MMSA$_3$ is $((x_1 \text{ AND } x_3 \text{ AND } x_5) \text{ OR } (x_2 \text{ AND } x_3))$ AND $((x_2 \text{ AND } x_4 \text{ AND } x_5 \text{ AND } x_6) \text{ OR } (x_1))$.

Given an instance of MMSA$_3$, our reduction defines the following UIBFF. For each variable x in conjunction c, which is, in turn, a part of disjunction d, define edge $e_{x,c,d}$ of capacity 1. Define also a bad flow b_x of value 1 and weight 1 that has all the edges $\{e_{x,c,d} | x \text{ appears in } c, \text{ a part of } d\}$ on its path, and no others of the above edges. The rest of the edges, if any, are arbitrary and unique for each bad flow. For each conjunction c of variables, which is a part of disjunction d, define a good flow $g_{c,d}$ that flows through the edges $\{e_{x,c,d} | x \text{ appears in } c, \text{ a part of } d\}$ and perhaps arbitrary other edges of capacity 1 without any bad flows that pass through them. For each disjunction d of conjunctions, let the respectively defined

---

[1] In our case, these are the incidence vectors of the bad flows that, if filtered, would allow assigning the good flows the maximum possible total value. Thus, we have $D \subseteq \mathbb{N}^n$.

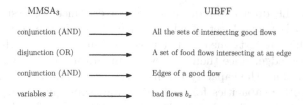

MMSA$_3$	$\longrightarrow$	UIBFF
conjunction (AND)	$\longrightarrow$	All the sets of intersecting good flows
disjunction (OR)	$\longrightarrow$	A set of food flows intersecting at an edge
conjunction (AND)	$\longrightarrow$	Edges of a good flow
variables x	$\longrightarrow$	bad flows b_x

Fig. 2. The approximation preserving reduction from MMSA$_3$ to UIBFF.

good flows $\{g_{c,d}|c \text{ appears in } d\}$ intersect at a single edge e_d where no bad flows pass. Let these be the only intersections of good flows among themselves. The reduction is illustrated in Fig. 2.

First, since we define the BFF instance such that every good flow intersects all the other good flows at a single edge, it is indeed a UIBFF. To prove validity of the reduction, note that the MMSA$_3$ instance is satisfied if and only if the main conjunction holds, which holds if and only if at least one conjunction in every disjunction holds. The last statement is equivalent to at least one good flow in all the intersecting sets of good flows has all its edges free and can thus be given the value of 1. Therefore, feasibility is transferred by the reduction. Since the costs are equivalent as well, the reduction preserves approximation.

Since MMSA$_3$ is not approximable within $2^{\log^{1-1/\log\log^c(n)}(n)}$, for any $c < 0.5$, as shown in [9], we infer that UIBFF is not approximable within the same ratio, when $n = |E| + |GF|$. This is because the size of the MMSA$_3$ formula translates to $|E| + |GF|$. ∎

We now prove that the decision version is indeed NP-complete. We first define the decision version of BBFF.

Definition 7. *The Decision-BBFF receives (x, l) in its input, where x is an instance of BBFF and l is a natural number. The question is whether there exists a feasible solution for BBFF with weight at most l.*

We finally prove that

Theorem 3. *Decision-BBFF is NP-complete, for any $k \geq 0$.*

Proof. First, we show that Decision-BBFF is in NP. Indeed, for a candidate solution S, filter all the flows there and maximize the good ones in the remaining network. The maximization can be done polynomially by solving the Linear Program (LP):

$$\max \sum_{i \in GF} x_i \qquad (1)$$

$$\text{such that}$$

$$\forall e \in E: \sum_{i \in GF: e \in P(i)} x_i + \sum_{b \in BF \setminus S: e \in P(i)} v(b) \leq c(e) \qquad (2)$$

$$\forall i \in GF: x_i \geq 0 \qquad (3)$$

Remark 1. This LP does not require an LP solver, since we assume that any good flow has a bottleneck set of edges which always is the constraint to any intersection. This property means that it is not important how the good flows divide a common edge, since their sum will remain the same. Consequently, we can effectively reduce the capacity taken by the bad flows in $O(|BF||E|)$ time, storing the effective capacities for each edge, and subsequently maximize each good flow one after another, by checking the bottlenecks of each good flow in $O(|E|)$ time and storing the effectively remaining capacities, amounting to the total time of $O((|GF| + |BF|)|E|) = O(|F||E|)$.

By comparing the maximum of this LP with the maximum when all the bad flows are filtered, i.e. when $S = BF$, we check whether S is feasible. If it is, then it constitutes a certificate if and only if $w(S) \leq l$. Therefore, the problem belongs to NP.

To prove the NP-hardness, notice that our reduction from MMSA$_3$ is also a Karp reduction for the decision versions, and since Decision-MMSA$_3$ is NP-hard [9], so is Decision-BBFF. However, to claim the NP-hardness for any $k \geq 0$, we now present a reduction from the decision version of Set Cover (SC) [12]. Let us remind the definition of SC.

Definition 8. *SC receives as input a universe U, a collection of its subsets $\{S_1, S_2, \ldots, S_m\}$, such that $\cup_{i=1}^m S_i = U$ and a natural number d. A solution is a subset of $\{S_1, S_2, \ldots, S_m\}$, while a solution $C \subset \{S_1, S_2, \ldots, S_m\}$ is called feasible or a cover if $\cup_{S \in C} S = U$. The question is whether there exists a cover C such that $|C| \leq d$.*[2]

Our reduction takes an input of SC, which is $U, S_1, S_2, \ldots, S_m, d$, and returns the following instance of Decision-BBFF (even Decision-UIBFF). First, we define a bad flow b_i for each set S_i, with $v(b_i) = 1$. For each element $x \in U$, let g_x be a good flow with a path consisting of the two edges: $e_x^{(1)}$ and $e_x^{(2)}$. We set the $c(e_x^{(1)})$ to be the number of sets S_i that contain x and we set $c(e_x^{(2)}) = 1$. For every S_i that contains x, let b_i intersect g_x at $e_x^{(1)}$. Besides these intersections, no more flow intersections take place. The weight of every b_i is defined to be 1.[3] The parameter l of the constructed Decision-UIBFF instance is defined to be d. The reduction is exemplified in Fig. 3.

We now prove that this reduction is valid. Indeed, for every element $x \in U$, any set cover includes at least one set that contains x, say S_i and this is transformed to filtering the corresponding bad flow b_i. This allows the good flow g_x to increase till it uses up all the capacity it can, i.e. 1. Note, that if more covering sets are selected as well, then the corresponding bad flows are filtered as well, but the value of flow g_x remains 1 because $c(e_x^{(2)}) = 1$; one filtering is enough to maximize $v(g_x)$, guaranteeing feasibility to Decision UIBFF.

---

[2] We use the unweighted set cover, where each set has the same importance, because it has the same hardness results as the weighted version.

[3] Did we reduce the weighted SC, we would define it to be the weight of the respective set.

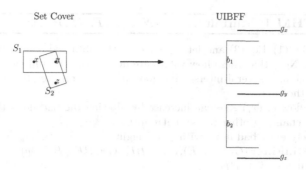

Fig. 3. The Karp reduction from Decision-SC to Decision-UIBFF.

In the other direction, a feasible solution to Decision UIBFF maximizes the sum of good flows. This requires filtering at least one bad flow from those that intersect every g_x, which means that a feasible solution to the constructed instance of Decision-UIBFF is obtained from a set cover. This completes the proof of the NP-hardness. ∎

4 Approximation

Consider the local ratio approximation Algorithm 1 for BBFF. We explain it now in the terms of Sect. 2.1. Line 1 finds the maximum total good flow that is available at the current invocation. The recursion basis appears at line 2 and line 3 removes the zero weight bad flows. We abuse notation by writing w in the recursive call, while we actually mean the restriction of w to $BF \setminus BF_0$. The central scene of the algorithm occurs at line 4. There, we pick a good flow that would benefit from filtering bad flows at line 4a and construct the set of all the bad flows we may need to filter at line 4b. This serves us to decompose the weights at line 4c.

When choosing the flows in H at line 4a and 4b, we take the minimum possible flow each time. The idea is to select all the possible good flows that can increase, to cover all the possibilities, and a smaller flow has more chances to increase.

We now prove that in polynomial time, this algorithm returns a feasible solution approximating the optimum within $b(k+1)$. This means, for example, that if any good flow intersects at most 1 another good flow ($k = 1$), and at most one bad flow contains a given edge of a good flow ($b = 1$), then the algorithm approximates the optimal solution within $1 \cdot (1 + 1) = 2$. And in case the good flows do not intersect one another ($k = 0$), the algorithm is optimal.

Another interesting particular case is the result of the reduction of Set Cover to BBFF from Theorem 3. In the outcome of the reduction, b is the maximum number of sets that can include a given element, and $k = 0$. Therefore, Algorithm 1, acting like the Algorithm 15.2 from [20], approximates set cover within the $b(0 + 1) = b$, which is the maximum number of sets that can include

---

ALGORITHM 1: MinFilter$(G = (N, E), c, F, GF, BF, w)$

---

1. Solve **LP** Eq. (1)–Eq. (3) and let $(x_i)_{i \in GF}$ be the obtained result (the current maximum). Note that these flow values are not necessarily unique, since we can sometimes change several intersecting good flows one on the other's expense, preserving the sum.

2. **If** no good flow in $(x_i)_{i \in GF}$ can increase by filtering the bad flows that intersect it (without changing other flows), **return** $\emptyset$.

3. **Else, if** there exist bad flows with zero weight BF_0,
   (a) $S' \leftarrow$ **MinFilter** $(G = (N, E), c, F \setminus BF_0, GF, BF \setminus BF_0, w)$.
   (b) **Return** $S \leftarrow S' \cup BF_0$.

4. **Else,**
   (a) Pick any $g \in GF$ that can be increased (without changing other flows) if we filter the bad flows that intersect it.
      This should be done by taking the minimum $v(g)$ that can be in a maximum total good flow. (Maximize the good flows that intersect g on g's expense.)
   (b) Consider all the other good flows $g_1, g_2, \ldots, g_p$ (by the definition of k, $p \leq k$) that intersect g and would grow after filtering some bad flows, if we take the minimum possible $v(g_i)$ in a maximum total good flow. Let H be the set of the considered good flows, i.e. $H \triangleq \{g, g_1, g_2, \ldots, g_p\}$ and let $D(H)$ be the set of their respective saturated edges, chosen one from a flow (may choose the same saturated edge from several good flows, so $|D(H)| \leq |H|$). Denote all the bad flows that contain edge(s) from $D(H)$ as $B(D(H))$.
   (c) Let $\delta > 0$ be the minimum total weight in $B(D(H))$,
      i.e. $\min_{b \in B(D(H))} \{w(b)\}$. Define the weight function on the bad flows:

$$w_1 \triangleq \begin{cases} \delta & \text{if } b \in B(D(H)), \\ 0 & \text{otherwise.} \end{cases}$$

   (d) **Return MinFilter** $(G = (N, E), c, F, GF, BF, w - w_1)$.

---

an element. Of course, the general BBFF is much harder than SC, being not easier than MMSA_3, as we show in Theorem 2. Approximating the general BBFF constitutes our main contribution.

First, we make a crucial observation, which guarantees that the algorithm always finds a required good flow g. This property requires the restriction of every good flow to have a bottleneck, introduced in Definition 2.

Observation 1. *A solution for an BBFF instance is infeasible if and only if in any maximal allocation of good flows for it there exists a good flow that can increase if we filter some bad flows that intersect it, without changing other flows.*

Proof. If a good flow can increase, the solution is infeasible by definition.

In the other direction, let S be an infeasible solution. This means that any allocation of good flows can increase if we filter some more bad flows. The only option where "no good flow exists that would increase if we filtered some bad flows that intersect it" needs good flows that can grow only at the expense of other good flows, like, for example, in Fig. 4. However, since any good flow in BBFF has a bottleneck, where all the intersecting good flows pass, (unlike shown in Fig. 4), increasing good flows at the expense of others would never increase the total good flow. ∎

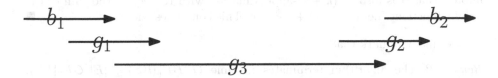

Fig. 4. Flow g_1 can grow if we filter b_1 and if g_3 decreases. Since decreasing g_3 can also help increasing g_2, if b_2 is filtered, the total good flow will increase.

As in any correctness proof for a local ratio algorithm, we show that the weight function w_1 is fully $b(k+1)$-effective, meaning that any feasible solution S is a $b(k+1)$ approximation to the optimum.

Lemma 1. *Let S be any feasible solution to the instance of the problem at some invocation of Algorithm 1. Then, $w_1(S) \leq b(k+1) \cdot w_1(S^*)$, where S^* is an optimal solution at that invocation.*

Proof. Any feasible solution will either allow g to grow by filtering at least one bad flow that contains its chosen saturated edge, or it will allow at least one of the good flows that intersect it to grow by filtering at least one of the bad flows that contain their respectively chosen saturated edges. Therefore, with respect to w_1, any feasible solution will cost at least δ. On the other hand, any solution costs at most $b(k+1) \cdot \delta$. Therefore, any feasible solution costs at most $b(k+1)$ times the minimum cost. ∎

We are finally set to prove the correctness and the approximation ratio of the algorithm.

Theorem 4. *Algorithm 1 always returns a feasible solution that approximates the optimal solution within the ratio of $b(k+1)$.*

Proof. We prove by induction on our recursive algorithm.

In the basis (line 2), the good flow is optimal and therefore, the empty set of bad flows is feasible. Since the empty set weighs zero, it is also optimal.

At a non-final stage, we need to prove that line 3 and line 4 both return feasible $b(k+1)$-approximation. In the case of line 3, S' is a feasible $b(k+1)$-approximation for the instance after removing BF_0, by induction on the algorithm. Now, a feasible solution for the instance after removing the flows in BF_0 remains feasible w.r.t. the original instance if we add the removed flows to the solution. Second, the approximation ratio keeps holding, since the optimum stays the same after this operation, and the solution cost remains the same as well.

Having said that, let us show that line 4 returns a feasible $b(k + 1)$-approximation. First, the recursive invocation at line 4d returns a feasible solution and a $b(k + 1)$-approximation with respect to the weight function $w - w_1$, by the induction hypothesis; call this solution $\hat{S}$. Set $\hat{S}$ is also a $b(k + 1)$-approximation with respect to w_1, by Lemma 1. The Local Ratio Theorem 1 implies that $\hat{S}$ is also a $b(k + 1)$-approximation with respect to the sum of the weight functions, i.e. $(w - w_1) + w_1 = w$. This completes the proof. ∎

We finally remark that

Remark 2. The algorithm terminates in time $O(|BF|(l(S) + |E||GF||F|))$, where $l(S)$ is the time taken to solve the LP that corresponds to instance S. As we explain in Remark 1, $l(S) = O(|F||E|)$, implying the total running time of $O(|E||BF||GF||F|)$.

Proof. The algorithm performs $O(|BF|)$ iterations, since at least one bad flow gets filtered at each invocation of line 4.

At each invocation of line 3, we filter the zero-weight bad flows and then add them back in $|BF|$ time.

At each invocation of line 4, we solve the LP in $l(S)$ time. Next, we go over all the good flows, checking for each good flow g in $O(|P(g)||F|) = O(|E||F|)$ time whether filtering bad flows can help increasing this flow, by passing through all the edges of the path of the flow and checking whether the good flows can be increased if no bad ones existed. If yes, we construct the sets H and $D(H)$ in $O(|E||GF||F|)$, construct $B(D(H))$ and define the weight function w_1 in $O(|E||BF|)$. This takes together $O(|E||GF||F| + l(S))$. Finally, we make the recursive invocation.

Summing up the above time bounds and multiplying by $|BF|$, we obtain $O(|BF|(l(S) + |E||GF||F|))$. ∎

Finally, we prove that the approximation ratio $b(k + 1)$ is tight for our algorithm, even on a UIBFF. To this end, we employ the following example, partially inspired by Example 15.4 from [20].

Example 3. The general idea is to reduce Example 15.4 from [20] for Set Cover by our reduction from Theorem 3 to UIBFF, while also translating the weights of sets to be the weights of the corresponding bad flows. This would produce a tight example, but the parameter k would be zero. To allow for any k, we consider several instances of such an example and make them intersect in a specific way.

Concretely, consider the following UIBFF instance, depicted in Fig. 5. We have the good flows $g_1, g_2, \ldots, g_{n+1}$, each g_i with the path of two edges $e_i^{(1)}$ and $e_i^{(2)}$, where $c(e_i^{(1)}) = 2$, for $i = 1, \ldots, n - 1$, $c(e_n^{(1)}) = n$, and $c(e_{n+1}^{(1)}) = 1$. For all $i = 1, \ldots, n + 1$, we have $c(e_i^{(2)}) = 1$. We have the bad flows $b_{\{1,n\}}, b_{\{2,n\}}, \ldots, b_{\{n-1,n\}}$ with weight 1 each and the bad flow $b_{\{1,2,\ldots,n+1\}}$ with weight $1 + \epsilon$ for a positive ϵ. Every bad flow has the value of 1. A bad flow b_S

intersects the good flows corresponding to the elements of S at their respective first edges, and, for now, no more intersections exist.

Next, consider $m + 1$ copies of the constructed problem instance. Let the distinct copies intersect only at the edges $e_i^{(2)}$, for $i = 1, \ldots, n+1$, where all the copies intersect.

Algorithm 1 can choose at its first recursive invocation any good edge that can increase from filtering the bad ones. Assume it chooses g_n of one of the m copies. The weights of each of the bad flows in all the copies decrease by 1 and in the next invocation we remove all the bad edges from all the copies, besides the bad edges $b_{\{1,2,\ldots,n+1\}}$. Their weights will now also go to zero and in the following invocation the empty set becomes feasible. In the unwinding of the recursion, we will add all the bad flows to the solution, accruing the total weight of $m+1$ times $\underbrace{1 + \ldots + 1}_{n-1} + 1 + \epsilon = (m+1)(n+\epsilon)$. Now, the optimal solution is just $b_{\{1,2,\ldots,n+1\}}$ of one of the copies, because the intersections among the good flows let only values 1 in every intersecting set. Therefore, the optimal weight is $1 + \epsilon$, and we can obtain an arbitrarily close to $n(m+1)$ ratio, for a sufficiently small ϵ. This is exactly $b(k+1) = n(m+1)$, demonstrating the tightness of $b(k+1)$.

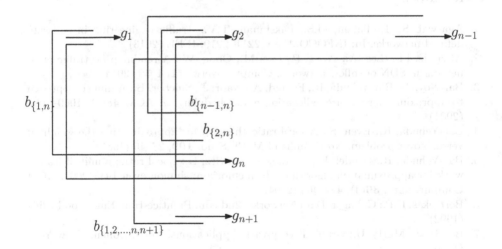

Fig. 5. The relevant part of the network. The bad flows are denoted by b with an index, while the good ones are denoted by g with an index. An optimal solution would be to filter $b_{\{1,2,\ldots,n+1\}}$, while our algorithm filters everyone.

5   Conclusion

Aiming to optimally mitigate DoS or unintended congestion, we study the BBFF problem of filtering the minimum number of undesirable (bad) flows so as to allow the desirable (good) flows to maximally utilize the network. We demonstrate that this practical problem is also very interesting theoretically. First, we reduce the MMSA$_3$ to BBFF while preserving approximation, proving that approximating within $2^{\log^{1-1/\log\log^c(n)}(n)}$, for $n = |$edges in the network$| + |$desirable flows$|$ and any $c < 0.5$ is NP-hard. We then provide a local ratio approximation algorithm for BBFF.

An interesting variation of the problem would be to assume that the flows are always allocated by a given protocol, for example, by the max-min fairness algorithm [6, Sect. 6.5.2]. This would render the problem of filtering non-monotonic with respect to inclusion, which would make many approximation techniques fail. Another point is that we are given a continuous ranking of the bad flows by weight, but the distinction between the bad and the good is binary. Exploring other rankings would allow modeling other congestion domains.

To summarize, we have modeled an important NP-complete problem, proven it be not easier than MMSA$_3$ and approximated it.

Acknowledgments. This research is funded by the Dutch Science Foundation project SARNET (grant no: CYBSEC.14.003/618.001.016).

References

1. Agarwal, S., Kodialam, M.S., Lakshman, T.V.: Traffic engineering in software defined networks. In: INFOCOM, pp. 2211–2219. IEEE (2013)
2. Akyildiz, I.F., Lee, A., Wang, P., Luo, M., Chou, W.: A roadmap for traffic engineering in SDN-openflow networks. Comput. Netw. **71**, 1–30 (2014)
3. Bar-Noy, A., Bar-Yehuda, R., Freund, A., Naor, J., Shieber, B.: A unified approach to approximating resource allocation and scheduling. J. ACM **48**(5), 1069–1090 (2001)
4. Bar-Yehuda, R., Even, S.: A local-ratio theorem for approximating the weighted vertex cover problem. North-Holland Math. Stud. **109**, 27–45 (1985)
5. Bar-Yehuda, R., Bendel, K., Freund, A., Rawitz, D.: Local ratio: a unified framework for approximation algorithms. In memoriam: shimon even 1935–2004. ACM Comput. Surv. **36**(4), 422–463 (2004)
6. Bertsekas, D.P.: Gallager: Data Networks, 2nd edn. Prentice-Hall, Englewood Cliffs (1992)
7. Bondy, J., Murty, U.: Graph Theory with Applications. North Holland, New York (1976)
8. Cormen, T., Leiserson, C., Rivest, R., Stein, C.: Introduction to Algorithms, 3rd edn. MIT Press, Cambridge (2009)
9. Dinur, I., Safra, S.: On the hardness of approximating label-cover. Inf. Process. Lett. **89**(5), 247–254 (2004)
10. Even, S., Itai, A., Shamir, A.: On the complexity of time table and multi-commodity flow problems. In: Proceedings of the 16th Annual Symposium on Foundations of Computer Science (SFCS 1975), pp. 184–193. IEEE Computer Society, Washington (1975)
11. Ferguson, P., Senie, D.: Network ingress filtering: defeating denial of service attacks which employ IP source address spoofing (1998)

12. Garey, M.R., Johnson, D.S.: Computers and Intractability: A Guide to the Theory of NP-Completeness. W. H. Freeman & Co., New York (1979)
13. Italiano, G.F.: Finding paths and deleting edges in directed acyclic graphs. Inf. Process. Lett. **28**(1), 5–11 (1988)
14. Khuller, S., Thurimella, R.: Approximation algorithms for graph augmentation. J. Algorithms **14**(2), 214–225 (1993)
15. Kleinberg, J., Tardos, E.: Algorithm Design. Addison-Wesley Longman Publishing Co., Inc., Boston (2005)
16. Koning, R., de Graaff, B., de Laat, C., Meijer, R., Grosso, P.: Interactive analysis of SDN-driven defence against distributed denial of service attacks. In: 2016 IEEE NetSoft Conference and Workshops (NetSoft), pp. 483–488, June 2016
17. Mirkovic, J., Dietrich, S., Dittrich, D., Reiher, P.: Internet Denial of Service: Attack and Defense Mechanisms (Radia Perlman Computer Networking and Security). Prentice Hall PTR, Upper Saddle River (2004)
18. Mirkovic, J., Reiher, P.: A taxonomy of DDOS attack and DDOS defense mechanisms. SIGCOMM Comput. Commun. Rev. **34**(2), 39–53 (2004)
19. Rodrigue, J.P.: The Geography of Transport Systems, 4th edn. Routledge, New York (2017)
20. Vazirani, V.: Approximation Algorithms. Springer (2001)
21. Yannakakis, M.: Node-and edge-deletion NP-complete problems. In: Proceedings of the Tenth Annual ACM Symposium on Theory of Computing (STOC 1978), pp. 253–264. ACM, New York (1978)
22. Zargar, S.T., Joshi, J., Tipper, D.: A survey of defense mechanisms against distributed denial of service (DDOS) flooding attacks. IEEE Commun. Surv. Tutor. **15**(4), 2046–2069 (2013)

A Framework for Overall Storage Overflow Problem to Maximize the Lifetime in WSNs

Guoliang Song, Chen Zhang$^{(\boxtimes)}$, Chuang Liu, and Yuna Chai

Harbin Institute of Technology Shenzhen Graduate School, Shenzhen, China
s_guoliang@foxmail.com, chenzhanghit@outlook.com, chuangliuhit@gmail.com,
chaiyuna@outlook.com

Abstract. Storage overflow problem in wireless sensor networks is a new and challenging issue, wherein data-collecting base station is not available while more data items are generated than available storage space in the entire network. In this paper, we consider overall storage overflow problem in WSNs, the goal of which is to maximize the minimum remaining energy of data node (the node with overflow data) in order to prolong the lifetime of the sensor network. For overall storage overflow problem, we propose a two-step solution. A degree-constrained data aggregation algorithm is presented, and then we further propose a data replication algorithm which is a unified method, integrating data aggregation and data redistribution. Extensive simulations show that our proposed algorithms significantly outperform than existing algorithms especially in extending the lifetime of the sensor network.

Keywords: Wireless sensor networks · Overall storage overflow · Data aggregation · Data redistribution

1 Introduction

In recent years, wireless sensor networks (WSNs) have been widely used in various fields. Many of them are deployed in remote area or challenging environments to collect large volumes of data for a long period of time, such as ocean monitoring, volcano eruption monitoring and climate change. Due to the inaccessible and hostile environments, it is not feasible to deploy long-term base station with power outlets. Therefore, the generated data is first stored inside the sensor network for a period of time, and then collected by periodic visit of the robots or data mules. In the challenging environment, however, uploading opportunities would be unpredictable and rare, a major problem is how to store the massive amount of data inside the network comprising of nodes with limited storage space and limited energy. In the sensor network, sensor nodes are randomly deployed in the area and then each node collects data independently. When events of the interest take place, sensor nodes close to them may collect data more frequently than nodes far away, therefore these nodes may run out of their storage space quickly than others. After a period of time, some nodes may deplete their storage space and generate overflow data, while other nodes may have available storage space. There are two level of data overflow in the sensor network.

X. Gao et al. (Eds.): COCOA 2017, Part I, LNCS 10627, pp. 18–32, 2017.
https://doi.org/10.1007/978-3-319-71150-8_2

1. Partial Storage Overflow: In this level of data overflow, some nodes (denoted as data nodes) in the sensor network deplete their own storage space while other nodes (denoted as storage nodes) still have available storage space. And the total size of available storage space is greater than or equal to the total size of overflow data. If uploading opportunities are not available, the newly generated data at data node can not be stored, causing data loss. In order to avoid data loss, data redistribution are proposed. The idea of data redistribution is that redistributing overflow data from data node to storage node, such that any data node does not have any overflow data.
2. Overall Storage Overflow: This is a more serious situation, where the total size of overflow data exceeds the total size of available storage space in the network. To overcome overall storage overflow problem, it needs two step: data aggregation [1] and data redistribution. Data aggregation for overall storage overflow problem is to reduce the size of overflow data, such that the overflow data can fit into the available storage space. After data aggregation, overall storage overflow problem becomes partial storage overflow problem and it can be solved by data redistribution.

Therefore overall storage overflow problem is more serious and complicated compared to partial storage overflow problem. In this paper, we focus on overall storage overflow problem. For this problem, we consider different sensor nodes may have different remaining energy, especially data node which usually has low remaining energy as collecting massive data consumes a lot of energy. The contributions of this paper are as follows:

1. We first study overall storage overflow problem in WSNs for maximizing the minimum remaining energy of data node. To our best knowledge, the problem has not been addressed by any of existing research.
2. We propose a data aggregation algorithm and a data replication algorithm. Data replication algorithm is a unified method which integrate data aggregation and data redistribution.
3. Extensive experiments have been conducted to verify that our algorithm achieves higher lifetime than existing approaches.

The rest of this paper is organized as follows. In Sect. 2, we present related work. In Sect. 3, we introduce overall storage overflow problem. Sections 4 and 5, we introduce data aggregation and unified method for overall storage overflow problem respectively. And we also give its corresponding algorithms. In Sect. 6, we compare the proposed algorithms with existing algorithms and discuss the performance. Section 7 concludes the paper with future work.

2   Related Work

Storage overflow problem in wireless sensor network is relatively new research topic. Tang et al. [1] study overall storage overflow problem in base station-less sensor networks. They solve the problem by data aggregation and data redistribution. And they address data aggregation for overall storage overflow is equivalent to multiple traveling salesman walks problem (MTSW). Alhakami et al. [2] proposed a unified method that is based upon data replication techniques for overall storage overflow problem. Both above work assume that the energy of each node is infinity, ignoring different nodes may have different remaining energy.

Tang et al. [3] also study how to minimize the total energy consumption in the process of data redistribution, and address it as a minimum cost flow problem. Hou et al. [4] study how to maximize the minimum remaining energy of the nodes after data preservation, such that the data can be preserved for maximum amount of time. And Takahashi et al. [5] try to preserve the data inside the network for maximum possible time, by distributing the data items from low energy nodes to high energy nodes. Xue et al. [6] consider different data may have different importance and priority, and study how to preserve data with maximum priority. They address the core of the problem is a maximum weighted flow problem and propose a time efficient heuristic algorithm. A network flow perspective of data preservation problem in sensor networks is given in [7]. All above work, however, do not address overall storage overflow problem and they just try to redistribute overflow data as much as possible. In this paper, we consider the different remaining energy of nodes, and try to maximize the minimum remaining energy of data node in the process of data aggregation in order to prolong the lifetime of the network.

There are active research that focused on data aggregation. Kuo et al. [8] studies how to construct a data aggregation tree that minimizes the total energy cost of data transmission, while Chen et al. [9] study the construction of a data gathering tree to maximize the network lifetime. Yan et al. [10] and Lee et al. [11] consider the aggregation delay in the process of data aggregation and propose data aggregation scheduling scheme to minimize latency in duty-cycled WSNs. Some other work use mobile base stations collect aggregated data [12,13]. However data aggregation for overall storage overflow problem significantly differs from above data aggregation. The above data aggregation in wireless sensor network is used to collect data items from different sensor nodes, in order to reduce number of transmissions and energy consumption. Data aggregation for overall storage overflow problem is to aggregate the overflow data, so that the overflow data can be stored in the available storage space. In Sect. 3, we introduce the process of data aggregation for overall storage overflow problem in detail.

3   Overall Storage Overflow Problem

The wireless sensor network consists of many nodes, we denote the node with overflow data as data node, and the node with available storage space as storage node. To aggregate data, one or more data nodes (called initiators) send their overflow data to other data nodes. When a data node (called an aggregator) receives the data, it aggregates its own overflow data, then forwards the initiators entire overflow data to another data node, which becomes an aggregators and aggregates its own overflow data, and so on so forth. This continues until enough aggregators are visited such that the total size of overflow data is equals to or is slightly less than total available storage in the network. Each aggregator can aggregate its own overflow data only once. If an aggregator receives another initiator's overflow data, it just transfers it to other data node. And if a storage node receives the initiator's overflow data, it simply relays it. After the aggregation, the initiators' overflow data become zero, and the last aggregator has both its own aggregated data and the entire overflow data from initiator. Some data nodes which neither an initiator nor an aggregator are not involved in data aggregation, they have original overflow data which is not aggregated.

Network Model. The sensor network can be modeled as an undirected graph $G = (V, E)$, where $V = \{1, 2, \cdots, |V|\}$ is set of $|V|$ sensor nodes, and E is set of $|E|$ edges. Every sensor node can transmit and receive data, but its transmission range is limited. $\forall v_i, v_j \in V$, there exists an edge $(v_i, v_j) \in E$ in graph G if and only if node v_i and v_j are in each other transmission range. Assume that each node has same transmission range and there are p data nodes, denoted as v_d. Thus the number of storage nodes (denoted as v_s) is $|V| - p$. We consider that each data node has same size of overflow data and each storage node has same available storage space. Let R denote the size of overflow data in bits at each data node, and let m denote the available storage space in bits at each storage node. For overall storage overflow problem, it satisfies the following equation.

$$p \times R > (|V| - p) \times m \tag{1}$$

Feasible Overall Storage Overflow. In order to reduce the overflow data to the size which can be stored by the available storage capacity, enough number of aggregators should be visited. Let q denotes the number of aggregators, and r represents the size of overflow data after data aggregation, which based on a spatial correlation model [14], indicating that the size of redundant overflow data between any two data nodes is $R - r$. The feasibility of data aggregation can be derived in [1].

$$q = \lceil \frac{p \times R - (|V| - p) \times m}{R - r} \rceil = \lceil \frac{p \times (R + m) - |V| \times m}{R - r} \rceil \tag{2}$$

There is at least one initiator, and the maximum number of aggregators is $p - 1$. Therefore, the valid range of p is

$$\frac{|V|m}{m + R} < p \leq \lfloor \frac{|V|m - R + r}{m + r} \rfloor \tag{3}$$

Example 1. Figure 1 is an example of overall storage overflow problem in a linear sensor network with five nodes. Figure 1(a) is data aggregation step for overall storage overflow problem. Node A, C and D are data nodes, while B and E are storage nodes. Each data node has 2 units overflow data, and each storage node has 2 units available storage space. There are total 6 units of overflow data while

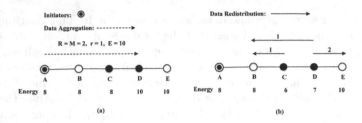

Fig. 1. A sensor network with overall storage overflow problem. (a) Data aggregation step. (b) Data redistribution step.

there are only 4 units of available storage space, causing overall storage overflow problem. We assume that $r = 1$. The number of aggregators q is calculated as 2 by using Eq. 2. Therefore the number of initiator is one. One possible aggregation walk is A, B, C, D and node A is the initiator. After data aggregation, the size of overflow data at A, C and D are 0, 1 and 3 respectively.

After data aggregation, the next step is data redistribution. Data redistribution is to decide how to redistribute overflow data from data node to storage node. This has been shown to be a minimum cost flow problem [3], which can be solved efficiently. Figure 1(b) shows data redistribution step post data aggregation. After data aggregation, the total size of overflow data is equal to available storage space. One possible data redistribution solution is redistributing C's 1 unit of data to B, D's 1 unit of data to B via C and D's 2 units of data to E. Finally, any data node does not have overflow data.

4   Data Aggregation for Overall Storage Overflow Problem

4.1   Data Aggregation Formulation

In the sensor network, all sensor nodes have a limited energy source, typically in the form of a battery. It is awkward and unreasonable to replace the energy source of node. For data node, it costs more energy than storage node as it collects and saves massive data items. The remaining energy of the data node is generally lower than the residual energy of the storage nodes. It is therefore significant to save the data node's energy for prolonging the lifetime of network (the time until the first node depletes its energy in the network).

Let $V_{DN} = \{DN_1, DN_2, \cdots, DN_p\}$ denotes the set of data nodes and there is a set of aggregation walks: $W = \{W_1, W_2, \cdots, W_a\}$, where each walk $W_i (1 \leq i \leq a)$ start from a distinct initiator. Each node have its own energy E_i. Let E_i' denote node's residual energy after data aggregation. Then,

$$E_i' = E_i - \sum_{j=1}^{a} C_{i,W_j} \tag{4}$$

where C_{i,W_j} is the energy cost of node i in the aggregation walk W_j by transmitting or receiving data items. If node i is not in the aggregation walk W_j, $C_{i,W_j} = 0$. The objective of data aggregation is to find a set of aggregation walk $W = \{W_1, W_2, \cdots, W_a\}$, such that the minimum energy among all data node V_{DN} is maximized post aggregation, while saving as much energy as possible.

$$\max_{W} \min_{1 \leq i \leq p} E'_{DN_i} \tag{5}$$

under the energy constraint that each node can not spend more energy than its own energy, $E'_i \geq 0, \forall i \in V$.

4.2 Data Aggregation Algorithm

Since the main participant in the process of data aggregation is the data node, we firstly transform the original sensor network $G(V, E)$ into an aggregation network $G'(V', E')$. In the aggregation network $G'(V', E')$, V' is set of p data node in V. For any two data node $v_i, v_j \in V'$, if there exists an edge $(v_i, v_j) \in E$, we add the same edge in $G'(V', E')$, thus $(v_i, v_j) \in E'$. Otherwise we find the shortest paths between node v_i and v_j, and add a new edge in the aggregation network. Its weight of the new added edge is the cost of the shortest path between those two nodes. Therefore the aggregation network is a complete graph. In this paper, we introduce a new variable to represent the weight of edge between data nodes in the network.

Definition 1 (Quality of Edge). *For any two data nodes, $v_i, v_j \in V$ and $e_{ij} = (v_i, v_j) \in E$, $Q(e_{ij}) = \frac{d_{ij}^2}{E_{ij}}$, where d_{ij} is the distance between node v_i and v_j, and $E_{ij} = \min(E_{v_i}, E_{v_j})$ is the minimum energy between node v_i and v_j.*

The quality of edge is proportional to the square of the distance between two data nodes in the edge. The idea behind the quality of the edge is that the larger quality of the edge, the two node in this edge will have less energy or longer distance, then less likely the edge will be selected as data aggregation path. We use the quality represent the weight of edge to help us to select the aggregation path. In the aggregation network, if two data nodes are not directly connected in the original sensor network, the distance d_{ij} is the cost of the shortest path between those two nodes, and E_{ij} is the minimum energy of node in the found

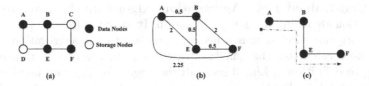

Fig. 2. (a) Original sensor network G. (b) Aggregation network G'. (c) Aggregation walk.

shortest path. Figure 2(a) shows the original wireless sensor network. Figure 2(b) is the corresponding aggregation network G'. In the original grid sensor network Fig. 2(a), we assume that the distance between any pair of connected nodes is 1 and the energy of every node is 2. The quality of edge is marked on every edge in Fig. 2(b) according to Definition 1.

Algorithm 1. Degree-Constrained Data Aggregation Algorithm

Input: $G(V, E)$ and the number of aggregators q
Output: The set of aggregation walks W and $E'_{min_{DN}}$
    Notations:
    e_i: the edge in the aggregation network graph;
    $v_{e'_i}$: one node of the edge e_i;
    $v_{e''_i}$: the other node of the edge e_i;
    $rf(v_i)$: the number of reference of the node v_i;
    $Q(e_i)$: the quality of the edge e_i;
    $E'_{min_{DN}}$: the minimum remaining energy of data node;
1: Transform $G(V, E)$ into $G'(V', E')$;
2: Calculate the quality of each edge in $G'(V', E')$;
3: Sort edges' quality in $G'(V', E')$, $Q(e_1) \leq Q(e_2) \leq \ldots \leq Q(e_N)$);
4: **for** $1 \leq j \leq |V'|$ **do**
5:     Initialize node $rf(v_j) = 2$;
6: **end for**
7: $W = \phi$, $count = i = 1$;
8: **while** $count \leq q$ **do**
9:     **if** $rf(v_{e'_i} > 0)$ and $rf(v_{e''_i} > 0)$ and (e_i in W will not induce cycle) **then**
10:         $W = W \cup \{e_i\}$;
11:         $rf(v_{e'_i}) - -$;
12:         $rf(v_{e''_i}) - -$;
13:         $count + +$;
14:     **end if**
15:     $i + +$;
16: **end while**
17: **for** $1 \leq j \leq |W|$ **do**
18:     Aggregate data along W_j from one end which has the smaller quality of edge;
19: **end for**
20: find the minimum remaining energy of data node $E'_{min_{DN}}$;
21: **return** W and $E'_{min_{DN}}$;

Degree-Constrained Data Aggregation Algorithm. Now we present an approximation algorithm for data aggregation. It works as follows. Line 1 transforms the original wireless sensor network graph into the aggregation network graph. Line 2 calculates the quality of each edge in the aggregation network according to definition 1. Line 3 sorts all the edges' quality into nondecreasing order. Lines 4–6 initialize the number of reference of each node to be 2. That is, the degree of each node in the set of aggregation walks will not exceed 2. The while loop in lines 8–16 check if each edge in W is cycleless and the number of reference of each node is greater than 0. If yes, add it into W. This continues until q edges are added into W. After that, it starts from one end which has the smaller quality of edge and aggregate overflow data via visiting the rest nodes. Figure 2(c) shows the aggregation walk which generated by the algorithm corresponding to aggregation network graph Fig. 2(b).

Time Complexity. Due to space constraints, the analysis is omitted. The time complexity of this algorithm is $O(|V|^3)$.

5   Integrating Data Aggregation and Data Redistribution

To overcome overall storage overflow problem in the sensor network, we have a two-step solution. But this solution does not necessarily achieve good performance. A unified method is proposed [2] which is based on data replication techniques. Data replication technology for overall storage overflow problem is that using storage node which is on the aggregation walk to replicate part or all overflow data of initiator in the process of data aggregation. However the total size of replicated data on any storage node along any aggregation walk cannot exceed this node's available storage space. As it does not consume extra energy, it saves a lot of energy in the step of data redistribution.

Example 2. Figure 1 shows the example of two-step solution for overall storage overflow problem. Considering that initial energy E of each node is 10, and the energy cost is 1 by transmitting 1 data item. Thus, data aggregation and redistribution cost is 6 and 5 respectively, and the residual energy of each node is marked under the node in Fig. 1. The total energy cost of two-step solution is 11 and the minimum residual energy of data node is 6. Figure 3 illustrate the data replication technology with the same sensor network which described in Fig. 1. It shows that when initiator node A sends its 2 units of data passing storage node B, it replicates 1 unit of the data (marked in parentheses) and stores at B. Therefore, next in data redistribution step, node D only needs redistribute 2 units of data to node E. Finally the total energy cost is 9 which has an 18% improvement compared to two-step solution, while the minimum residual energy of data node is 7, having a 17% improvement.

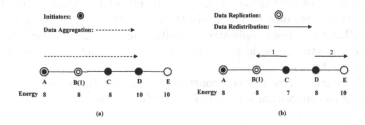

Fig. 3. Data replication technology for same sensor network in Fig. 1.

Since our Degree-Constrained Data Aggregation Algorithm can find the set of aggregation walks and data redistribution can be solved by minimum cost flow algorithm, the challenges of data replication technology are how to select initiator for each aggregation walk and how many units of data to replicate at storage nodes which in the aggregation walk. In each aggregation walk, the initiator has two choices. For example, in Fig. 3(a) the initiator can be node A or node D which can lead to different energy consumption. Observing that the last aggregator has more overflow data than other data nodes after data aggregation and having more available storage nodes around the last aggregator would make the data redistribution more energy-efficient. Therefore we select the node which surrounded by less storage nodes as the initiator in each aggregation walk. For the second challenge, we give below definition.

Definition 2 (Demand Number of Storage Node). *For any storage node u on any aggregation walk, let $N(u)$ be all its one-hop neighbor nodes. For each data node $v \in N(u) \cap V_{DN}$, let $D_{u,v}$ represent the distance between node u and node v, and $V_{SN} = V - V_{DN}$ denotes the set of storage nodes. The demand number $d(u)$ of storage node u,* $d(u) = \sum_{v \in N(u) \cap V_{DN}} \dfrac{1}{\sum_{q \in N(v) \cap V_{SN}} \frac{D_{u,v}^2}{D_{q,v}^2}}$.

Noted that $\sum_{q \in N(v) \cap V_{SN}} \frac{D_{u,v}^2}{D_{q,v}^2} > 0$ since node v has at least one neighboring storage node u. And if each node v just has one neighboring storage node u, the value of $d(u)$ is equal to the number of data nodes which surround node u. The idea behind $d(u)$ is that the less number of data nodes surrounding u and the more number of storage nodes surrounding such data nodes with shorter distance, the more unites of data items should be replicated at storage node u. Next we give a data replication algorithm, it works as follows. In each aggregation walk, the initiator sends all its overflow data to the next node along the walk. If a data node receives the overflow data, it just aggregates its own data and then sends the received data to the next node. For a storage node u receiving the data, firstly it calculates its own demand number $d(u)$. And then calculate the amount of data to be replicated as $\min\left(\frac{z}{d(u)}, z, s\right)$, where z is the rest of overflow data which has not been replicated and s represents the available storage space of this node. Finally node u replicates the calculated units of data in its storage space and relays the entire overflow data to the next node along the aggregation walk. This continues until the last aggregator receives the overflow data. The last aggregator aggregates its data and keeps the rest units (may be zero) of overflow data which have not been replicated. Finally, the aggregated overflow data is redistributed to storage node which has available storage space by minimum cost flow algorithm [3].

Time Complexity. Due to space constraints, the analysis is omitted. The time complexity of this algorithm is $O(|V|^2 |E| \log(|V|C))$, where $C = \max\{R+r, m\}$.

Algorithm 2. Data Replication Algorithm

Input: The sensor network G, and the set of aggregation walk W
Output: Minimum remaining energy of data node $E''_{min_{DN}}$
    Notations:
    $|W_i|$: the number of nodes on the aggregation walk W_i;
    $mcfa$: minimum cost flow algorithm;
    $V_j.space$: available storage space of storage node V_j;
1: $a = |W|$;
2: **for** $1 \leq i \leq a$ **do**
3:    Let V_{ini} and V_{agg} be the initiator and the last aggregator on the aggregation walk W_i respectively;
4:    V_{ini} sends all its overflow data to the next node along W_i;
5:    $z = R, b = |W_i|$;
6:    **for** $2 \leq j \leq (b-1)$ **do**
7:      **if** ($V_j \in W_i$ is data node) **then**
8:        V_j aggregates its own data with overflow data of V_{ini};
9:      **else**
10:        $s = V_j.space$;
11:        **if** ($s > 0$ and $z > 0$) **then**
12:          Calculate $d(V_j)$;
13:          $t = \min\left(\frac{z}{d(V_j)}, z, s\right)$;
14:          Replicate t units of overflow data on V_j;
15:          $z = z - t$;
16:        **end if**
17:      **end if**
18:      V_j sends the entire overflow data of V_{ini} to the next node along W_i;
19:    **end for**
20:    V_{agg} aggregates its own data and keeps the rest z units of data of V_{ini};
21: **end for**
22: $E''_{min_{DN}} = mcfa(G)$;
23: **return** $E''_{min_{DN}}$;

---

6   Performance Evaluation

This section presents the effectiveness of our proposed algorithms for overall storage overflow problem. Extensive experiments were performed in Java. In our experiment, we adopt first order radio model [15]. For node u send R-bit data to its neighbor v over their distance d, the transmission energy cost at u is $E_t(R, d) = E_{elec} \times R + \epsilon_{amp} \times R \times d^2$, and the receiving energy cost at v is $E_r(R) = E_{elec} \times R$, where $E_{elec} = 100\,\text{nJ/bit}$ and $\epsilon_{amp} = 100\,\text{pJ/bit/m}^2$. 50 and 100 sensor nodes are scattered randomly across a $1000 \times 1000\,\text{m}^2$ network, in which no two nodes can be in the same location. The transmission range of each node is $250\,\text{m}$. For data node, the initial energy is randomly around $600\,\text{J}$–$700\,\text{J}$, while the initial energy of each storage node is randomly around $900\,\text{J}$–$1000\,\text{J}$. Unless otherwise mentioned, the sensor network consists of 50 nodes, and $R = m = 1\,\text{MB}$. To eliminate the impact of randomness, each experiment scenario is repeated 100 times.

6.1   Performance of Data Aggregation Algorithm

For data aggregation algorithm, we compare the performance of our Degree-Constrained Data Aggregation algorithm (denoted as DCDA) with STF-Walk [1] and LP-Walk [1] algorithm. STF-Walk algorithm is a $(2 - \frac{1}{q})$-approximation data aggregation algorithm, while LP-Walk is a novel heuristic algorithm.

We compare DCDA algorithm with STF-Walk algorithm where considering $r/R = 0.5$ and the whole valid range of $p \in [26, 33]$. Figure 4(a) shows the total aggregation cost of STF-Walk and DCDA algorithms. With the increase of the number of data nodes p, total aggregation costs of both STF-Walk and DCDA increase. It's obviously that the DCDA algorithm yields less cost than STF-Walk. This is because STF-Walk algorithm visits some edges twice in the process of data aggregation, while DCDA algorithm tries to visit some edges which has smaller weight instead of edges' second visiting. Figure 4(b) shows the minimum remaining energy of data node after data aggregation corresponding to Fig. 4(a). The minimum remaining energy of data node decrease with increase p in both DCDA and STF-Walk algorithms. As DCDA algorithm considers different node with different remaining energy and selects the data node with higher priority which has higher remaining energy to participate in the process of data aggregation, the minimum remaining energy of data node in DCDA algorithm is always higher than the remaining energy in STF-Walk algorithm. And the performance difference between DCDA and STF-Walk algorithm gets bigger with the increase of number of data nodes.

LP-Walk algorithm which is a novel heuristic algorithm outperforms STF-Walk algorithm in total energy consumption. We adopt $r/R = 0.3$ and 0.7, for $r/R = 0.3$, the valid range of p is from 26 to 37, while $p \in [26, 29]$ for $r/R = 0.7$. Figure 5(a) is the aggregation energy cost by varying r/R and p. And Fig. 5(b) is corresponding the minimum remaining of data node after data aggregation. It shows that for the same p, with the increase of r/R, the total aggregation cost for both DCDA and LP-Walk algorithm increase and the minimum remaining energy of data node post aggregation decrease. The reason is that less redundant

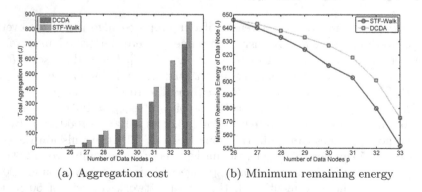

(a) Aggregation cost	(b) Minimum remaining energy

Fig. 4. Comparing DCDA with STF-Walk by varying p where $r/R = 0.5$.

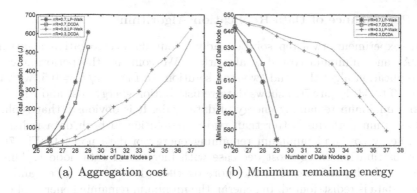

(a) Aggregation cost   (b) Minimum remaining energy

Fig. 5. Comparing DCDA with LP-Walk by varying p and r/R.

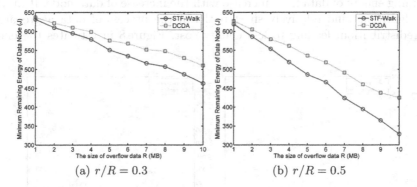

(a) $r/R = 0.3$ (b) $r/R = 0.5$

Fig. 6. Comparing DCDA with LP-Walk in minimum remaining energy of data node by varying R where $p = 30$.

data between data nodes leads to more number of aggregator are visited, thus increasing aggregation energy cost and reducing the minimum remaining energy of data node. However DCDA algorithm outperforms LP-Walk algorithm in both aggregation cost and minimum remaining energy of data node for the same reason. In Fig. 6(a), $p = 30$ and $r/R = 0.3$, we vary R from 1 MB to 10 MB while in Fig. 6(b) $p = 30$ and $r/m = 0.5$. It's obviously that as the increase of the size of overflow data, the minimum remaining energy of data node post aggregation in both two algorithm decrease, and the minimum remaining energy in Fig. 6(b) decrease faster than that in Fig. 6(a). This is because with the increase of r/R, the more number of aggregators should be visited, it costs more energy for data nodes. However the minimum remaining energy of data node in DCDA algorithm is still higher than that in LP-Walk algorithm and the performance difference get bigger with the increase of R. Therefore DCDA algorithm is more energy efficient and can prolong the lifetime of the sensor network compared to LP-Walk algorithm.

6.2 Performance of Data Replication Algorithm

In this experiment, two-step solution adopts our data aggregation algorithm (DCDA) and minimum cost flow algorithm. We compare the performance of data replication algorithm and two-step solution. In Fig. 7, $r/R = 0.5$, we vary p from 26 to 33. Figure 7(a) shows data redistribution energy cost and Fig. 7(b) presents minimum remaining energy of data node. It is obviously that replication algorithm performs better than two-step solution in both data redistribution energy cost and minimum remaining energy of data node. In Fig. 7(a), data redistribution energy cost decrease with increase of data node p. This is because with the increase of p, the more overflow data is aggregated and less overflow data is redistributed. In general, the minimum remaining energy of data node decrease with increase of data node. However, in some case, the minimum remaining energy of data node increase with the increase of data node. It maybe that algorithms find relatively short paths in the process of data aggregation and redistribution, leading to less energy cost. Figure 8 investigates the effect

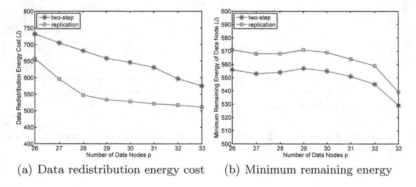

(a) Data redistribution energy cost   (b) Minimum remaining energy

Fig. 7. Comparing data replication algorithm with two-step solution by varying p where $r/R = 0.5$.

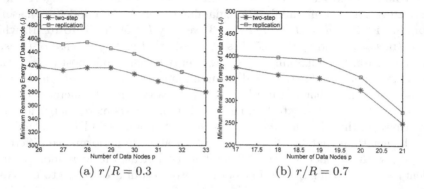

(a) $r/R = 0.3$ (b) $r/R = 0.7$

Fig. 8. Comparing replication algorithm with two-step solution in minimum remaining energy of data node by varying p where $R/m = 5$.

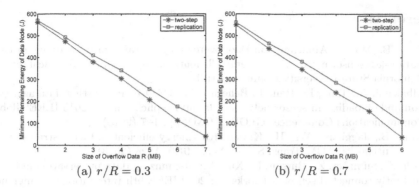

(a) $r/R = 0.3$ (b) $r/R = 0.7$

Fig. 9. Comparing replication algorithm with two-step solution in minimum remaining energy of data node by varying R where $p = 28$.

of $R/m = 5$ and $m = 1\,\text{MB}$. Figure 8(a) shows the performance of data replication algorithm and two-step solution in minimum remaining energy of data node where $r/R = 0.3$ while in Fig. 8(b) $r/R = 0.7$. It shows the same trend as Fig. 7(b). For data replication algorithm, it replicates data items at storage nodes, saving a lot of energy of sensor nodes during data redistribution. And it still has advantage over two-step solution.

Figure 9 presents the effect of the size of overflow data where $p = 28$. We vary R from 1 MB to 7 MB. With the increase of R, it becomes more challenging since there are more overflow data. In Fig. 9(a), $r/R = 0.3$ while $r/R = 0.7$ in Fig. 8(b). We observe that with the increase of R, the minimum remaining energy of data node decrease linearly. And with the same R, the minimum remaining energy of data node in $r/R = 0.7$ is lower than that in $r/R = 0.3$. This is because with the increase of r/R, more aggregators are visited, it costs more energy. In Fig. 8(b), the sensor network can not finish data aggregation and data redistribution work as some data node delept their energy when $R = 7\,\text{MB}$. However, data replication algorithm outperforms than two-step solutioin. And the performance difference gets larger with the increase of R. This again demonstrates the effectiveness of replication algorithm.

7 Conclusions and Future Work

In this paper, we study overall storage overflow problem in wireless sensor network, the goal of which is to maximize the minimum energy of data node. To our best knowledge, the problem has not been addressed by any of existing research. And we propose energy-efficient data aggregation and data replication algorithms. Via extensive simulations, it shows that our algorithms can effectively prolong the lifetime of sensor network compared with existing algorithms. As for future work, we will consider that different data nodes may have different size of overflow data. And in order to adapt to large scale sensor networks, we will design distributed algorithms.

References

1. Tang, B., Ma, Y., Alhakami, B.: Dao: overcoming overall storage overflow in base station-less sensor networks via energy efficient data aggregation. Technical report, California State University Dominguez Hills
2. Alhakami, B., Tang, B., Han, J., Beheshti, M.: DAO-R: integrating data aggregation and offloading in sensor networks via data replication. In: 2015 IEEE Global Communications Conference (GLOBECOM), pp. 1–7 (2015)
3. Tang, B., Jaggi, N., Wu, H., Kurkal, R.: Energy-efficient data redistribution in sensor networks. ACM Trans. Sensor Netw. 9(2), 1–28 (2013)
4. Hou, X., Sumpter, Z., Burson, L., Xue, X.: Maximizing data preservation in intermittently connected sensor networks. In: 2012 IEEE 9th International Conference on Mobile Ad-Hoc and Sensor Systems (MASS), pp. 448–452 (2012)
5. Takahashi, M., Tang, B., Jaggi, N.: Energy-efficient data preservation in intermittently connected sensor networks. In: 2011 IEEE Conference on Computer Communications Workshops (INFOCOM WKSHPS), pp. 590–595 (2011)
6. Xue, X., Hou, X., Tang, B., Bagai, R.: Data preservation in intermittently connected sensor networks with data priority. In: 2013 10th Annual IEEE Communications Society Conference on Sensor, Mesh and Ad Hoc Communications and Networks (SECON), pp. 122–130 (2013)
7. Tang, B., Bagai, R., Nilofar, F., Yildirim, M.B.: A generalized data preservation problem in sensor networks–a network flow perspective. In: International Conference on Ad-Hoc Networks and Wireless (ADHOC-NOW), pp. 275–289 (2014)
8. Kuo, T.W., Lin, K.C.J., Tsai, M.J.: On the construction of data aggregation tree with minimum energy cost in wireless sensor networks: NP-completeness and approximation algorithms. IEEE Trans. Comput. 65(10), 3109–3121 (2016)
9. Chen, Z., Yang, G., Chen, L., Xu, J.: Constructing maximum-lifetime data-gathering tree in wsns based on compressed sensing. Int. J. Distrib. Sensor Netw. 12(5), 1–11 (2016)
10. Yan, X., Du, H., Ye, Q., Song, G.: Minimum-delay data aggregation schedule in duty-cycled sensor networks. In: Yang, Q., Yu, W., Challal, Y. (eds.) WASA 2016. LNCS, vol. 9798, pp. 305–317. Springer, Cham (2016). https://doi.org/10.1007/978-3-319-42836-9_28
11. Lee, T., Kim, D.S., Choo, H., Kim, M.: A delay-aware scheduling for data aggregation in duty-cycled wireless sensor networks. In: 2013 IEEE 9th International Conference on Mobile Ad-Hoc and Sensor Networks (MSN), pp. 254–261 (2013)
12. Liu, C., Du, H., Ye, Q.: Sweep coverage with return time constraint. In: 2016 IEEE Global Communications Conference (GLOBECOM), pp. 1–6 (2016)
13. Alia, O.M.: Dynamic relocation of mobile base station in wireless sensor networks using a cluster-based harmony search algorithm. Inf. Sci. 385–386, 76–95 (2017)
14. Cristescu, R., Beferull-Lozano, B., Vetterli, M., Wattenhofer, R.: Network correlated data gathering with explicit communication: NP-completeness and algorithms. IEEE/ACM Trans. Netw. 14(1), 41–54 (2006)
15. Heinzelman, W.R., Chandrakasan, A., Balakrishnan, H.: Energy-efficient communication protocol for wireless microsensor networks. In: 33rd Annual Hawaii International Conference on System Sciences (2000)

Floorplans with Columns

Katsuhisa Yamanaka[1], Md. Saidur Rahman[2], and Shin-Ichi Nakano[3]([envelope])

[1] Iwate University, Morioka, Japan
[2] Bangladesh University of Engineering and Technology, Dhaka, Bangladesh
[3] Gunma University, Kiryu, Japan
nakano@cs.gunma-u.ac.jp

Abstract. Given an axis-aligned rectangle R and a set P of n points in the proper inside of R we wish to partition R into a set S of $n + 1$ rectangles so that each point in P is on the common boundary between two rectangles in S. We call such a partition of R a feasible floorplan of R with respect to P. Intuitively P is the locations of columns and a feasible floorplan is a floorplan in which no column is in the proper inside of a room, i.e., columns are allowed to be placed only on the partition walls between rooms. In this paper we give an efficient algorithm to enumerate all feasible floorplans of R with respect to P. The algorithm is based on the reverse search method, and enumerates all feasible floorplans in $O(|S_P|)$ time using $O(n)$ space where S_P is the set of the feasible floorplans of R with respect to P, while the known algorithms need either $O(n|S_P|)$ time and $O(n)$ space or $O(\log n|S_P|)$ time and $O(n^3)$ space.

Keywords: Enumeration · Floorplan · Algorithm

1 Introduction

Given an axis-aligned rectangle R and a set P of n points in the proper inside of R we wish to partite R into a set S of $n + 1$ rectangles so that each point in P is on the common boundary between two rectangles in S. We call such a partition of R a feasible floorplan of R with respect to P. Figure 1(b) illustrates the 22 feasible floorplans of R with respect to the point set P in Fig. 1(a). For simplicity we assume no two points have the same x-coordinate, and no two points have the same y-coordinate. Intuitively P is the locations of columns and a feasible floorplan is a floorplan in which no column is in the proper inside of a room, i.e. columns are allowed to be placed only on the partition walls between rooms.

Ackerman et al. [ABP1, ABP2] gave an algorithm to enumerate all feasible floorplans with respect to P. The algorithm is based on the reverse search method [A1, AF1] and enumerates all feasible floorplans in either $O(n|S_P|)$ time using $O(n)$ space or $O(\log n|S_P|)$ time using $O(n^3)$ space, where S_P is the set of feasible floorplans with respect to P.

In this paper we design a faster algorithm, which is also based on the reverse search method. Our algorithm is simple and uses only $O(n)$ space, and enumerates all feasible floorplans in $(|S_P|)$ time. Using a similar method we have designed efficient enumeration algorithms [N1, N2, LN1].

© Springer International Publishing AG 2017
X. Gao et al. (Eds.): COCOA 2017, Part I, LNCS 10627, pp. 33–40, 2017.
https://doi.org/10.1007/978-3-319-71150-8_3

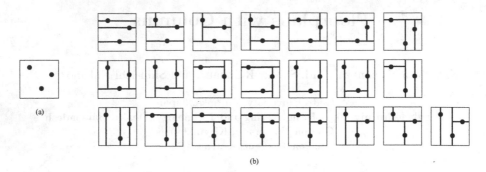

Fig. 1. (a) An example of a rectangle R and a set P of three points in R, (b) all feasible floorplans of R with respect to P.

The rest of the paper is organized as follows. Section 2 gives some definitions. Section 3 defines a tree structure among the feasible floorplans. Section 4 gives our enumeration algorithm. Finally Sect. 5 is a conclusion.

2   Preliminaries

In this section we give some definitions.

A *floorplan* is a partition of an axis-aligned rectangle R into a set S of rectangles. In this paper we have typical assumption for floorplans that no four rectangles in S share a common corner point in a floorplan. We call R the outer rectangle and each rectangle in S *a face*. Given an axis-aligned rectangle R and a set P of n points in R *a feasible floorplan of R with respect to P* is a partition of R into a set S of $n + 1$ rectangles (or faces) so that each point in P is on the common boundary between two rectangles (or faces) in S. We assume that no two points have the same x-coordinate, and no two points have the same y-coordinate. Then one can observe that every maximal line segment not on R contains exactly one point in P in any feasible floorplan of R with respect to P. Let S_P be the set of all feasible floorplans with respect to P. For R and P in Fig. 1(a) all feasible floorplans in S_P is illustrated in Fig. 1(b).

Let Q be a feasible floorplan of R with respect to P. A maximal line segment containing no end vertex of other line segment is called *a basic line segment*. Each maximal line segment consists of one or more basic line segments. A maximal vertical line segment contains a point $p \in P$ is $type(u, d)$ if it contains u end points of maximal horizontal line segments above p and d end points of maximal horizontal line segment below p. Thus a basic vertical line segment is $type(0, 0)$.

3   Family Tree

In this section we define a tree structure among the feasible floorplans of R with respect to P.

Let Q_r be the feasible floorplan of R with respect to P such that every point in P is on a horizontal line segment.

Given a feasible floorplans $Q \neq Q_r$ of R with respect to P, we define the parent feasible floorplan $P(Q)$ of R with respect to P, as follows. Let s be the leftmost maximal vertical line segment in Q except the left vertical line segment of R, and $p \in P$ be the point on s. We have the following two cases to consider.

Case 1 s is $type(0,0)$

In this case we (1) remove s from Q and then (2) append a horizontal line segment containing p as a basic horizontal line segment as illustrated in Fig. 2. Intuitively this is a rotation of s.

Case 2 Otherwise

In this case we (1) remove s from Q, then (2) extend to left each maximal horizontal line segment having left end on s so that it has the same number of basic line segments as it was, then (3) extend to right each maximal horizontal line segment having right end on s so that it has the same number of basic line segments as it was, and finally (4) append a horizontal line segment containing p as a basic horizontal line segment. Intuitively this is a rotation of s after shrinking. See Fig. 3.

Note that the number of maximal vertical line segments of $P(Q)$ is decreased by one from the one of Q, and the rotated line segment is always basic in $P(Q)$. We have defined $P(Q)$ for each feasible floorplan Q except Q_r. We say $P(Q)$ is the *parent* of Q and Q is a *child* of $P(Q)$.

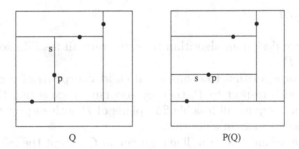

$$Q \qquad\qquad\qquad P(Q)$$

Fig. 2. The parent floorplan $P(Q)$ of Q in Case 1.

Given a feasible floorplan Q in S_P, by repeatedly computing its parent, we can have the unique sequence $Q, P(Q), P(P(Q)), \cdots$ of feasible floorplans with respect to P which eventually ends with Q_r. See an example of such sequence in Fig. 4. We call the sequence *the removing sequence* of Q.

By merging those sequences we define the family tree T_P of S_P such that the vertices of T_P correspond to feasible floorplans of R with respect to P, and each edge corresponds to each relation between some Q and $P(Q)$. See Fig. 5. Note that each vertex of depth i in T_P corresponds to a feasible floorplan with i

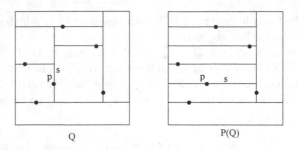

Fig. 3. The parent floorplan $P(Q)$ of Q in Case 2.

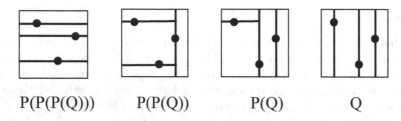

Fig. 4. The removing sequence.

vertical line segments except the two vertical line segments on R, and the height of the family tree is n.

4   Algorithm

In this section we design an algorithm to enumerate all feasible floorplans of R with respect to P.

If we have an algorithm to compute all child floorplans of a given feasible floorplan of R with respect to P, then by recursively executing the algorithm from Q_r, we can compute all feasible floorplans of R with respect to P. We are now going to design such an algorithm.

Let s' be the leftmost vertical line segment in Q except the left vertical line segment of R, and $p' \in P$ be the point on s'. (Thus if $Q = Q_r$ then s' is the right vertical line segment of R, and we regard any point on s as p'.) One can observe that each child floorplan of Q is one of the following two types. Let s be a basic horizontal line segment containing a point $p \in P$ locating left of p'.

Type 1: $C(s, 0, 0)$

$C(s, 0, 0)$ is the floorplan constructed from Q by (1) removing s from Q then (2) appending a vertical line segment containing p as a basic line segment.

Note that $C(s, 0, 0)$ is also a feasible floorplan with respect to P. Intuitively this is the child floorplan derived from Q by rotation of s.

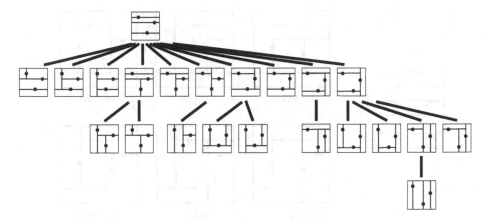

Fig. 5. The family tree.

Type 2: $C(s, u, d)$

Let u' be the number of maximal horizontal line segments above p and d' the number of maximal horizontal line segments below p in Q. For example for the floorplan in Fig. 6, $u' = 3$ and $d' = 2$. For two integers $u < u'$ and $d < d'$, $C(s, u, d)$ is the floorplan constructed from Q by (1) removing s from Q then (2) appending a vertical line segment s' containing p as a basic line segment then (3) extending s' upward and downward so that it becomes $type(u, d)$ then (4) shrinking each maximal horizontal line segment intersecting s' so that it has an end point on s'. See examples in Fig. 6. Intuitively this is the child floorplan derived from Q by rotation then extension of s.

Lemma 1. $C(s, u, d)$ *is a child of Q if and only if every shrinked horizontal line segment having a point of P on the left of p consists of exactly one basic line segment in Q. (We do not care each shrinked horizontal line segment having a point of P on the right of p.)*

Proof (sketch). Assume that some shrinked horizontal line segment has a point of P on the left of p but consists of two or more basic line segments in Q. For example see Fig. 7. The horizontal line segment t just above p has a point of P on the left of p but consists of three basic line segments in Q. We need to shrink t in $C(s, 1, 0)$. For any suitable definition of $C(s, 1, 0)$ each basic line segments of t except the leftmost one cannot exists in $C(s, 1, 0)$ (as depicted as dashed line segments) since we have cut t at s. Thus $P(C(s, 1, 0))$ is not Q and $C(s, 1, 0)$ is not a child of Q. □

Note that if s is a basic horizontal line segment containing a point in P locating right of p' then resulting floorplan $C(s, u, d)$ is not a child of Q, since the leftmost vertical line segment of the resulting floorplan is s' not s. Thus we do not need check $C(s, u, d)$ with such s.

Based on the above observation we can enumerate all child floorplan of given Q.

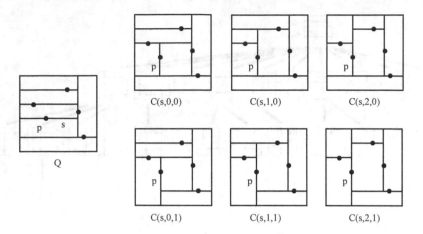

Fig. 6. The children.

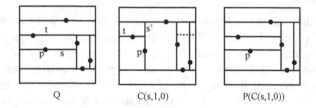

Fig. 7. $C(s, 1, 0)$ is not a child of Q.

We now explain data structures required for our algorithm above. We regard each corner of a rectangle as a vertex and each basic line segment as an edge and a floorplan as a graph. We store and maintain the current floorplan using some standard data structure for plane graphs during the execution of our enumeration algorithm. This part needs $O(n)$ space. We can efficiently trace the basic segments on the boundary of each face. Also given a vertex and a direction (up/down/left/right) we can find the neighbour vertex in constant time.

We also maintain the list of the basic horizontal line segments located left of the leftmost vertical line segment. We assume the basic horizontal line segments are sorted in the list by the x-coordinates of the points in P on the basic horizontal line segments. For Q_r such list can be constructed in $O(n \log n)$ time. For any feasible floorplan with respect to P the list is a prefix of the list of Q_r. Thus we need $O(n)$ space for the list and can update it efficiently.

For each recursive call we need a constant amount of memory and the depth of the call is at most n so this part needs $O(n)$ space in total. Thus we need $O(n)$ space in total.

We have the following lemma.

Lemma 2. *Given a child floorplan $C(s, u, d)$ of Q one can check if $C(s, u+1, d)$ is a child floorplan of Q or not, and if it is a child floorplan of Q one can generate $C(s, u + 1, d)$ in constant time.*

Proof. Let t be the maximal horizontal line segment containing the upper end point of s in $C(s, u, d)$. Now t consists of two or more basic line segments in $C(s, u, d)$. We have the following three cases.

If t has a point in P on the left of s and t consists of exactly two basic line segments in $C(s, u, d)$, then removing the right basic line segment of t from $C(s, u, d)$ then extending s upward so that it has one more basic line segment results in $C(s, u + 1, d)$ and it is a child of Q. The number of different segments between them is clearly a constant.

If t has a point in P on the left of s and t consists of three or more basic line segments in $C(s, u, d)$, then $C(s, u + 1, d)$ is not a child of Q, as explained just after the definition of $Type(s, u, d)$.

If t has a point in P on the right of s then removing the leftmost basic line segment of t from $C(s, u, d)$ then extending s upward so that it has one more basic line segment results in $C(s, u + 1, d)$ and it is a child of Q. The number of different segments between them is also a constant.

Thus in constant time we can check if $C(s, u + 1, d)$ is a child of Q or not, and if it is a child one can generate $C(s, u + 1, d)$ from $C(s, u, d)$. $\square$

Intuitively we can generate $C(s, u + 1, d)$ from $C(s, u, d)$ by removing a suitable basic horizontal line segment having an end point at the upper end point of s then appending the basic vertical line segment having lower end point at the upper end point of s. See Fig. 6.

Similarly given $C(s, u, d)$ one can check if $C(s, u, d + 1)$ is a child or not, and if it is a child one can generate $C(s, u, d + 1)$ from $C(s, u, d)$ in constant time. Thus we have the following lemma.

Lemma 3. *One can enumerate all child floorplans of a given feasible floorplan Q with respect to P in $O(k)$ time, where k is the number of child floorplans of Q.*

Since we need $O(k)$ time for each vertex of the family tree, where k is the number of children of the floorplan corresponding to the vertex, the algorithm above runs in $O(|S_P|)$ time, where S_P is the set of feasible floorplans with respect to P.

We have the following theorem.

Theorem 1. *After $O(n \log n)$ time preprocessing one can enumerate all feasible floorplans with respect to P in $O(|S_P|)$ time and $O(n)$ space.*

5   Conclusion

In this paper we have designed a simple and efficient algorithm to enumerate all feasible floorplans with respect to a given set P of points. Our algorithm enumerate all such floorplans in $O(|S_P|)$ time and $O(n)$ space after $O(n \log n)$

time preprocessing, where S_P is the set of floorplans with respect to P, and $|P| = n$.

Can we enumerate all feasible floorplans with respect to P when some walls are fixed? Can we enumerate all feasible floorplans with respect to P when some rooms are fixed? Can we enumerate all feasible floorplans with respect to P so that each room has at least one window, which means each room must share some part of the boundary of R?

Acknowledgment. This work is partially supported by JSPS KAKENHI Grant Number JP16K00002 and JP17K00003.

References

[ABP1] Ackerman, E., Barequet, G., Pinter, R.Y.: On the number of rectangulations. In: Proceedings of SODA, pp. 729–738 (2004)

[ABP2] Ackerman, E., Barequet, G., Pinter, R.Y.: On the number of rectangulations of a planar point set. J. Comb. Theor. Ser. A **113**, 1072–1091 (2006)

[A1] Avis, D.: Generating rooted triangulations without repetitions. Algorithmica **16**, 618–632 (1996)

[AF1] Avis, D., Fukuda, K.: Reverse search for enumeration. Discrete Appl. Math. **65**, 21–46 (1996)

[N1] Nakano, S.: Enumerating floorplans with n rooms. In: Eades, P., Takaoka, T. (eds.) ISAAC 2001. LNCS, vol. 2223, pp. 107–115. Springer, Heidelberg (2001). https://doi.org/10.1007/3-540-45678-3_10

[N2] Nakano, S.: Efficient generation of plane trees. Inf. Process. Lett. **84**, 167–172 (2002)

[LN1] Li, Z., Nakano, S.: Efficient generation of plane triangulations without repetitions. In: Orejas, F., Spirakis, P.G., Leeuwen, J. (eds.) ICALP 2001. LNCS, vol. 2076, pp. 433–443. Springer, Heidelberg (2001). https://doi.org/10.1007/3-540-48224-5_36

A Parallel Construction of Vertex-Disjoint Spanning Trees with Optimal Heights in Star Networks

Shih-Shun Kao[1], Jou-Ming Chang[1]([✉]), Kung-Jui Pai[2], Jinn-Shyong Yang[3], Shyue-Ming Tang[4], and Ro-Yu Wu[5]

[1] Institute of Information and Decision Sciences,
National Taipei University of Business, Taipei, Taiwan
spade@ntub.edu.tw
[2] Department of Industrial Engineering and Management,
Ming Chi University of Technology, New Taipei City, Taiwan
[3] Department of Information Management,
National Taipei University of Business, Taipei, Taiwan
[4] Department of Psychology and Social Work,
National Defense University, Taipei, Taiwan
[5] Department of Industrial Management, Lunghwa University of Science
and Technology, Taoyuan, Taiwan

Abstract. Constructing vertex-disjoint spanning trees (VDSTs for short) of a given network is an important issue in the research of network fault-tolerance and security. The star network was proposed as an attractive interconnection network model for competing with n-cube. Accordingly, Rescigno in [Inform. Sci. 137 (2001) 259–276] proposed an algorithm to construct $n - 1$ VDSTs rooted at a common node in an n-dimensional star network S_n. In this paper, we point out that there exists a flaw in Rescigno's algorithm, and thus the spanning trees constructed by this algorithm may not be vertex-disjoint. As a result, a correct scheme of constructing $n - 1$ VDSTs on S_n is presented. Moreover, based on the reversing rule of building certain paths of VDSTs in the amendatory scheme, we propose a new algorithm to construct $n - 1$ VDSTs with optimal heights on S_n. In particular, the proposed algorithm is more efficient and can easily be implemented in parallel.

Keywords: Vertex-disjoint spanning trees · Interconnection networks · Star networks · Fault-tolerance · Network security

1 Introduction

As usual, the underlying topology of an interconnection network is modeled as a graph G, where the vertex set $V(G)$ represents the set of processing elements and the edge set $E(G)$ represents the set of connection links between processing elements. In this paper, for convenience, the terms "networks" and "graphs", "nodes" and "vertices", "links" and "edges" are often used interchangeably. An algorithm

© Springer International Publishing AG 2017
X. Gao et al. (Eds.): COCOA 2017, Part I, LNCS 10627, pp. 41–55, 2017.
https://doi.org/10.1007/978-3-319-71150-8_4

developed for an interconnection network is said to be *fully parallelized* if it could make use of all nodes of the network as processors for computation [6,20].

Two paths in a graph G are said to be *vertex-disjoint* (resp., *edge-disjoint*) if they have no common vertices except for the two extreme vertices (resp., if they have no common edges). A set of spanning trees of G are *vertex-disjoint* (resp., *edge-disjoint*) if they are rooted at the same vertex r such that, for any pair of them and for each vertex $v(\neq r)$ in G, the two different paths from v to r, one path in each tree, are vertex-disjoint (resp., edge-disjoint). Hereafter, vertex-disjoint and edge-disjoint spanning trees are referred as VDSTs and EDSTs for short, respectively. Constructing multiple spanning trees in interconnection networks has many practical applications such as fault-tolerant broadcasting and secure message distribution [5,13,15,17].

The star network, proposed by Akers and Krishnamurty [1], is one of the most popular architectures for interconnecting a large number of nodes in a parallel computing system. One advantage of the star network is that it is able to accommodate more nodes with less connection links and less communication delay than the hypercube, and thus is a viable alternative to the hypercube [2,8]. In addition, the star network enjoys a number of properties desirable in interconnection networks. The illustrious features of the star network include low degree of nodes, small diameter, vertex- and edge-symmetry, recursive decomposition, maximal fault-tolerance, and strong resilience. For more topological properties of star networks, we suggest readers to refer [1,14,16].

Based on these good features, many algorithms for solving diverse problems on star networks have been developed (e.g., see [3,4,9,10]). In particular, the construction of multiple spanning trees on star networks has been studied in several papers. For example, Fragopoulou and Akl [10] first considered some fundamental communication problems on star networks, and showed that a common approach to implement algorithms for solving these problems is to design communication protocols based on spanning trees with some special properties of the networks. At a later time, for the n-dimensional star graph S_n, Fragopoulou and Akl [11] proposed an algorithm to construct $n - 1$ EDSTs of S_n, and Bao et al. [5] proposed an algorithm to construct $n - 2$ VDSTs of S_n. Shortly after, an improvement of $n - 1$ VDSTs rooted at arbitrary node was provided by Rescigno [15]. Although this result is optimal in the sense that the number of VDSTs is maximized (i.e., the number of VDSTs has reached to the vertex-connectivity of S_n which is an upper bound), it is highly regrettable that we recently found a flaw in Rescigno's algorithm, and thus the proposed algorithm needs to be repaired.

In this paper, we revisit the problem of constructing VDSTs on S_n. For Rescigno's algorithm, we provide an amendatory scheme to recover the fault. In addition, based on the reversing rule of building certain paths of VDSTs in the amendatory scheme, we propose a new algorithm for constructing VDSTs on S_n. In particular, our approach can be fully parallelized such that each node can determine its parent in every spanning tree directly by only referring its own label and the index of a tree. Consequently, the proposed algorithm for constructing $n-1$ VDSTs on S_n can be run in $\mathcal{O}(n)$ time using $n!$ nodes of S_n as processors.

The remaining sections are organized as follows. Section 2 formally gives the definition of S_n and introduces a particular shortest path routing scheme of S_n. Section 3 presents Rescigno's algorithm for constructing VDSTs of S_n, and points out the existence of a critical flaw in this algorithm. Section 4 provides our amendatory scheme and shows its correctness. Section 5 proposes a newly parallel algorithm to construct $n-1$ VDSTs of S_n. The final section contains our concluding remarks.

2 The Star Graphs

Let Σ_n be the set of all permutations of symbols $\{1, 2, \ldots, n\}$. For a node $\mathbf{x} = x_1 \cdots x_n \in \Sigma_n$ and $i \in \{2, \ldots, n\}$, we denote $\mathbf{x}\langle i \rangle = x_i \cdots x_1 \cdots x_n$ as the element of Σ_n obtained from $\mathbf{x}$ by swapping the first symbol with the ith symbol. The n-star graph, denoted by S_n, has the vertex set $V(S_n) = \Sigma_n$ and edge set $E(S_n) = \{(\mathbf{x}, \mathbf{x}\langle i \rangle) : \mathbf{x} \in \Sigma_n, 2 \leqslant i \leqslant n\}$. Note that S_n is a Cayley graph with $n!$ nodes, is both vertex- and edge-symmetric, and is regular of degree $n-1$. Moreover, it has connectivity $\kappa(S_n) = n-1$ and diameter $D(S_n) = \lfloor 3(n-1)/2 \rfloor$ (see [1,2]). For $k, p \in \{1, \ldots, n\}$, we denote by $\mathscr{C}_p^k$ the subgraph of S_n induced by the vertices having symbol k in position p. Clearly, $\mathscr{C}_p^k$ is isomorphic to S_{n-1} if $p \neq 1$, and is a set of isolated vertices of S_n otherwise.

A permutation $\mathbf{x} \in \Sigma_n$ can be expressed by the representation of cyclic decomposition, i.e., the product of $c(\mathbf{x})$ disjoint cycles $K_1, K_2, \ldots, K_{c(\mathbf{x})}$ and $\psi(\mathbf{x})$ invariances, where a cycle $K_i = (k_1, \ldots, k_\ell)$ contains $\ell \geqslant 2$ distinct symbols such that the desired symbol of $\mathbf{x}$ at position k_j, $1 \leqslant j \leqslant \ell$, is that occupied by the next symbol k_{j+1} (where the index $j+1$ takes modulo ℓ), and an invariance is a symbol x_j such that $x_j = j$. In particular, all cycles $K_1, K_2, \ldots, K_{c(\mathbf{x})}$ are chosen in lexicographic order. Also, let $\Psi(\mathbf{x}) = \{j : x_j = j\}$ be the set of invariances of $\mathbf{x}$. Hence, K_1 always contains the symbol 1 if $1 \notin \Psi(\mathbf{x})$. In [15], a cycle K_i is called the good cycle of $\mathbf{x}$ provided i is the smallest index such that K_i does not contain the symbol 1.

Let $\mathbf{1} = 1 \cdots n$ be the identity permutation. For any vertices $\mathbf{x}, \mathbf{y} \in V(S_n)$, we use $P[\mathbf{x}, \mathbf{y}]$ to denote a path joining $\mathbf{x}$ and $\mathbf{y}$. Due to the vertex-symmetry of S_n, all paths from $\mathbf{x}$ to $\mathbf{y}$ are one-to-one correspondence with the paths from the vertex $\mathbf{x}\mathbf{y}^{-1}$ to $\mathbf{1}$ by multiplying each vertex in the latter path with the specific permutation $\mathbf{y}$. Hence, Day and Tripathi [8] gave the following lemma to characterize shortest paths between pair of vertices in S_n.

Lemma 1 (See [8]). A path $P[\mathbf{x}^0, \mathbf{1}] = (\mathbf{x}^0, \mathbf{x}^1, \ldots, \mathbf{x}^{h-1}, \mathbf{x}^h = \mathbf{1})$ joining vertices $\mathbf{x}^0 (\neq \mathbf{1})$ and $\mathbf{1}$ is a shortest path in S_n if and only if, for $i = 0, \ldots, h-1$, the vertex $\mathbf{x}^{i+1}$ is obtained from $\mathbf{x}^i$ by swapping symbols x_1^i and x_s^i (i.e., $x_1^{i+1} = x_s^i$ and $x_s^{i+1} = x_1^i$) such that s fulfills the condition: if $x_1^i = 1$, then s is any position with $x_s^i \neq s$; otherwise, either $s = x_1^i$ or s is any position belonging to a cycle of $\mathbf{x}^i$ such that the cycle does not contain the symbol 1.

From Lemma 1, Rescigno in [15] adopted a particular shortest path called the *basic path*, hereafter denoted by $\hat{P}[\mathbf{x}^0, \mathbf{1}]$, which is defined as follows.

Definition 1 (See [15]). A path $\hat{P}[\mathbf{x}^0, \mathbf{1}] = (\mathbf{x}^0, \mathbf{x}^1, \ldots, \mathbf{x}^{h-1}, \mathbf{x}^h = \mathbf{1})$ is called the *basic path* from the vertex $\mathbf{x}^0 (\neq \mathbf{1})$ to $\mathbf{1}$ if, for $i = 0, \ldots, h-1$, the vertex $\mathbf{x}^{i+1}$ is obtained from $\mathbf{x}^i$ by swapping symbols x_1^i and x_s^i such that s fulfills the condition: if $x_1^i \neq 1$ and $c(\mathbf{x}^i) = 1$, then $s = x_1^i$; otherwise, s is the smallest position of the good cycle of $\mathbf{x}^i$.

By Definition 1, it is easy to see that any subpath $(\mathbf{x}^i, \mathbf{x}^{i+1}, \ldots, \mathbf{x}^{h-1}, \mathbf{x}^h)$ with $i \geqslant 1$ is still a basic path from $\mathbf{x}^i$ to $\mathbf{1}$. Moreover, if $\mathbf{x}^0 \in \mathscr{C}_r^1$ and $r \neq 1$, it implies that $\mathbf{x}^i \in \mathscr{C}_r^1$ for all $i \in \{1, \ldots, h-1\}$.

3   Rescigno's Algorithm for Constructing VDSTs of S_n

In this section, we will introduce Rescigno's algorithm for constructing VDSTs of S_n. As the vertex-symmetry of S_n, without loss of generality, all spanning trees of S_n constructed by Rescigno are considered to be rooted at the vertex $\mathbf{1}$, and assume that $T_2, \ldots, T_n$ are such spanning trees in which the common root $\mathbf{1}$ of each spanning tree T_i for $i \in \{2, \ldots, n\}$ has a unique child $\mathbf{1}\langle i \rangle$. For convenience, a structure that contains all spanning tree by identifying the common root is called the *multiple spanning tree rooted at* $\mathbf{1}$ in [15]. Throughout this paper, we also use the following notation. The *length* of a path P, denoted by LEN(P), is the number of edges passing through P. The *distance* between two vertices $\mathbf{x}$ and $\mathbf{y}$, denoted by $d(\mathbf{x}, \mathbf{y})$, is the length of a shortest path from $\mathbf{x}$ to $\mathbf{y}$. Let $P \oplus Q$ denote the concatenation of two paths P and Q. The parent of a node $\mathbf{u}(\neq \mathbf{1})$ in T_i is denoted by PAR($\mathbf{u}, i$).

For a vertex $\mathbf{u}(= u_1 \cdots u_n) \in V(\mathscr{C}_r^1)$ where $r \in \{2, \ldots, n\}$, let t be the index such that $u_t = r$. Clearly, $t \neq r$. Also, let f be the smallest symbol in K_2 if $c(\mathbf{u}) \geqslant 2$, and $f = u_1$ otherwise. According to Rescigno's algorithm, a set of $n-1$ paths, denoted by $I_2(\mathbf{u}), \ldots, I_n(\mathbf{u})$, from $\mathbf{u}$ to a vertex of $V(\mathscr{C}_1^1)$ are described as follows. For $j \in \{2, \ldots, n\}$, define

$$I_j(\mathbf{u}) = \begin{cases} (\mathbf{u}, \mathbf{u}\langle j \rangle) \oplus \hat{P}[\mathbf{u}\langle j \rangle, \mathbf{1}] & \text{if } j = f, \\ (\mathbf{u}, \mathbf{u}\langle j \rangle) & \text{if } j = r, \\ (\mathbf{u}, \mathbf{u}\langle j \rangle, \mathbf{u}\langle j \rangle\langle f \rangle, \mathbf{u}\langle j \rangle\langle f \rangle\langle r \rangle) & \text{if } j = t, \\ (\mathbf{u}, \mathbf{u}\langle j \rangle, \mathbf{u}\langle j \rangle\langle r \rangle) & \text{if } j \neq r, f, t. \end{cases} \tag{1}$$

Note that $I_f(\mathbf{u})$ matches the basic path defined in Definition 1. Thus, LEN($I_j(\mathbf{u})$) = $d(\mathbf{u}, \mathbf{1})$ if $j = f$; and LEN($I_j(\mathbf{u})$) $\leqslant 3$ otherwise. Also, a function that assigns an index to each path for constructing spanning trees is given as follows:

$$sp(I_j(\mathbf{u})) = \begin{cases} r & \text{if } j = f, \\ u_f & \text{if } (j = r = u_1) \text{ or } (j = t \text{ and } u_1 \neq r), \\ u_1 & \text{if } j = r \neq u_1, \\ u_j & \text{otherwise.} \end{cases} \tag{2}$$

Note that, in Eq. (2), if $u_1 = r$ then $K_1 = (1, r)$, otherwise, $|K_1| \geqslant 3$. Moreover, for $j, j' \in \{2, \ldots, n\}$, $sp(I_j(\mathbf{u})) \neq sp(I_{j'}(\mathbf{u}))$ if and only if $j \neq j'$. To construct T_i for $i = 2, \ldots, n$, the algorithm proposed in [15] builds a multiple spanning tree rooted at $\mathbf{1}$ that takes $\mathbf{1}\langle i \rangle$ as child as follows.

Algorithm 1. Rescigno's algorithm

---

1. Build a spanning tree of $\mathscr{C}_i^1$ rooted at $\mathbf{1}\langle i \rangle$ in the following way and denote it by T_i^1:

   for each vertex $\mathbf{u}(= u_1 \cdots u_n) \in V(\mathscr{C}_i^1) \setminus \{\mathbf{1}\langle i \rangle\}$ **do**

   > Let PAR$(\mathbf{u}, i) = \mathbf{u}\langle s \rangle$ such that
   > **if** $c(\mathbf{u}) = 1$ **then** s is the position for which $u_1 = s$;
   > **if** $c(\mathbf{u}) \geqslant 2$ **then** s is the smallest position of the good cycle of $\mathbf{u}$;

2. Add vertices $V(\mathscr{C}_1^1) \setminus \{\mathbf{1}\}$ to T_i^1 and denote the resulting tree by $T_{i,1}^1$ as follows:

   for each vertex $\mathbf{u}(= 1u_2 \cdots u_n) \in V(\mathscr{C}_1^1) \setminus \{\mathbf{1}\}$ **do**

   > Let PAR$(\mathbf{u}, i) = \mathbf{u}\langle i \rangle$;

3. Complete T_i by adding to $T_{i,1}^1$ the vertices $V(\mathscr{C}_r^1)$ for $r \in \{2, \ldots, n\} \setminus \{i\}$ as follows:

   for each vertex $\mathbf{u} \in V(\mathscr{C}_r^1)$ **do**

   > Construct $n - 1$ vertex-disjoint paths $I_2(\mathbf{u}), \ldots, I_n(\mathbf{u})$ defined by Eq. (1);
   > Compute $sp(I_j(\mathbf{u}))$ for each $j \in \{2, \ldots, n\}$ defined by Eq. (2);

   for each vertex $\mathbf{u} \in V(\mathscr{C}_r^1)$ with $r \in \{2, \ldots, n\} \setminus \{i\}$ **do**

   > Let PAR$(\mathbf{u}, i)$ be the successor of $\mathbf{u}$ on the path $I_j(\mathbf{u})$ where $i = sp(I_j(\mathbf{u}))$;

For example, a multiple spanning tree of S_4 rooted at $\mathbf{1}(= 1234)$ that contains T_2, T_3 and T_4 is presented in [15]. Here, we reproduce the multiple spanning tree as shown in Fig. 1 to illustrate some faults occurred in [15]. In this figure, solid, dashed and bold lines represent edges produced by Step 1, 2 and 3, respectively. In particular, we note that a path with edges of solid lines is a basic path (see Definition 1). We first consider the vertex $\mathbf{u} = 3142 \in V(\mathscr{C}_2^1)$. Clearly, $r = 2$, $t = 4$ and $f = u_1 = 3$. By Eq. (1), we have

$$I_2(\mathbf{u}) = (\mathbf{u}, \mathbf{u}\langle 2 \rangle) = (3142, 1342),$$
$$I_3(\mathbf{u}) = (\mathbf{u}, \mathbf{u}\langle 3 \rangle) \oplus \hat{P}[\mathbf{u}\langle 3 \rangle, 1] = (3142, 4132, 2134, 1234), \text{ and}$$
$$I_4(\mathbf{u}) = (\mathbf{u}, \mathbf{u}\langle 4 \rangle, \mathbf{u}\langle 4 \rangle\langle 3 \rangle, \mathbf{u}\langle 4 \rangle\langle 3 \rangle\langle 2 \rangle) = (3142, 2143, 4123, 1423).$$

Moreover, by Eq. (2), we obtain $sp(I_2(\mathbf{u})) = u_1 = 3$, $sp(I_3(\mathbf{u})) = r = 2$, and $sp(I_4(\mathbf{u})) = u_3 = 4$. Thus, the paths $I_2(\mathbf{u}), I_3(\mathbf{u})$ and $I_4(\mathbf{u})$ are contained in spanning trees T_3, T_2 and T_4, respectively.

Next, we consider the vertex $\mathbf{u} = 2143 \in V(\mathscr{C}_2^1)$. In this case, $r = 2$, $t = 1$ and $f = 3$ (i.e., the smallest symbol of K_2). By Eq. (1), we have

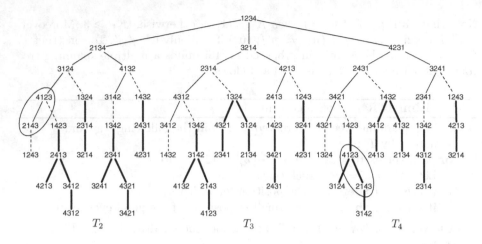

Fig. 1. A multiple spanning tree of S_4 constructed by Rescigno's algorithm. (See Example 3 of [15])

$$I_2(\mathbf{u}) = (\mathbf{u}, \mathbf{u}\langle 2 \rangle) = (2143, 1243),$$
$$I_3(\mathbf{u}) = (\mathbf{u}, \mathbf{u}\langle 3 \rangle) \oplus \hat{P}[\mathbf{u}\langle 3 \rangle, 1] = (2143, 4123, 3124, 2134, 1234), \text{ and}$$
$$I_4(\mathbf{u}) = (\mathbf{u}, \mathbf{u}\langle 4 \rangle, \mathbf{u}\langle 4 \rangle\langle 2 \rangle) = (2143, 3142, 1342).$$

Again, by Eq. (2), we obtain $sp(I_2(\mathbf{u})) = u_3 = 4$, $sp(I_3(\mathbf{u})) = r = 2$, and $sp(I_4(\mathbf{u})) = u_4 = 3$. Thus, the paths $I_2(\mathbf{u}), I_3(\mathbf{u})$ and $I_4(\mathbf{u})$ are contained in spanning trees T_4, T_2 and T_3, respectively.

We are now at a position to show that there exists a conflict in the structure of the multiple spanning tree. Since T_4 contains the path $I_4(3142) = (3142, 2143, 4123, 1423)$, we have PAR$(2143, 4) = 4123$. Since T_2 contains the path $I_3(2143) = (2143, 4123, 3124, 2134, 1234)$, we have PAR$(2143, 2) = 4123$. This shows that the edge $(2143, 4123)$ is shared by T_2 and T_4, and thus they are not vertex-disjoint (see vertices surrounded by the two ovals in Fig. 1). By Eq. (1), the two paths $I_4(3142) = (3142, 2143, 4123, 1423)$ and $I_2(2143) = (2143, 1243)$ in T_4 are indeed inconsistent. Moreover, the function $I_j(\mathbf{u})$ is not well-defined in Eq. (1), e.g., consider the vertex $\mathbf{u} = 4213 \in V(\mathscr{C}_3^1)$ or $\mathbf{u} = 4231 \in V(\mathscr{C}_4^1)$. For the former, we have $r = 3$, $f = u_1 = 4$ and $t = 4$, and the latter we have $r = 4$, $f = u_1 = 4$ and $t = 1$. Thus, the situation $f = t$ or $f = r$ lead to a conflict in Eq. (1).

4   An Amendatory Scheme

In this section, we will repair Rescigno's algorithm. For a vertex $\mathbf{u}\ (= u_1 \cdots u_n) \in V(\mathscr{C}_r^1)$ where $r \in \{2, \ldots, n\}$, let t be the index such that $u_t = r$. Also, let $f_{\mathbf{u}}$ be the smallest symbol in K_2 if $c(\mathbf{u}) \geqslant 2$, and $f_{\mathbf{u}} = u_1$ otherwise. For $j \in \{2, \ldots, n\}$, we define

$$
I_j(\mathbf{u}) = \begin{cases}
(\mathbf{u}, \mathbf{u}\langle j \rangle) \oplus \hat{P}[\mathbf{u}\langle j \rangle, 1] & \text{if } j = f_{\mathbf{u}}, \\
(\mathbf{u}, \mathbf{u}\langle j \rangle) & \text{if } j = r \neq f_{\mathbf{u}}, \\
(\mathbf{u}, \mathbf{u}\langle j \rangle, \mathbf{u}\langle j \rangle\langle f_{\mathbf{u}} \rangle, \mathbf{u}\langle j \rangle\langle f_{\mathbf{u}} \rangle\langle r \rangle) & \text{if } j = t \neq f_{\mathbf{u}} \text{ and } f_{\mathbf{u}} \neq f_{\mathbf{u}\langle j \rangle} \\
(\mathbf{u}, \mathbf{u}\langle j \rangle, \mathbf{u}\langle j \rangle\langle r \rangle) & \text{otherwise.}
\end{cases}
\tag{3}
$$

In addition, a function $sp(I_j(\mathbf{u}))$ is defined in a similar way to Eq. (2) as follows.

$$
sp(I_j(\mathbf{u})) = \begin{cases}
r & \text{if } j = f_{\mathbf{u}}, \\
u_f & \text{if } (j = r = u_1 \neq f_{\mathbf{u}}) \text{ or } (j = t \neq f_{\mathbf{u}} \text{ and } u_1 \neq r), \\
u_1 & \text{if } j = r \notin \{u_1, f_{\mathbf{u}}\}, \\
u_j & \text{otherwise.}
\end{cases}
\tag{4}
$$

Therefore, an amendatory scheme can be obtained from Algorithm 1 (as shown in the previous section) by modifying the following statement in Step 3:

---

for each vertex $\mathbf{u} \in V(\mathscr{C}_r^1)$ **do**
    Construct $n - 1$ vertex-disjoint paths $I_2(\mathbf{u}), \ldots, I_n(\mathbf{u})$ defined by Eq. (3);
    Compute $sp(I_j(\mathbf{u}))$ for each $j \in \{2, \ldots, n\}$ defined by Eq. (4);

As a consequence, for each $i \in \{2, \ldots, n\}$, the result that T_i is a spanning tree of S_n can be proved by a similar way as that in [15], and thus we have the following lemma.

Lemma 2. *For each* $i \in \{2, \ldots, n\}$, *the construction of* T_i *is a spanning tree of* S_n.

To guarantee that every path in a spanning tree constructed above is correct, we say that two paths $I_j(\mathbf{u})$ and $I_{j'}(\mathbf{v})$ are *consistent* in T_i if $i = sp(I_j(\mathbf{u})) = sp(I_{j'}(\mathbf{v}))$ for some $j, j' \in \{2, \ldots, n\}$ and $\mathbf{v}$ is the parent of $\mathbf{u}$ in T_i such that $I_{j'}(\mathbf{v})$ is a subpath of $I_j(\mathbf{u})$. As we have mentioned earlier, any subpath of a basic path is still a basic path. In the following, we give a proof to show that any non-basic path and its subpath in a spanning tree constructed by Algorithm 1 with the amendatory scheme are consistent.

Lemma 3. *For any vertex* $\mathbf{u}(= u_1 \cdots u_n) \in V(\mathscr{C}_r^1)$ *where* $r \in \{2, \ldots, n\} \setminus \{i\}$, *let* $\mathbf{v} = \mathbf{u}\langle j \rangle = v_1 \cdots v_n$ *be the successor of* $\mathbf{u}$ *on the path* $I_j(\mathbf{u})$ *for some* $j \in \{2, \ldots, n\} \setminus \{f_{\mathbf{u}}, r\}$ *such that* $i = sp(I_j(\mathbf{u}))$, *and let* $I_{j'}(\mathbf{v})$ *be the path for some* $j' \in \{2, \ldots, n\} \setminus \{f_{\mathbf{v}}\}$ *such that* $i = sp(I_{j'}(\mathbf{v}))$. *Then, the two paths* $I_j(\mathbf{u})$ *and* $I_{j'}(\mathbf{v})$ *are consistent in* T_i.

Proof. Since $\mathbf{v} = \mathbf{u}\langle j \rangle$, it is clear that $v_j = u_1$, $v_1 = u_j$ and $v_k = u_k$ for $k \in \{2, \ldots, n\} \setminus \{j\}$. Since $r \notin \{1, j\}$, we have $\mathbf{v} \in V(\mathscr{C}_r^1)$. By definition, $f_{\mathbf{u}} \neq 1$

and $f_{\mathbf{v}} \neq 1$. To avoid confusion, we denote by $t_{\mathbf{u}}$ and $t_{\mathbf{v}}$ to mean the indices such that $u_{t_{\mathbf{u}}} = v_{t_{\mathbf{v}}} = r$. We consider the following cases to build the paths $I_j(\mathbf{u})$ and $I_{j'}(\mathbf{v})$ for proving the consistency.

Case 1: $j \neq t_{\mathbf{u}}$. By Eq. (3) with the condition $j \notin \{f_{\mathbf{u}}, r, t_{\mathbf{u}}\}$, we obtain the path $I_j(\mathbf{u}) = (\mathbf{u}, \mathbf{v}, \mathbf{v}\langle r\rangle)$. Also, by Eq. (4) with the condition $j \notin \{f_{\mathbf{u}}, r, t_{\mathbf{u}}\}$, we have $i = sp(I_j(\mathbf{u})) = u_j$. Since $i = sp(I_{j'}(\mathbf{v}))$ and $u_j = v_1$, it follows that $sp(I_{j'}(\mathbf{v})) = v_1$. Again by Eq. (4), since $j' \neq f_{\mathbf{v}}$, we have $sp(I_{j'}(\mathbf{v})) \neq r$. Also, since $f_{\mathbf{v}} \neq 1$, we have $sp(I_{j'}(\mathbf{v})) \neq v_{f_{\mathbf{v}}}$. Thus, $sp(I_{j'}(\mathbf{v})) = v_1$ implies $j' = r \notin \{v_1, f_{\mathbf{v}}\}$. Hence, by Eq. (3) with the condition $j' = r \neq f_{\mathbf{v}}$, we obtain the path $I_{j'}(\mathbf{v}) = (\mathbf{v}, \mathbf{v}\langle j'\rangle) = (\mathbf{v}, \mathbf{v}\langle r\rangle)$, which is a subpath of $I_j(\mathbf{u})$.

Case 2: $j = t_{\mathbf{u}}$ and $f_{\mathbf{u}} \neq f_{\mathbf{v}}$. By Eq. (3) with conditions $j = t_{\mathbf{u}} \neq f_{\mathbf{u}}$ and $f_{\mathbf{u}} \neq f_{\mathbf{v}}$, we obtain $I_j(\mathbf{u}) = (\mathbf{u}, \mathbf{v}, \mathbf{v}\langle f_{\mathbf{u}}\rangle, \mathbf{v}\langle f_{\mathbf{u}}\rangle\langle r\rangle)$. Since $r = u_{t_{\mathbf{u}}} = u_j$ and $j \neq 1$, it implies $u_1 \neq r$. By Eq. (4) with conditions $j = t_{\mathbf{u}} \neq f_{\mathbf{u}}$ and $u_1 \neq r$, we have $i = sp(I_j(\mathbf{u})) = u_{f_{\mathbf{u}}}$. Since $i = sp(I_{j'}(\mathbf{v}))$ and $f_{\mathbf{u}} \notin \{1, j\}$, it follows that $sp(I_{j'}(\mathbf{v})) = i = u_{f_{\mathbf{u}}} = v_{f_{\mathbf{u}}}$. Moreover, since $f_{\mathbf{u}} \notin \{1, f_{\mathbf{v}}\}$ and $i \neq r$, we have $sp(I_{j'}(\mathbf{v})) \notin \{r, v_1, v_{f_{\mathbf{v}}}\}$. Hence, by Eq. (4) again, only the case $sp(I_{j'}(\mathbf{v})) = v_{j'}$ is possible, and it further implies that $j' = f_{\mathbf{u}} \notin \{f_{\mathbf{v}}, r, t_{\mathbf{v}}\}$. Therefore, by Eq. (3) with the last condition, we obtain the path $I_{j'}(\mathbf{v}) = (\mathbf{v}, \mathbf{v}\langle j'\rangle, \mathbf{v}\langle j'\rangle\langle r\rangle) = (\mathbf{v}, \mathbf{v}\langle f_{\mathbf{u}}\rangle, \mathbf{v}\langle f_{\mathbf{u}}\rangle\langle r\rangle)$, which is a subpath of $I_j(\mathbf{u})$.

Case 3: $j = t_{\mathbf{u}}$ and $f_{\mathbf{u}} = f_{\mathbf{v}}$. By Eq. (3) with conditions $j = t_{\mathbf{u}} \notin \{f_{\mathbf{u}}, r\}$ and $f_{\mathbf{u}} = f_{\mathbf{v}}$, we obtain $I_j(\mathbf{u}) = (\mathbf{u}, \mathbf{v}, \mathbf{v}\langle r\rangle)$. Since $v_1 = u_j = u_{t_{\mathbf{u}}} = r = v_{t_{\mathbf{v}}}$ and $j \neq 1$, it implies $u_1 \neq r$ and $t_{\mathbf{v}} = 1$. By Eq. (4) with conditions $j = t_{\mathbf{u}} \neq f_{\mathbf{u}}$ and $u_1 \neq r$, we have $i = sp(I_j(\mathbf{u})) = u_{f_{\mathbf{u}}}$. Since $i = sp(I_{j'}(\mathbf{v}))$ and $f_{\mathbf{v}} = f_{\mathbf{u}} \notin \{1, j\}$, it follows that $sp(I_{j'}(\mathbf{v})) = i = u_{f_{\mathbf{u}}} = u_{f_{\mathbf{v}}} = v_{f_{\mathbf{v}}}$. Again by Eq. (4), since $j' \neq f_{\mathbf{v}}$, we have $sp(I_{j'}(\mathbf{v})) \neq r$. Also, since $f_{\mathbf{v}} \neq 1$, we have $sp(I_{j'}(\mathbf{v})) \neq v_1$. Thus, if $sp(I_{j'}(\mathbf{v})) = v_{f_{\mathbf{v}}}$, then it implies either (i) $j' = r = v_1 \neq f_{\mathbf{v}}$ or (ii) $j' = t_{\mathbf{v}} \neq f_{\mathbf{v}}$ and $v_1 \neq r$. However, the latter is impossible since $j' \neq 1$ and $t_{\mathbf{v}} = 1$. Thus, if (i) holds, by Eq. (3) with the condition $j' = r \neq f_{\mathbf{v}}$, we obtain $I_{j'}(\mathbf{v}) = (\mathbf{v}, \mathbf{v}\langle j'\rangle) = (\mathbf{v}, \mathbf{v}\langle r\rangle)$, which is a subpath of $I_j(\mathbf{u})$. □

For instance, we consider the vertex $\mathbf{u} = 3142 \in V(\mathscr{C}_2^1)$, where $r = 2$, $t = 4$ and $f_{\mathbf{u}} = 3$. We check $I_4(\mathbf{u})$ as follows. Clearly, $\mathbf{u}\langle 4\rangle = 2143 \in V(\mathscr{C}_2^1)$ and $f_{\mathbf{u}\langle 4\rangle} = 3 = f_{\mathbf{u}}$. Thus, by Eq. (3), we have $I_4(\mathbf{u}) = (\mathbf{u}, \mathbf{u}\langle 4\rangle, \mathbf{u}\langle 4\rangle\langle 2\rangle) = (3142, 2143, 1243)$. Further, $sp(I_4(\mathbf{u})) = u_{f_{\mathbf{u}}} = u_3 = 4$. Next, we consider the vertex $\mathbf{v} = 2143 \in V(\mathscr{C}_2^1)$, where $r = 2$, $t = 1$ and $f_{\mathbf{v}} = 3$. In this case, we have $I_2(\mathbf{v}) = (\mathbf{v}, \mathbf{v}\langle 2\rangle) = (2143, 1243)$ and $sp(I_2(\mathbf{v})) = v_{f_{\mathbf{v}}} = v_3 = 4$. It is obvious that $I_2(2143)$ is a subpath of $I_4(3142)$, and the two paths are consistent in T_4. According to Algorithm 1 with the amendatory scheme, we can build a multiple spanning tree of S_4 rooted at $\mathbf{1}(= 1234)$ that contains T_2, T_3 and T_4 as shown in Fig. 2.

In what follows, we show the correctness that all spanning trees $T_2, T_3, \ldots, T_n$ in the multiple spanning tree of S_n are VDSTs.

Lemma 4. *For any vertex* $\mathbf{u}$ $(= u_1 \cdots u_n) \in V(\mathscr{C}_r^1)$ *where* $r \in \{2, \ldots, n\}$, *the paths* $I_j(\mathbf{u})$ *for* $j \in \{2, \ldots, n\}$ *constructed by Eq. (3) are pairwise vertex-disjoint.*

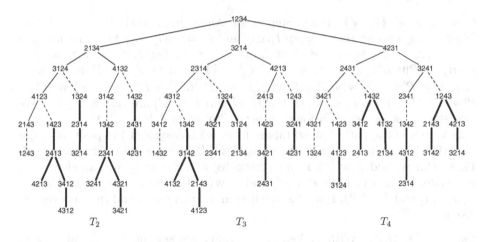

Fig. 2. A multiple spanning tree of S_4 constructed by Algorithm 1 with the amendatory scheme.

Proof. For $c(\mathbf{u}) \geqslant 2$, $f_{\mathbf{u}}$ is the smallest symbol in K_2. For each $j \in \{2, \ldots, n\} \setminus \{f_{\mathbf{u}}\}$, the symbol u_1 appears in position j along the path $I_j(\mathbf{u})$. Hence, for any $j, j' \in \{2, \ldots, n\} \setminus \{f_{\mathbf{u}}\}$ with $j \neq j'$, the two paths $I_j(\mathbf{u})$ and $I_{j'}(\mathbf{u})$ are vertex-disjoint. Furthermore, for the basic path $I_{f_{\mathbf{u}}}(\mathbf{u})$, the symbol u_1 appears in positions $f_{\mathbf{u}}$, 1 and u_1, successively. By Eq. (3) with the condition $j \neq f_{\mathbf{u}}$ (i.e., the last three conditions), it guarantees that $I_{f_{\mathbf{u}}}(\mathbf{u})$ and $I_j(\mathbf{u})$ for $j \in \{2, \ldots, n\} \setminus \{f_{\mathbf{u}}, u_1\}$ are vertex-disjoint. Finally, let us consider paths $I_{f_{\mathbf{u}}}(\mathbf{u})$ and $I_{u_1}(\mathbf{u})$. Since $I_{f_{\mathbf{u}}}(\mathbf{u})$ matches the basic path, if a vertex $\mathbf{w}(= w_1 \cdots w_n) \in I_{f_{\mathbf{u}}}(\mathbf{u})$ has symbol u_1 in position u_1, then every symbol $k \in K_2 \cup \cdots \cup K_{c(\mathbf{u})}$ appears in the right position (i.e., $w_k = k$). By contrast, every symbol $k \in K_2 \cup \cdots \cup K_{c(\mathbf{u})}$ in a vertex $\mathbf{w} \in I_{u_1}(\mathbf{u})$ appears in the same position as that in $\mathbf{u}$. It follows that $I_{f_{\mathbf{u}}}(\mathbf{u})$ and $I_{u_1}(\mathbf{u})$ are vertex-disjoint.

For $c(\mathbf{u}) = 1$, every vertex on the path $I_j(\mathbf{u})$ for $j \in \{2, \ldots, n\}$ has symbol u_1 in position j. Hence, for any $j, j' \in \{2, \ldots, n\} \setminus \{f_{\mathbf{u}}\}$ with $j \neq j'$, the two paths $I_j(\mathbf{u})$ and $I_{j'}(\mathbf{u})$ are vertex-disjoint. $\qquad \square$

Theorem 1. *For each vertex $\mathbf{u} \in V(S_n) \setminus \{\mathbf{1}\}$ and any pair of integers $i, i' \in \{2, \ldots, n\}$ with $i \neq i'$, the two paths $P[\mathbf{u}, \mathbf{1}]$, respectively, in T_i and $T_{i'}$, are vertex-disjoint.*

Proof. Let $\mathbf{u} \in V(\mathscr{C}_r^1)$ where $r \in \{1, \ldots, n\}$. We use $P_i[\mathbf{u}, \mathbf{1}]$ and $P_{i'}[\mathbf{u}, \mathbf{1}]$ to distinguish the two paths $P[\mathbf{u}, \mathbf{1}]$ in T_i and $T_{i'}$, respectively. Without loss of generality, we may consider the following cases:

Case 1: $r = 1$. From Step 2 of Algorithm 1, $\mathrm{PAR}(T_i, \mathbf{u}) = \mathbf{u}\langle i \rangle$ and $\mathrm{PAR}(T_{i'}, \mathbf{u}) = \mathbf{u}\langle i' \rangle$. Clearly, $P_i[\mathbf{u}, \mathbf{1}] = (\mathbf{u}, \mathbf{u}\langle i \rangle) \oplus \hat{P}[\mathbf{u}\langle i \rangle, \mathbf{1}]$ and $P_{i'}[\mathbf{u}, \mathbf{1}] = (\mathbf{u}, \mathbf{u}\langle i' \rangle) \oplus \hat{P}[\mathbf{u}\langle i' \rangle, \mathbf{1}]$. Since $\hat{P}[\mathbf{u}\langle i \rangle, \mathbf{1}]$ and $\hat{P}[\mathbf{u}\langle i' \rangle, \mathbf{1}]$ are basic paths in distinct trees, vertices in the two paths (apart from $\mathbf{1}$) keep symbol 1 in position i and i', respectively. Hence, $P_i[\mathbf{u}, \mathbf{1}]$ and $P_{i'}[\mathbf{u}, \mathbf{1}]$ are vertex-disjoint.

Case 2: $r \notin \{1, i, i'\}$. From Step 3 of Algorithm 1 with the amendatory scheme, we suppose that $i = sp(I_j(\mathbf{u}))$ and $i' = sp(I_{j'}(\mathbf{u}))$ for some integers $j, j' \in \{2, \ldots, n\}$. Since $i \neq i'$, it implies $j \neq j'$. Let $I_j(\mathbf{u}) = P[\mathbf{u}, \mathbf{w}]$ and $I_{j'}(\mathbf{u}) = P[\mathbf{u}, \mathbf{w}']$, where $\mathbf{w}, \mathbf{w}' \in V(\mathscr{C}_1^1) \setminus \{\mathbf{1}\}$ and $\mathbf{w} \neq \mathbf{w}'$. Thus, we have $P_i[\mathbf{u}, \mathbf{1}] = P[\mathbf{u}, \mathbf{w}] \oplus (\mathbf{w}, \mathbf{w}\langle i \rangle) \oplus \hat{P}[\mathbf{w}\langle i \rangle, \mathbf{1}]$ and $P_{i'}[\mathbf{u}, \mathbf{1}] = P[\mathbf{u}, \mathbf{w}'] \oplus (\mathbf{w}', \mathbf{w}'\langle i' \rangle) \oplus \hat{P}[\mathbf{w}'\langle i' \rangle, \mathbf{1}]$. Lemma 4 assures that $P[\mathbf{u}, \mathbf{w}]$ and $P[\mathbf{u}, \mathbf{w}']$ are vertex-disjoint. We now show that no vertex belongs to both $P[\mathbf{u}, \mathbf{w}]$ and $\hat{P}[\mathbf{w}'\langle i' \rangle, \mathbf{1}]$. By Eq. (3), each vertex in the path $P[\mathbf{u}, \mathbf{w}]$ (apart from $\mathbf{w}$) has symbol 1 in position r. On the other hand, each vertex in the path $\hat{P}[\mathbf{w}'\langle i' \rangle, \mathbf{1}]$ has symbol 1 in position i'. Thus, $P[\mathbf{u}, \mathbf{w}]$ and $\hat{P}[\mathbf{w}'\langle i' \rangle, \mathbf{1}]$ are vertex-disjoint. Similarly, we can show that no vertex belongs to both $P[\mathbf{u}, \mathbf{w}']$ and $\hat{P}[\mathbf{w}\langle i \rangle, \mathbf{1}]$. Finally, similar to Case 1, $\hat{P}[\mathbf{w}\langle i \rangle, \mathbf{1}]$ and $\hat{P}[\mathbf{w}'\langle i' \rangle, \mathbf{1}]$ are basic paths in distinct trees, and thus are vertex-disjoint.

Case 3: $r \in \{i, i'\}$. Without loss of generality we assume $r \neq i$ and $r = i'$. From Step 3 of Algorithm 1 with the amendatory scheme, we suppose that $i = sp(I_j(\mathbf{u}))$ and $i' = sp(I_{j'}(\mathbf{u}))$ for some integers $j, j' \in \{2, \ldots, n\}$ and $j \neq j'$. Since $r \neq i$, we have $P_i[\mathbf{u}, \mathbf{1}] = P[\mathbf{u}, \mathbf{w}] \oplus (\mathbf{w}, \mathbf{w}\langle i \rangle) \oplus \hat{P}[\mathbf{w}\langle i \rangle, \mathbf{1}]$, where $P[\mathbf{u}, \mathbf{w}] = I_j(\mathbf{u})$ for $\mathbf{w} \in V(\mathscr{C}_1^1) \setminus \{\mathbf{1}\}$. On the other hand, since $r = i'$, we have $P_{i'}[\mathbf{u}, \mathbf{1}] = I_{j'}(\mathbf{u}) = \hat{P}[\mathbf{u}, \mathbf{1}]$. By Lemma 4, $P[\mathbf{u}, \mathbf{w}]$ and $P_{i'}[\mathbf{u}, \mathbf{1}]$ are vertex-disjoint. Also, $\hat{P}[\mathbf{w}\langle i \rangle, \mathbf{1}]$ and $P_{i'}[\mathbf{u}, \mathbf{1}]$ are basic paths in distinct trees, and thus are vertex-disjoint.   □

In [15], a detail of analysis for the lengths of paths in the constructed VDSTs was given. Here we omit the analysis because, from the proof of Theorem 1, it is easy to check that the length of $P[\mathbf{u}, \mathbf{1}]$ for any $\mathbf{u} \in V(S_n) \setminus \{\mathbf{1}\}$ is at most the length of a certain shortest path plus 4.

5   A Fully Parallelized Algorithm for Constructing VDSTs of S_n

In this section, for the purpose of designing a more efficient way to construct VDSTs on S_n, we propose an alternative algorithm. Technically, the newly proposed algorithm relies on a function that can determine the parent of any vertex in a spanning tree directly. Because such a function only needs to refer the label of a vertex and the index of a tree as parameters, the advantage of this algorithm is that it can easily be parallelized.

For $i \in \{2, \ldots, n\}$ and $\mathbf{u}(= u_1 \cdots u_n) \in V(S_n) \setminus \{\mathbf{1}\}$, let r be the index such that $u_r = 1$, let t be the index such that $u_t = r$, and let j be the index such that $u_j = i$. Also, let f be the smallest symbol in K_2 if $c(\mathbf{u}) \geqslant 2$, and $f = u_1$ otherwise. In the following, we give the function to determine the parent of a vertex $\mathbf{u}$ in T_i as follows.

$$
\text{PAR}(\mathbf{u}, i) = \begin{cases}
\mathbf{u}\langle f\rangle & \text{if } i = r \text{ (i.e., } u_i = 1), & (5.1)\\
\mathbf{u}\langle i\rangle & \text{if } u_1 = 1, & (5.2)\\
\mathbf{u}\langle r\rangle & \text{if } (i = u_f \neq r \text{ and } u_1 = r \neq 1) \text{ or } i = u_1 \notin \{1, r\}, & (5.3)\\
\mathbf{u}\langle t\rangle & \text{if } i = u_f \neq r \text{ and } u_1 \notin \{1, r\}, & (5.4)\\
\mathbf{u}\langle j\rangle & \text{if } i \notin \{r, u_1, u_f\} \text{ and } u_1 \neq 1. & (5.5)
\end{cases}
$$

Obviously, to determine the parent of a vertex $\mathbf{u}$ in T_i, we only need to test two variables i and u_1. Perhaps the reader will be interested in how to derive the above function. The basic idea is that we make a composition of two functions $sp(I_j(\mathbf{u}))$ and $I_j(\mathbf{u})$ defined by Eqs. (4) and (3), respectively. Then, such a composition can assign a certain path $I_j(\mathbf{u})$ (i.e., a path starting from $\mathbf{u}$ to a vertex with symbol 1 in the right position) to the tree T_i. Hence, reversing the composition, we obtain the desired function that takes the vertex $\mathbf{u}$ and T_i as parameters to compute a certain path starting from $\mathbf{u}$ with the successor on this path as its parent. The following lemma shows that the function $\text{PAR}(\mathbf{u}, i)$ is well-defined.

Lemma 5. *For any $\mathbf{u}(= u_1 \cdots u_n) \in V(S_n) \setminus \{\mathbf{1}\}$ and $i \in \{2, \ldots, n\}$, the function $\text{PAR}(\mathbf{u}, i)$ matches exactly one condition of Eqs. (5.1)–(5.5).*

Proof. We first show that the union of all conditions in Eq. (5) is a tautology. To avoid lengthy depiction, in the following we write Con(5.x) to mean the condition of an equation labeled by (5.x). Consider the following compound statement in disjunctive normal form:

$$
\begin{aligned}
& [(i = u_f) \wedge (i \neq r) \wedge (u_1 = r) \wedge (u_1 \neq 1)] && \text{// The front part of Con(5.3)}\\
\vee\ & \quad [(i = u_1) \wedge (u_1 \neq 1) \wedge (i \neq r)] && \text{// The rear part of Con(5.3)}\\
\vee\ & [(i = u_f) \wedge (i \neq r) \wedge (u_1 \neq 1) \wedge (u_1 \neq r)] && \text{// Con(5.4)}\\
\vee\ & [(i \neq r) \wedge (i \neq u_1) \wedge (i \neq u_f) \wedge (u_1 \neq 1)]. && \text{// Con(5.5)}
\end{aligned}
$$

Clearly, it can be reduced to $i \neq r$ and $u_1 \neq 1$. Hence, the union of Con(5.3), Con(5.4) and Con(5.5) is logically equivalent to the complement of the union of Con(5.1) and Con(5.2).

To complete the proof, we need to prove that all conditions in Eq. (5) are pairwise disjoint. Clearly, Con(5.1) and Con(5.2) are disjoint because $i \geqslant 2$. From above, any two conditions, one choosing from the front two and the other choosing from the last three, are disjoint. Since the front part of Con(5.3) includes the literal $u_1 = r$ and Con(5.4) includes the literal $u_1 \neq r$, they are disjoint. We now show that the rear part of Con(5.3) and Con(5.4) are disjoint. Suppose not, i.e., there exists some vertex $\mathbf{u}$ such that both conditions hold. Then, $u_f = i = u_1$, and so $f = 1$. Since $f = 1$ cannot be the smallest symbol of K_2 for $\mathbf{u}$, by the definition of f, we have $f = u_1$. Hence, $u_1 = 1$, which leads to a contradiction since $u_1 \notin \{1, r\}$ in both condition. Finally, it is easy to see that Con(5.5) and each of Con(5.3) and Con(5.4) are disjoint, respectively. $\square$

Algorithm 2. A fully parallelized algorithm

---

for each vertex $\mathbf{u}(= u_1 \cdots u_n) \in V(S_n) \setminus \{1\}$ do in parallel
    Compute r, t, f and $j[i]$ for $i = 1, \ldots, n$;
    for $i \leftarrow 2$ to n do
        Let $j \leftarrow j[i]$;
        Compute PAR$(\mathbf{u}, i)$ defined by Eq. (5);

Table 1. Computing the parent of every vertex $\mathbf{u} \in V(S_4) \setminus \{1\}$ in T_i for $i \in \{2, 3, 4\}$.

u	i	r	t	f	j	u_1	u_f	rule	PAR(u,i)
1234	-	-	-	-	-	-	-	-	-
1243	2	1	1	1	2	1	1	(5.2)	2143
	3				4			(5.2)	4213
	4				3			(5.2)	3241
1324	2	1	1	1	3	1	1	(5.2)	3124
	3				2			(5.2)	2314
	4				4			(5.2)	4321
1342	2	1	1	1	4	1	1	(5.2)	3142
	3				2			(5.2)	4312
	4				3			(5.2)	2341
1423	2	1	1	1	3	1	1	(5.2)	4123
	3				4			(5.2)	2413
	4				2			(5.2)	3421
1432	2	1	1	1	4	1	1	(5.2)	4132
	3				3			(5.2)	3412
	4				2			(5.2)	2431
2134	2	2	1	2	1	2	1	(5.1)	1234
	3				3			(5.5)	3124
	4				4			(5.5)	4132
2143	2	2	1	3	1	2	4	(5.1)	4123
	3				4			(5.5)	3142
	4				3			(5.3)	1243
2314	2	3	2	2	1	2	3	(5.3)	1324
	3				2			(5.1)	3214
	4				4			(5.5)	4312
2341	2	4	3	2	1	2	3	(5.3)	1342
	3				2			(5.4)	4321
	4				3			(5.1)	3241
2413	2	3	4	2	1	2	4	(5.3)	1423
	3				4			(5.1)	4213
	4				2			(5.4)	3412
2431	2	4	2	2	1	2	4	(5.3)	1432
	3				3			(5.5)	3421
	4				2			(5.1)	4231
3124	2	2	3	3	3	3	2	(5.1)	2134
	3				1			(5.3)	1324
	4				4			(5.5)	4123
3142	2	2	4	3	4	3	4	(5.1)	4132
	3				1			(5.3)	1342
	4				3			(5.4)	2143
3214	2	3	1	3	2	3	1	(5.5)	2314
	3				1			(5.1)	1234
	4				4			(5.5)	4213
3241	2	4	3	3	2	3	4	(5.5)	2341
	3				1			(5.3)	1243
	4				3			(5.1)	4231
3412	2	3	1	2	4	3	4	(5.5)	2413
	3				1			(5.1)	4312
	4				2			(5.3)	1432
3421	2	4	2	3	3	3	2	(5.4)	4321
	3				1			(5.3)	1423
	4				2			(5.1)	2431
4123	2	2	3	4	3	4	3	(5.1)	3124
	3				4			(5.4)	2143
	4				1			(5.3)	1423
4132	2	2	4	4	4	4	2	(5.1)	2134
	3				3			(5.5)	3142
	4				1			(5.3)	1432
4213	2	3	4	4	2	4	3	(5.5)	2413
	3				4			(5.1)	3214
	4				1			(5.3)	1243
4231	2	4	1	4	2	4	1	(5.5)	2431
	3				3			(5.5)	3241
	4				1			(5.1)	1234
4312	2	3	2	4	4	4	2	(5.4)	3412
	3				2			(5.1)	2314
	4				1			(5.3)	1342
4321	2	4	1	2	3	4	3	(5.5)	2341
	3				2			(5.3)	1324
	4				1			(5.1)	3421

Now, we present a fully parallelized algorithm for constructing $n - 1$ VDSTs of S_n with $\mathbf{1}$ as the common root. For easily parallel implementation, every node $\mathbf{u}(= u_1 \cdots u_n) \in V(S_n) \setminus \{1\}$ has its own private variables r, t, f and j. Since j is dependent on i, an auxiliary array $j[1..n]$ is also available for storing the index where the symbol i occurs in the label of $\mathbf{u}$ (i.e., set $u_{j[i]} = i$).

For example, we summarize the result of calculating the parent of every vertex $\mathbf{u} \in V(S_4) \setminus \{1\}$ in T_i for $i \in \{2, 3, 4\}$ in Table 1. According to this table, a multiple spanning tree rooted at **1** constructed by Algorithm 2 is the same as that in the previous section (see Fig. 2).

Since the algorithm for constructing VDSTs in this section is followed by the reversing rule of the amendatory scheme described in the previous section, the correctness of Algorithm 2 directly follows from Lemma 5 and Theorem 1. Clearly, the computation of r, t, f and $j[1..n]$ for each vertex $\mathbf{u} \in V(S_n) \setminus \{\mathbf{1}\}$ in Algorithm 2 can be done in $\mathcal{O}(n)$ time. Also, determining the parent of each vertex $\mathbf{u}$ in a spanning tree using the function PAR$(\mathbf{u}, i)$ only requires a constant time. Therefore, the total complexity of the proposed algorithm is $\mathcal{O}(n \cdot n!)$ time. Note that, by the vertex-symmetry of S_n, the root of VDSTs constructed in Algorithm 2 can be changed to any vertex of S_n. In addition, all VDSTs constructed by Algorithm 2 are indeed isomorphic to those in Algorithm 1 with the amendatory scheme, and thus the height of each spanning tree is bounded by a shortest path of S_n plus four. As a result, we have the following theorem.

Theorem 2. *Let $D(S_n)$ stands for the diameter of S_n and $\mathbf{r} \in V(S_n)$ be an arbitrary vertex. Then, Algorithm 2 can correctly construct $n - 1$ VDSTs of S_n rooted at $\mathbf{r}$ in $\mathcal{O}(n \cdot n!)$ time and the height of each spanning tree is at most $D(S_n) + 4$. In particular, the algorithm can be parallelized to run in $\mathcal{O}(n)$ time by using $n!$ vertices of S_n as processors.*

6 Concluding Remarks

The star network is an important architecture on the research of interconnection networks and has been studied in the last three decades. Constructing VDSTs has many applications in interconnection networks such as fault-tolerant broadcasting and secure message distribution. In this paper, we mainly point out that there exists a flaw in the previous algorithm for constructing VDSTs on star networks and then provide an amendatory scheme to correct it. Moreover, we propose another alternative algorithm that is suitable and efficient for parallel implementation.

We close this paper by giving the following two remarks. Since there are many interconnection networks are defined by using permutations of symbols as vertices such that the adjacency between vertices can be described by a transposition set. For instance, bubble-sort graphs, alternating group graphs and arrangement graphs are such kind of networks. However, to the best of our knowledge, so far does not exist fully parallelized approaches for constructing VDSTs on these networks, which is worthwhile to be addressed in a future research.

Also, a challenge of constructing spanning trees in a network as broadcasting schemes is to pursue the goal of reducing the heights of spanning trees (e.g., see [12, 18, 19]). An instinctive question is to ask whether the heights of the constructed VDSTs are optimal? Since the maximum height of $\kappa(= \kappa(G))$ VDSTs is bounded below by the κ-wide distance of G (i.e., the length of a longest path in an optimal parallel routing of width κ between any two vertices u and v in G)

and a result in [7] showed that the κ-wide distance of S_n is $D(S_n) + 4$, we conclude that the constructed VDSTs are optimal in the sense that the maximum height of VDSTs is minimized.

Acknowledgments. This research was partially supported by MOST grants 104-2221-E-141-002-MY3 (Jou-Ming Chang), 105-2221-E-131-027 (Kung-Jui Pai), 106-2221-E-141-001 (Jinn-Shyong Yang) and 104-2221-E-262-005 (Ro-Yu Wu) from the Ministry of Science and Technology, Taiwan.

References

1. Akers, S.B., Krishnamurty, B.: A group theoretic model for symmetric interconnection networks. IEEE Trans. Comput. **28**, 555–566 (1989)
2. Akers, S.B., Harel, D., Krishnamurty, B.: The star graph: an attractive alternative to the n-cube. In: Proceedings of the International Conference on Parallel Processing (ICPP 1987), University Park, pp. 393–400 (1987)
3. Akl, S.G., Qiu, K., Stojmenović, I.: Fundamental algorithms for the star and pancake interconnection networks with applications to computational geometry. Networks **23**, 215–226 (1993)
4. Akl, S.G., Wolff, T.: Efficient sorting on the star graph interconnection network. Telcom. Syst. **10**, 3–20 (1998)
5. Bao, F., Funyu, Y., Hamada, Y., Igarashi, Y.: Reliable broadcasting and secure distributing in channel networks. In: Proceedings of 3rd International Symposium on Parallel Architectures, Algorithms and Networks (ISPAN 1997), Taipei, pp. 472–478 (1997)
6. Chang, J.-M., Yang, T.-J., Yang, J.-S.: A parallel algorithm for constructing independent spanning trees in twisted cubes. Discrete Appl. Math. **219**, 74–82 (2017)
7. Chen, C.-C., Chen, J.: Optimal parallel routing in star networks. IEEE Trans. Comput. **46**, 1293–1303 (1997)
8. Day, K., Tripathi, A.: A comparative study of topologies properties of hypercubes and star networks. IEEE Trans. Parallel Distrib. Syst. **5**, 31–38 (1994)
9. Fragopoulou, P., Akl, S.G.: A parallel algorithm for computing Fourier transforms on the star graph. IEEE Trans. Parallel Distrib. Syst. **5**, 525–531 (1994)
10. Fragopoulou, P., Akl, S.G.: Optimal communication algorithms on star graphs using spanning tree constructions. J. Parallel Distrib. Comput. **24**, 55–71 (1995)
11. Fragopoulou, P., Akl, S.G.: Edge-disjoint spanning trees on the star network with applications to fault tolerance. IEEE Trans. Comput. **45**, 174–185 (1996)
12. Hasunuma, T., Nagamochi, H.: Independent spanning trees with small depths in iterated line digraphs. Discrete Appl. Math. **110**, 189–211 (2001)
13. Itai, A., Rodeh, M.: The multi-tree approach to reliability in distributed networks. Inform. Comput. **79**, 43–59 (1988)
14. Qiu, K., Akl, S.G., Meijer, H.: On some properties and algorithms for the star and pancake interconnection networks. J. Parallel Distrib. Comput. **22**, 16–25 (1994)
15. Rescigno, A.A.: Vertex-disjoint spanning trees of the star network with applications to fault-tolerance and security. Inform. Sci. **137**, 259–276 (2001)
16. Sur, S., Srimani, P.K.: Topological properties of star graph. Comput. Math. Appl. **25**, 87–98 (1993)
17. Yang, J.-S., Chan, H.-C., Chang, J.-M.: Broadcasting secure messages via optimal independent spanning trees in folded hypercubes. Discrete Appl. Math. **159**, 1254–1263 (2011)

18. Yang, J.-S., Chang, J.-M., Tang, S.-M., Wang, Y.-L.: Reducing the height of independent spanning trees in chordal rings. IEEE Trans. Parallel Distrib. Syst. **18**, 644–657 (2007)
19. Yang, J.-S., Luo, S.-S., Chang, J.-M.: Pruning longer branches of independent spanning trees on folded hyper-stars. Comput. J. **58**, 2979–2981 (2015)
20. Yang, J.-S., Wu, M.-R., Chang, J.-M., Chang, Y.-H.: A fully parallelized scheme of constructing independent spanning trees on Möbius cubes. J. Supercomput. **71**, 952–965 (2015)

Protein Mover's Distance: A Geometric Framework for Solving Global Alignment of PPI Networks

Manni Liu and Hu Ding[✉]

Department of Computer Science and Engineering, Michigan State University,
East Lansing, USA
{liumanni,huding}@msu.edu

Abstract. A protein-protein interaction (PPI) network is an unweighted and undirected graph representing the interactions among proteins, where each node denotes a protein and each edge connecting two nodes indicates their interaction. Given two PPI networks, finding their alignment is a fundamental problem and has many important applications in bioinformatics. However, it often needs to solve some generalized version of subgraph isomorphism problem which is challenging and NP-hard. Following our previous geometric approach [21], we propose a unified algorithmic framework for PPI networks alignment. We first define a general concept called "Protein Mover's Distance (PMD)" to evaluate the alignment of two PPI networks. PMD is similar to the well known "Earth Mover's Distance"; however, we also incorporate some other information, e.g., the functional annotation of proteins. Our algorithmic framework consists of two steps, Embedding and Matching. For the embedding step, we apply three different graph embedding techniques to preserve the topological structures of the original PPI networks. For the matching step, we compute a rigid transformation for one of the embedded PPI networks so as to minimize its PMD to the other PPI network; by using the flow values of the resulting PMD as the matching scores, we are able to obtain the desired alignment. Also, our framework can be easily extended to joint alignment of multiple PPI networks. The experimental results on two popular benchmark datasets suggest that our method outperforms existing approaches in terms of the quality of alignment.

1 Introduction

Proteins are essential parts of organisms and participate in virtually every process within cells [36]. *Protein-Protein Interaction (PPI) networks* provides effective tools for studying protein complexes and understanding their functional

---

The research of this work was supported in part by NSF through grant CCF-1656905 and a start-up fund from Michigan State University. Ding also wants to thank Profs. Bonnie Berger and Roded Sharan for their helpful discussions at Simons Institute, UC Berkeley.

© Springer International Publishing AG 2017
X. Gao et al. (Eds.): COCOA 2017, Part I, LNCS 10627, pp. 56–69, 2017.
https://doi.org/10.1007/978-3-319-71150-8_5

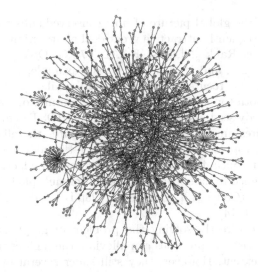

Fig. 1. An example of yeast PPI network [35].

interactions, modules, and pathways, in many cellular processes. A **PPI network** is a graph that describes the interaction of proteins, where a node represents a protein and an edge means that the two corresponding proteins interact with each other [35]. See Fig. 1.

The current research on PPI networks mainly focus on two directions: (1) knowledge discovery inside each individual network and (2) comparison and integration of different networks. The first direction includes the problems of link prediction (i.e., adding new interactions) and modules/pathways detection, while the second one often targets finding the similarity or distinction between two or more networks. Actually these two directions are closely related with each other, e.g., better knowledge discovery inside each network could lead to more accurate comparison between networks, and the integrated analysis on different networks could improve the knowledge discovery inside each individual network. In this paper, we focus on a fundamental problem in the latter direction, **PPI networks alignment**, which is often modeled as the problem of mapping two undirected graphs:

Let two undirected graphs $G_1 = (V_1, E_1)$ and $G_2 = (V_2, E_2)$ denote two PPI networks. An alignment of G_1 and G_2 is to compute a mapping between V_1 and V_2 satisfying some given criteria, where the mapping could be one-to-one or many-to-many.

Since it is usually a generalized NP-hard subgraph isomorphism problem, most of the existing algorithms on PPI networks alignment are heuristic and aimed at achieving good practical efficiency. Current research includes local and global alignment. Local alignment algorithms are designed to find isomorphic subgraphs of two or more PPI networks, where the popular ones include Mawish [15] and AlignNemo [5]. Comparing with local alignment, global alignment can better capture the global picture of how conserved substructure motifs are organized, and consequently attracts a great deal of attentions. The well known algorithms include IsoRank [32], MI-GRAAL [17], GHOST [25], MAGNA [29], Prob [34], NETAL [23], and HubAlign [11]. For example, IsoRank defines the similarity of two nodes recursively based on the similarity of their neighbors; MI-GRAAL uses both topological and biological information, and generates the alignment by a greedy seed-and-extend approach; GHOST defines the difference of spectral signatures among the nodes and generates the alignment greedily; NETAL defines the topological similarity between the nodes in a similar way to IsoRank and tries to optimize the number of conserved edges. Moreover, some algorithms are designed to handle joint alignment of multiple PPI networks, such as IsoRankN [20], NetCoffee [13], SMETANA [28], BEAMS [2], ConvexAlign [9], and NetworkBlast-M [14].

Comparing with directly solving the problem of graph isomorphism, the aforementioned heuristic approaches can alleviate the high computational complexity to certain extent. However, they still suffer several unavoidable drawbacks. For example, their time complexities could still be relatively high (e.g., $O(n^3 \log n)$ where n is the number of vertices [10]). Moreover, the available PPI networks are often very sparse, and thus the alignment based on the local topology of each vertex is not quite reliable. One way to solve this issue is to first make use of the fact that biological networks can often be embedded into Euclidean space (due to their intrinsic nature [12,18]; recently, Cho et al. [4] propose a new algorithm for low-dimensional geometric representation of biological networks called *Diffusion Component Analysis*), and then convert the alignment problem from graph domain to geometry domain. Besides the lower computational complexity, the geometric representations of PPI networks can also remedy the issue caused by the sparse and noisy interactions of PPI networks [18].

Inspired by this observation, our previous work [21] provides a geometric embedding based algorithm "GeoAlign". Roughly speaking, GeoAlign first embeds the given two PPI networks into a Euclidean space via the method of structure preserving embedding [31], and then computes their alignment in the space.

1.1   Our Contributions

The goal of this paper is twofold. First, we follow and generalize our previous work [21] to a unified algorithmic framework for PPI networks alignment. Second, we study and compare the experimental performance of our framework with other popular methods on two benchmark datasets.

Our algorithmic framework includes two steps: **(1) embedding** and **(2) matching**. Given two PPI networks, we first use a graph embedding technique to represent them in some Euclidean space. As a consequence, each network is transformed to a point set and the local topological properties (such as the connectivity and length of shortest path between nodes) are well preserved in a geometric form. We adopt three different embedding methods, the recent popular deep learning based approach *node2vec* [8], the well studied *multi-dimensional* *scaling (MDS)* [16], and *structure preserving embedding (SPE)* [31] which was used in [21] (see Sect. 2 for details). Then, we use both the geometric information and given sequence similarity scores of the proteins to establish the matching. Note that the matching should also take into account of certain transformations in Euclidean space, such as rigid transformation. To realize this idea, we propose a novel concept **"Protein Mover's Distance (PMD)"** to measure the matching cost between two PPI networks. Moreover, our framework can be naturally extended for joint alignment of multiple PPI networks.

Note: of course, the embedding method should not be limited to the aforementioned three algorithms in our general algorithmic framework, and we expect a more extensive experimental study on different embedding methods in future work.

2   Embedding Methods

In this section, we introduce three different methods for embedding PPI networks in our framework.

2.1   Node2vec

Recently, Grover and Leskovec [8] present a new algorithm called *node2vec* for feature learning. Given a graph, the key idea of node2vec is to define a novel random walk procedure to generate the neighborhood of each node (vertex) and maximize the likelihood for maintaining the interactions among the neighbors; eventually, it obtains a representation of the nodes in Euclidean space. For the sake of completeness, we briefly introduce the method below.

Let $G = (V, E)$ be a given unweighted and undirected graph and $f : V \to \mathbb{R}^d$ be the (to be learned) mapping function from the nodes to a d-dimensional space where d is a parameter that can be specified as the input. For each node $u \in V$, node2vec defines its neighborhood $N_S(u)$ based on two classic sampling strategies, *Breadth First Sampling (BFS)* and *Depth First Sampling (DFS)*. In BFS, the neighborhood $N_S(u)$ covers the nodes which are directly connected with the source node u. Differently, DFS defines $N_S(u)$ to contain the nodes which may have indirected interactions (by depth first search) with the source node u.

Node2vec applies random walk to make a balance between BFS and DFS. For a source node u and a given positive integer l, node2vec runs the fixed l steps of random walk and the neighborhood $N_S(u)$ consists of all the passed nodes. After generating the neighborhood $N_S(u)$ for each node u, node2vec is to optimize the following objective function inspired by the Skip-gram Model [22]:

$$\max_f \sum_{u \in V} \log Prob(N_S(u)|f(u)). \tag{1}$$

With the standard assumptions of conditional independence and symmetry of feature space, the objective function (1) can be further simplified to be:

$$\max_f \sum_{u \in V} [-\log Z_u + \sum_{n_i \in N_S(u)} f(n_i) \times f(u)] \tag{2}$$

where $Z_u = \sum_{v \in V} \exp(f(u) \times f(v))$ (see [8] for the omitted details). Finally, the objective function (1) is optimized by stochastic gradient descent (SGD) on single hidden-layer feedforward neural networks.

2.2 Multi-dimensional Scaling

Multi-dimensional Scaling (MDS) is a widely used tool for embedding graph into Euclidean space [16]. In particular, Higham et al. [12] and Kuchaiev et al. [18] introduce the ideas based on MDS to tackle the problems of de-noising and link prediction for PPI networks.

The input of MDS is the matrix of the $n \times n$ pairwise distances (suppose the number of nodes is n in the given graph). To define the pairwise distance, [12,18] adopt the length of the shortest path between each pair of nodes in the graph (in case that the PPI network is not connected, they handle the connected components separately). Obviously, computing the whole distance matrix could be very costly if using *Dijkstra's* or other shortest path algorithms [6]. However, since PPI networks are unweighted and usually sparse, we can directly run breadth first search n times to obtain the n^2 pairwise distances, and the total running time is only $O(n^2)$ (also Higham et al. [12] set an upper bound for the distances which makes the method even more practical).

Let the obtained distance between node i and j be d_{ij} and the dimension of the desired embedding space be d. The goal of MDS is to find n points $x_i \in \mathbb{R}^d$, $i = 1, \cdots, n$, such that the distance between each pair (x_i, x_j) is roughly equal to d_{ij}. First, we generate a positive semi-definite matrix A where each

$$a_{ij} = -\frac{1}{2}(d_{ij}^2 - \frac{1}{n}\sum_{k=1}^{n} d_{ij}^2 - \frac{1}{n}\sum_{k=1}^{n} d_{kj}^2 + \frac{1}{n^2}\sum_{k=1}^{n}\sum_{l=1}^{n} d_{kl}^2). \tag{3}$$

Consequently, we know that

$$X^T X \approx A. \tag{4}$$

Further, we decompose the matrix A to be $U^T \Sigma U$ where the rows of U are the eigenvectors of A and the diagonal entries of Σ are the eigenvalues ordered decreasingly. Finally, MDS lets $X = \sqrt{\Sigma_d} U$ be the embedding solution where Σ_d contains only the top d eigenvalues.

2.3 Structure Preserving Embedding

Given the adjacency matrix of a graph, traditional graph embedding algorithms often need to employ a spectral decomposition of the Laplacian and take the top eigenvectors as the embedding coordinates. However, a drawback of such embedding algorithms is that they cannot efficiently preserve the topology of the input graph. To remedy this issue, Shaw and Jebara propose a novel embedding algorithm called structure preserving embedding (SPE) [31]. Different from the previous spectral embedding methods, SPE learns a new positive semi-definite kernel matrix K whose spectral decomposition can preserve the topology exactly; moreover, the problem can be modeled as a semi-definite programming with a set of linear constraints. For more detailed explanation on SPE, we refer the readers to [31].

Due to the advantage on preserving topological structure, our previous work [21] adopts SPE to embed the given PPI networks into Euclidean space for computing their alignment.

3 Protein Mover's Distance

Since the given PPI networks become point sets in Euclidean space after embedding, the next question is how to measure their similarity. Actually, our idea comes from the well known concept **earth mover's distance (EMD)** in computational geometry which has been extensively studied in many areas [19, 26, 27].

*Given two point sets $A = \{p_1, p_2, \cdots, p_n\}$ and $B = \{q_1, q_2, \cdots, q_m\}$ in $\mathbb{R}^d$ with nonnegative weights α_i and β_j for each $p_i \in A$ and $q_j \in B$ respectively, define the **ground distance** $D(p_i, q_j) \geq 0$ for each pair of p_i and q_j (normally, the ground distance is simply their (squared) Euclidean distance). The EMD between A and B is:*

$$EMD(A, B) = \frac{\min_F \sum_{i=1}^{n} \sum_{j=1}^{m} f_{ij} \cdot D(p_i, q_j)}{\min \left\{ \sum_{i=1}^{n} \alpha_i, \sum_{j=1}^{m} \beta_j \right\}}, \tag{5}$$

where $F = \{f_{ij}\}$ is a feasible flow from A to B, such that $\forall i, j,\ f_{ij} \geq 0$, $\sum_{j=1}^{m} f_{ij} \leq \alpha_i$, $\sum_{i=1}^{n} f_{ij} \leq \beta_j$, and $\sum_{i=1}^{n} \sum_{j=1}^{m} f_{ij} = \min\{\sum_{i=1}^{n} \alpha_i, \sum_{j=1}^{m} \beta_j\}$.

Intuitively, EMD can be viewed as the minimum transportation cost between A and B, where the weights of A and B are the "supplies" and "capacities" respectively, and the cost of an edge between any pair of points from A to B is their ground distance (see Fig. 2(a)). Also, since EMD is associated with an

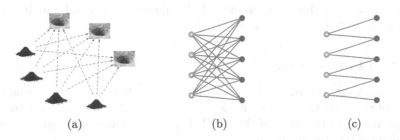

(a)   (b)   (c)

Fig. 2. (a) An illustration for earth mover's distance; (b) min-cost max flow for computing EMD; (c) the simplified min-cost max flow via FastEMD.

underlying flow F, a many-to-many matching is naturally generated via simply matching the points that have a positive flow between them. More importantly, EMD is based on a global optimization. That is, instead of greedily matching local points that are close to each other, EMD finds a matching that is able to capture the global relationship between them.

For the sake of simplicity, we also use A and B to denote the point sets, i.e., the two embedded PPI networks, respectively; each point p_i (q_j) indicates one protein. For normalization, we let each $\alpha_i = m$ and $\beta_j = n$, and thus both the total weights $\sum_{i=1}^{n} \alpha_i$ and $\sum_{j=1}^{m} \beta_j$ are equal to nm. To measure their similarity, a significant difference to EMD is that we have to consider both local topology and biological information. We introduce the following definition.

Definition 1 (Protein Mover's Distance (PMD)). *Given a parameter* $\lambda \in [0, 1]$,

$$PMD(A, B) = \lambda EMD_t(A, B) + (1 - \lambda)EMD_b(A, B), \qquad (6)$$

where $EMD_t(A, B)$ is simply the EMD between A and B with the ground distance D_t being the squared Euclidean distance, while $EMD_b(A, B)$ is the EMD between A and B with the ground distance D_b being some decreasing function on the given sequence similarity scores of the proteins.

Due to the embedding procedure, we know that $EMD_t(A, B)$ reveals the similarity of local topology between A and B. Meanwhile, $EMD_b(A, B)$ shows the similarity based on biological information, where the ground distance D_b could have different forms depending on the setting in practice. In our experiment, we simply use the inverse of the similarity score as the ground distance; if the similarity score of a pair of proteins does not exist, their ground distance is $+\infty$.

We can see that the parameter λ allocates the importances of local topology and biological information in PMD. Namely, the higher (lower) λ, the more important the local topology (biological information).

4   Our Algorithms

We first introduce our algorithm for pairwise alignment of two PPI networks in Sect. 4.1, and then show how to extend the algorithm to handle multiple PPI networks in Sect. 4.2.

4.1   Two PPI Networks

After embedding, the main idea of our alignment algorithm is to compute the PMD between the two PPI networks and generate the matching between the proteins based on the flows of the PMD. For this purpose, we need to consider the following two technical issues.

(1) Registration. Note that the embedding only preserves the pairwise distances of the nodes, thus each network actually becomes a rigid structure in the space. Consequently, we need to consider the registration between A and B under rigid transformation. Before computing the PMD, we fix A and apply the widely used *Iterative Closest Point (ICP)* [3] algorithm to find an appropriate position for B. ICP algorithm is an alternating minimization procedure that each iteration fixes either the matching or the current transformation and modifies the other to minimize the difference. ICP algorithm is guaranteed to converge and performs quite well in practice.

(2) The computation of EMD. From Definition 1, we know that both EMD_t and EMD_b need to compute the EMD between A and B but with different ground distances. Actually, optimizing the objective function of EMD is a typical instance of min-cost max flow problem which can be solved by linear programming (Fig. 2(b)). However, the numbers of points (nodes) in the PPI networks A and B are often thousands which make the computation complexity of linear programming extremely high. To resolve this issue, we use the approximate algorithm FastEMD [26] instead. Roughly speaking, FastEMD deletes the flows which have large ground distances, where the intuition is that the flows with large ground distances are more likely to be small or even zero. In practice, FastEMD makes the connecting graph of EMD much more sparse (Fig. 2(c)) and thus reduces the running time significantly.

Overall, our algorithm is shown in Algorithm 1.

Algorithm 1. Pairwise alignment

Input: two PPI networks $G_1 = (V_1, E_1)$ and $G_2 = (V_2, E_2)$, three parameters $d \in \mathbb{Z}^+$, $0 \leq \lambda \leq 1$, and $\mu > 0$.
Output: An alignment between G_1 and G_2.

1. Embed G_1 and G_2 into d-dimensional Euclidean space as A and B (by node2vec, MDS, or SPE).
2. Fix A, and run ICP to registrate B to A (with a little abuse of notations, we still use B to denote the transformed B).
3. Apply FastEMD to compute $PMD(A, B) = \lambda EMD_t(A, B) + (1 - \lambda)EMD_b(A, B)$.
4. Match protein i in A to protein j in B, if the flow between them in the PMD is larger than μ.

4.2 Multiple PPI Networks

Our method in Sect. 4.1 can be easily extended to the case with multiple networks. Given N PPI networks $G_i = (V_i, E_i), i = 1, \ldots, N$, we aim to find the alignment among all of them jointly. First, we use Algorithm 1 (step 1–3) to compute the PMD between each pair of networks, and build a N-partite graph

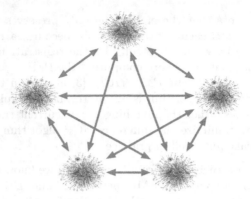

Fig. 3. Five PPI networks: we compute the PMD between each pair of networks, and build a 5-partite graph where each network is denoted as a column of vertices and the weight of each edge connecting two vertices from different columns is the corresponding value of PMD flow.

(see Fig. 3 as an example); then we apply the recent proposed convex optimization model by Hashemifar et al. [9] on the N-partite graph to find the joint alignment.

Let X_{ij} be the binary variable matrix indicating the alignment between V_i and V_j, that is, $X_{i,j}(u,v) = 1$ if $u \in V_i$ and $v \in V_j$ are aligned with each other; otherwise $X_{i,j}(u,v) = 0$. By using our obtained PMD between each pair of networks, we modify the objective function from [9] to be

$$F = \sum_{1 \leq i < j \leq N} \sum_{u \in V_i, v \in V_j} f_{PMD}(u,v) X_{ij}(u,v) \tag{7}$$

where $f_{PMD}(u,v)$ indicates the PMD flow from u to v. To make the optimization convex, according to [9] each binary variable matrix X_{ij} is relaxed to satisfy the following constraints: (i) X_{ii} is an identity matrix; (ii) X_{ij} is positive semi-definite. Finally, we use the alternating direction of multiplier method (ADMM) [9] to find the solution.

5   Experiments

For pairwise alignment, we compare our algorithm with IsoRank [32], MI-GRAAL [17], GHOST [25], and NETAL [23]; for joint alignment of multiple networks, we compare our algorithm with IsoRankN [20], NetCoffee [13], SMETANA [28], and BEAMS [2]. In our algorithms, we try the three embedding methods node2vec, MDS, and SPE, where the algorithms are denoted as **Geo-node2vec**, **Geo-mds**, and **Geo-spe** respectively. All of the experimental results are obtained on a Windows workstation with 2.4 GHz Intel Xeon E5-2630 v3 CPU and 32 GB DDR4 2133 MHz Memory.

5.1 Datasets

First, We use the popular benchmark dataset NAPAbench [30] to test the algorithms for pairwise alignment. NAPAbench has three children datasets which are generated through crystal growth (CG), duplication-mutation-complementation (DMC), and duplication-with-random-mutation (DMR); each dataset is composed of 10 pairs of PPI networks, where each pair includes a 3000-node and a 4000-node PPI network. NAPAbench also provides the sequence similarity scores among the proteins.

To further test the algorithms for joint alignment, we use another benchmark dataset Isobase [24] which contains multiple PPI networks. Isobase is a database of functionally related orthologs developed from five major eukaryotic PPI networks; it contains five species, including H.sapiens (human), S.cerevisiae (yeast), Drosophila melanogaster (fly), Caenorhabditis elegans (worm), and Mus musculus (mouse). We use BLAST bit scores [33] as the given sequence similarity scores for Isobase. See Table 1.

Table 1. a_1: number of the proteins having interaction with other proteins; a_2: number of the proteins having BLAST bit scores with other proteins; a_3: number of interactions in the network.

	a_1	a_2	a_3
Homo sapiens (human)	10403	20313	105232
Saccharomyces cerevisiae (yeast)	5524	3764	164718
Drosophila melanogaster (fly)	7396	10336	49467
Caenorhabditis elegans (worm)	2995	10945	8639
Mus musculus (mouse)	623	21856	776

To evaluate the alignment results, we compare the obtained matchings with the annotations gene ontology (GO) terms [1]. GO terms describe the roles of proteins in terms of their associated biological process, molecular function, and cellular component (CC). We exclude CC because it only annotates a small percentage of the proteins, and moreover, the proteins with matched CC are not usually considered to be functionally similar.

5.2 Evaluation Metrics

We use the following evaluation metrics which are widely used in the previous articles to measure the alignment qualities.

1. Induced Conserved Structure (ICS). Let the two PPI networks be $G_1 = (V_1, E_1)$ and $G_2 = (V_2, E_2)$, and the resulting matching be $\mathcal{M}$. We denote the subgraph induced by $\mathcal{M}$ in G_2 as $G_2(\mathcal{M}(V_1))$ and the corresponding edge sets as $E_2(\mathcal{M}(V_1))$. Also, the set of the edges conserved in the alignment is denoted as

$\mathcal{M}(E_1, E_2)$. Then the induced conserved structure score $ICS = \frac{|\mathcal{M}(E_1,E_2)|}{|E_2(\mathcal{M}(V_1))|}$ [25]. ICS is a topological measurement, because it only takes into account the graph topology.

2. Specificity. We call each connected component of the matching a *cluster*. A cluster is annotated if at least two of the proteins are annotated, and we call a cluster *correct* if all the annotated proteins share the same annotation. Specificity [7] measures the ratio of correct clusters to annotated clusters. Obviously, the higher Specificity an alignment has, the more functional consistent it is.

3. Mean Normalized Entropy (MNE). The mean normalized entropy [20] is also a measure of the consistency of the alignment. The smaller MNE an alignment has, the more functionally coherent it is. For a cluster $\mathcal{C}$ induced by the matching, the normalized entropy (NE) is defined as $NE(\mathcal{C}) = -\frac{1}{\log t} \cdot \sum_{i=1}^{t} p_i \cdot \log p_i$, where t is the number of annotations in $\mathcal{C}$ and p_i is the fraction of proteins with annotation i. Then the mean normalized entropy (MNE) is simply the average normalized entropy for all annotated clusters. We can see that a cluster that consists of proteins with higher functional consistency will have lower normalized entropy.

4. Conserved Orthologous Interactions (COI). COI is recently introduced by Hashemifar et al. [9] which only considers the total number of interactions between all pairwise correct clusters. Here we modify it to be the ratio of the total number of interactions between all pairwise correct clusters to the total number of aligned interactions. It measures the alignment algorithm's ability of detecting conserved interactions between orthologous proteins.

The latter three metrics, Specificity, MNE, and COI, are all biological measurements, since they take into account the functional annotation of each protein.

5.3   Results

In our experiments, we determine the values of λ and μ (see Algorithm 1) through optimizing Specificity score over a 10-fold cross-validation on the NAPAbench CG dataset. For simplicity, we always set the dimensionality $d = 3$ in all the embedding methods.

The average results (over 10 pairs of networks in each dataset) on pairwise alignment are shown in Table 2, where the best results are labeled in black (**for ICS, Specificity, and COI, the higher the better; for MNE, the lower the better**). Because ICS and COI are only for pairwise alignment, we use Specificity and MNE for joint alignment and the results are shown in Table 3.

We can see that Geo-spe always achieves the best for ICS, where we believe that it is due to the advantage of SPE on preserving topological structure (note that ICS is a topological measurement); for the other three evaluation metrics, Geo-node2vec often achieves the best and significantly outperforms the second best. For joint alignment, Geo-node2vec achieves the second best for Specificity which is slightly lower than the best one by NetCoffee.

Table 2. Pairwise alignment for three NAPAbench datasets CG, DMC, and DMR.

CG	IsoRank	GHOST	MI-GRAAL	NETAL	Geo-spe	Geo-node2vec	Geo-mds
ICS	0.58	0.81	0.76	0.52	**0.90**	0.72	0.66
Specificity	0.78	0.83	0.80	0.21	0.82	**0.85**	0.80
MNE	0.21	0.17	0.20	0.79	0.17	**0.15**	0.19
COI	0.42	0.51	0.53	0.49	0.72	**0.95**	0.94
DMC	IsoRank	GHOST	MI-GRAAL	NETAL	Geo-spe	Geo-node2vec	Geo-mds
ICS	0.47	0.69	0.55	0.51	**0.87**	0.56	0.50
Specificity	0.76	0.81	0.78	0.33	0.79	**0.86**	0.80
MNE	0.23	0.19	0.22	0.67	0.17	**0.14**	0.19
COI	0.45	0.58	0.60	0.48	0.68	**0.92**	0.90
DMR	IsoRank	GHOST	MI-GRAAL	NETAL	Geo-spe	Geo-node2vec	Geo-mds
ICS	0.56	0.79	0.62	0.55	**0.85**	0.62	0.57
Specificity	0.79	0.82	0.81	0.38	0.81	**0.86**	0.81
MNE	0.20	0.18	0.19	0.62	0.16	**0.14**	0.19
COI	0.44	0.55	0.59	0.46	0.71	**0.94**	0.93

Table 3. Joint alignment of the five PPI networks from Isobase

	IsoRankN	SMETANA	NetCoffee	BEAMS	Geo-spe	Geo-node2vec	Geo-mds
Specificity	0.74	0.54	**0.77**	0.73	0.73	0.75	0.71
MNE	0.83	0.99	0.95	0.81	0.81	**0.79**	0.82

6 Conclusion

In this paper, we generalize our previous work [21] and propose a unified algorithmic framework for PPI networks alignment. Different from previous methods, our framework is a geometric approach which consists of embedding and matching steps. The embedding step transforms the input PPI networks from graph domain to Euclidean space, and the matching step yields the final solution for the alignment. To efficiently solve the matching step, we define the general objective function "protein mover's distance". Moreover, our framework can be naturally extended to joint alignment of multiple PPI networks. The experimental results suggest that our method outperforms previous methods in terms of accuracy to certain extent.

To enrich the experimental study of our framework, it is deserved to explore more embedding methods instead of the three that are studied in this paper. Also, we hope that our framework can be applied to a broader range of network problems (e.g., social network) in future.

References

1. Aladağ, A.E., Erten, C.: Spinal: scalable protein interaction network alignment. Bioinformatics **29**(7), 917–924 (2013)
2. Alkan, F., Erten, C.: Beams: backbone extraction and merge strategy for the global many-to-many alignment of multiple PPI networks. Bioinformatics **30**(4), 531–539 (2013)
3. Besl, P.J., McKay, N.D.: Method for registration of 3-D shapes. In: Robotics-DL Tentative, pp. 586–606. International Society for Optics and Photonics (1992)
4. Cho, H., Berger, B., Peng, J.: Diffusion component analysis: unraveling functional topology in biological networks. In: Przytycka, T.M. (ed.) RECOMB 2015. LNCS, vol. 9029, pp. 62–64. Springer, Cham (2015). https://doi.org/10.1007/978-3-319-16706-0_9
5. Ciriello, G., Mina, M., Guzzi, P.H., Cannataro, M., Guerra, C.: Alignnemo: a local network alignment method to integrate homology and topology. PLoS ONE **7**(6), e38107 (2012)
6. Cormen, T.H., Stein, C., Rivest, R.L., Leiserson, C.E.: Introduction to Algorithms, 2nd edn. McGraw-Hill Higher Education, New York (2001)
7. Flannick, J., Novak, A., Do, C.B., Srinivasan, B.S., Batzoglou, S.: Automatic parameter learning for multiple network alignment. In: Vingron, M., Wong, L. (eds.) RECOMB 2008. LNCS, vol. 4955, pp. 214–231. Springer, Heidelberg (2008). https://doi.org/10.1007/978-3-540-78839-3_19
8. Grover, A., Leskovec, J.: node2vec: scalable feature learning for networks. In: Proceedings of the 22nd ACM SIGKDD International Conference on Knowledge Discovery and Data Mining, pp. 855–864. ACM (2016)
9. Hashemifar, S., Huang, Q., Xu, J.: Joint alignment of multiple protein-protein interaction networks via convex optimization. J. Comput. Biol. **23**(11), 903–911 (2016)
10. Hashemifar, S., Ma, J., Naveed, H., Canzar, S., Xu, J.: Modulealign: module-based global alignment of protein-protein interaction networks. Bioinformatics **32**(17), 658–664 (2016)
11. Hashemifar, S., Xu, J.: HubAlign: an accurate and efficient method for global alignment of protein-protein interaction networks. Bioinformatics **30**(17), i438–i444 (2014)
12. Higham, D.J., Rasajski, M., Przulj, N.: Fitting a geometric graph to a protein-protein interaction network. Bioinformatics **24**(8), 1093–1099 (2008)
13. Hu, J., Kehr, B., Reinert, K.: NetCoffee: a fast and accurate global alignment approach to identify functionally conserved proteins in multiple networks. Bioinformatics **30**(4), 540–548 (2013)
14. Kalaev, M., Bafna, V., Sharan, R.: Fast and accurate alignment of multiple protein networks. In: Vingron, M., Wong, L. (eds.) RECOMB 2008. LNCS, vol. 4955, pp. 246–256. Springer, Heidelberg (2008). https://doi.org/10.1007/978-3-540-78839-3_21
15. Koyutürk, M., Kim, Y., Topkara, U., Subramaniam, S., Szpankowski, W., Grama, A.: Pairwise alignment of protein interaction networks. J. Comput. Biol. **13**(2), 182–199 (2006)
16. Kruskal, J.B., Wish, M.: Multidimensional Scaling, vol. 11. Sage, Newbury Park (1978)
17. Kuchaiev, O., Pržulj, N.: Integrative network alignment reveals large regions of global network similarity in yeast and human. Bioinformatics **27**(10), 1390–1396 (2011)

18. Kuchaiev, O., Rasajski, M., Higham, D.J., Przulj, N.: Geometric de-noising of protein-protein interaction networks. PLoS Comput. Biol. **5**(8), e1000454 (2009)
19. Kusner, M., Sun, Y., Kolkin, N., Weinberger, K.: From word embeddings to document distances. In: International Conference on Machine Learning, pp. 957–966 (2015)
20. Liao, C.-S., Lu, K., Baym, M., Singh, R., Berger, B.: IsoRankN: spectral methods for global alignment of multiple protein networks. Bioinformatics **25**(12), i253–i258 (2009)
21. Liu, Y., Ding, H., Chen, D., Xu, J.: Novel geometric approach for global alignment of PPI networks. In: Proceedings of the Thirty-First AAAI Conference on Artificial Intelligence, 4–9 February 2017, San Francisco, pp. 31–37 (2017)
22. Mikolov, T., Chen, K., Corrado, G., Dean, J.: Efficient estimation of word representations in vector space. arXiv preprint arXiv:1301.3781 (2013)
23. Neyshabur, B., Khadem, A., Hashemifar, S., Arab, S.S.: NETAL: a new graph-based method for global alignment of protein-protein interaction networks. Bioinformatics **29**(13), 1654–1662 (2013)
24. Park, D., Singh, R., Baym, M., Liao, C.-S., Berger, B.: IsoBase: a database of functionally related proteins across PPI networks. Nucl. Acids Res. **39**(suppl 1), D295–D300 (2011)
25. Patro, R., Kingsford, C.: Global network alignment using multiscale spectral signatures. Bioinformatics **28**(23), 3105–3114 (2012)
26. Pele, O., Werman, M.: Fast and robust earth mover's distances. In: 2009 IEEE 12th International Conference on Computer Vision, pp. 460–467 (2009)
27. Rubner, Y., Tomasi, C., Guibas, L.J.: The earth mover's distance as a metric for image retrieval. Int. J. Comput. Vis. **40**(2), 99–121 (2000)
28. Sahraeian, S.M.E., Yoon, B.-J.: SMETANA: accurate and scalable algorithm for probabilistic alignment of large-scale biological networks. PLoS ONE **8**(7), e67995 (2013)
29. Saraph, V., Milenković, T.: MAGNA: maximizing accuracy in global network alignment. Bioinformatics **30**(20), 2931–2940 (2014)
30. Sayed Mohammad, E.S., Yoon, B.-J.: A network synthesis model for generating protein interaction network families. PLoS ONE **7**, e41474 (2012)
31. Shaw, B., Jebara, T.: Structure preserving embedding. In: Proceedings of the 26th Annual International Conference on Machine Learning, pp. 937–944. ACM (2009)
32. Singh, R., Xu, J., Berger, B.: Global alignment of multiple protein interaction networks with application to functional orthology detection. Proc. Natl. Acad. Sci. **105**(35), 12763–12768 (2008)
33. Tatusova, T.A., Madden, T.L.: BLAST 2 sequences, a new tool for comparing protein and nucleotide sequences. FEMS Microbiol. Lett. **174**(2), 247–250 (1999)
34. Todor, A., Dobra, A., Kahveci, T.: Probabilistic biological network alignment. IEEE/ACM Trans. Comput. Biol. Bioinform. (TCBB) **10**(1), 109–121 (2013)
35. Online Computational Biology Textbook. http://compbio.pbworks.com/w/page/16252899/Mass
36. Protein. http://www.hemostasis.com/protein/

On the Profit-Maximizing for Transaction Platforms in Crowd Sensing

Xi Luo[1](✉), Jialiang Lu[1](✉), Guangshuo Chen[2], Linghe Kong[1],
and Min-You Wu[1]

[1] Shanghai Jiao Tong University, Shanghai, China
luoxiqqrenren@sjtu.edu.cn, jialiang.lu@sjtu.edu.cn
[2] INRIA, University Paris Saclay, Paris, France

Abstract. Crowd sensing is a novel sensing paradigm, in which a challenging task is to balance benefits of various participants, *i.e.*, data requesters, data providers and transaction platforms for attracting sufficient participants. Little attention in literature has been paid to the transaction platform's profit which is one of the major issues for maintaining a crowd sensing system consistently. In this paper, we aim to propose a mechanism design for optimizing the platform's profit. For that, we first model the interactions in crowd sensing by leveraging tools of game theory, and then we prove the best strategy for maximizing the benefit of transaction platforms with satisfying individual rationality constraint and incentive compatibility constraint. Finally, we propose two practical algorithms based on the best strategy. Our simulations show that the algorithms are effective in terms of keeping the platform's profit and time efficiency.

Keywords: Crowd sensing · Bayesian game · Incentive mechanism design · Assignment problem · Hungarian algorithm

1 Introduction

In recent years, there has been an increasing amount of literature on crowd sensing [1]. As a new compelling paradigm, crowd sensing is to perform large-scale sensing applications using sensors of mobile devices based on the cloud, and allows mobile devices to be utilized not only for providing information to their users, but also for sensing tasks. Fruitful applications have been implemented according to the idea of crowd sensing [2–4].

One of important keys to the success of a crowd sensing system is to motivate more agents to participate crowd sensing platform through incentive mechanisms. In recent years, two main streams of incentive mechanism designs have been investigated: auction-based mechanism and reputation-based mechanism. In addition to achieve this goal, the platform is obliged to earn enough benefit through transactions in order to keep the system working consistently. For that, the design of crowd sensing incentive mechanism needs to take care of

© Springer International Publishing AG 2017
X. Gao et al. (Eds.): COCOA 2017, Part I, LNCS 10627, pp. 70–84, 2017.
https://doi.org/10.1007/978-3-319-71150-8_6

not only the social welfare and revenue but also the profit of transaction platform. The formers have been frequently addressed in the literature [5–7]. Still, a limited literature has investigated the profit maximization problem in crowd sensing systems. Han *et al.* [8] construct a multistage model to maximize platform's profit by using Lyapunov optimization techniques in stochastic process. Shah-Mansouri *et al.* [9] propose a profit maximizing truthful auction mechanism for mobile crowd sensing systems while providing satisfactory rewards to the smartphone users. Luo *et al.* [10] apply all-pay auction approach to design incentive mechanisms, emphasizing on incomplete information, risk-averse agent, and stochastic population.

In the above studies, the platform's profit is formulated as the difference between the revenue and cost of tasks. In this formulation, the profit is actually shared by the platform and data requesters having different motivations and behaviors. In contrast to these studies, we propose a novel profit model by leveraging the information asymmetry of data requesters and data providers, where we model the platform's own profit via commission charges.

In this paper, we model the interactions among data providers, data requesters, and platform. We define the profit-maximizing problem and design an incentive mechanism based on algorithms, which meets the constraints of incentive compatibility, individual rationality, and computational efficiency. Our contributions are summarized as follows:

- To the best of our knowledge, this is the first model taking profit of platforms into consideration in crowd sensing system from the perspective of an intermediary.
- We apply the auction-based design pattern and formalize the interactions between data requesters, data providers, and transaction platforms as a Bayesian game.
- Discrete mathematical deduction is given to formalize profit-maximizing problem.

The rest of the paper is organized as follows. In Sect. 2, we model the interactions of crowd sensing system. In Sect. 3, we define and formulate the maximization problem of platform's profit. In Sect. 4, we propose two algorithms to maximize platform's profit. In Sect. 5, we evaluate our algorithms based on simulations. Finally, we conclude the paper with future work in Sect. 6.

2   System Model

In this section, we start with presenting the participants in crowd sensing system. We then model the crowd sensing system from the perspective of Bayesian game. Finally, we propose the auction-based incentive mechanism of crowd sensing system.

2.1 Crowd Sensing System

Crowd sensing involves a large group of participants. Categorized by function, there are three roles in a crowd sensing system: data requesters, data providers, and a platform. A data requester publishes its tasks on the platform and is willing to pay for the sensing data. A data provider decides whether to bid for a task regarding its cost. A platform serves as an intermediary between data providers and data requesters.

2.2 Basic Definitions

Suppose a crowd sensing system with m data requesters and n data providers handled by a platform. We assume that a crowd sensing system contains only one platform. Let $M = \{1, 2, \cdots, m\}$ and $N = \{1, 2, \cdots, n\}$ denote the sets of data requesters and providers, respectively.

We model and formalize the auction mechanism between data requesters, data providers, and the platform as a Bayesian game [11,12]. In a Bayesian game, every player knows its own type, $i.e.$, the private information of itself, but only probabilistically of others' type. For platform, it has neither access to data requesters' task value, nor access to data providers' self-cost.

First, we present our model from the perspective of platform. Let $v_j \in [0, \bar{v}_j]$ denote the value of task T_j brought by data requester j to the platform, which is given by probability density function (PDF) $f_j^R : [0, \bar{v}_j] \to \mathbb{R}^+$. We assume that $\forall v_j \in [0, \bar{v}_j],\ f_j^R(v_j) > 0$ where f_j^R represents a continuous function on $[0, \bar{v}_j]$. The cumulative distribution function (CDF) $F_j^R : [0, \bar{v}_j] \to [0, 1]$ corresponding to f_j^R is given by $F_j^R(v_j) = \int_0^{\bar{v}_j} f_j^R(u) du$.

Each data provider i evaluates costs for all tasks in the system, which is represented by a vector $\mathbf{c_i} : (c_{i,1}, c_{i,2}, \cdots, c_{i,j}, \cdots, c_{i,m})$, where $c_{i,j}$ is the cost evaluation for task T_j by provider i.

We assume that each cost evaluation is independent and follows the PDF $f_{i,j}^W : [0, c_{i,j}^-] \to \mathbb{R}^+$, so that $f_{i,1}^W, f_{i,2}^W, \cdots, f_{i,m}^W$ are mutually independent, where f_i^W denotes the joint density function for the vector costs. Let V_i denote the space that $\mathbf{c_i}$ belongs to, defined as $V_i = [0, c_{i,1}^-] \times [0, c_{i,2}^-] \times \cdots \times [0, c_{i,m}^-]$. Since all tasks are independent, the joint density function is $f_i^W = f_{i,1}^W \times f_{i,2}^W \times \cdots \times f_{i,m}^W$ and the CDF $F_i^W : V_i \to [0, 1]$ corresponding to data provider i is $F_i^W(\mathbf{c_i}) = \int_{V_{c_i}} f_i^W(\mathbf{x}) d\mathbf{x}$ where $V_{c_i} = [0, c_{i,1}] \times [0, c_{i,2}] \times \cdots \times [0, c_{i,m}]$. The CDF for the cost of a single task j as $F_{i,j}^W : [0, c_{i,j}^-] \to [0, 1]$ is defined as $F_{i,j}^W(c_{i,j}) = \int_0^{c_{i,j}} f_{i,j}^W(u) du$. With this formula and independence assumption, we are able to derive the following results, $F_i^W(\mathbf{c_i}) = \prod_{j=1}^m F_{i,j}^W(c_{i,j})$.

Second, we proceed with descriptions related to data requesters and data providers. For a data requester j, its valuation for the task T_j is only known to itself. Meanwhile, it does not know others' valuations or costs, but has prior knowledge to this game which is others' distribution functions. Let V_{R_j} represent the space of incomplete information for data requester: $V_{R_j} = \prod_{k=1, k\neq j}^m [0, \bar{v}_k] \times \prod_{i=1}^n V_i$. We also assume that all estimated values and costs of participants are independent random variables. Thus, the joint density distribution function $f_{R_j} = \prod_{k=1, k \neq j}^{m} f_k^R \times \prod_{i=1}^{n} f_i^W$.

Similarly, for each data provider i, the vector $\mathbf{c_i}$ is the cost of task from every data requester, only known to the data provider itself. A data provider knows density distribution functions of all the data requesters and the other data providers, but does not know their costs or valuations. We use V_{W_i} to denote the incomplete information for data provider i: $V_{W_i} = \prod_{j=1}^{j=m} [0, \bar{v}_j] \times \prod_{k=1, k \neq i}^{n} V_k$. Based on the independence assumption, the joint density distribution for data provider i is: $f_{W_i} = \prod_{j=1}^{m} f_j^R \times \prod_{k=1, k \neq i}^{n} f_k^W$.

Let V denote the entire space for data requesters' valuations and data providers' costs: $V = \prod_{j=1}^{j=m} [0, \bar{v}_j] \times \prod_{k=1}^{n} V_k$.

2.3 Auction Mechanism

Before defining utility functions for data requesters and data providers, we describe the procedure of auction. Data requester j brings a task T_j to the platform and offers detailed information about the task and reports its bid $\tilde{v}_j$ to the platform according to its valuation v_j of the task. After that, data provider i sees the information of tasks on the platform, submits its bid for task $\tilde{c}_{i,j}$ to the platform according to its practical cost $c_{i,j}$. Notice that data provider i only needs to submit its practical cost to those tasks it is interested in. In the next step, the platform automatically sets $c_{i,k}^-$ to close infinite, and the distribution for $c_{i,k}$ to Dirac distribution for data provider i to those task T_k that it has not submitted.

Now the platform collects all bids from data requesters and data providers, which is the profile of the game, represented by

$$\mathbf{v} = (\tilde{v}_1, \tilde{v}_2, \cdots, \tilde{v}_j, \cdots, \tilde{v}_m, \tilde{\mathbf{c}}_1, \tilde{\mathbf{c}}_2, \cdots, \tilde{\mathbf{c}}_i, \cdots, \tilde{\mathbf{c}}_n) \qquad (1)$$

For data requester j, we use $\mathbf{v}_j$ to denote the vector $\mathbf{v}$ except its report $\tilde{v}_j$ to platform. For data provider i, $\mathbf{v}_j$ denotes the vector $\mathbf{v}$ except its report $\tilde{\mathbf{c}}_i$.

Having collected all reports, the platform refreshes allocation rules and payment rules.

Let $\mathbf{x}$ represent the allocation rules, which is a matrix of binary variables of every participants.

$$\mathbf{x} = \begin{bmatrix} x_1^R & x_2^R & \cdots & x_j^R & \cdots & x_m^R \\ x_{1,1}^W & x_{1,2}^W & \cdots & x_{1,j}^W & \cdots & x_{1,m}^W \\ x_{2,1}^W & x_{2,2}^W & \cdots & x_{2,j}^W & \cdots & x_{2,m}^W \\ \vdots & \vdots & \cdots & \vdots & \cdots & \vdots \\ x_{i,1}^W & x_{i,2}^W & \cdots & x_{i,j}^W & \cdots & x_{i,m}^W \\ \vdots & \vdots & & & & \\ x_{n,1}^W & x_{n,2}^W & \cdots & x_{n,j}^W & \cdots & x_{n,m}^W \end{bmatrix} \qquad (2)$$

We denote x_j^R as the binary variable for data requester j, x_j^R equals to 1 means that data requester j gets the task T_j after the auction, which implies it does not successfully allocate its task to some data providers. Let x_j^R equal to zero: it means that the data requester j does not get the task T_j after the auction and successfully allocate its task. We use $x_{i,j}^W$ to represent the binary variable for data provider i of task T_j. Let $x_{i,j}^W$ equal to 1 means that data provider i is allocated to task T_j after the auction. While $x_{i,j}^W$ equals to 0 means that data provider i is not allocated to task T_j after the auction.

Let $\mathbf{p}$ represent the payment to every participants.

$$\mathbf{p} = (p_1^W, p_2^W, \cdots, p_j^W, \cdots, p_n^W, p_1^R, p_2^R, \cdots, p_j^R, \cdots, p_m^R) \tag{3}$$

where p_i^W stands for expected payment from the platform to data provider i from the platform, and p_j^R stands for the expected payment from data requester j to the platform.

As a result, the pair $(\mathbf{p}, \mathbf{x})$ represents the auction mechanism for platform in a Bayesian game.

3   Problem Formulation

In this section, we develop participants' utility functions and clarify the constraints for formulating the platform profit maximization problem as an optimization problem. Then, we deduce the final expression of the problem by using mathematical methods.

3.1   Participants' Utility Functions

For data requester j, it can get a profit of v_j if it successfully allocates its task. So the utility for data requester is

$$U_j^R(v_j, \tilde{v}_j) = \mathbb{E}_{V_{R_j}} \{v_j(1 - x_j^R(\mathbf{v}_j, \tilde{v}_j)) - p_j^R(\mathbf{v})\} \tag{4}$$

For the data provider i, it will cost it $c_{i,j}$ when it is allocated to the task T_j. Its utility is the difference between platform's payment and its cost,

$$U_i^W(\mathbf{c_i}, \tilde{\mathbf{c_i}}) = \mathbb{E}_{V_{W_i}} \{p_i^R(\mathbf{v}_i, \tilde{\mathbf{c_i}}) - \sum_{j=1}^{m} x_{i,j}^W c_{i,j}\} \tag{5}$$

3.2   Platform Profit Maximization Problem

Once the platform has collected reports from all participants, it decides the auction mechanism $(\mathbf{p}, \mathbf{x})$. Platform's profit is hence the difference between payment from data requesters to the platform and payment from the platform to data providers.

$$U_p(\mathbf{p}, \mathbf{x}) = \mathbb{E}_V\{\sum_{j=1}^{m} p_j^R(\mathbf{v}) - \sum_{i=1}^{n} p_i^W(\mathbf{v})\} \tag{6}$$

According to the revelation principle, for any Bayesian Nash equilibrium of an auction mechanism $(\mathbf{p}, \mathbf{x})$, there exists an incentive compatible, individually rational, direct mechanism that yields the seller and bidders to the same expected utilities as in the original auction mechanism.

Without loss of generality, it means that we can only consider direct truthful revelation mechanism, in which every participant reports its true type in equilibrium. Under this circumstance, our auction mechanism design problem can be greatly simplified. Yet, we still need to ensure that our mechanism follows basic constraints of auction mechanism design, i.e. incentive compatibility constraint, individual rationality constraint and allocation constraint.

Incentive compatibility constraint. At the equilibrium, the utility of data requesters and data providers when reporting their true type is no less than when not reporting their true types. Therefore, the constraints are

$$\forall i \in N, \forall \mathbf{u} \in V_i, U_i^W(\mathbf{c_i}) \geq U_i^W(\mathbf{u}) \tag{7}$$

$$\forall j \in M, \forall v \in [0, \bar{v}_j], U_j^R(v_j) \geq U_j^R(v) \tag{8}$$

Individual rationality constraint. The second constraint is that the platform needs to assure that every participant receives non-negative utility. For data requesters and the data providers, respectively

$$\forall i \in N, U_i^W(\mathbf{c_i}) \geq 0 \tag{9}$$

$$\forall j \in M, U_j^R(v_j) \geq 0 \tag{10}$$

Allocation constraint. The allocation constraint means that task T_j can be assigned only to either data requester j or all other data providers.

$$\begin{cases} 1 = x_1^R + \sum_{i \in N} x_{i,1}^W \\ 1 = x_2^R + \sum_{i \in N} x_{i,2}^W \\ \vdots \\ 1 = x_m^R + \sum_{i \in N} x_{i,n}^W \end{cases} \tag{11}$$

Task constraint. In order to increase the chance that a data provider is allocated to a task, we assume that every data provider can be assigned at most one task.

$$\begin{cases} \sum_{j=1}^m x_{1,j}^W \leq 1 \\ \sum_{j=1}^m x_{2,j}^W \leq 1 \\ \vdots \\ \sum_{j=1}^m x_{n,j}^W \leq 1 \end{cases} \tag{12}$$

Now, we can transform Platform's Profit Maximization Problem into a Bayesian Nash equilibrium as an optimization problem. The mathematical expression is given as follows:

$$\max U_p(\mathbf{p}, \mathbf{x}) = \mathbb{E}_V\{\sum_{j=1}^{m} p_j^R(\mathbf{v}) - \sum_{i=1}^{n} p_i^W(\mathbf{v})\} \tag{13}$$

$$s.t. \quad (7), (8), (9), (10), (11), (12)$$

3.3 Mathematical Deduction

Our deduction begins with the incentive compatibility constraints, namely (7) and (8). A truth-telling agent will get the maximum profit, which is described as $U_j^R(v_j) = \max_{\tilde{v}_j} U_j^R(v_j, \tilde{v}_j) = \max_{\tilde{v}_j} \mathbb{E}_{V_{R_j}}\{v_j(1 - x_j^R(\tilde{v}_j, \mathbf{v}_j) - p_j^R(\tilde{v}_j, \mathbf{v}_j)\}$.

We discover that $U_j^R(v_j, \tilde{v}_j)$ is continuous of v_j. Consequently, we can have its derivation by applying envelop theorem: $\frac{d}{dv_j} U_j^R(v_j) = \frac{\partial}{\partial v_j} U_j^R(v_j, \tilde{v}_j = v_j) = \mathbb{E}_{V_{R_j}}\{(1 - x_j^R(\mathbf{v}))\}$. Notice that the above term is always non-negative, so the utility for data requester $U_j^R(v_j)$ is non-decreasing with v_j. Integrating the above equations, we have: $U_j^R(v_j) = U_j^R(0) + \int_0^{v_j} \mathbb{E}_{V_{-0}}\{(1 - x_j^R(u, \mathbf{v}))\}du$.

As data provider i can submit its costs to several tasks, it is necessary to respectively take costs $c_{i,j}$ for each task T_j in partial derivatives. The expression of data provider i's utility is: $U_i^W(\mathbf{c_i}) = U_i^W(\bar{\mathbf{c}}_{\mathbf{i}}) + \sum_{j=1}^{m} \int_{c_{i,j}}^{\bar{c}_{i,j}} \mathbb{E}_{V_{W_i}}\{x_{i,j}^W(u, \mathbf{v}_i)\}du$ where $\bar{\mathbf{c}}_{\mathbf{i}} = (\bar{c_{i,1}}, \bar{c_{i,2}}, \cdots, \bar{c_{i,m}})$.

From now on, we apply the integration expression of utility function so as to satisfy incentive compatibility constraint. In addition, notice that the integrand terms are always non-negative, the utility of data provider i and data requester j are consequently both non-negative if the following conditions are met,

$$\forall i \in N, \quad U_i^W(\bar{\mathbf{c}}_{\mathbf{i}}) \geq 0 \tag{14}$$

$$\forall j \in M, \quad U_j^R(0) \geq 0 \tag{15}$$

Lemma 1. *Suppose that* $\boldsymbol{x} : V \to \mathbb{R}^{\kappa}$ *maximizes*

$$\max \mathbb{E}_V\{\sum_{j=1}^{m}\{Q_j^R(v_j) - x_j^R(v)Q_j^R(v_j)\} - \sum_{i=1}^{n}\{\sum_{j=1}^{m} Q_{i,j}^W(c_{i,j})x_{i,j}^W\}\}$$

$$s.t. \quad (11), (12)$$

with the definition of $Q_j^R(v_j)$ *and* $Q_{i,j}^W(c_{i,j})$ *as follows, then we have maximized the platform's profit.*

$$Q_j^R(v_j) = v_j - \frac{1 - F_j^R(v_j)}{f_j^R(v_j)} \tag{16}$$

$$Q_{i,j}^W(c_{i,j}) = \frac{F_{i,j}^W(c_{i,j})}{f_{i,j}^W(c_{i,j})} + c_{i,j} \tag{17}$$

$Q_j^R(v_j)$ and $Q_{i,j}^W(c_{i,j})$ are interpreted as virtual valuations of data requester j and data provider i, which are essential role of Bayesian auction. We can discover that the virtual valuations $Q_j^R(v_j)$ and $Q_{i,j}^W(c_{i,j})$ consist of two terms: participant's cost or valuation of task, and a positive term $\frac{1-F_j^R(v_j)}{f_j^R(v_j)}$ or $\frac{F_{i,j}^W(c_{i,j})}{f_{i,j}^W(c_{i,j})}$, which is only related to participant's distribution. For platform, it should have charged v_j from data requester j and paid $c_{i,j}$ for data provider i for task T_j, if it had known their true valuation or cost in advance. Nevertheless, the platform does not have access to participants' true types, which gives rise to that the platform can only charge $Q_j^R(v_j)$ from data requester j, and have to pay $Q_{i,j}^W(c_{i,j})$ for the data provider i, and we have $Q_j^R(v_j) \leq v_j$ and $Q_{i,j}^W(c_{i,j}) \geq c_{i,j}$.

Consequently, data requester j is charged less and data provider i is paid more in the formulation of Bayesian auction, because of incomplete information. Lemma 1 shows that the platform decides who wins the auction according to $Q_j^R(v_j)$ and $Q_{i,j}^W(c_{i,j})$, not v_j and $c_{i,j}$.

4   Strategies and Algorithms for Platforms

In this section, we present solutions to the optimization problem (Lemma 1). It indicates how the transaction platform can insure its maximum profit. First of all, we study the problem where only one data requester in the system. We then extend it to the general case of multiple data requesters. For the general case, we propose the global optimum strategy based on Hungarian algorithm, and a local optimum strategy with greedy algorithm. Compared with Hungarian algorithm, the greedy algorithm achieves almost equivalent platform profit, but has much lower time consumption.

4.1   Basic Case: One Requester and Multiple Providers

When there is only one data requester in the crowd sensing system, the objective in Lemma 1 is simplified as follows:

$$\max \mathbb{E}_V \{ Q_1^R(v_1) - x_1^R(\mathbf{v}) Q_1^R(v_1) - \sum_{i \in N} x_{i,1}^W(\mathbf{v}) Q_{i,1}^W(c_{i,1}) \} \tag{18}$$

Before solving this problem, we introduce the *rejection*, in order to ensure that the platform obtains non-negative profit. In the *rejection*, a requester with task v_1 is rejected if $Q_1^R(v_1) < 0$, since in the case, the platform cannot achieve non-negative profit.

To solve the problem is equivalent to find the most competitive providers for the task, mathematically as follows:

$$j = \arg \min_{i \in \{R_1, W_1, \cdots, W_n\}} \{ Q_1^R(v_1), Q_{1,1}^W(c_{1,1}), \cdots, Q_{n,1}^W(c_{n,1}) \} \tag{19}$$

If $j = R_1$, the platform should not delegate the task to any providers since it has no profit to the platform. If $j \neq R_1$, the platform should assign the task to data provider j so it is guaranteed to get non-negative profit because $Q_1^R(v_1) \geq Q_{i,1}^W(c_{i,1})$.

4.2 General Case: Multiple Requesters and Multiple Providers

We consider the case in which there are multiple requesters as well as tasks. Similar to the basic case, it is reasonable to apply the *rejection* on tasks, in order to ensure the profit of platform. Similarly, in the rejection, a task of data requester j, which leads to $Q_j^R(v_j) < 0$ is refused. After the rejection, we can assure that $\sum_{j=1}^m Q_j^R(v_j)$ is non-negative.

In this case, we have the simplified equivalent problem of Lemma 1 as follows:

$$\min \mathbb{E}_V \{ \sum_{j=1}^m x_j^R(\mathbf{v}) Q_j^R(v_j) + \sum_{i=1}^n \sum_{j=1}^m Q_{i,j}^W(c_{i,j}) x_{i,j}^W \} \tag{20}$$

$$s.t. \quad \forall j \in M, x_j^R + \sum_{i=1}^n x_{i,j}^W = 1$$

$$\forall i \in N, \sum_{j=1}^m x_{i,j}^W \leq 1, \quad \forall j \in M, x_j^R \in \{0,1\}$$

$$\forall i \in N, j \in M, x_{i,j}^W \in \{0,1\}$$

As $x_j^R(\mathbf{v})$ and $x_{i,j}^W$ are binary variables for data requester j and data provider i, they can only take the value 0 or 1, our problem can be regarded as a 0–1 integer programming problem [13]. The 0–1 integer programming problem is a mathematical optimization problem in which some or all variables are restricted to the set $\{0, 1\}$. It is proved to be NP-hard, so that we can hardly expect a polynomial algorithm [13].

In order to find a polynomial solution, we convert our problem to an assignment problem [14]. An assignment problem is to choose an optimal assignment of n workers to m jobs assuming that numerical costs are given for each worker on each job. The optimal assignment is one which minimizes the sum of worker's cost for their assigned task [14]. We discover that our optimization problem is very similar to the assignment problem. Without the term $\sum_{j=1}^m x_j^R(\mathbf{v}) Q_j^R(v_j)$ and interpret the $Q_{i,j}^W(c_{i,j})$ as cost for each data provider on each task, our optimization problem is exactly the assignment problem. This term $\sum_{j=1}^m x_j^R(\mathbf{v}) Q_j^R(v_j)$ can be interpreted as the sum of costs for data requester to take the tasks. Thus, we can also treat data providers as workers in the scenario of assignment problem. As data requester j only has cost $Q_j^R(v_j)$ for task T_j, we need to add virtual costs as infinite to the task other than task T_j. The following matrix is the input to the assignment problem, in which each entry of the matrix represents the cost for each participant on each job.

$$
\begin{bmatrix}
Q_1^R(v_1) & \cdots & \text{inf} & \cdots & \text{inf} \\
\text{inf} & \cdots & \text{inf} & \cdots & \text{inf} \\
\vdots & \cdots & \vdots & \cdots & \vdots \\
\text{inf} & \cdots & Q_j^R(v_j) & \cdots & \text{inf} \\
\vdots & \cdots & \vdots & \cdots & \vdots \\
\text{inf} & \cdots & \text{inf} & \cdots & Q_m^R(v_m) \\
Q_{1,1}^W(c_{1,1}) & \cdots & Q_{1,j}^W(c_{1,j}) & \cdots & Q_{1,m}^W(c_{1,m}) \\
Q_{2,1}^W(c_{2,1}) & \cdots & Q_{2,j}^W(c_{2,j}) & \cdots & Q_{2,m}^W(c_{2,m}) \\
\vdots & \cdots & \vdots & \cdots & \vdots \\
Q_{i,1}^W(c_{i,1}) & \cdots & Q_{i,j}^W(c_{i,j}) & \cdots & Q_{i,m}^W(c_{i,m}) \\
\vdots & \cdots & \vdots & \cdots & \vdots \\
Q_{n,1}^W(c_{n,1}) & \cdots & Q_{n,j}^W(c_{n_j}) & \cdots & Q_{n,m}^W(c_{n,m})
\end{bmatrix}
\tag{21}
$$

4.3 Solutions for the General Case

Hereby we propose two solutions for the problem Lemma 1 in the general case. Recall that the objective is to assign tasks to providers while keeping the maximum profit of platform.

(1) **Hungarian algorithm**: As we convert our problem to an assignment problem, the Hungarian algorithm [15] can be utilized, which is a polynomial algorithm to solve the linear assignment problem. The key point of Hungarian algorithm is trying to find maximum matching in the bipartite graph by using zero edges. The time complexity of Hungarian algorithm is $O(n^4)$. See [15] for the detail of Hungarian algorithm.

(2) **Greedy algorithm**: We present an greedy algorithm in order to reduce the time consumption. As an approximation algorithm, this greedy algorithm has lower time complexity while not producing the maximum profit of platform.

Steps of the greedy algorithm greedy approach are illustrated in Algorithm 1 as pseudo-code. The general idea is we continuously find the minimum cost in the input matrix for those pairs who are not matched.

Although not optimal, this greedy algorithm consumes lower time than the Hungarian algorithm does. We can prove that the time complexity of greedy algorithm is $O(n^2 \log n)$ based on quick sort or merge sort. We present the performance of this two algorithms in Sect. 5.

5 Simulations

To evaluate performance of our proposed algorithms to the platform profit maximization problem, we implemented Hungarian algorithm and greedy approach

---

Algorithm 1. Greedy Approach for **Lemma 1**

Input:
   $\mathbb{M}$: Cost matrix
   W: data provider set
   R: data requester set
Output:
   A: matching set
   C: total minimum cost
Main Procedure
   1: $A \leftarrow \varnothing$
   2: $C \leftarrow 0$
   3: **while** $R \neq \varnothing$ and $W \neq \varnothing$ **do**
   4:    $(i,j) \leftarrow \arg\min \mathbb{M}$
   5:    **if** $r_i \in R$ and $W_j \in W$ **then**
   6:       $A \leftarrow A \cup \{r_j, w_j\}$
   7:       $R \leftarrow R \setminus r_i$
   8:       $W \leftarrow W \setminus w_j$
   9:       $C \leftarrow C + \mathbb{M}(i,j)$
   10:      $\mathbb{M}(i,j) \leftarrow \infty$
   11:   **end if**
   12: **end while**

---

against the random assignment. The performance metrics include platform profit ratio, which is profit calculated by different algorithms or strategies divides optimal profit; running time ratio, which is running time by greedy algorithm divides that of Hungarian algorithm.

We perform three experiments based on simulations. In the experiments, we randomly generate data collection tasks for data requesters. For each task, its real cost $c_{i,j}$ follows the uniform distribution unif$(10, 30)$. For all costs, we have $\forall i \in N, f_i^W(c_{i,j}) = \frac{1}{20}, F_i^W(c_{i,j}) = \frac{c_{i,j}-10}{20}$. According to Lemma 1, their virtual costs $Q_j^W(c_{i,j})$ considered by the platform are $\forall j \in M, Q_j^W(c_{i,j}) = 2c_{i,j} - 10$. For each task, we set its value v_j declaimed by its requester follows unif$(10, 100)$.

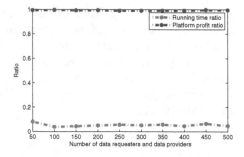

Fig. 1. The running time ratio and the platform profit

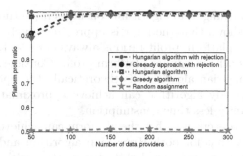

Fig. 2. The platform profit ratio regarding data providers

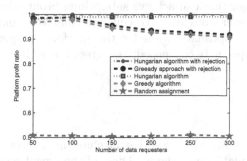

Fig. 3. The platform profit ratio regarding data requesters.

For all values, we have $\forall j \in M, f_j^R(v_j) = \frac{1}{90}, F_j^R(v_j) = \frac{v_j - 10}{90}$. Similarly, their virtual values $Q_j^R(v_j)$ considered by the platform are $\forall j \in M, Q_j^R(v_j) = 2v_j - 100$. Note that each task is generated independently. The experiments are as follows:

- **Experiment 1**: This experiment evaluates the Hungarian algorithm and the greedy algorithm in terms of time consumption. We apply the two algorithms on participants from 50/50 to 500/500 data requesters/providers, and compute the running time ratio and platform profit ratio.
- **Experiment 2**: This experiments evaluates the two algorithms that we implement in terms of data providers. In the experiment, we fix the number of data requesters as 200, and set the number of data providers ranging from 50 to 300. We apply four approaches: Hungarian algorithm with rejection, Hungarian algorithm, greedy approach with rejection, and greedy approach. We compute the platform profit ratio, as well as by the random algorithm in which each task is assigned randomly.
- **Experiment 3**: This experiments evaluates the two algorithms that we implement in terms of data requesters. Opposite to Experiment 2, we keep the number of data provider invariant as 200, and set the number of data requesters from 50 to 300. We also use the four approaches and compute the platform profit ratio.

Theoretically, the Hungarian algorithm consumes more time than the greedy algorithm. Figure 1 shows that the later is approximately 10 times faster than the former, while the platform profit ratio is always close to 1.0, indicating that these two algorithms achieve similar platform profit. Considering that the profit obtained by the Hungarian algorithm is theoretically the optimal solution, we conclude that the greedy algorithm can achieve approximate optimal performance with significantly less time consumption.

The rejection process ensures that the platform can achieve positive benefits. Figures 2 and 3 show that for both Hungarian algorithm and our greedy algorithm, one with rejection performs slightly better than without rejection. The Hungarian algorithm achieves the best performance, ignoring time consumption, while the greedy algorithm can achieve suboptimal profit when the numbers of providers and requester are close. We conclude that the Hungarian algorithm with rejection is the optimal option for the platform to get maximal profit. When time consumption is concerned, the greedy algorithm with rejection is preferred.

6   Conclusion

In our paper, we studied the incentive mechanism in crowd sensing system. Because the role of platform in the workflow of the crowd sensing system needs to be considered separately from requesters, we built a platform's profit model by formulating the transactions between data requesters and providers as a Bayesian game. After mathematical deductions, we formalized an 0–1 integer programming problem. Realizing the similarity of assignment problem, we transformed the programming program and proposed two solutions, $i.e.$, the optimal solution and the greedy solution (suboptimal and faster).

Future work will be extending the model in terms of long-term profit. The time period concept would be taken in to account to describe every "round" of transactions in long-term mode.

Acknowledgements. This research was supported by 863 under grant No. 2015AA015802, NSF of China under grant No. U1401253 and No. 61373155.

Appendix: Proof of Lemma 1

Proof. The original problem that we formulated is as follows,

$$\max U_p(\mathbf{p}, \mathbf{x}) = \mathbb{E}_V\{\sum_{j=1}^{m} p_j^R(\mathbf{v}) - \sum_{i=1}^{n} p_i^W(\mathbf{v})\} \tag{22}$$

$$s.t. \quad (7), (8), (9), (10), (11), (12)$$

We first have $\mathbb{E}_V\{\sum_{j=1}^{m} p_j^R(\mathbf{v})\} = \sum_{j=1}^{m} \int_0^{\bar{v}_j} \mathbb{E}_{V_{R_j}}\{p_j^R(u, \mathbf{v}_j)\} f_j^R(u) du$.

For data requester j, we discover $\mathbb{E}_{V_{R_j}}\{p_j^R(u, \mathbf{v}_j\} = \mathbb{E}_{V_{R_j}}\{u(1 - x_j^R(u, \mathbf{v}_j))\} - U_j^R(u)$ according to its utility function. Recalling in incentive compatibility constraint, we manage to transform utility function in to following integration form, $U_j^R(u) = U_j^R(0) + \int_0^u \mathbb{E}_{V_{R_j}}\{(1 - x_j^R(u, \mathbf{v}_j))\}du$

We then substitute the term in the integration with the above two equations.

$$\mathbb{E}_V\{\sum_{j=1}^m p_j^R(\mathbf{v})\} = \sum_{j=1}^m \int_0^{\bar{v}_j} \mathbb{E}_{V_{R_j}}\{u(1 - x_j^R(u, \mathbf{v}_j) - U_j^R(u)\}f_j^R(u)du$$

$$= \sum_{j=1}^m \mathbb{E}_V\{u(1 - x_j^R(u, \mathbf{v}_j) - U_j^R(0)\} \tag{23}$$

$$- \sum_{j=1}^m \int_0^{\bar{v}_j} \int_0^u \mathbb{E}_{V_{R_j}}\{(1 - x_j^R(x, \mathbf{v}_j))\}dx f_j^R(u)du$$

We can apply integration by parts ($f' = f_j^R(u)$, $g = \int_0^u \mathbb{E}_{V_{R_j}}\{(1 - x_j^R(x, \mathbf{v}_j))\}dx$) to the second term and rewrite the above equation, $\sum_{j=1}^m \int_0^{\bar{v}_j} \int_0^u \mathbb{E}_{V_{R_j}}\{(1 - x_j^R(x, \mathbf{v}_j))\}dx f_j^R(u)du = \sum_{j=1}^m \mathbb{E}_V\{\frac{1-F_j^R(u)}{f_j^R(u)}(1 - x_j^R(u, \mathbf{v}_j)\}$.

So, we have rewrite the payment of data requesters with definition of $Q_j^R(v_j) = v_j - \frac{1-F_j^R(v_j)}{f_j^R(v_j)}$: $\mathbb{E}_V\{\sum_{j=1}^m p_j^R(\mathbf{v})\} = \mathbb{E}_V\{\sum_{j=1}^m\{Q_j^R(v_j) - x_j^R(\mathbf{v})Q_j^R(v_j) - U_j^R(0)\}$.

The deduction of data providers' payment is almost the same. We first have: $\mathbb{E}_V\{\sum_{i=1}^n p_i^W(\mathbf{v})\} = \sum_{i=1}^n \int_{V_i} \mathbb{E}_{V_{W_i}}\{p_i^W(\mathbf{v})\}f_i^W(\mathbf{u})du$. Apply again integration form of utility function in individual rationality constraint to substitute the term in the integration.

$$\mathbb{E}_V\{\sum_{i=1}^n p_i^W(\mathbf{v})\} = \sum_{i=1}^n \int_{V_i}\{U_i^W(\mathbf{u}) + \mathbb{E}_{V_{W_i}}\{\sum_{j=1}^m x_{i,j}^W c_{i,j}\}\}f_i^W(\mathbf{u})du$$

$$= \sum_{i=1}^n \mathbb{E}_V\{\sum_{j=1}^m x_{i,j}^W c_{i,j} + U_i^W(\bar{\mathbf{c}}_i)\} \tag{24}$$

$$- \sum_{i=1}^n \int_{V_i} \sum_{j=1}^m \int_{u_{i,j}}^{c_{i,j}} \mathbb{E}_{V_{W_i}}\{x_{i,j}^W(\mathbf{x}, \mathbf{v})\}dx\}f_i^W(\mathbf{u})du$$

The second term needs to simplify further. Replace V_i and f_i^W by their definition and apply integration by parts for every components in summation and $f_{i,j}^W$, and the results is as follows: $\mathbb{E}_V\{\sum_{i=1}^n p_i^W(\mathbf{v})\} = \sum_{i=1}^n \mathbb{E}_V\{\sum_{j=1}^m \frac{F_{i,j}^W(c_j)}{f_{i,j}^W(c_j)}x_{i,j}^W + \sum_{j=1}^m x_{i,j}^W c_{i,j} + U_i^W(\bar{\mathbf{c}}_i)\}$. Let $Q_{i,j}^W(c_{i,j})$ represent $\frac{F_{i,j}^W(c_{i,j})}{f_{i,j}^W(c_{i,j})} + c_{i,j}$, we have $\mathbb{E}_V\{\sum_{i=1}^n p_i^W(\mathbf{v})\} = \sum_{i=1}^n \mathbb{E}_V\{\sum_{j=1}^m Q_{i,j}^W(c_{i,j})x_{i,j}^W + U_i^W(\bar{\mathbf{c}}_i)\}$. In order to satisfy the individual rationality constraint and maximizing the profit of platform, our best choice is set all $U_i^W(\mathbf{c_i})$ and $U_j^R(0)$ equal to zero according to Eqs. (14) and (15).

The final problem of the platform's profit is as follows.

$$\max \mathbb{E}_V\{\sum_{j=1}^{m}\{Q_j^R(v_j) - x_j^R(\mathbf{v})Q_j^R(v_j)\} - \sum_{i=1}^{n}\{\sum_{j=1}^{m} Q_{i,j}^W(c_{i,j})x_{i,j}^W\}\} \qquad (25)$$

$$s.t. \quad (11), (12)$$

References

1. Ganti, R.K., Ye, F., Lei, H.: Mobile crowdsensing: current state and future challenges. IEEE Commun. Mag. **49**(11), 32–39 (2011)
2. Capponi, A., Fiandrino, C., Kliazovich, D., et al.: A cost-effective distributed framework for data collection in cloud-based mobile crowd sensing architectures. IEEE Trans. Sustain. Comput. **2**(1), 3–16 (2017)
3. Kalejaiye, G.B., Orefice, H.R., Moura, T.A., et al.: Frugal crowd sensing for bus arrival time prediction in developing regions. In: 2017 IEEE/ACM Second International Conference on Internet-of-Things Design and Implementation (IoTDI), pp. 355–356. IEEE (2017)
4. Aly, H., Basalamah, A., Youssef, M.: Automatic rich map semantics identification through smartphone-based crowd-sensing. IEEE Trans. Mob. Comput. **16**, 2712–2725 (2016)
5. Fan, Y., Sun, H., Liu, X.: Truthful incentive mechanisms for dynamic and heterogeneous tasks in mobile crowdsourcing. In: 2015 IEEE 27th International Conference on Tools with Artificial Intelligence (ICTAI), pp. 881–888. IEEE (2015)
6. Wei, Y., et al.: Truthful online double auctions for dynamic mobile crowdsourcing. In: 2015 IEEE Conference on Computer Communications (INFOCOM). IEEE (2015)
7. Wen, Y., Shi, J., Zhang, Q., et al.: Quality-driven auction-based incentive mechanism for mobile crowd sensing. IEEE Trans. Veh. Technol. **64**(9), 4203–4214 (2015)
8. Han, Y., Zhu, Y.: Profit-maximizing stochastic control for mobile crowd sensing platforms. In: 2014 IEEE 11th International Conference on Mobile Ad Hoc and Sensor Systems (MASS), pp. 145–153. IEEE (2014)
9. Shah-Mansouri, H., Wong, V.W.S.: Profit maximization in mobile crowdsourcing: a truthful auction mechanism. In: 2015 IEEE International Conference on Communications (ICC), pp. 3216–3221. IEEE (2015)
10. Luo, T., Tan, H.P., Xia, L.: Profit-maximizing incentive for participatory sensing. In: 2014 Proceedings of IEEE INFOCOM, pp. 127–135. IEEE (2014)
11. Harsanyi, J.C.: Games with incomplete information played by Bayesian players, I–III: Part I. The basic model. Manag. Sci. **50**(12 supplement), 1804–1817 (2004). MLA
12. Myerson, R.B.: Optimal auction design. Math. Oper. Res. **6**(1), 58–73 (1981)
13. Johnson, E.L., Kostreva, M.M., Suhl, U.H.: Solving 0–1 integer programming problems arising from large scale planning models. Oper. Res. **33**(4), 803–819 (1985)
14. Munkres, J.: Algorithms for the assignment and transportation problems. J. Soci. Ind. Appl. Math. **5**(1), 32–38 (1957)
15. Kuhn, H.W.: The Hungarian method for the assignment problem. Nav. Res. Logist. (NRL) **2**(12), 83–97 (1955)

A New Approximation Algorithm
for the Maximum Stacking Base Pairs Problem
from RNA Secondary Structures Prediction

Aizhong Zhou[1], Haitao Jiang[2(✉)], Jiong Guo[1], and Daming Zhu[2]

[1] School of Computer Science and Technology and School of Mathematics
and System Science, Shandong University, Jinan, People's Republic of China
398239146@qq.com, jguo@sdu.edu.cn
[2] School of Computer Science and Technology, Shandong University,
Jinan, People's Republic of China
{htjiang,dmzhu}@sdu.edu.cn

Abstract. This paper investigates the problem of maximum stacking base pairs from RNA secondary structure prediction. The basic version of maximum stacking base pairs problem as: given an RNA sequence, to find a maximum number of base pairs where each base pair is involved in a stacking. Ieong et al. showed this problem to be NP-hard, where the candidate base pairs follow some biology principle and are given implicitly. In this paper, we study the version of this problem that the candidate base pairs are given explicitly as input, and present a new approximation algorithm for this problem by the local search method, improving the approximation factor from 5/2 to 7/3. The time complexity is within $O(n^{14})$, since we adopt 1-substitution and special 2-substitutions in the local improvement steps.

1 Introduction

RNA are versatile molecules. To understand the functions of RNAs in biological processes, we first need to understand their structures. An RNA folds into a three dimensional structure by forming hydrogen bonds between nonconsecutive bases that are complementary, such as the Watson-Crick pairs *G-C* and *A-U* and the wobble pair G-U. The primary structure of an RNA is the sequence of nucleotides in its single-stranded polymer. The collection of base pairs in the tertiary structure is the *secondary* structure. The three-dimensional arrangement of the atoms in the folded RNA molecule is the *tertiary* structure. The primary structure can be used to establish the secondary structure through the use of simple, robust rules of secondary folding. Next, the secondary structure can be used to predict the tertiary contacts in the structure, following again simple tertiary folding rules.

Actually, the secondary structure can tell us where there are additional connections between the bases, and where the RNA molecule could be folded. The folding of RNA is hierarchical, since secondary structure is much more stable

© Springer International Publishing AG 2017
X. Gao et al. (Eds.): COCOA 2017, Part I, LNCS 10627, pp. 85–92, 2017.
https://doi.org/10.1007/978-3-319-71150-8_7

than tertiary folding structure [1], which means the tertiary folding would obey the secondary structure mostly. Since the 3-dimensional structure determines the function of the RNA to some extent, predicting the secondary structure of RNA becomes a key problem to study RNA in a larger and deeper scope.

The computational study of RNA secondary structure prediction began in 1978 by Nussinov et al. [2], but this problem is still not well solved now. It can be very hard to predict the secondary structure when pseudoknots do exist in some RNAs. Lyngsø and Pedersen [8] have proven that determining the optimal secondary structure possibly with pseudoknots is NP-hard under special energy functions, and Akutsu [9] has shown that it remains NP-hard, even if the secondary structure is required to be planar. The pseudoknot is composed of two interleaving base pairs that we arrange the RNA sequence in a linear order. There are a lot of positive works where there are no pseudoknot. [2–7] have computed the optimal RNA secondary structure in $O(n^3)$ time and $O(n^2)$ space by the method of dynamic programming. And even when the size of evaluated internal loops is bounded by k (a commonly used heuristic), the time complexity of internal loop evaluation is $O(kn^2)$ in [7]. Akutsu in [9], Rivas and Eddy in [10], and Uemura et al. in [11] have presented polynomial-time algorithm when the types of pseudoknots are limited.

As an alternative, the set of candidate base pairs may be given explicitly as input, because there could be additional conditions from comparative analysis which prevent two bases forming a pair. It would generalize the maximum stacking base pairs problem with explicit base pairs, so the problem remains NP-hard. Jiang [15] improved the approximation factor for the maximum stacking base pairs problem with explicit base pairs to $\frac{5}{2}$. The problem is similar to the maximum base pairs stackings problem in Zhou [17]. The difference between the two problems is that this problem calculates the number of base pairs but the problem in [17] calculates the number of stackings. Both of these two problems use the technique of local search, but the performance analysis are different.

In this paper, we devise a new approximation algorithm for the maximum stacking base pairs problems with explicit base pairs. The approximation factor reaches to $\frac{7}{3}$, and the time complexity is $O(n^{14})$.

2   Preliminaries

Let $S = s_1 s_2 \cdots s_n$ be an RNA sequence of n bases. A secondary structure of S is a set of base pairs $(s_{i_1}, s_{j_1}), (s_{i_2}, s_{j_2}), \ldots, (s_{i_r}, s_{j_r})$, where $s_{i_k} + 2 \leq s_{j_k}$ for all $k = 1, \cdots, r$. Two base pairs, such as (s_i, s_j) and (s_{i+1}, s_{j-1}) with $i + 4 \leq j$, are *adjacent*. A base pair stacking is constitute by two adjacent base pairs. A *helix* of length q, H_q, is composed of q consecutive base pairs $(s_i, s_j), (s_{i+1}, s_{j-1}), \ldots,$ (s_{i+q-1}, s_{j-q+1}), denoted by $H_q = (s_i, s_{i+1}, \ldots, s_{i+q-1}; s_{j-q+1}, s_{j-q}, \ldots, s_j)$. Let H_α be the segment of $\{s_i, s_{i+1}, \ldots, s_{i+q-1}\}$ of H, symmetrically, let H_β be the segment of $\{s_{j-q+1}, s_{j-q}, \ldots, s_j\}$ of H. Let B_H be the set of bases which can construct the helix H, and let B'_H be the set of bases $\{ s_{i-1}, s_{i+q}, s_{j-q}, s_{j+1} \}$. Also denote the base pair (s_i, s_j) and (s_{i+q-1}, s_{j-q+1}) be the two *terminal* base pair of helix H_q. Since each base pair should be in stacking, the length of each helix is at least 2.

Now we present the formal definition of the problem studied in this paper.

Problem Description: Maximum Stacking Base Pairs.
Input: An RNA sequence S, and a set of candidate base pairs BP.
Output: A set of chosen base pairs to constitute stackings with maximum number of base pairs and no two base pairs share one common base (An example shows in Fig. 1).

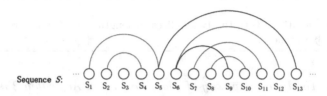

Sequence S: $\quad S_1 \ S_2 \ S_3 \ S_4 \ S_5 \ S_6 \ S_7 \ S_8 \ S_9 \ S_{10} \ S_{11} \ S_{12} \ S_{13}$

Fig. 1. We can choose the base pairs of $(s_1, s_5), (s_2, s_4), (s_6, s_{12}), (s_7, s_{11}), (s_8, s_{10})$, and the maximum stacking base pairs is 5. The base pair (s_5, s_{13}) can not be chosen because the base s_5 has been chosen. We also can not choose base pair (s_5, s_{13}) instead of (s_1, s_5) since the base pair (s_2, s_4) should be in a stacking.

3 Algorithm Description

In this section, we will show the details of our algorithm. The main idea of our algorithm is a local search method. Firstly, we need to obtain an initial solution. After that, we will perform the following 3 operations of some 1-substitutions and some special 2-substitutions to obtain more base pairs in stackings. A base is *free* if it is not involved in any base pairs we have chosen, otherwise, it is *occupied*. The algorithm keeps T, a set of base pairs, as the current solution.

- *Operation* ①: Find a helix H_q, replace the q base pairs in T by other $q'(q' > q)$ base pairs, all of which are in stackings.
- *Operation* ②: Find two helices H_2 and H_q, replace the 2 base pairs of the helix H_2 and one terminal base pair of H_q by other $q'(q' > 3)$ base pairs, all of which are in stackings.
- *Operation* ③: Find two helices H_2^1 and H_2^2 of length 2, replace the 4 base pairs of helices H_2^1 and H_2^2 by other $q'(q' > 4)$ base pairs, all of which are in stackings.

Algorithm 1. Preprocessing

1: Repeat whenever possible: Find any helix H of length q ($4 \leq q \leq 11$) where all bases in B_H are free and put H in T.

2: Repeat whenever possible: Find any helix H of length 3 where all bases in B_H are free and put H in T.

3: Repeat whenever possible: Find any helix H of length 2 where all bases in B_H are free and put H in T.

Algorithm 2. local improvement

1: Apply the operation ① to a helix H in T until we can not do operation ① any more.

2: Apply the operation ② to the two helices in T until we can not do operation ② any more.

3: Apply the operation ③ to the two helices of length 2 in T until we can not do operation ③ any more.

Theorem 1. *The time complexity of Algorithm 1 and Algorithm 2 is $O(n^{14})$.*

Proof. In the algorithm 1, to generate an initial feasible solution, we search for long helices of length at most 11. There are at most $O(n)$ such helices and it takes $O(n)$ time to find each helix. So the time complexity of algorithm 1 is $O(n^2)$.

In the algorithm 2, we will do 3 types of operations to the initial feasible solution from algorithm 1. Since a helix H of length q will destroy at most $2*q+4$ base pairs. Applying the operation ① needs to find how many helices can be formed by these $2q+4$ bases at the same time. These $2q+4$ bases at most constitute $(2q+4)/2$ helices, and it takes $O(n)$ time to constitute each helix. So the time complexity of operation ① is $O(n^{14})$ time. Do the operation ② to an H_2 and an $H_q(q > 2)$, since we just think one terminal base pair of H_q, so it will at most constitute 4 helices. Also it will have at most $O(n^2)$ such two helices. the time complexity of operation ② is $O(n^6)$ Do the operation ③ to two H_2s, there are 8 bases in the B_H of the two helices and other 8 bases adjacent to them may be useful. These 16 bases can at most constitute 8 helices and there are at most $O(n^2)$ such two helices. and the time complexity of operation ③ is $O(n^{10})$.

Therefore the time complexity of our algorithm is $O(n^{14})$.

4 Performance Analysis

To analyze the performance of our algorithm, we need to compare the output of our algorithm to the optimal solution. Let $B^* = (s_x, s_y)$ be one base pair in the optimal solution. We say the base pair B^* is *destroyed* by helix H if B_H contains one base of B_{T^*}, that the T^* is the stacking in optimal solution and contains the base of s_x or s_y. Even if the base pair B^* is also in our solution. B^* can be destroyed by at most 4 helices of our algorithm. Since a base pair (s_i, s_j) can be in two continuous stackings of $(s_{i-1}, s_i; s_j, s_{j+1})$ and $(s_i, s_{i+1}; s_{j-1}, s_j)$.

4.1   The Analysis of Approximation Performance Ratio

Lemma 1. *A helix H of length q, it will be at most $2*q+4$ base pairs destroyed by H.*

Proof. The proof is same to the Lemma 1 in [17], we omit the proof here.

Let WS_H be the total weight of base pairs in optimal solution destroyed by helix H. B^* is *singly* destroyed if B^* just be destroyed by one helix H of our solution and then we assign the weight 1 to the WS_H. B^* is *multiply* destroyed if B^* is multiply destroyed by $k(k > 1)$ helices H^1, H^2,..., and H^k, then we assign a weight of $1/k$ to WS_{H^1}, WS_{H^2},..., and WS_{H^k}. From our algorithm, B^* will be chosen in our solution or be destroyed by our solution. We say a helix H_q is *safe*, if $WS_{H_q} \leq \frac{7}{3}*q$; otherwise it is *unsafe*.

Lemma 2. *A helix $H_q (q > 2)$, we can get from our algorithm finally, then H_q must be safe.*

Proof. From Lemma 1, there are at most $2*(q+2)$ base pairs destroyed by the helix H_q. Let k be the number of base pairs singly destroyed by H_q. Since we can not do the Operation ①, so it must be $k \leq q$. So we can get that:

$$WS_H \leq \frac{k*1 + (2*(q+2)-k)}{2} \leq q+2+\frac{k}{2} \leq \frac{3}{2}*q+2 < \frac{7}{3}*q; \quad (1)$$

So each helix $H_q (q > 2)$ must be safe.

We can also get that: When $q = 2$ and $k \leq 1$, $\frac{WS_{H_2}}{2} = \frac{8+k}{4} \leq \frac{9}{4} < \frac{7}{3}$; and when $q = 2$ and $k = 2$, $\frac{WS_{H_2}}{2} = \frac{8+k}{4} \leq \frac{5}{2} > \frac{7}{3}$.

Lemma 3. *A helix $H_q (q \geq 2)$, if two base pairs multiply destroyed by H_q and other two helices, then this helix H_q must be safe.*

Proof. Let B^1 and B^2 be the two base pairs multiply destroyed by H_q and other two helices. And B^1 and B^2 will both distribute weight $\frac{1}{3}$ to WS_H.

$$WS_H \leq k*1 + \frac{2*1}{3} + \frac{(2*(q+2)-2-k)}{2} \leq q+\frac{5}{3}+\frac{k}{2}$$
$$\leq \frac{3}{2}*q+\frac{5}{3} \leq \frac{7}{3}*q; \quad (2)$$

So in this situation, the H_q must be safe.

Lemma 4. *A helix H_q, if one base in $B_{H_q} \bigcup B'_{H_q}$ is not in optimal solution, then H_q must be safe.*

Proof. Because one base in $B_{H_q} \bigcup B'_{H_q}$ is not in optimal solution, then at most $2*q+3$ base pairs in optimal solution destroyed by H_q. Let $k(k \leq q)$ be the number of base pairs simply destroyed by H_q. Then

$$WS_{H_q} \leq k + \frac{2*q+3-k}{2} \leq q+\frac{3}{2}+\frac{k}{2} \leq \frac{3}{2}*q+\frac{3}{2} \leq \frac{7}{3}*q; \quad (3)$$

So the H_q is safe.

From Lemmas 2 and 3, only the helix H_2 that at most one base pair can assign weight $w(w \leq 1/3)$ to the WS_{H_2} may be unsafe.

- *Unsafe helix H_2'*: One base pair assigns weight w to the $WS_{H_2} \leq \frac{29}{6}$ and the H_2 is unsafe.
- *Unsafe helix H_2''*: No base pair assigns weight w to the $WS_{H_2} \leq 5$ and the H_2 is unsafe.

Lemma 5. *Helices H_2^1 and H_2^2 in our solution finally are both unsafe, then H_2^1 and H_2^2 can not both destroy one base pair B^*.*

Proof. Because H_2^1 and H_2^2 are unsafe, then there will be 2 base pairs simply destroyed by H_2^1 and 2 base pairs simply destroyed by H_2^2. And the base pair B^* must just be destroyed by H_2^1 and H_2^2 (from Lemma 3). Then we can do the operation ② to the H_2^1 and H_2^2. This contradicts to the H_2^1 and H_2^2 in our solution finally.

To deal with the unsafe helices, we think the unsafe helices with other helices together as a *whole*, denote as W. Let AW_H be the aggregate of helix H' that H and H' both destroy one base pair B^*.

We will constitute the each whole as follows:

- *Whole type* ①: Helix H and all the helices in AW_H are safe, then helix H is a whole.
- *Whole type* ②: Helix H is safe and some helices in AW_H are unsafe, then these unsafe helices and H together is a whole.

Let the safe helix be the *center helix* of the whole. From Lemmas 4 and 5, one safe helix is definitely in one whole and one unsafe helix can be in one or more wholes. Similarly, let W_i be the whole containing i base pairs and WS_W be the total weight of the whole W.

Lemma 6. *For a W_q, if the length of center helix $j \geq 4$ in W_q, then $WS_{W_q} \leq \frac{7}{3} * q$.*

Proof. If the W_q is whole type ①, from Lemma 2, $WS_{W_q} = WS_{H_j} \leq 7/3 * q$. If the W_q is whole type ②, then the key is to know how many unsafe helices are in the whole W_q. Since 3 base pairs in optimal solution concerned with W_q will be destroyed by at most two unsafe helices H_2', and each such 2 base pairs will be destroyed by at most one helix H_2''. Let r_1 be the number of H_2', r_2 be the number of H_2'' and $k(k \leq j)$ be the number of base pairs simply destroyed by H_j, then

$$r_1 * \frac{3}{2} + r_2 * 2 \leq 2 * j + 4 - k; \tag{4}$$

$$q = j + 2 * (r_1 + r_2); \tag{5}$$

So to the whole W_j, we can get:

$$WS_{W_q} \leq \frac{29r_1}{6} + 5r_2 + k + \frac{2j + 4 - k - \frac{r_1}{2}}{2} + \frac{r_1}{2} * \frac{1}{3}$$

$$\leq \frac{29r_1}{6} + 5r_2 + j + 2 + \frac{k}{2} - \frac{r_1}{12}; \tag{6}$$

From (4), (5), (6) and $j \geq 4$, we can get $WS_{W_q} \leq \frac{7}{3} * q$.

Lemma 7. *A whole W_q, if the length of center helix is 3 in W_q, then $WS_{W_j} \leq \frac{7}{3} * q$.*

Proof. If the W_q is whole type ①, from Lemma 2, $WS_{W_q} = WS_{H_3} \leq \frac{7}{3} * q$. If the W_j is whole type ②, then the key is to know how many unsafe helices are in the whole W_q. Denote the center helix $H_3 = (s_i, s_{i+1}, s_{i+2}; s_{j-2}, s_{j-1}, s_j)$, and let the $\{s_{i-1}, s_i, s_{i+1}, s - i + 2\}$ be the α segment of H_3, similarly, $\{s_{j-3}, s_{j-2}, s_{j-1}, s_j, s_{j+1}\}$ be the β segment of H_3. Let $k(k \leq 3)$ be the number of base pairs simply destroyed by H_3, since some unsafe helices are in W_j, then k at most be 1.

(1) $k = 1$ and assume the base in α segment, then at most 3 base pairs can be multiply destroyed by other helices from α segment, and at most 5 base pairs can be multiply destroyed by other helices from β segment. Then there are at most one H_2'' or two H_2' from α segment. And there are at most two H_2'' or one H_2'' and two H_2' from β segment.

$$\frac{WS_{W_q}}{q} \leq \frac{5 + 5 * 2 + 1 + 8 * \frac{1}{2}}{2 + 2 * 2 + 3} = \frac{20}{9} < \frac{7}{3}; \tag{7}$$

(2) $k = 0$ and both of the α segment and β segment are at most 5 base pairs multiply destroyed by H_3 and other helices. Both of the two segments at most two H_2'' or one H_2'' and two H_2'.

$$\frac{WS_{W_q}}{q} \leq \frac{5 * 2 + 5 * 2 + 10 * \frac{1}{2}}{2 * 2 + 2 * 2 + 3} = \frac{25}{11} < \frac{7}{3}; \tag{8}$$

So for a W_q, if the length of center helix is 3 in W_q, then $WS_{W_q} \leq \frac{7}{3} * q$.

Lemma 8. *A whole W_q, if all the length of helices in the whole is 2, then $WS_{W_q} \leq \frac{7}{3} * q$.*

Proof. The proof is also similar to the Lemma 6, we omit due to space constraint.

Theorem 2. *Our algorithm approximates the maximum stacking base pairs within a factor $\frac{7}{3}$.*

Proof. From Lemmas 6, 7 and 8, all the whole W_q satisfy $WS_{W_q} \leq \frac{7}{3} * q$. So each helix satisfies $WS_{H_{q'}} \leq \frac{7}{3} * q'$. Then we are done.

References

1. Tinoco Jr., I., Bustamante, C.: How RNA folds. J. Mol. Biol. **293**, 271–281 (1999)
2. Nussinov, R., Pieczenik, G., Griggs, J.R., Kleitman, D.J.: Algorithms for loop matchings. SIAM J. Appl. Math. **35**(1), 68–82 (1978)
3. Nussinov, R., Jacobson, A.B.: Fast algorithm for predicting the secondary structure of single-stranded RNA. Proc. Natl. Acad. Sci. USA **77**, 6309–6313 (1980)
4. Zuker, M., Stiegler, P.: Optimal computer folding of large RNA sequences using thermodynamics and auxiliary information. Nucleic Acids Res. **9**, 133–148 (1981)
5. Zuker, M., Sankoff, D.: RNA secondary structures and their prediction. Bull. Math. Biol. **46**, 591–621 (1984)
6. Sankoff, D.: Simultaneous solution of the RNA folding, alignment and protosequence problems. SIAM J. Appl. Math. **45**, 810–825 (1985)
7. Lyngsø, R.B., Zuker, M., Pedersen, C.N.S.: Fast evaluation of interval loops in RNA secondary structure prediction. Bioinformatics **15**, 440–445 (1999)
8. Lyngsø, R.B., Pedersen, C.N.S.: RNA pseudoknot prediction in energy based models. J. Comput. Biol. **7**(3/4), 409–428 (2000)
9. Akutsu, T.: Dynamic programming algorithms for RNA secondary structure prediction with pseudoknots. Discrete Appl. Math. **104**(1–3), 45–62 (2000)
10. Rivas, E., Eddy, S.R.: A dynamic programming algorithm for RNA structure prediction including pseudoknots. J. Mol. Biol. **285**(5), 2053–2068 (1999)
11. Uemura, Y., Hasegawa, A., Kobayashi, S., Yokomori, T.: Tree adjoining grammars for RNA structure prediction. Theoret. Comput. Sci. **210**(2), 277–303 (1999)
12. Tinoco Jr., I., Borer, P.N., Dengler, B., Levine, M.D., Uhlenbeck, O.C., Crothers, D.M., Gralla, J.: Improved estimation of secondary structure in ribonucleic acids. Nature New Biol. **246**, 40–42 (1973)
13. Ieong, S., Kao, M.-Y., Lam, T.-W., Sung, W.-K., Yiu, S.-M.: Predicting RNA secondary structure with arbitrary pseudoknots by maximizing the number of stacking pairs. J. Comput. Biol. **10**, 981–995 (2003)
14. Lyngsø, R.B.: Complexity of pseudoknot prediction in simple models. In: Díaz, J., Karhumäki, J., Lepistö, A., Sannella, D. (eds.) ICALP 2004. LNCS, vol. 3142, pp. 919–931. Springer, Heidelberg (2004). https://doi.org/10.1007/978-3-540-27836-8_77
15. Jiang, M.: Approximation algorithms for predicting RNA secondary structures with arbitrary pseudoknots. IEEE/ACM Trans. Comput. Biol. Bioinform. **7**(2), 323–332 (2010)
16. Berman, P.: A $d/2$ approximation for maximum weight independent set in d-Claw Free Graphs. Nordic J. Comput. **7**, 178–184 (2000)
17. Zhou, A., Jiang, H., Guo, J., Feng, H., Liu, N., Zhu, B.: Improved Approximation algorithm for the maximum base pair stackings problem in RNA secondary structures prediction. In: Cao, Y., Chen, J. (eds.) COCOON 2017. LNCS, vol. 10392, pp. 575–587. Springer, Cham (2017). https://doi.org/10.1007/978-3-319-62389-4_48

Approximation Algorithm and Graph Theory

Approximation Algorithms for the Generalized Stacker Crane Problem

Jianping Li$^{(\boxtimes)}$, Xiaofei Liu, Weidong Li, Li Guan, and Junran Lichen

Department of Mathematics, Yunnan University,
Kunming 650091, People's Republic of China
{jianping,weidong,guanli}@ynu.edu.cn, lxfjl2016@163.com,
hebe_jie@sina.com

Abstract. The stacker crane problem is treated as one modified arc routing problem. This problem is to find some route for stacker cranes on a construction site such that all arcs in a mixed graph $G = (V, E \cup A; w)$ must be traversed at least once. In the real literature, since many different building materials must be handled, we consider the generalized stacker crane (GSC) problem, and the objective of this new problem is to determine a minimum weighted tour C traversing each arc e (in A) a number of times between the lower demand and upper demand.

In this paper, we design two approximation algorithms for the GSC problem. The first algorithm uses some exact algorithm to solve the integral circulation problem, and the second algorithm uses some approximation algorithm to solve the metric traveling salesman problem. Combining these two approximation algorithms, we can design a 9/5-approximation algorithm to solve the GSC problem.

Keywords: Approximation algorithm · Stacker crane problem · Lower/upper demands

1 Introduction

The arc routing problem determines a minimum weighted traversal of a set of required edges and/or arcs of a graph. Such problems are encountered in a variety of practical situations, such as road or street maintenance, garbage collection, mail delivery, school bus routing, and meter reading. Guan [4] was the first to propose the Chinese postman problem (CPP), and the CPP is tasked with finding a minimum weighted tour traversing all edges in a weighted graph at least once. Two close problems related to the CPP are the directed Chinese postman problem (DCPP) and the mixed Chinese postman problem (MCPP), respectively. The first one is a variant of the CPP on digraphs, and the second one is a generalization of the CPP. Even though Edmonds and Johnson [1] designed two combinatorial algorithms to solve the CPP and the DCPP, respectively, Papadimitriou [7] showed that the MCPP becomes *NP*-complete. As far as we have known, the best approximation algorithm to solve the MCPP is a 3/2-approximation algorithm [8].

© Springer International Publishing AG 2017
X. Gao et al. (Eds.): COCOA 2017, Part I, LNCS 10627, pp. 95–102, 2017.
https://doi.org/10.1007/978-3-319-71150-8_8

Another practical generalization of the CPP is the rural postman problem (RPP) that was first presented by Orloff [6], which is tasked with finding a minimum weighted tour traversing each of the required edges in a weighted graph at least once. Lenstra and Rinnooy Kan [5] proved that the RPP is *NP*-hard, and then Frederickson [3] designed a 3/2-approximation algorithm to solve it. The stacker crane problem which was originally addressed by Frederickson et al. [2] is a modified version of the RPP that requires that a set of edges be traversed at least once in a given direction. Frederickson et al. [2] proved the fact that the stacker crane problem is *NP*-hard and then presented a 9/5-approximation algorithm to solve it.

In the real literature, each workplace of construction site needs many different kinds of building materials for performing construction mission. This means that the stacker crane must travel in certain directions at least several times to finish the hauling task. At the same time, if the building materials are not used up on a single day, then these unused materials may be stored in the workplace and then used the next day. However, for security reasons, the construction site workplace cannot store the building materials, so the stacker crane cannot travel in certain directions more than a certain number of times.

Motivated by the proceeding problems, we address the generalized stacker crane problem (GSC) defined as follows. Given a weighted mixed graph $G = (V, E \cup A; w; l, u)$, we have a weight function $w : E \cup A \to R^+$ and two integral functions $l, u : A \to Z^+$, satisfying $l(a) \leq u(a)$ for each arc $a \in A$. If an arc $a = \langle x, y \rangle \in A$, then there is an edge $e = (x, y) \in E$ to satisfy $w(\langle x, y \rangle) = w(x, y)$. We are asked to find a tour C traversing each arc a in A at least $l(a)$ and at most $u(a)$ times, and the objective is to minimize the total weight $w(C) = \sum_{e \in E} t(e) \cdot w(e) + \sum_{a \in A} t(a) \cdot w(a)$, where $t(e)$ and $t(a)$ are the numbers of times that the tour C traverses an edge e and an arc a, respectively. Especially, when $l(a) = 1$ holds for each arc a in the mixed graph G, the GSC problem becomes the stacker crane problem.

If there is no path from vertex x_i to vertex x_j on G, where any two arcs $\langle x_i, y_i \rangle, \langle x_j, y_j \rangle \in A$, then there is no feasible tour of mixed graph G. In this paper, we may assume that a mixed graph G is strongly connected. An illustration as shown in Fig. 1, we define a triple $(w(a), l(a), u(a))$ to represent the weight, lower bound and upper bound for each arc $a \in A$ and $w(e)$ for each edge $e \in E$.

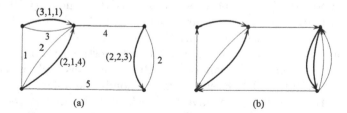

Fig. 1. (a) a graph G of the GSC problem and (b) a feasible tour of G.

Our paper is structured as follows. In Sect. 2, using an exact algorithm to solve the integral circulation problem, we construct an approximation algorithm GSC_1 to solve the GSC problem. In Sect. 3, using an approximation algorithm to solve the metric traveling salesman problem, we construct another approximation algorithm GSC_2 to solve the GSC problem. In Sect. 4, combining the two approximation algorithms GSC_1 and GSC_2 to produce a better tour better tour, we design a 9/5-approximation algorithm $GSC_{9/5}$ for the GSC problem as expected. In Sect. 5, we state our conclusion and direction for future work.

2 Algorithm GSC_1

In this section, using an exact algorithm to solve the integral flow problem, we design an approximation algorithm GSC_1 to solve the GSC problem. In detail, for a mixed graph G as an instance of the GSC problem, using the GSC_1 algorithm, we construct a feasible tour C to satisfy $w(C) \leq 3 \cdot OPT - 2 \cdot \sum_{a \in A} w(a) \cdot l(a)$, where OPT is the value of an optimal tour in G.

We use the following strategy in designing an approximation algorithm GSC_1. First, construct an auxiliary network $D = (V; A_D; l_D, u_D; w_D)$ from a mixed graph $G = (V, E \cup A; w; l, u)$, where $A_D = \{\langle x, y \rangle, \langle y, x \rangle | (x, y) \in E\}$, and then define lower bound function $l_D(\cdot)$, upper bound function $u_D(\cdot)$ and length weight function $w_D(\cdot)$ on A_D as follows:

$$l_D(a) = \begin{cases} l(a) & \text{for an arc } a \in A, \\ 0 & \text{for an arc } a \in A_D - A. \end{cases}$$
$$u_D(a) = \infty, \quad \text{for an arc } a \in A_D.$$
$$w_D(a) = w(x, y), \quad \text{for an arc } a = \langle x, y \rangle \in A_D - A.$$

Using some algorithm to find an integral circulation flow f in D, we consider a reduced digraph $D[f]$. If the reduced digraph $D[f]$ is not strongly connected, then we may find an edge subset E_1 from E with minimum-weight such that reduced mixed graph $G[f \cup E_1]$ is strongly connected. Finally, construct a mixed Eulerian tour C from the mixed graph $G[f \cup E_1]$. Algorithm GSC_1 is formally described in Algorithm 1.

For convenience, let C^* be an optimal tour of G in the proof of following theorem, where the C^* traverses each arc a in A at least $l(a)$ times and $w(C^*) = OPT$ the value of an optimal tour in the mixed graph G. And let $C(m, n)$ be the running time of the algorithm to solve the minimum cost integral circulation problem.

Theorem 1. *For a mixed graph $G = (V, E \cup A; w; l, u)$ for the GSC problem, Algorithm GSC_1 can produce a feasible tour C, satisfying*

$$w(C) \leq 3 \cdot OPT - 2 \cdot \sum_{a \in A} w(a) \cdot l(a) \tag{1}$$

and Algorithm GSC_1 runs in time $C(m, n)$.

Algorithm 1. GSC$_1$

Input: A mixed graph $G = (V, E \cup A; w; l, u)$, satisfying $l(a) \leq u(a)$ for each arc $a \in A$.

Output: A feasible tour C.

1: Construct a new network $D = (V; A_D; l_D, u_D; w_D)$ from a mixed graph $G = (V, E \cup A; w; l, u)$ as mentioned-above.
2: Determine a minimum-cost integral circulation flow f on D.
3: If digraph $D[f]$ is strongly connected, we can construct an Eulerian tour C from the mixed graph $D[f]$, and output C, stop.
4: Find an edge subset E_1 from E with minimum-weight such that the reduced graph $G[f \cup E_1]$ is strongly connected.
5: Determine a directed Eulerian tour C_i on each strongly connected components G_i of $G[A_1]$, where $A_1 = \{\langle x, y \rangle, \langle y, x \rangle | (x, y) \in E_1\}$.
6: Using the integral circulation f and these directed Eulerian tours C_i, we construct an Eulerian tour C, and output C, stop.

Proof. Since C^* is an optimal tour of G, where C^* traverses a at least $l(a)$ and at most $u(a)$ times for each arc $a \in A$, *i.e.* we can construct an integral circulation f^* on D from C^*. Since f is a minimum-cost integral circulation on D, we have

$$w_D(f) \leq OPT. \tag{2}$$

If Algorithm GSC$_1$ stops at Step 3, then we can construct a tour C from the minimum-cost integral circulation f, and we have $w(C) = w_D(f) \leq OPT$ that follows from the inequality (2), thus C is an optimal solution.

If Algorithm GSC$_1$ stops at Step 6, then we can obtain an output tour C by the integral circulation f and these Eulerian tours C_i.

Because C^* is an optimal tour on G, then $C^* - A$ can become a tour C_1^* traversing all strongly connected components of $G[f]$. Thus, we can use set C_1^* to construct a set E_1^* on G, where the graph $G[f \cup E_1^*]$ is strongly connected. Since E_1 is an edge subset of minimum weight, we thus have

$$w(E_1) = \sum_{e \in E_1} w(e) \leq \sum_{e \in E_1^*} w(e) = w(E_1^*) \leq OPT - \sum_{a \in A} l(a) \cdot w(a). \tag{3}$$

According to the structure of A_1 at Step 5, we have

$$w(A_1) = \sum_{a \in A_1} w(a) = 2 \cdot \sum_{e \in E_1} w(e) = 2 \cdot w(E_1). \tag{4}$$

By the inequalities (2)–(4), we have

$$w(C) = w_D(f) + w(A_1) \leq OPT + 2 \cdot w(E_1)$$
$$\leq 3 \cdot OPT - 2 \cdot \sum_{a \in A} w(a) \cdot l(a).$$

It is obvious to see that the running time of Algorithm GSC$_1$ depends on the running time of the integral circulation flow algorithm. Consequently, the running time of Algorithm GSC$_1$ is $C(m, n)$. □

3   Algorithm GSC$_2$

In this section, we consider a feasible tour C of a mixed graph G for the GSC problem and let $E' = \{(x,y) \in E \mid \langle x,y \rangle \in A \text{ or } \langle y,x \rangle \in A\}$. Since a feasible tour C of the GSC problem needs to traverse each connected component of $G[E']$, motivating the algorithm to solve the metric traveling salesman problem (MTSP), we design an approximation algorithm GSC$_2$ to solve the GSC problem. In detail, for a mixed graph G as an instance of the GSC problem, we find a feasible tour C by using the GSC$_2$ algorithm to satisfy $w(C) \le 3/2 \cdot OPT + 1/2 \cdot \sum_{a \in A} w(a) \cdot l(a)$.

We use the following notations. Let D_i be a directed cycle on a mixed graph G, define A_i^+ as the arc set in A that is traversed in the same direction on D_i' and let A_i^- be the arc set in A that is traversed in the opposite direction on D_i'. A T-join of V' $(\subseteq V)$ in a graph $G = (V,E)$ is a set $J \subseteq E$ such that $|J \cap \delta_G(v)|$ is odd if and only if $v \in V'$.

We denote $G_1, \ldots, G_k$ to be the connected components of $G[E']$. (In fact, if $k = 1$, using the algorithm GSC$_1$, we can have an optimal tour of G for the GSC problem.) We may suppose that $k > 1$, and then construct a contracted graph $G_1 = (V_1, E_1; w_1)$ from a mixed graph $G = (V, E \cup A; w; l, u)$. Let $P_{v_i', v_j'}$ be a shortest path from v_i' to v_j' on a graph G_1 and $w_1(P_{v_i', v_j'}) = \sum_{e \in P_{v_i', v_j'}} w_1(e)$. Contracting G_i to a single vertex v_i' in V_1, we shall use the following complete auxiliary graph $H = (V', E_H; w_H)$, where $V' = \{x_1, x_2, \ldots, x_k\}$ and $w_H(x_i, x_j) = w_1(P_{v_i', v_j'})$.

The algorithm GSC$_2$ is formally described in Algorithm 2.

Theorem 2. *For a mixed graph $G = (V, E \cup A; w; l, u)$ for the GSC problem, Algorithm GSC$_2$ can produce a feasible tour C, satisfying*

$$w(C) \le 3/2 \cdot OPT + 1/2 \cdot \sum_{a \in A} l(a) \cdot w(a), \tag{6}$$

and Algorithm GSC$_2$ runs in time $\mathcal{O}(n^3)$.

Proof. Using Algorithm GSC$_2$, we can obtain an output tour C by directed cycle C_i and these rings $r_{(x,y)}$.

Since C^* is an optimal tour of G, then $C^* - A$ becomes a tour C_1^* traversing all connected components of $G[E']$. Thus, we can construct a tour C_H^* by using the set C_1^*, where tour C_H^* traverses each vertex v in V' at least once. By Step 3, a graph H is a complete graph to satisfy the triangle inequality. For convenience, let C_H' be the optimal tour on H for the MTSP. It is obvious that C_H^* is a feasible tour on H for the MTSP. Thus, we have

$$w_H(C_H') = \sum_{e \in C_H'} w_H(e) \le \sum_{e \in C_H^*} w_H(e) = w_h(C_H^*) \tag{7}$$

By the theorem of MTSP, we have

$$w_H(C_H) = \sum_{e \in C_H} w_H(e) \le 3/2 \cdot w_H(C_H') \tag{8}$$

Algorithm 2. GSC_2

Input: A mixed graph $G = (V, E \cup A; w; l, u)$, satisfying $l(a) \le u(a)$ for each arc $a \in A$.

Output: A constrained stacker crane tour C.

1: Construct a complete auxiliary graph $H = (V', E_H; w_H)$ as mentioned-above.
2: Find a minimum cost tour C_H in H for the metric traveling salesman problem.
3: For each edge $x_i x_j$ in C_H, denote by E_1 a set consisting of such edges on these shortest paths P_{v_i, v_j} in G. Let $G_H = G[E_1]$. And denote by V_{H_o} the set of odd-degree vertices in G_H.
4: Determine a minimum T-join E_T on V_{H_o} in $G[E']$. Set $G' = G[E_1 \cup E_T]$. And let $G'_1, \ldots, G'_q$ be the connected components of G'.
5: Determine a directed Eulerian tour D_i on the graph G'_i, $i = 1, 2, \ldots, q$. If

$$\sum_{a \in A_i^-} w(a) > \sum_{a \in A_i^+} w(a), \tag{5}$$

we let the directed tour D_i be the opposite direction. And denote $A^+ = \bigcup_{i=1}^q A_i^+$ and $A^- = \bigcup_{i=1}^q A_i^-$. If an arc $a \in A^+$, denote $l'(a) := l(a) - 1$, and otherwise, if an arc $a \in A - A^+$, denote $l'(a) := l(a)$.
6: For each arc $\langle x, y \rangle \in A$, construct $l'(\langle x, y \rangle)$ rings $r_{(x,y)}$ composed an arc $\langle x, y \rangle$ and an edge (x, y). By these rings and directed cycle C_i, merge them into an Eulerian tour C, and output tour C.

By the inequalities (7) and (8), we have

$$w(E_1) = w_H(C_H) \le 3/2 \cdot w_H(C'_H) \le 3/2 \cdot w_H(C_H^*)$$
$$\le 3/2 \cdot (OPT - \sum_{\langle x,y \rangle \in A} l(\langle x, y \rangle) \cdot w(\langle x, y \rangle)) \tag{9}$$

By Step (5) and inequality (9), we have

$$w(C) = \sum_i w(C_i) + \sum_{\langle x,y \rangle \in A} (l'(\langle x, y \rangle) \cdot (w(\langle x, y \rangle) + w(x, y)))$$

$$= w(E_1) + \sum_{a \in A^+} w(a) + \sum_{a \in A^-} w(a) + 2 \cdot \sum_{\langle x,y \rangle \in A} l'(\langle x, y \rangle) \cdot w(\langle x, y \rangle)$$

$$\le w(E_1) + 2 \cdot \sum_{a \in A^+} w(a) + 2 \cdot (\sum_{\langle x,y \rangle \in A} l(\langle x, y \rangle) \cdot w(\langle x, y \rangle) - \sum_{a \in A^+} w(a))$$

$$\le w(E_1) + 2 \cdot \sum_{\langle x,y \rangle \in A} l(\langle x, y \rangle) \cdot w(\langle x, y \rangle)$$

$$\le 3/2 \cdot (OPT - \sum_{\langle x,y \rangle \in A} l(\langle x, y \rangle) \cdot w(\langle x, y \rangle)) + 2 \cdot \sum_{\langle x,y \rangle \in A} l(\langle x, y \rangle) \cdot w(\langle x, y \rangle)$$

$$\le 3/2 \cdot OPT + 1/2 \cdot \sum_{\langle x,y \rangle \in A} l(\langle x, y \rangle) \cdot w(\langle x, y \rangle).$$

It is obvious to see that the running time of Algorithm GSC$_2$ is dependent on the running time of the T-join algorithm [1]. We know that the running time of Algorithm GSC$_2$ is $\mathcal{O}(n^3)$. □

4 Algorithm GSC$_{9/5}$

By Theorems 1 and 2, we show that Algorithm GSC$_1$ is more efficient than Algorithm GSC$_2$ if the value $\sum_{a \in A} l(a) \cdot w(a)$ is larger than $3/5 \cdot OPT$, meanwhile Algorithm GSC$_2$ is more efficient than Algorithm GSC$_1$ if the value $\sum_{a \in A} l(a) \cdot w(a)$ is smaller than $3/5 \cdot OPT$. Combining these two algorithms, we would obtain an 9/5-approximation algorithm to solve the GSC problem.

In fact, we design the following algorithm to solve the GSC problem.

Algorithm 3. GSC$_{9/5}$

Input: A mixed graph $G = (V, E \cup A; w; l, u)$, satisfying $l(a) \leq u(a)$ for each arc $a \in A$.
Output: A constrained arc routing C.
1: Determine two tours C_1 and C_2 by means of Algorithm GSC$_1$ and Algorithm GSC$_2$, respectively.
2: Select the lowest cost tour between two tours C_1 and C_2, and output the better one.

Theorem 3. *For a mixed graph $G = (V, E \cup A; w; l, u)$ for the GSC problem, Algorithm GSC$_{9/5}$ is a 9/5-approximation algorithm to solve the GSC problem, and its running time is $\max\{C(m, n), n^3\}$.*

Proof. By Algorithm GSC$_{9/5}$, C_1 and C_2 are the output tours produced by Algorithm GSC$_1$ and Algorithm GSC$_2$, respectively, then we have

$$w(C_1) \leq 3 \cdot OPT - 2 \cdot \sum_{a \in A} w(a) \cdot l(a)$$

and

$$w(C_2) \leq \frac{3}{2} \cdot OPT + \frac{1}{2} \cdot \sum_{a \in A} w(a) \cdot l(a)$$

If $\sum_{a \in A} w(a) \cdot l(a) \geq 3/5 \cdot OPT$, then Algorithm GSC$_1$ provides a tour C_1 with its cost at most

$$w(C_1) \leq 3 \cdot OPT - 2 \cdot \sum_{a \in A} w(a) \cdot l(a) \leq 3 \cdot OPT - 2 \cdot (\frac{3}{5} \cdot OPT) = \frac{9}{5} \cdot OPT.$$

If $\sum_{a \in A} w(a) \cdot l(a) < 3/5 \cdot OPT$, then Algorithm GSC$_2$ provides a tour C_2 with its cost at most

$$w(C_2) \leq \frac{3}{2} \cdot OPT + \frac{1}{2} \cdot \sum_{a \in A} w(a) \cdot l(a) < \frac{3}{2} \cdot OPT + \frac{1}{2} \cdot (\frac{3}{5} \cdot OPT) = \frac{9}{5} \cdot OPT.$$

It follows that Algorithm $GSC_{9/5}$ has a worst-case ratio of 9/5.

It is obvious to see that the running time of Algorithm $GSC_{9/5}$ is dependent on the running time of both Algorithm GSC_1 and Algorithm GSC_2. Thus, we know that the running time of the $GSC_{9/5}$ algorithm is $\max\{C(m,n), n^3\}$.   □

5   Conclusion and Future Work

In this paper, we address a generalization of the stacker crane problem, where each arc a in a mixed graph is traversed by a tour at least $l(a)$ and at most $u(a)$ times. And we design a 9/5-approximation algorithm to solve the GSC problem.

In some future work, we would like to consider some optimization problems in which the times that an edge or arc is traversed should be given a reasonable range.

Acknowledgments. The work is supported in part by the National Natural Science Foundation of China [Nos. 11461081, 61662088, 11761078] and the Natural Science Foundation of Education Department of Yunnan Province [No. 2017ZZX235].

References

1. Edmonds, J., Johnson, E.L.: Matching, Euler tours and the Chinese postman. Math. Program. **5**, 88–124 (1973)
2. Frederickson, G.N., Hecht, M.S., Kim, C.E.: Approximation algorithms for some routing problems. SIAM J. Comput. **7**(2), 178–193 (1978)
3. Frederickson, G.N.: Approximation algorithms for some postman problems. J. ACM **26**, 538–554 (1979)
4. Guan, M.G.: Graphic programming using odd and even points (in Chinese). Acta Mathematica Sinica **10**, 263–266 (1960). [English translation: Chinese Mathematics, 1, 273–277 (1962)]
5. Lenstra, J.K., Rinnooy Kan, A.H.G.: On general routing problems. Networks **6**, 273–280 (1976)
6. Orloff, C.S.: A fundamental problem in vehicle routing. Networks **4**, 35–64 (1974)
7. Papadimitriou, C.H.: On the complexity of edge traversing. J. ACM **23**(3), 544–554 (1976)
8. Raghavachari, B., Veerasamy, J.: A 3/2-approximation algorithm for the mixed postman problem. SIAM J. Discrete Math. **12**(4), 425–433 (1999)

Fast Approximation Algorithms for Computing Constrained Minimum Spanning Trees

Pei Yao and Longkun Guo[✉]

College of Mathematics and Computer Science, Fuzhou University,
Fuzhou 350116, People's Republic of China
lkguo@fzu.edu.cn

Abstract. Given an integer $L \in \mathbb{Z}^+$ and an undirected graph with a weight and a length associated with every edge, the constrained minimum spanning tree (CMST) problem is to compute a minimum weight spanning tree with total length bounded by L. The problem was shown weakly $\mathcal{NP}$-hard in [1], admitting a PTAS with a runtime $O(n^{O(\frac{1}{\epsilon})}(m \log^2 n + n \log^3 n))$ due to Ravi and Goemans [13]. In the paper, we present an exact algorithm for CMST, based on our developed bicameral edge replacement which improves a feasible solution of CMST towards an optimal solution. By applying the classical rounding and scaling technique to the exact algorithm, we can obtain a fully polynomial-time approximation scheme (FPTAS), i.e. an approximation algorithm with a ratio $(1 + \epsilon)$ and a runtime $O(mn^5 \frac{1}{\epsilon^2})$, where $\epsilon > 0$ is any fixed real number.

Keywords: Constrained minimum spanning tree · Bicameral edge replacement · Approximation algorithm · FPTAS

1 Introduction

Broadcasting has become a fundamental method for public information dissemination in nowaday networks because of its advantages in high throughput, energy saving, efficiency, etc. Most data broadcasting applications require to minimize the occupied resources while guarantee customer experience simultaneously, which is typically to minimize the waiting time of the clients between proposing a request and receiving the data. In the context, a link has a length as the delay of data transmission over the link, and a weight as its occupied resource. Then the constrained minimum spanning tree problem arises, which is to compute a tree spanning all the nodes in the network, such that the total edge weight of the tree is minimized and the total length is bounded by a given threshold. Formally, we have the following definition:

Definition 1 (*The Constrained Minimum Spanning Tree problem, CMST*). *Given an undirected graph $G = (V, E)$, a weight function $w : E \rightarrow \mathbb{Z}_0^+$, a length function $l : E \rightarrow \mathbb{Z}_0^+$, and a length upper bound $L \in \mathbb{Z}_0^+$, the CMST problem is to calculate a spanning tree T with its weight sum minimized and length sum bounded by L, i.e. to minimize $\sum_{e \in T} w(e)$ subject to $\sum_{e \in T} l(e) \leq L$.*

© Springer International Publishing AG 2017
X. Gao et al. (Eds.): COCOA 2017, Part I, LNCS 10627, pp. 103–110, 2017.
https://doi.org/10.1007/978-3-319-71150-8_9

1.1   Related Works

The CMST problem is a basic theoretical problem attracting interest from both research community and industry, because it has broad applications in transportation networks, power grids, telephone networks, information dissemination in network, etc. It was put forward by Aggarwal, Aneja and Nair for the first time in the paper [1], where the problem was shown weakly $\mathcal{NP}$-hard by reducing from the knapsack problem. Then Marathe and Ravi etc. gave a (2, 2)-approximation algorithm for the CMST problem based on Hassin's approximation approach [9]; later Ravi and Goemans presented a (2, 1)-approximation algorithm based on Lagrangean relaxation [13].

Some special cases of CMST have been well studied. The minimum spanning tree (MST) problem, CMST when all edges are with length 0, is one of the most famous optimization problems. Kruskal [7], Prim [11], Dijkstra [5] and Sollin [14] have designed fast algorithms for the MST problem. Meanwhile, there were also some other interesting constrained minimum spanning tree problems with different constraints, such as the degree constrained spanning tree problem, the delay constrained minimum spanning tree problem (or namely shallow-light minimum spanning tree problem) and the hop constrained minimum spanning tree problem etc. The degree constrained minimum spanning tree problem, proposed in paper [10] by Subhash C. and Cesar A. for the first time, is to compute minimum spanning tree connecting all the vertices, with the sum of degree bounded by a given integer. Then the problem was shown $\mathcal{NP}$-complete by reducing from the Hamiltonian path problem [3]. Later, an approximation algorithm with a ratio $(1 + \epsilon)$ and a run-time $O(\log_{1+\epsilon} n)$ was developed [12]. At last, for the delay constrained minimum spanning tree problem, Leggieri, Haouari and Triki gave an exact algorithm based on Branch-and-Cut method [8]. In particular, it is shown that even the 2-hop constrained minimum spanning tree problem, i.e. the delay constrained minimum spanning tree problem when every edge is with delay 1 and the given delay constraint is 2, is $\mathcal{NP}$-Hard [4]. For the special case, Alfandari and Paschos presented a $\frac{5}{4}$-approximation algorithm when all edge costs are within $\{1, 2\}$ [2].

1.2   Our Results

In this paper, we develop an exact algorithm for CMST based on our proposed bicameral edge exchange method. The algorithm runs in time $O(mn^3 l_{max}^2)$, where $l_{max} = \max\{l(e)|e \in G \& l(e) \leq L\}$. By employing the classical rounding and scaling technique, the algorithm can be improved into a bi-factor FPTAS, i.e. an approximation algorithm with ratio $(1 + \epsilon)$, with a run-time $O(mn^5 \frac{1}{\epsilon^2})$. This improves the run-time $O(n^{O(\frac{1}{\epsilon})}(m \log^2 n + n \log^3 n))$ of the previous PTAS that has a factor $\frac{1}{\epsilon}$ on the shoulder of n.

2 Algorithms for CMST via Bicameral Edge Replacement

The key idea of our exact algorithm for CMST is similar to local search: initially compute a spanning tree T and then repeatedly improve it towards an optimal solution of CMST. Without loss of generality, we assume that the computed tree T violates the length constraint, i.e. $l(T) > L$.

Actually, a naive idea of directly employing local search can be simply as: repeat swapping an edge of T and edge out of T (with comparative smaller length), such that the length of T decreases until the length constraint is satisfied. Formally, for a pair of edges e and e', $e \in T$, $e' \in G \setminus T$, we say (e', e) is a *tree edge replacement (TER)* iff $T \setminus \{e\} \cup \{e'\}$ remains a tree. For notation briefness, we denote $l(e', e) = l(e') - l(e)$ and $w(e', e) = w(e') - w(e)$. Then obviously, $\frac{l(e', e)}{w(e', e)}$ is the exchanging rate between length and weight, when using (e', e) to improve the length of T. The edge swap simply processes as: (1) Compute a tree edge replacement (e', e), such that $l(e', e) < 0$ and $\frac{l(e', e)}{w(e', e)}$ attains minimum; (2) Set $T := T \setminus \{e\} \cup \{e'\}$. However, such a naive local search method might not output an optimal solution when it terminates.

The traditional local search method is bad since it only swaps edges to decrease the length of T but never allows increasing the length of T. So different to traditional local search method, our algorithm allows necessary increment over the length of T, and in general acts like a crafty merchant who sells or buys items depending on whichever produces better profit. That is, the algorithm decides to decrease or increase the length (and the weight of T will increase or decrease accordingly) depending on whichever produces better "profit".

2.1 Bicameral Edge Replacement

A question remains how to decide which tree edge replacement would benefit better. Intuitively, $r(e', e) = \frac{l(e', e)}{w(e', e)}$ is the exchanging rate wrt length and weight while swapping e and e' for T. Note that when $l(e', e) < 0$ and $w(e', e) > 0$, we want to minimize $r(e', e)$ as it means maximizing the rate of length decrement over weight increment; Constractly, when $l(e', e) > 0$ and $w(e', e) < 0$, we want to maximize $r(e', e)$. For briefness, we say a TER (e', e) is positive if $l(e', e) < 0$ and $w(e', e) > 0$; and the TER is negative, if $l(e', e) > 0$ and $w(e', e) < 0$. It remains to choose between the best negative and positive tree edge replacement, i.e. choose between the positive *TER* with minimum $r(e', e)$ and the negative *TER* with maximum $r(e', e)$. For the task, inspired by the bicameral cycle cancelation in [6], we introduce bicameral edge replacement formally as below:

Definition 2. *(Bicameral Edge Replacement, BER) Let T be a spanning tree and $(e'\ e)$ be a TER wrt T. Then we have three types of bicameral edge replacements as in the following:*

1. *Let (e', e) be a tree edge replacement. If $l(e', e) < 0$ and $w(e', e) \leq 0$ or $l(e', e) \leq 0$ and $w(e', e) < 0$, then it is a type-I bicameral edge replacement (BER-I);*

2. Let (e'_1, e_1) be a tree edge replacement with $l(e'_1, e_1) < 0$ and $w(e'_1, e_1) > 0$, such that

$$r(e'_1, e_1) = \min\left\{ \left.\frac{l(e', e)}{w(e', e)}\right| (e', e) \text{ is a positive TER}\right\};$$

Similarly, let (e'_2, e_2) be a tree edge replacement with $l(e'_2, e_2) > 0$ and $w(e'_2, e_2) < 0$, such that

$$r(e'_2, e_2) = \max\left\{ \left.\frac{l(e', e)}{w(e', e)}\right| (e', e) \text{ is a negative TER}\right\}.$$

Then if $r(e'_1, e_1) \geq r(e'_2, e_2)$, (e'_1, e_1) is a type-II bicameral edge replacement (BER-II); Otherwise, (e'_2, e_2) is a type-III bicameral edge replacement (BER-III).

Theorem 3. Let $\triangle L = L - \sum_{e \in T} l(e)$ and $\triangle W = W_{OPT} - \sum_{e \in T} w(e)$. Assume there exists no BER-I. Then if (e', e) is BER-II, we have $r(e', e) \leq \frac{\triangle L}{\triangle W}$; if (e', e) is BER-III, we have $r(e', e) \geq \frac{\triangle L}{\triangle W}$.

Let T be the current spanning tree and T^* be an optimal solution to CMST. The key observation of the proof of Theorem 3 is that the edges of $T \setminus T^*$ can pair with the edges of $T^* \setminus T$ to compose a set of disjoint edge pairs, each of which is a tree edge replacement (TER). Then, we can show BER is better than the average of the disjoint TERs that are composed by the set of disjoint edge pairs regarding $T \setminus T^*$ and $T^* \setminus T$. So first of all, we will show the edges of $T \setminus T^*$ can pair with the edges of $T^* \setminus T$ to compose a set of disjoint TERs. Actually, we have a more general property over the relationship of the two sets of different edges of any two distinct spanning trees, as stated below:

Lemma 4. Let T, T' be two distinct spanning trees in graph that $E_1 = E(T) \setminus E(T')$ and $E_2 = E(T') \setminus E(T)$. Let $H = (U, V, E)$ be a bipartite graph whose vertices are $U = E_1$ and $V = E_2$. Let $E(H) = \{(e', e) | e \in E_1, e' \in E_2, T \setminus \{e\} \cup \{e'\} \text{ is a tree}\}$. Then there exists a perfect matching in H.

Proof. The proof is omitted due to the length limitation.   □

Following the above lemma, we immediately have the following corollary:

Corollary 5. There exists a perfect matching $\mathcal{M}$ between the edges of $T \setminus T^*$ and $T^* \setminus T$, such that: $(1)|\mathcal{M}| = |T \setminus T^*| = |T^* \setminus T|$; $(2) x \cap y = \emptyset$ for any distinct $x \neq y \in \mathcal{M}$; (3) for any $(e', e) \in \mathcal{P}$, $e \in T \setminus T^*$ and $e' \in T^* \setminus T$ both hold, and $T \setminus \{e\} \cup \{e'\}$ is a tree.

Lemma 6. If there exists no BER-I, then one of the following two cases must hold:

1. There exists a positive TER $(e', e) \in \mathcal{M}$, such that $\frac{l(e', e)}{w(e', e)} \leq \frac{\triangle L}{\triangle W}$; OR
2. There exist a negative TER $(e', e) \in \mathcal{M}$, such that $\frac{l(e', e)}{w(e', e)} \geq \frac{\triangle L}{\triangle W}$.

Proof. The proof is omitted due to the length limitation.   □

From Lemma 6, we immediately have the correctness of Theorem 3.

Algorithm 1. An algorithm for CMST.

Input: An undirected graph $G = (V, E)$, a weight function $w : E \to Z_0^+$, a length function $l : E \to Z_0^+$, and a length bound $L \in \mathbb{R}^+$;
Output: An approximation solution to CMST.
 0: Calculate a minimum weight spanning tree T, without considering edge length;
 /*T can be computed by Prim's algorithm [11], etc. */
 1: **If** $l(T) = \sum_{e \in T} l(e) \leq L$ **then** return T;
 /*Otherwise $l(T) > L$, decrease $l(T)$ as in the following.*/
 2: Set $i = 1$, $T_1 := T$, $\mathcal{T} := \{T\}$;
 3: **While** *true* **do**
 4:    Set $R_{min}^i := \emptyset$ and $R_{max}^i := \emptyset$;
       /*R_{min}^i and R_{max}^i are respectively candidates of BER-II and BER-III for T_i.*/
 5:    **For each** $e' \in G \setminus T_i$ **do**
 6:       **For each** $e \in T_i$ **do**
 7:          **If** $T_i \setminus \{e\} \cup \{e'\}$ is a tree **then** /* Note that no BER-I exists. */
 8:             **If** $l(e', e) < 0$, $w(e', e) > 0$ **then**
 9:                $R_{min}^i = R_{min}^i \cup \{(e', e)\}$;
10:             **If** $l(e', e) > 0$, $w(e', e) < 0$ **then**
11:                $R_{max}^i = R_{max}^i \cup \{(e', e)\}$;
12:       **Endfor**
13:    **Endfor**
14:    **If** $R_{min}^j = R_{max}^j = \emptyset$ for each $j \in [i]^+$ **then** go to Step 34; /*Terminate. */
15:    **For** $j = 1$ to i **do**
16:       Set $R_1^j = arg\min_{(e', e) \in R_{min}^j}\{r(e', e)\}$, $R_2^j = arg\max_{(e', e) \in R_{max}^j}\{r(e', e)\}$;
          /*The rate is $r(e', e) := \frac{l(e', e)}{w(e', e)}$.*/
17:    **Endfor**
18:    Set $\alpha := arg\min_j\{R_1^j | j = 1, \ldots, i\}$ and $\beta := arg\max_j\{R_2^j | j = 1, \ldots, i\}$;
19:    **If** $r(R_1^\alpha) > r(R_2^\beta)$ **then**
20:       Set $T_{i+1} := T_\alpha \setminus \{e\} \cup \{e'\}$, where $R_1^\alpha = (e', e)$; /*R is the desired BER-II.*/
21:       **If** T_{i+1} is not new for $\mathcal{T}$ **then**
22:          Set $R_{min}^\alpha := R_{min}^\alpha \setminus R_1^\alpha$;
23:          Go to Step 14;
24:       **Endif**
25:    **Else**
26:       Set $T_{i+1} := T_\beta \setminus \{e\} \cup \{e'\}$, where $R_2^\beta = (e', e)$; /*R is BER-III.*/
27:       **If** T_{i+1} is not new for $\mathcal{T}$ **then**
28:          Set $R_{max}^\beta := R_{max}^\beta \setminus R_2^\beta$;
29:          Go to Step 14;
30:       **Endif**
31:    **Endif**
32:    Set $\mathcal{T} := \mathcal{T} \cup \{T_{i+1}\}$ and $i := i + 1$;
33: **Endwhile**
34: Return $arg\min_T\{w(T) | T \in \mathcal{T} \,\&\, l(T) \leq L\}$.

2.2   The Exact Algorithm

The key idea of our algorithm is first to compute a minimum weight spanning tree T, and then compute a set of trees $\mathcal{T}$ based on the idea of dynamic programming, such that each tree $T_i \in \mathcal{T}$ attains minimum weight under the length bound $l(T_i)$. The set $\mathcal{T}$ initially contains only T, and its size increases one in each iteration of our algorithm, until an optimal solution to CMST is obtained. More precisely, if $l(T) > L$, we will compute bicameral edge replacements (BERs) to construct a new tree for $\mathcal{T}$. According to the definition of BER, obviously BER-I is the first choice as it will decrease weight (or length) without any increment on length (or weight). However, since every $T_i \in \mathcal{T}$ is a minimum weight tree under the length bound $l(T_i)$, there actually exists no BER-I for every tree in $\mathcal{T}$. Therefore, our algorithm will compute *active* BER-II or BER-III for all the trees in $\mathcal{T}$, and choose the best one among them for further improvement as to generate a new element of $\mathcal{T}$.

To find an *active* BER-II or BER-III for T_i, we will find both an *active* positive TER (e_1', e_1) with

$$r(e_1', e_1) = \min \left\{ \left. \frac{l(e', e)}{w(e', e)} \right| (e', e) \text{ is an active positive TER} \right\},$$

and an *active* negative TER (e_2', e_2) with

$$r(e_2', e_2) = \max \left\{ \left. \frac{l(e', e)}{w(e', e)} \right| (e', e) \text{ is an active negative TER} \right\},$$

where a TER (e', e) is *active* if and only if $T_i \setminus \{e\} \cup \{e'\}$ is new to $\mathcal{T}$, i.e. $\mathcal{T}$ does not contain a tree with the same weight and length as $T_i \setminus \{e\} \cup \{e'\}$. Then, the algorithm will choose (e_1', e_1) as an *active* BER-II if $r(e_1', e_1) \geq r(e_2', e_2)$; and (e_2', e_2) as an *active* BER-III otherwise. The detailed algorithm is as stated in Algorithm 1.

For the time complexity and correctness of Algorithm 1, we have:

Theorem 7. *Algorithm 1 runs in time $O(mn^3 l_{max}^2)$ and produces an optimal solution to CMST, or determines the instance of CMST is infeasible, where $l_{max} = \max\{l(e)|e \in G \& l(e) \leq L\}$.*

The proof of the lemma will be given in the next section, as it needs some more properties on the relationship between T and an optimum solution of CMST. We note that Algorithm 1 is a pseudo polynomial time algorithm, as the time complexity contains L. However, by employing the classical scaling and rounding technique, we can immediately obtain an FPTAS for CMST. The key idea hereby is to set the length of every edge e as $\left\lfloor l(e) \frac{n}{\epsilon l_{max}} \right\rfloor$, where $\epsilon > 0$ is any fixed number. Then we have the improved time complexity and the compromised quality of the output of algorithm as follows:

Corollary 8. *The CMST problem admits an approximation algorithm with a ratio $(1 + \epsilon)$ and a runtime $O(mn^5 \frac{1}{\epsilon^2})$.*

We omitted the proof of the corollary due to the length limitation of the paper.

3 Proof of Theorem 7

In this section, we will show Algorithm 1 eventually outputs an optimal solution in pseudo polynomial time. First we will prove Algorithm 1 will terminate in finite steps. The key observation is that the BER we used in algorithm is not worse than the average rate of exchanging all the edges of $T \setminus T^*$ and $T^* \setminus T$. Then, our algorithm will either improve the average rate or keep the average rate unchanged but decrease the length of the tree.

Lemma 9. *In the ith iteration in Algorithm 1, $\forall i$, $w(T_i)$ is the minimum spanning tree with length bounded by $l(T_i)$ in with respect to the CMST instance.*

Proof. The proof is omitted due to the length limitation.   □

From the above lemma, we know that Algorithm 1 computes at most $O(nl_{max})$ entries, where $l_{max} = \max\{l(e)|e \in G \& l(e) \leq L\}$. That is, the algorithm iterates for at most $O(nl_{max})$ times, each of which takes $O(mn^2 l_{max})$ time to find a bicameral edge replacement. Hence, we have:

Lemma 10. *Algorithm 1 terminates in $O(mn^3 l_{max}^2)$ time.*

Due to the length limitation of the paper, we omitted the proof of the lemma above. Then it remains only to show Algorithm 1 will return an optimal solution satisfying the length bound.

Lemma 11. *If the CMST problem is feasible, then Algorithm 1 outputs a solution with length bounded by L and weight exactly being W_{OPT}.*

Proof. The proof is omitted due to the length limitation.   □

Combining Lemmas 10 and 11, we immediately have the correctness of Theorem 7.

4 Conclusion

Based on an enhanced local search method, this paper gave an exact algorithm for the constrained minimum spanning tree (CMST) problem with a pseudo polynomial time $O(mn^3 l_{max}^2)$, where m, n are respectively the number of edges, the number of vertices and $l_{max} = \max\{l(e)|e \in G \& l(e) \leq L\}$. By the rounding and scaling technique, the time complexity can be improved to $O(mn^5 \frac{1}{\epsilon^2})$, and an FPTAS is obtained. We are currently investigating a further improved FPTAS which is with an even better time complexity.

Acknowledgements. The research of the first author is supported by Natural Science Foundation of China (Nos. 61772005, 61300025) and Natural Science Foundation of Fujian Province (No. 2017J01753).

References

1. Aggarwal, V., Aneja, Y.P., Nair, K.P.K.: Minimal spanning tree subject to a side constraint. Comput. Operat. Res. **9**(4), 287–296 (1982)
2. Alfandari, L., Paschos, V.T.: Approximating minimum spanning tree of depth 2. Int. Trans. Oper. Res. **6**(6), 607–622 (1999)
3. Boldon, B., Deo, N., Kumar, N.: Minimum-weight degree-constrained spanning tree problem: Heuristics and implementation on an simd parallel machine. Parallel Comput. **22**(3), 369–382 (1996)
4. Dahl, G.: The 2-hop spanning tree problem. Oper. Res. Lett. **23**(1), 21–26 (1998)
5. Dijkstra, E.W.: A note on two problems in connexion with graphs. Numer. Math. **1**(1), 269–271 (1959)
6. Guo, L., Liao, K., Shen, H., Li, P.: Brief announcement: efficient approximation algorithms for computing k disjoint restricted shortest paths. In: Proceedings of the 27th ACM on Symposium on Parallelism in Algorithms and Architectures, SPAA 2015, Portland, OR, USA, 13–15 June 2015, pp. 62–64 (2015)
7. Kruskal, J.B.: On the shortest spanning subtree of a graph and the traveling salesman problem. Proc. Am. Math. Soc. **7**(1), 48–50 (1956)
8. Leggieri, V., Haouari, M., Triki, C.: An exact algorithm for the steiner tree problem with delays. Electron. Notes Discrete Math. **36**, 223–230 (2010)
9. Marathe, M., Ravi, R., Sundaram, R., Ravi, S., Rosenkrantz, D., Hunt, H.: Bicriteria network design problems. In: Automata, Languages and Programming, pp. 487–49 (1995)
10. Narula, S.C., Ho, C.A.: Degree-constrained minimum spanning tree. Comput. Oper. Res. **7**(4), 239–249 (1980)
11. Prim, R.C.: Shortest connection networks and some generalizations. Bell Labs Tech. J. **36**(6), 1389–1401 (1957)
12. Ravi, R., Marathe, M.V., Ravi, S.S., Rosenkrantz, D.J., Hunt III, H.B.: Approximation algorithms for degree-constrained minimum-cost network-design problems. Algorithmica **31**(1), 58–78 (2001)
13. Ravi, R., Goemans, M.X.: The constrained minimum spanning tree problem. In: Karlsson, R., Lingas, A. (eds.) SWAT 1996. LNCS, vol. 1097, pp. 66–75. Springer, Heidelberg (1996). https://doi.org/10.1007/3-540-61422-2_121
14. Sollin, M.: Le trace de canalisation. In: Berge, C., Ghouilla-Houri, A. (eds.) Programming, Games, and Transportation Networks. Wiley, New York (1965)

Trajectory-Based Multi-hop Relay Deployment in Wireless Networks

Shilei Tian, Haotian Wang, Sha Li, Fan Wu, and Guihai Chen[✉]

Shanghai Key Laboratory of Scalable Computing and Systems,
Department of Computer Science and Engineering,
Shanghai Jiao Tong University, Shanghai 200240, China
{tianshilei,ltwbwht,Zoey.Lee}@sjtu.edu.cn,
{fwu,gchen}@cs.sjtu.edu.cn

Abstract. In this paper, we identify a novel problem *Trajectory-Based Relay Deployment* (TBRD) which aims at maximizing user connection time as the users roam through the target area while complying with relay resource constraints. To solve the TBRD, we first propose the concept *Demand Nodes* (DNs). Next, we design a Demand Node Generation (DNG) algorithm that transforms the continuous historical user trajectory into a number of discrete DNs. By generating DNs, we convert the TBRD problem into a Demand Node Coverage (DNC) problem, which is NP-complete. After that, we design an approximation algorithm, named *Submodular Iterative Deployment Algorithm* (SIDA), to solve the DNC problem with the approximation factor $1 - \frac{1}{\sqrt{e \cdot (1 - 1/k)}}$. The simulation on five real datasets shows that our algorithm can obtain high coverage for users in motion, leading to better user experience.

1 Introduction

With the explosive growth of mobile users, wireless coverage has become an increasingly challenging problem. However, due to transmission distance, limited coverage of Access Point (AP), path loss, and so forth, the signal quality at some locations fails to provide satisfactory Internet access [3]. Deploying relays in multi-hop networks has become an effective method to improve the wireless coverage and service quality [6].

In this paper, we investigate the relay deployment problem under the non-stationary user setting. Due to limited transmission power and path loss, the base station (BS) may fail to cover all users at all times. We hope to deploy a limited number of relays to keep users connected to the Internet as long as possible when they are wandering.

This work was supported in part by Program of International S&T Cooperation (2016YFE0100300), the State Key Development Program for Basic Research of China (973 project 2014CB340303), China NSF Projects (Nos. 61672348, 61672353, 61422208, and 61472252), and CCF-Tencent Open Research Fund.

X. Gao et al. (Eds.): COCOA 2017, Part I, LNCS 10627, pp. 111–118, 2017.
https://doi.org/10.1007/978-3-319-71150-8_10

Existing works designed algorithms according to the exact user locations, they all assumed that users are stationary, which is not realistic in practice. As a result, once a user location changes, the network performance will be affected.

In fact, the movements of users within an area are not completely random. They are strongly affected by people's social demands [4]. Therefore, some *hot spots*, which mean locations where users often pass, or linger around, can be inferred from the user trajectory. Therefore, we consider utilizing the historical user trajectory to infer the tendency of the user movement and deploy the relays.

In this paper, we first define the connectivity of the network. Since relays cannot access the Internet directly, each relay must have a path to the BS. Then we define the *Trajectory-Based Relay Deployment* (TBRD) problem, which aims at maximizing user connection time as the users roam through the target area while complying with relay resource constraints. We introduce a concept *Demand Nodes* (DNs), which are *virtual weighted nodes* representing locations where users often pass or stay for a long time. Next, we propose a matrix-based trajectory representation and design the *Demand Node Generation* (DNG) algorithm. After that, the original TBRD problem is converted to a new problem called *Demand Node Coverage* (DNC). We claim that a DN is covered if its distance to an AP is less than the coverage radius of the AP. The DNC problem is to maximize the total weight of DNs covered by deployed relays and BS. The DNC is NP-complete, which can be reduced from a known NP-complete problem named *budget set cover* (BSC) [2]. To tackle this problem, we propose an approximation algorithm, named *Submodular Iterative Deployment Algorithm* (SIDA), which has an approximation ratio of $1 - \frac{1}{\sqrt{e \cdot (1-1/k)}}$, where e is the mathematical constant, and k is the relay number constraint. Finally, we use real datasets to evaluate our algorithm. The simulation results indicate that our algorithm can perform well.

The paper is organized as follows. The problem statement is given in Sect. 2. Section 3 describes the DNG algorithm. Section 4 presents the SIDA. Simulations are demonstrated in Sect. 5. Finally, Sect. 6 concludes this paper.

2 Problem Statement

2.1 System Model

In this model, the user can either communicate with BS directly, or connect to BS with the help of relays. Since too many hops will lead to a high delay, we limit the number of communication hops to 2.

2.2 Problem Definition

The user trajectory set is denoted by T. P_B and P_R represent the set of BS and relay candidate positions respectively. k is the number of relays we can deploy.

Definition 1 (Communication Radius). *Two APs can communicate with each other within a communication radius. We use d_B and d_R to denote the communication radius of BS and relay respectively.*

Definition 2 (2-Hop Relay Connectivity). *Given the AP candidate position set $P = P_B \cup P_R$, we generate a weighted graph $G = (P, E)$, where $(p_i, p_j) \in E$ if the distance between these two locations is less than the corresponding communication radius. The weight of each edge is set to 1. 2-hop relay connectivity means that in the induced graph $G[F]$, there always exists a path between any selected relay node and the selected BS node, while its distance is less than or equal to 2.*

Definition 3 (TBRD Problem). *Given a set of trajectories T, BS candidate locations P_B, relay candidate locations P_R, relay number constraint k, the TBRD problem is to find a BS location $p_B \in P_B$ and relay locations $P_S \subset P_R$ to maximize user connection time. P_S must be subject to $|P_S| = k$, and the induced subgraph $G[\{p_B\} \cup P_S]$ has 2-hop relay connectivity.*

As we mentioned before, hot spots can be inferred from historical user trajectory. We introduce a novel concept called *Demand Node* (DN) to represent them.

Definition 4 (Demand Node). *Demand Nodes (DNs) are virtual weighted nodes representing the locations where users often pass or stay for a long time. They are at the center of the grids which are generated by the division of the target area. The weight is the probability of user's appearance in the corresponding location. The larger the weight is, it is more possible that users will pass through or stay at the corresponding location.*

Definition 5 (Coverage Radius). *Coverage radius, denoted by r_B for BS and r_R for a relay, is the distance threshold for the BS or relay. Only DNs whose distance to an AP is less than its coverage radius can ensure Internet connection for users. We say that the DN is covered by the corresponding AP.*

Before we introduce the Demand Node Coverage (DNC) problem, we first give some definitions that are used throughout this paper. We use D to denote the DNs set, and W for the weight set of DNs.

Definition 6 (Covered DNs Set). *The covered DNs set $C(\cdot)$ is the set of DNs covered by a given AP. For a BS candidate location $p_B^i \in P_B$, $C(p_B^i) = \{d_j | dist(d_j, p_B^i) \leq r_B\}$ where $dist(\cdot)$ denotes the Euclidean distance. For a relay candidate location $p_R^i \in P_R$, $C(p_R^i) = \{d_j | dist(d_j, p_R^i) \leq r_R\}$.*

Definition 7 (Weight Function). *The weight function $w(\cdot)$ is the sum of weights of the covered DNs set. For an AP candidate location $p \in P_B \cup P_R$, $w(p) = \sum_{s_i \in C(p)} w_{s_i}$. For an AP candidate location set P, the DNs covered by P are represented as $D_C = \cup_{p_i \in P} C(p_i)$, $w(P) = \sum_{s_i \in D_C} w_{s_i}$.*

Definition 8 (Residual Weight). *Considering a selected AP candidate location set S_A, when we continue to select a AP candidate location set S_B, the residual weight of S_B based on S_A is defined as $w_R(S_A, S_B) = w(S_B) - w(S_A \cap S_B)$.*

Assume the width of the target area is w, and the height is h. There is also a filter threshold θ, which constrains the weight of each generated DN to be larger than θ. Now we can define the DNG problem.

Definition 9 (DNG Problem). *Given a user trajectory set T, the width w and height h of the target area, and a filter threshold parameter θ, the DNG problem is to generate a set of DNs D and a relative weight set W. The weight of each DN is in the range of $[\theta, 1]$.*

Now we can define the Demand Node Coverage (DNC) problem.

Definition 10 (DNC Problem). *Given a set of DNs D and the corresponding weight set W, BS candidate locations P_B, relay candidate locations P_R, relay number constraint k, the DNC problem is to find a location $p_B \in P_B$, and relay candidate locations subset $P_S \subseteq P_R$ to maximize $w(F)$, where $F = \{p_B\} \cup P_S$ while $|P_S| = k$. The induced subgraph complies with the 2-hop relay connectivity constraint.*

3   Demand Node Generation

In this section, we show how to extract "hotspots" which we refer to as Demand Nodes (DNs) from user trajectories. The Demand Node Generation (DNG) algorithm consists of three major steps: (1) trajectory matrix generation; (2) prediction matrix; (3) filtering.

3.1   Trajectory Matrix Generation

Since the DNs depend on both the temporal and spatial information of the user trajectory, we segment each trajectory according to a fixed time span t and record the location of each segment where the user appears in the target area by a binary matrix. Figure 1 illustrates the details of converting a trajectory into a binary matrix. Figure 1(a) shows a trajectory in the area.

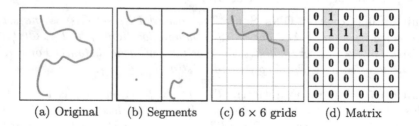

(a) Original   (b) Segments   (c) 6 × 6 grids   (d) Matrix

Fig. 1. An illustration of the process of a trajectory.

Firstly, we divide the trajectory into a number of segments, and each segment shows the trajectory of a user at the corresponding time span t, as shown in Fig. 1(b). Then, the target area is further partitioned into small sizes of grids which are the candidate locations for the demand nodes. Figure 1(c) shows the distribution of the upper left segment. Lastly, Fig. 1(d) shows the binary matrix of the trajectory at one time segment. The whole trajectory area is seen as a matrix and entries of the matrix represent the partitioned grids. If the trajectory passes through the grid, the corresponding entry of the matrix is set to 1.

After the conversion, we obtain numerous binary matrices for the target area.

3.2   Prediction Matrix

Since the value of a grid x_{ij} is 0 or 1, we assume the probability distributions of these grids are independent Bernoulli distributions, which can be written as $x_{ij} \sim p(x_{ij}|\mu_{ij}) = \mu_{ij}^{x_{ij}}(1 - \mu_{ij})^{1-x_{ij}}$, where the parameter $\mu_{ij} \in [0,1]$ is the probability of $x_{ij} = 1$.

We can estimate the μ_{ij} by maximizing likelihood estimation. However, this may lead to over-fitted results for small datasets [1]. In order to alleviate this problem, we first introduce a prior distribution $p(\mu_{ij}|a_{ij}, b_{ij})$, *beta* distribution, over the parameter μ_{ij}, which is easy to interpret while having some properties. The posterior distribution of μ_{ij} is now obtained by Bayesian theorem

$$p(\mu_{ij}|X_{ij}) = \frac{p(X_{ij}|\mu_{ij})p(\mu_{ij}|a_{ij}, b_{ij})}{\int p(X_{ij}|\mu_{ij})p(\mu_{ij}|a_{ij}, b_{ij})d\mu_{ij}}. \tag{1}$$

Then, we estimate the value of μ_{ij} by maximizing the posterior distribution $p(\mu_{ij}|x_{ij})$. We see that this posterior distribution has the form

$$p(\mu_{ij}|X_{ij}) \propto \mu_{ij}^{m+a_{ij}-1}(1 - \mu_{ij})^{n-m+b_{ij}-1}. \tag{2}$$

Finally, maximizing Eq. (2) with respect to μ_{ij}, we obtain the maximum posterior solution given by $\mu_{ij} = \frac{m+a_{ij}}{n+a_{ij}+b_{ij}}$.

3.3   Filtering

After the prediction matrix of the target area is determined, the DNs are at the center of those grids with higher probabilities for 1. In our model, a threshold θ is set, and the grids whose probabilities for 1 are not less than θ are DNs.

4   Submodular Iterative Deployment Algorithm (SIDA)

We now focus on selecting the locations for APs from the candidate location set. It is clear that the weight function $w(\cdot)$ is a submodular function.

4.1   The SIDA

The main idea of SIDA is as follows. First, we construct an undirected graph $G = (P, E)$, where $P = P_B \cup P_R$. For any two nodes $p_i, p_j \in P$, $(p_i, p_j) \in E$ if $dist(p_i, p_j)$ is less than the corresponding communication radius. Then, we scan each p_B^i sequentially, and generate a subgraph with its 2-hop neighbors. The following operations are taken within this subgraph.

Algorithm 1. SIDA

Input: An instance of DNC problem, $\langle P_B, P_R, k, w(\cdot), w_R(\cdot) \rangle$
Output: The final solution F

1 $D \leftarrow \emptyset$;
2 **for** $b \in P_B$ **do**
3      $k' \leftarrow k$; $S \leftarrow b$; $V_t \leftarrow \{v : hop(v, b) \leq 2, v \in P_R\}$; // $hop(v, b)$ is the least hop number from v to b, the same as below.
4   **while** $k' > 0$ *and* $V_t \neq S$ **do**
5          $j \leftarrow 0$;
6   **while** $j \leq \lfloor k'/2 \rfloor$ **do**
7              $\quad$ Find $\max\{w_R(S, S \cup \{v\}) : v \in V_t\}$; $S \leftarrow S \cup \{v\}$; $j \leftarrow j + 1$;
8   **for** $v \in S$ **do**
9   **if** v *is not connected with* b **then**
10                 $\quad$ $V_d \leftarrow \{u|u$ is one hop neighbor of v that also one hop neighbor of $b\}$; Find $\max\{w_R(S, S \cup \{u\}) : u \in V_d\}$; $S \leftarrow S \cup \{u\}$;
11         $k' \leftarrow k - |S|$;
12   **if** $w(F) \leq w(S)$ **then**
13         $\quad$ $F \leftarrow S$;

14 **return** F;

Next, we repeatedly select $\lfloor k/2 \rfloor$ candidate locations with maximum residual weight in the subgraph. For each selected candidate location p_i, check whether it is the 1-hop or 2-hop neighbor of the BS. If it is a 2-hop neighbor, then we check whether those selected locations can construct a path from p_i to the BS. If not, we need to select another one p_j from the 1-hop neighbors of p_i that brings the maximum residual weight while ensuring that $p_i \to p_j \to BS$ is a path. In this way, the number of all selected locations is at most $\lfloor k/2 \rfloor \times 2 \leq k$. It is very likely that we still have available relays. Therefore, assume that we have selected g relays, and $g < k$, then we run the same procedure on this subgraph with $k = k - g$. Repeat this procedure and use S to record all the selected locations, and it will terminate once $|S| = k$ and S is a feasible solution.

Finally, choose the solution with the maximum total weight. The details of SIDA are shown in Algorithm 1.

4.2 Performance Analysis

In this subsection, we analyze the performance guarantee of SIDA. We consider a BS location, its 2-hop neighbors and the generated subgraph. We propose two lemma for this subgraph.

Lemma 1. *After each greedy iteration l_i, $i = 2, \ldots, t$, $t \leq \lfloor k/2 \rfloor$, the inequality $w(G_i) - w(G_{i-1}) \geq \frac{1}{k} \left[w(OPT') - w(G_{i-1}) \right]$ holds, where G_i is the selected set after i-th iteration, and OPT' is the optimal solution within the current subgraph.*

Proof. First, we denote $w(G_i) - w(G_{i-1})$ as W_i', which is the maximum residual weight in ith iteration according to the greedy strategy. Clearly, $w(OPT') - w(G_{i-1})$ is no more than the weight of the elements covered by OPT', but not covered by G_{i-1}, i.e. $w(OPT') - w(G_{i-1}) \leq w(OPT'\backslash G_{i-1})$. Since the size of the set $OPT'\backslash G_{i-1}$ is bounded by the budget k, the total weight of DNs covered by $OPT'\backslash G_{i-1}$ and not covered by G_{i-1}, is at most kW_i'. Hence we get $w(OPT') - w(G_{i-1}) \leq kW_i'$. Substituting $w(G_i) - w(G_{i-1})$ for W_i', and multiplying both sides by $1/k$, we get the required inequality.

Lemma 2. *After each iteration l_i, $i = 2,\ldots,t$, $t \leq \lfloor k/2 \rfloor$, the inequality $w(G_i) \geq [1 - (1 - 1/k)^i]w(OPT')$ holds.*

Proof. According to Lemma 1, we have:

$$k(w(G_i) - w(G_{i-1})) \geq w(OPT') - w(G_{i-1}) \Rightarrow \frac{w(G_i) - w(OPT')}{w(G_{i-1}) - w(OPT')} \leq 1 - 1/k.$$

Therefore, let $j = 1, 2, \ldots, i$, and multiply those inqualities, we can get:

$$\prod_{j=1}^{i} \frac{w(G_j) - w(OPT')}{w(G_{j-1}) - w(OPT')} \leq (1 - 1/k)^i \Rightarrow \frac{w(G_i) - w(OPT')}{w(G_0) - w(OPT')} \leq (1 - 1/k)^i.$$

Since $G_0 = \emptyset$, thus $w(G_0) = 0$, then we have $w(G_i) \geq [1 - (1 - 1/k)^i]w(OPT')$.

Theorem 1. *SIDA achieves an approximation factor of $1 - \frac{1}{\sqrt{e \cdot (1-1/k)}}$ for the DNC problem.*

Proof. For each BS candidate location, the algorithm iterates for at least $\lfloor k/2 \rfloor$ times. We suppose OPT is the optimal solution of the DNC problem. For the subgraph which contains OPT, we denote the set of locations selected by SIDA as F_{OPT}. Then in the light of Lemma 2, we could get:

$$w(F_{OPT}) \geq [1 - (1 - 1/k)^{\lfloor k/2 \rfloor}]w(OPT) \geq \left[1 - \frac{1}{\sqrt{e \cdot (1 - 1/k)}}\right]w(OPT).$$

5   Simulations

In this section, we conduct extensive simulation experiments to evaluate our algorithm via C++. We evaluate our entire procedure including trajectory processing, DNG, and SIDA on five real GPS data [5] from CRAWDAD: NCSU and KAIST, New York City, Orlando, and North Carolina state fair. We randomly divide each dataset into training and validation group.

The parameters of each dataset are shown in Table 1. We divide the map into grids of $g_B \times g_B$ and set the candidate BS locations to the center of these grids. Similarity, the candidate relay locations are set to the center of $g_R \times g_R$ grids. d_B is set to $r_R + r_B$, and d_R is set to $2 \times r_R$. The number of relays we can deploy

Table 1. Parameters of each dataset

Dataset	s	r_B	r_R	g_B	g_R	θ
KAIST	200	1200	600	3000	500	0.14
NCSU	200	1200	600	3000	500	0.21
New York	400	2400	1200	6000	1000	0.21
Orlando	300	1500	1000	5000	1000	0.20
Statefair	20	150	75	350	50	0.35

is $k = 5$. For simplicity, both the two parameters a_{ij} and b_{ij} of beta distribution are set to 5. The time slot is set to $t = 200$.

We repeated the partition of the validation set and ran the procedure for 1000 times, and then took the average. The coverage performance for the five datasets are 95.10%, 85.83%, 62.52%, 85.44%, and 60.70%, respectively.

6   Conclusion

In this work, we have proposed the *Trajectory-Based Relay Deployment* (TBRD) problem in wireless networks, which aims at maximizing user connection time as the users roam through the target area while complying with relay resource constraints. We first transform the trajectories into a number of virtual weighted discrete Demand Nodes (DNs). In this way, the original TBRD problem is converted to an NP-complete problem called Demand Node Coverage (DNC) problem, which is to maximize total covered DN weight. Then, we design an approximation algorithm named Submodular Iterative Deployment Algorithm (SIDA) to solve the DNC problem, with an approximation ratio of $1 - \dfrac{1}{e \cdot \sqrt{(1-1/k)}}$. The simulation on five real datasets results show that our algorithm can obtain high coverage performance and thus significantly improve the user experience. To the best of our knowledge, we are the first to consider user trajectories for relay deployment.

References

1. Bishop, C.M.: Pattern Recognition and Machine Learning. Springer, New York (2006)
2. Hochba, D.S.: Approximation algorithms for np-hard problems. ACM SIGACT News **28**(2), 40–52 (1997)
3. Ma, L., Teymorian, A.Y., Cheng, X.: A hybrid rogue access point protection framework for commodity wi-fi networks. In: IEEE International Conference on Computer Communications (INFOCOM) (2008)
4. Musolesi, M., Hailes, S., Mascolo, C.: An ad hoc mobility model founded on social network theory. In: MSWiM, pp. 20–24 (2004)
5. Rhee, I., Shin, M., Hong, S., Lee, K., Kim, S., Chong, S.: CRAWDAD dataset ncsu/mobilitymodels (v. 2009-07-23), July 2009. http://crawdad.org/ncsu/mobilitymodels/20090723/GPS
6. Wang, H., Tian, S., Gao, X., Wu, L., Chen, G.: Approximation designs for cooperative relay deployment in wireless networks. In: ICDCS, pp. 2270–2275 (2017)

A Local Search Approximation Algorithm for a Squared Metric k-Facility Location Problem

Dongmei Zhang[1], Dachuan Xu[2(✉)], Yishui Wang[2], Peng Zhang[3],
and Zhenning Zhang[2]

[1] School of Computer Science and Technology, Shandong Jianzhu University,
Jinan 250101, People's Republic of China
[2] Department of Information and Operations Research, College of Applied Sciences,
Beijing University of Technology, Beijing 100124, People's Republic of China
xudc@bjut.edu.cn
[3] School of Computer Science and Technology, Shandong University, Jinan 250101,
People's Republic of China

Abstract. In this paper, we introduce a squared metric k-facility location problem (SM-k-FLP) which is a generalization of the squared metric facility location problem (SMFLP) and k-facility location problem (k-FLP). In the SM-k-FLP, we are given a client set $\mathcal{C}$ and a facility set $\mathcal{F}$ from a metric space, a facility opening cost $f_i \geq 0$ for each $i \in \mathcal{F}$, and an integer k. The goal is to open a facility subset $F \subseteq \mathcal{F}$ with $|F| \leq k$ and to connect each client to the nearest open facility such that the total cost (including facility opening cost and the sum of squares of distances) is minimized. Using local search and scaling techniques, we offer a constant approximation algorithm for the SM-k-FLP.

Keywords: Approximation algorithm · Facility location · Local search

1 Introduction

The facility location problem (FLP) and its variants are studied widely in the society of Theoretical Computer Science. The problem is defined as follows. Given a client set $\mathcal{C}$, a facility set $\mathcal{F}$ with a nonnegative opening cost for each facility, and a connection cost for each facility-client pair, we want to choose a subset of $\mathcal{F}$ to open and connect each client to the nearest open facility such that the total cost including opening costs of facilities and connection costs is minimized. We assume that the connection cost is a metric which satisfies nonnegativity, symmetry, and the so-called triangle inequality.

It is well-known that the FLP is NP-hard with a lower bound 1.463 assuming $NP \nsubseteq DTIME[n^{O(\log \log n)}]$ [10]. We briefly review the approximation algorithms for the FLP as follows. Shmoys et al. [15] give the first constant 3.16-approximation algorithm which is further improved by Chudak and Shmoys [4], Charikar and Guha [4], Jain and Vazirani [9], Jain et al. [8], Mahdian et al. [14], and Byrka and Aardal [2]. The currently best ratio 1.488 is due to Li [12]

© Springer International Publishing AG 2017
X. Gao et al. (Eds.): COCOA 2017, Part I, LNCS 10627, pp. 119–124, 2017.
https://doi.org/10.1007/978-3-319-71150-8_11

which combines the LP-rounding and dual-fitting techniques. Fernandes et al. [7] study the squared metric facility location problem (SMFLP) which generalizes the FLP. In the SMFLP, the square root of distance satisfies the so-called triangle inequality comparing with that in metric space. Fernandes et al. [7] prove that the lower bound for the SMFLP is 2.04 assuming $P \neq NP$. Furthermore, they obtain a 2.04-approximation which achieves the lower bound by adapting the techniques of Li [12].

The k-median problem is another important variant of the FLP. In this problem, each facility opening cost is zero and there is a upper bound the number of opening facilities. The approximation algorithms for the metric k-median problem are listed below. The first constant approximation algorithm is offered by Charikar et al. [5] using LP rounding technique. Arya et al. [1] give a local search $(3+\varepsilon)$-approximation algorithm. Li and Svensson [13] present an improved $(1+\sqrt{3}+\varepsilon)$-approximation algorithm via a novel pseudo-approximation approach. The currently best approximation ratio $2.675 + \varepsilon$ is due to Byrka et al. [3]. If each facility has an opening cost in the metric k-median problem, we obtain the k-facility location problem (k-FLP). Zhang [16] gives a local search $(2 + \sqrt{3} + \varepsilon)$-approximation algorithm for the k-FLP.

In this paper, we introduce a squared metric k-facility location problem (SM-k-FLP) which is a generalization of the SMFLP and k-FLP. In the SM-k-FLP, we are given a client set $\mathcal{C}$ and a facility set $\mathcal{F}$ from a metric space, a facility opening cost $f_i \geq 0$ for each $i \in \mathcal{F}$, and an integer k. The goal is to open a facility subset $F \subseteq \mathcal{F}$ with $|F| \leq k$ and to connect each client to the nearest open facility such that the total cost (including facility opening cost and the sum of squares of distances) is minimized.

The contributions of our paper are summarized as follows.

– Introduce firstly the SM-k-FLP which generalizes the SMFLP and k-FLP;
– Offer a constant approximation algorithm for the SM-k-FLP by using local search and scaling techniques.

The organization of this paper is as follows. In Sect. 2, we present a local search $\left(22 + \sqrt{505} + \varepsilon\right)$-approximation algorithm. We give some discussions in Sect. 3.

All the proofs are deferred to the journal version.

2   Approximation Algorithm for the SM-k-FLP

We first give some preliminaries for the SM-k-FLP in Subsect. 2.1. Then we present a local search algorithm for the SM-k-FLP in Subsect. 2.2 along with its analysis in Subsect. 2.3. Finally, we obtain the improved approximation ratio $\left(22 + \sqrt{505} + \varepsilon\right)$ using scaling technique in Subsect. 2.4.

2.1   Preliminaries

Given any two points $a, b \in \mathcal{C} \cup \mathcal{F}$, we define $\Delta(a, b) := \text{dist}^2(a, b)$. For a client subset $C \subseteq \mathcal{C}$, a point $i \in \mathcal{F}$, let us define the total sum of squares distances of C with respect to i as follows, $\Delta(i, C) := \sum_{j \in C} \Delta(i, j) = \sum_{j \in C} \text{dist}^2(i, j)$.

Let S be a feasible solution to the SM-k-FLP. We still use S to denote the open facility set in this solution. For each client $j \in \mathcal{C}$, we denote $S_j := \min_{s \in S} \Delta(j, s)$ and $s_j := \arg\min_{s \in S} \Delta(j, s)$. For each $s \in S$, let $N_S(s)$ be the client subset in which each client is closer to s than to other facilities. If a client j is at the same distance from several facilities, we arbitrarily choose one facility. The cost of S is denoted by $\mathrm{cost}(S) := C_f + C_s$ where $C_s := \sum_{j \in \mathcal{C}} S_j = \sum_{s \in S} \Delta(j, N_S(s)) = \sum_{s \in S} \sum_{j \in N_S(s)} \Delta(j, s)$ and $C_f := \sum_{s \in S} f_s$.

Let O be a global optimal solution to the SM-k-FLP. Similarly as the above notations for S, we introduces the following notations for O. For each client $j \in \mathcal{C}$, we denote $O_j := \min_{o \in O} \Delta(j, o)$, $o_j := \arg\min_{o \in O} \Delta(j, o)$, and $s_{o_j} := \arg\min_{s \in S} \Delta(o_j, s)$. For each $o \in O$, we denote $N_O(o) := \{j \in \mathcal{C} | o_j = o\}$. Denote $\mathrm{cost}(O) := C_f^* + C_s^*$ where $C_s^* := \sum_{j \in \mathcal{C}} O_j = \sum_{o \in O} \Delta(j, N_O(o)) = \sum_{o \in O} \sum_{j \in N_O(o)} \Delta(j, o)$; and $C_f^* := \sum_{o \in O} f_o$.

2.2   Local Search Algorithm

For any feasible solution S, we define the following three local operations.

(1) add(b). In add operation, a facility $b \in \mathcal{C} \setminus S$ is added to S if $|S| < k$.
(2) drop(a). In drop operation, a facility $a \in S$ is dropped.
(3) swap(A, B). In swap operation, we are given two subsets $A \subseteq S$ and $B \subseteq \mathcal{C} \setminus S$ with $|A| = |B| = p$, where p is a fixed integer. All facilities in A are dropped out of S. Meanwhile, all facilities in B are added to S.

We define the neighborhood of S associated with the above operations as follows,

$$
\mathrm{Ngh}(S) := \begin{cases}
\begin{aligned}
& \{S \cup \{b\} | b \in \mathcal{C} \setminus S\} \cup \\
& \{S \setminus \{a\} | a \in S\} \cup \\
& \{S \setminus A \cup B | A \subseteq S, B \subseteq \mathcal{C} \setminus S, |A| = |B| = p\}, \text{ if } |S| < k;
\end{aligned} \\[1em]
\begin{aligned}
& \{S \setminus \{a\} | a \in S\} \cup \\
& \{S \setminus A \cup B | A \subseteq S, B \subseteq \mathcal{C} \setminus S, |A| = |B| = p\}, \text{ if } |S| = k.
\end{aligned}
\end{cases}
$$

Alogrithm 1

Step 0. (Initialization) *Arbitrarily choose a feasible solution S.*
Step 1. (Local search) *Compute the best solution in the neighborhood of S,*

$$
S_{\min} := \arg \min_{S' \in \mathrm{Ngh}(S)} \mathrm{cost}(S').
$$

Step 2. (Stop criterion) *If $\mathrm{cost}(S_{\min}) \geq \mathrm{cost}(S)$, output S. Otherwise, set $S := S_{\min}$ and go to Step 1.*

2.3   Analysis

In order to bound the facility/connection cost, we need the following technical lemma.

Lemma 1 ([11]). *Let S and O be a local optimal solution and a global optimal solution to the SM-k-FLP, respectively. We have,*

$$\sum_{j \in C} \sqrt{S_j O_j} \le \sqrt{C_s C_s^*},$$

$$\sqrt{\Delta(s_{o_j}, j)} \le \sqrt{S_j} + 2\sqrt{O_j}, \qquad \forall j \in \mathcal{D}.$$

The facility opening cost for S is estimated in the following lemma.

Lemma 2. *The local optimal solution S satisfies that*

$$C_f \le C_f^* + 6C_s^* + 2C_s + 8\sqrt{C_s^* C_s}.$$

We will bound connection cost for S in the following two lemmas.

Lemma 3. *Suppose that $|S| \ge |O|$. The local optimal solution S satisfies that*

$$C_s \le C_f^* + \left(5 + \frac{4}{p}\right) C_s^* + 4\left(1 + \frac{1}{p}\right)\sqrt{C_s^* C_s}.$$

Lemma 4. *Suppose that $|S| < |O|$. The local optimal solution S satisfies that*

$$C_s \le C_f^* + C_s^*.$$

Lemma 5. *The local optimal solution S satisfies that*

$$\sqrt{C_s} \le 2\left(1 + \frac{1}{p}\right)\sqrt{C_s^*} + \sqrt{C_f^* + \left(3 + \frac{2}{p}\right)^2 C_s^*}.$$

Then, we estimate the cost of S in the following theorem.

Theorem 6. *The local optimal solution S satisfies that*

$$C_f + C_s \le \max\left\{14 + \frac{6}{p}, 161 + \frac{256}{p} + \frac{136}{p^2} + \frac{24}{p^3}\right\}(C_f^* + C_s^*).$$

2.4   Further Improvement by Scaling

Noting that Lemmas 2 and 5 hold for arbitrary feasible solution U of arbitrary instance I of SM-k-FLP, we have

$$C_f(I, S) \le C_f(I, U) + 6C_s(I, U) + 2C_s(I, S) + 8\sqrt{C_s(I, U)C_s(I, S)},$$

$$\sqrt{C_s(I, S)} \le 2\left(1 + \frac{1}{p}\right)\sqrt{C_s(I, U)} + \sqrt{C_f(I, U) + \left(3 + \frac{2}{p}\right)^2 C_s(I, U)}.$$

For any given integer $p > 0$, denote

$$\delta_0 := \frac{11 + \frac{30}{p} + \frac{20}{p^2} + \frac{4}{p^3} + \sqrt{\left(11 + \frac{30}{p} + \frac{20}{p^2} + \frac{4}{p^3}\right)^2 + 16\left(3 + \frac{2}{p}\right)\left(8 + \frac{12}{p} + \frac{6}{p^2} + \frac{1}{p^3}\right)}}{3 + \frac{2}{p}}.$$

Using the standard scaling technique [4], we present the following algorithm.

Alogrithm 2

Step 0. *Set $\delta := \delta_0$.*
Step 1. *For any given instance I, set $f' := \delta f$ which results in a modified instance I'.*
Step 2. *Run Algorithm 1 on I' to obtain a local optimal solution S.*
Step 3. *Output S as the solution of I.*

Theorem 7. *The local optimal solution S produced by Algorithm 2 satisfies that*

$$C_f + C_s \le \left\{11 + 3\delta_0 + 2\left(2 + \delta_0\right)\frac{1}{p}\right\}\left(C_f^* + C_s^*\right).$$

Using standard technique to obtain a polynomial time local search based approximation algorithm (cf. [1,16]), we have

Theorem 8. *For any fixed constant $\varepsilon > 0$, there is a $\left(22 + \sqrt{505} + \varepsilon\right)$-approximation algorithm for the SM-k-FLP when p is large enough.*

3   Discussions

In this paper, we introduce the SM-k-FLP and present a local search $(22 + \sqrt{505} + \varepsilon)$-approximation algorithm. There are several future research directions. First, it will be interesting to further improve our approximation ratio especially using LP rounding instead of local search technique. Second, since there are many variants for the k-means problem, it is natural to study the corresponding variants for the SM-k-FLP. Third, more applications of our model and its algorithm are worth to be further investigated.

Acknowledgements. The research of the first author is supported by Higher Educational Science and Technology Program of Shandong Province (No. J15LN22). The second author is supported by Natural Science Foundation of China (No. 11531014). The fourth author is supported by Natural Science Foundation of China (No. 61672323) and Natural Science Foundation of Shandong Province (ZR2016AM28). The fifth author is supported by Beijing Excellent Talents Funding (No. 2014000020124G046).

References

1. Arya, V., Garg, N., Khandekar, R., Meyerson, A., Munagala, K., Pandit, V.: Local search heuristics for k-median and facility location problems. SIAM J. Comput. **33**, 544–562 (2004)
2. Byrka, J., Aardal, K.: An optimal bifactor approximation algorithm for the metric uncapacitated facility location problem. SIAM J. Comput. **39**, 2212–2231 (2010)
3. Byrka, J., Pensyl, T., Rybicki, B., Srinivasan, A., Trinh, K.: An improved approximation for k-median, and positive correlation in budgeted optimization. In: Proceedings of SODA, pp. 737–756 (2014)
4. Charikar, M., Guha, S.: Improved combinatorial algorithms for facility location problems. SIAM J. Comput. **34**, 803–824 (2005)
5. Charikar, M., Guha, S., Tardos, É., Shmoys, D.B.: A constant-factor approximation algorithm for the k-median problem. In: Proceedings of STOC, pp. 1–10 (1999)
6. Chudak, F.A., Shmoys, D.B.: Improved approximation algorithms for the uncapacitated facility location problem. SIAM J. Comput. **33**, 1–25 (2003)
7. Fernandes, C.G., Meira, L.A., Miyazawa, F.K., Pedrosa, L.L.: A systematic approach to bound factor-revealing LPs and its application to the metric and squared metric facility location problems. Math. Program. **153**, 655–685 (2015)
8. Jain, K., Mahdian, M., Markakis, E., Saberi, A., Vazirani, V.V.: Greedy facility location algorithms analyzed using dual fitting with factor-revealing LP. J. ACM **50**, 795–824 (2003)
9. Jain, K., Vazirani, V.V.: Approximation algorithms for metric facility location and k-median problems using the primal-dual schema and Lagrangian relaxation. J. ACM **48**, 274–296 (2001)
10. Guha, S., Khuller, S.: Greedy strikes back: improved facility location algorithms. J. Algorithms **31**, 228–248 (1999)
11. Kanungoa, T., Mountb, D.M., Netanyahuc, N.S., Piatkoe, C.D., Silvermand, R., Wu, A.Y.: A local search approximation algorithm for k-means clustering. Comput. Geometry Theory Appl. **2**, 89–112 (2004)
12. Li, S.: A 1.488 approximation algorithm for the uncapacitated facility location problem. Inf. Comput. **222**, 45–58 (2013)
13. Li, S., Svensson, O.: Approximating k-median via pseudo-approximation. In: Proceedings of STOC, pp. 901–910 (2016)
14. Mahdian, M., Ye, Y., Zhang, J.: Approximation algorithms for metric facility location problems. SIAM J. Comput. **36**, 411–432 (2006)
15. Shmoys, D.B., Tardos, É., Aardal, K.: Approximation algorithms for facility location problems. In: Proceedings of STOC, pp. 265–274 (1997)
16. Zhang, P.: A new approximation algorithm for the k-facility location problem. Theor. Comput. Sci. **384**, 126–135 (2007)

Combinatorial Approximation Algorithms for Spectrum Assignment Problem in Chain and Ring Networks

Guangting Chen[1,2], Lei Zhang[2], An Zhang[2], and Yong Chen[2(✉)]

[1] Taizhou University, Taizhou 317000, Zhejiang, People's Republic of China
[2] Department of Mathematics, Hangzhou Dianzi University, Hangzhou 310018, Zhejiang, People's Republic of China
{gtchen,anzhang,chenyong}@hdu.edu.cn

Abstract. In this paper, we investigate the spectrum assignment (SA) problem in chain and ring topologies, which is the key network design and control problem in spectrum sliced elastic optical path network. Improved algorithms with guaranteed performance ratios are provided for several NP-hard scenarios of the SA problem. Concretely, we develop $\frac{4}{3}$-approximation algorithms for the SA problem in chain networks with five or six nodes, and for the SA problem in the clockwise direction of a bidirectional ring networks with five nodes. For the latter problem with six nodes, we propose a $\frac{3}{2}$-approximation algorithm. All the algorithms are combinatorial and constructive, whose performance ratios are strictly smaller than the best known ones to date.

Keywords: Network design · Spectrum assignment · Approximation algorithm · Worst-case performance ratio

Mathematics Subject Classification (2010): 68W25 · 90B35 · 90C27

1 Introduction

The orthogonal frequency division multiplexing (OFDM) technology [2,9,12] is the foundation of a spectrum efficient and scalable optical transport network called *spectrum-sliced elastic optical path network* (SLICE) [7]. The target of SLICE architecture is to allocate variable sized optical bandwidths that meet

G. Chen–Supported by the National Natural Science Foundation of China (11571252).

A. Zhang–Supported by the National Natural Science Foundation of China (11771114, 11201105) and the Zhejiang Provincial Natural Science Foundation of China (LY16A010015).

Y. Chen–Supported by the National Natural Science Foundation of China (11401149).

© Springer International Publishing AG 2017
X. Gao et al. (Eds.): COCOA 2017, Part I, LNCS 10627, pp. 125–132, 2017.
https://doi.org/10.1007/978-3-319-71150-8_12

a range of user traffic demands. The flexibility and scalability of the OFDM technology offers an opportunity for efficient resource utilization in SLICE. In order to minimize the utilized spectrum, the routing and spectrum assignment (RSA) problem [3, 6, 10] has emerged as the key network design and control problem in SLICE.

The routing and spectrum assignment (RSA) problem can be informally defined as follows [13, 15]. Given a directed graph $G = (V, A)$, where V is the set of nodes and A is the set of arcs (directed links), a spectrum demand matrix $M = [t_{sd}]$, where t_{sd} is the amount of spectrum required to carry the traffic from source node s to destination node d, and k alternate routes, $r_{sd}^1, r_{sd}^2, ..., r_{sd}^k$, from node s to node d, our goal is to assign a route and spectrum slots to each demand so as to minimize the total amount of spectrum used on any link in the network, under the following three constraints: (1) each demand is assigned contiguous spectrum (spectrum contiguity constraint), (2) each demand is assigned the same spectrum along all links of its path (spectrum continuity constraint), and (3) demands that share a link are assigned nonoverlapping parts of the available spectrum (nonoverlapping spectrum constraint). If a single route is provided for each source-destination pair (i.e., $k = 1$) and each traffic demand must follow the given route, then the RSA problem reduces to the spectrum assignment (SA) problem.

As we know, RSA problem was first proposed in [3, 6, 10]. After that only a few other papers, e.g., [8, 17], have done further studies on this problem. And in general, most of these studies [8, 10, 17] focuses on searching integer linear programming based solutions (for small network sizes) or heuristic solutions (for medium and large network sizes). Approximation results for RSA were rarely found in the literature, even when the optical network topology is as simple as a chain or a ring, which is of particular importance in the optical domain because of its application in metro networks and in some long haul networks.

Recently, Shirazipourazad et al. [13] proved that the RSA problem is NP-hard even when the optical network topology is a chain or a ring. Using results from graph coloring theory, it was shown in [13] that there exist a $2 + \epsilon$ and a $4 + \epsilon$-approximation algorithm for the SA problem in the chain and ring networks, respectively. From the perspective of scheduling theory, soon afterwards Talebi et al. [15] showed that the SA problem in (mesh) networks of general topology can be viewed as a parallel machine scheduling problem in which jobs may require more than one machine simultaneously, which is denoted by $P|fix_j|C_{max}$ and has been studied extensively in the literature [1, 5]. Note that in chain (linear) networks, the route r_{sd} of each traffic demand is uniquely determined by its source and destination nodes. Therefore, Talebi et al. [15] proved that the SA problem on a graph G that is a chain with four nodes can be solved in polynomial time and with five nodes is NP-hard. A $\frac{3}{2}$-approximation algorithm was proposed for $|V| = 5$ or $|V| = 6$ in [15], respectively. Based on the scheduling insight, a recent paper [16] studied the spectrum assignment (SA) problem in ring networks with shortest path routing. Under shortest path routing, they proved the SA subproblem defined in the clockwise direction of a bidirectional ring with four nodes is solvable in polynomial time and with five nodes is NP-hard. Moreover, a $\frac{3}{2}$-approximation algorithm for $|V| = 5$, a 2-approximation algorithm for $|V| = 6, 7$ and a $3 + \epsilon$-approximation algorithm for $|V| \geq 8$ are constructed respectively, where the $3 + \epsilon$-approximation algorithm for $|V| \geq 8$ is strictly smaller than the best known $4 + \epsilon$-approximation algorithm [13] to date. Besides, several variants of the RSA problem have been studied in the literature [4, 11], for a survey of spectrum management techniques and classification of solution approaches, the reader is referred to [14].

In this paper, we study the SA problem in chain and ring networks and propose combinatorial approximation algorithms with better guaranteed performance ratios. More concretely, we develop $\frac{4}{3}$-approximation algorithms for the SA problem in chains when the number of nodes is five or six, a $\frac{4}{3}$-approximation algorithm for the SA problem in the clockwise direction of a bidirectional ring networks with five nodes with shortest path routing and a $\frac{3}{2}$-approximation algorithm for six nodes. All our algorithm performance ratios are strictly smaller than the best known ones to date. The rest of the paper is organized as follows. In Sect. 2, we provide some preliminaries, including the terminologies and notations to be used throughout the paper. In Sect. 3, we present improved approximation algorithm and perform its worst case analysis for SA problem in chain and ring networks. We make conclusion in Sect. 4.

2   Preliminaries

For a SA problem in chain and ring topologies, we are given a directed graph $G = (V, A)$ and a spectrum demand matrix $M = [t_{sd}]$. Note that in chain networks, the route r_{sd} of the traffic demand t_{sd} is uniquely determined by its source node s and destination node t. Let D be the set of all the demands. It is clear that either the chain network or the ring network under clockwise direction with $|V| = 5$ serves 10 types of demands. For simplicity, we also use d_i to denote the demand that requires only link (arc) i, i.e., $d_i = t_{i,i+1}$. Similarly let d_{ij} denote the demands that require link i and j and let d_{ijk} denote the demands that require link i, j and k. Besides, we denote $d_{i\cdot}$ as the demands that require at least two links with respect to link i and D_i to denote the total demands on the ith link, i.e., $D_i = d_i + d_{i\cdot}$. For example, we have $d_{1\cdot} = d_{12} + d_{123}$, $d_{2\cdot} = d_{12} + d_{23} + d_{123} + d_{234}$, $d_{3\cdot} = d_{23} + d_{34} + d_{123} + d_{234}$, $d_{4\cdot} = d_{34} + d_{234}$ in the chain networks with $|V| = 5$, while $d_{1\cdot} = d_{12} + d_{51}$, $d_{2\cdot} = d_{12} + d_{23}$, $d_{3\cdot} = d_{23} + d_{34}$, $d_{4\cdot} = d_{34} + d_{45}$, $d_{5\cdot} = d_{45} + d_{51}$ in the ring networks with $|V| = 5$. The same notations can be similarly defined in the chain networks with $|V| = 6$ and in the ring networks with $|V| = 6$.

Let $LB = max\{D_1, D_2, D_3, D_4, D_5\}$, then the optimal value of the problem must satisfy that $OPT \geq LB$. In general, our algorithm consists of two phases. We start with a feasible assignment σ_0 of partial demands $D^0 \subset D$ such that the lower bound LB is attained in the first phase, i.e., the maximum amount of spectrum of σ_0 equals exactly $S^0_{max} = LB$, while the second phase inserts other demands $D \setminus D^0$ to σ_0.

Given a feasible assignment σ_0 of D^0, denote the only idle time period (gap) on link $L_i(i = 1, 2, 3, 4, 5)$ as G_i. If there are two idle time periods (gaps) on link L_i, denote the first one as G_{i1} and the second one as G_{i2}. We say $L_i(i = 1, 2, 3, 4, 5)$ is *Safe* in σ_0 if one of the following two conditions is satisfied: (1) there is only one idle time period on link L_i and $G_i + d_i. \geq LB$; (2) there are two idle time periods on link L_i and $G_{i1} + d_i. \geq LB$ or $G_{i2} + d_i. \geq LB$. Otherwise, L_i is *Unsafe* in σ_0. Obviously, by the definition we know that (1) L_i is *Safe* in σ_0 means that d_i can fit into the idle time period; (2) L_i is *Unsafe* in σ_0 means that d_i can not fit into the current idle time period, then we need make more space for d_i. For the latter case, we use the *Dragging Technique* [15] to make space, which allows us to expand the idle time periods for single-link demands until they fit into the demands exactly.

For easy understanding of the *Dragging Technique*, considering a feasible assignment σ_0 which is represented in Fig. 1(a) (Note that 'AOS' is the abbreviation of 'Amount of Spectrum'). Let S^0_{max} and S^A_{max} be the maximum amount of spectrum of the assignment σ_0 and the algorithm assignment σ_A, respectively. Clearly, L_2 and L_3 are *Safe* and L_4 might be *Unsafe*. If L_4 is *Unsafe*, then *Dragging Technique* is used to make enough space for d_4 and we have a new assignment σ^4_0 (Fig. 1(b)) with maximum amount of spectrum defined as $S^4_{max} = D_4 + s^4_0$, where $s^4_0 \leq LB - d_4. - G_4 = d_{123}$. Therefore, we have the following lemma.

Lemma 2.1. *Given a feasible assignment σ_0 of D^0 and $LB = D_i$, if $L_{i'}$ is Unsafe, then Dragging Technique is used to make enough space for $d_{i'}$ and we have a new assignment $\sigma^{i'}_0$ with maximum amount of spectrum defined as $S^{i'}_{max} = D_{i'} + s^{i'}_0$, where $s^{i'}_0 \leq LB - d_{i'}. - G_{i'}$.*

If two links $L_{i'}$, $L_{i''}$ are *Unsafe* in a given assignment σ_0 of D^0 and $LB = D_i$, then we have to use *Dragging Technique* twice and have the following lemma.

Lemma 2.2. *Given a feasible assignment σ_0 of D^0 and $LB = D_i$, if $L_{i'}$, $L_{i''}$ are Unsafe, then Dragging Technique is used twice to make enough space for $d_{i'}$, $d_{i''}$ and finally we have a new assignment σ_A with maximum amount of spectrum defined as $S^A_{max} = max\{D_{i'} + s^{i'}_0, D_{i''} + s^{i''}_0\}$, where $s^{i'}_0 \leq LB - d_{i'}. - G_{i'}$ and $s^{i''}_0 \leq LB - d_{i''}. - G_{i''}$.*

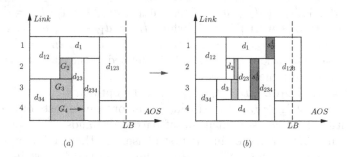

(a)   (b)

Fig. 1. Dragging technique.

3 Approximation Algorithm and Its Performance Ratio Analysis

We should note that our algorithm and analysis of its performance ratio use some ideas of Talebi et al. [15,16]. The main idea behind the algorithm is as follows: First of all, we construct an initial feasible assignment σ_0 based on the larger demand first rule. Secondly, check whether all the links are *Safe*, if so, output assignment σ_0 and stop. Otherwise, *Dragging Technique* is called to expand the constructed gaps in σ_0 for single-link demands until they fit into them exactly.

As our approximation algorithm is combinatorial and constructive, we do not give a detailed description of the algorithm. The following theorem shows that there exists a $\frac{4}{3}$-approximation algorithm for the SA problem in chain with $|V| = 5$ and the proof is by construction.

Theorem 3.1. *There exists a $\frac{4}{3}$-approximation algorithm for the SA problem in chain with $|V| = 5$.*

Proof. For the sake of simplicity of the proof, it is assumed that the multi-link demand d_{1234} is zero. This assumption does not impact the proof, in that d_{1234} can be assigned first. Moreover, as this problem is symmetric, we only

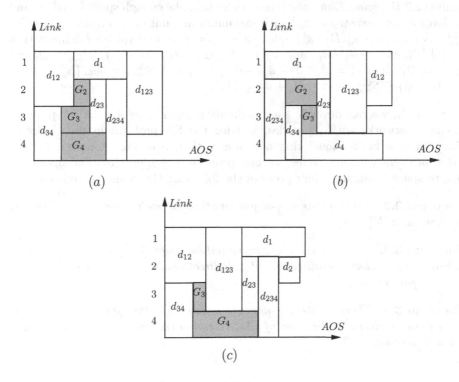

Fig. 2. Initial Assignment σ_0.

discuss the cases in which links 1 and 2 have the maximum amount of spectrum and the results can easily be generalized to the cases where links 3 and 4 are the busiest ones. Two possible cases are distinguished according to link 1 or 2 has the maximum amount of spectrum. Here, we only consider $LB = D_1$ and $LB = D_2$ can be similarly discussed.

Let $C_1 = d_1 + d_{12}$ and $C_2 = d_{123}$, then $D_1 = C_1 + C_2$. We discuss three cases, which are distinguished primarily by the value of C_1.

Case 1: $C_1 \geq \frac{2}{3}LB$. Clearly, $C_2 = d_{123} < \frac{1}{3}LB$. In the initial assignment σ_0 (See Fig. 2(a)), L_2 and L_3 are *Safe* and L_4 might be *Unsafe*. If L_4 is *Unsafe*, then by Lemma 2.1 *Dragging Technique* is used to make enough space for d_4 and we have a new assignment σ_0^4 with maximum amount of spectrum defined as $S_{max}^4 = D_4 + s_0^4$. As $G_4 + d_{4.} + d_{123} = D_1$ and $S_{max}^0 = D_1$, then $s_0^4 \leq d_{123}$ and $S_{max}^A = max\{S_{max}^0, D_4 + s_0^4\} \leq S_{max}^0 + s_0^4 \leq D_1 + d_{123} \leq \frac{4}{3}LB$.

Case 2: $C_1 < \frac{2}{3}LB$ and $d_{12} \leq \frac{1}{3}LB$. In the initial assignment σ_0 (See Fig. 2(b)), L_2 and L_4 are *Safe* and L_3 might be *Unsafe*. If L_3 is *Unsafe*, then by Lemma 2.1 *Dragging Technique* is used to make enough space for d_3 and we have a new assignment σ_0^3 with the maximum amount of spectrum modified as $S_{max}^3 = D_3 + s_0^3$. As $G_3 + d_{3.} + d_{12} = D_1$ and $S_{max}^0 = D_1$, then $s_0^3 \leq d_{12}$ and $S_{max}^A = max\{S_{max}^0, D_3 + s_0^3\} \leq S_{max}^0 + s_0^3 \leq D_1 + d_{12} \leq \frac{4}{3}LB$.

Case 3: $C_1 < \frac{2}{3}LB$ and $d_{12} > \frac{1}{3}LB$. In the initial assignment σ_0 (See Fig. 2(c)), L_3 and L_4 might be *Unsafe*. If L_3 and L_4 are *Unsafe*, then by Lemma 2.2 *Dragging Technique* is used twice to make enough space for d_3, d_4 and we have a new assignment σ_0^{34} with maximum amount of spectrum updated as $S_{max}^{34} = max\{D_3 + s_0^3, D_4 + s_0^4\}$. Obviously, $C_1 < \frac{2}{3}LB$ and $d_{12} > \frac{1}{3}LB$ imply that $d_1 \leq \frac{1}{3}LB$. As $G_3 + d_{3.} + d_1 - d_{23} - d_{234} = D_1$, $G_4 + d_{4.} + d_1 - d_{23} - d_{234} = D_1$ and $S_{max}^0 = D_1$, then $s_0^3 = s_0^4 \leq d_1 - d_{23} - d_{234} \leq d_1$ and $S_{max}^A = max\{S_{max}^0, D_3 + s_0^3, D_4 + s_0^4\} \leq S_{max}^0 + s_0^3 \leq D_1 + d_1 \leq \frac{4}{3}LB$.   □

Similarly, we can develop $\frac{4}{3}$-approximation algorithms for the SA problem in chain networks with six nodes, and for the SA problem in the clockwise direction of a bidirectional ring networks with five nodes. Moreover, for the latter problem with six nodes, we can propose a $\frac{3}{2}$-approximation algorithm. Due to space constraint, their proof of the following theorems are omitted.

Theorem 3.2. *There exists a $\frac{4}{3}$-approximation algorithm for the SA problem in chain with $|V| = 6$.*

Theorem 3.3. *There exists a $\frac{4}{3}$-approximation algorithm for the SA problem defined on the clockwise direction of a bidirectional ring with five nodes and shortest path routing.*

Theorem 3.4. *There exists a $\frac{3}{2}$-approximation algorithm for the SA problem defined on the clockwise direction of a bidirectional ring with six nodes and shortest path routing.*

4   Conclusion

In this paper, we study the SA problem in chain and ring networks and propose combinatorial approximation algorithms with better guaranteed performance ratio. More concretely, we develop $\frac{4}{3}$-approximation algorithms for the SA problem in chains when the number of nodes is five or six, a $\frac{4}{3}$-approximation algorithm for the SA problem in the clockwise direction of a bidirectional ring networks with five nodes with shortest path routing and a $\frac{3}{2}$-approximation algorithm for six nodes. All our algorithm performance ratios are strictly smaller than the best known ones to date. In future work, we plan to design better approximation algorithm to tackle the RSA problem in (mesh) networks of general topology.

References

1. Bampis, E., Caramia, M., Fiala, J., Fishkin, A.V., Iovanella, A.: Scheduling of independent dedicated multiprocessor tasks. In: Bose, P., Morin, P. (eds.) ISAAC 2002. LNCS, vol. 2518, pp. 391–402. Springer, Heidelberg (2002). https://doi.org/10.1007/3-540-36136-7_35
2. Chang, R.W.: Synthesis of band-limited orthogonal signals for multichannel data transmission. Ell Syst. Tech. J. **45**, 1775–1796 (1966)
3. Christodoulopoulos, K., Tomkos, I., Varvarigos, E.A.: Routing and spectrum allocation in OFDM-based optical networks with elastic bandwidth allocation. In: GLOBECOM, pp. 1–6 (2011)
4. Christodoulopoulos, K., Tomkos, I., Varvarigos, E.A.: Elastic bandwidth allocation in flexible OFDM - based optical networks. J. Lightwave Technol. **29**(9), 1354–1366 (2011)
5. Hoogeveen, J.A., Van de Velde, S.L., Veltman, B.: Complexity of scheduling multiprocessor tasks with prespecified processor allocations. Discr. Appl. Math. **55**, 259–272 (1994)
6. Jinno, M., Kozicki, B., Takara, H., Watanabe, A., Sone, Y., Tanaka, T., Hirano, A.: Distance-adaptive spectrum resource allocation in spectrumsliced elastic optical path network [topics in optical communications]. IEEE Commun. Magaz. **48**(8), 138–145 (2010)
7. Jinno, M., Takara, H., Kozicki, B., Tsukishima, Y., Sone, Y., Matsuoka, S.: Spectrum-efficient and scalable elastic optical path network: architecture, benefits, and enabling technologies. IEEE Commun. Magaz. **47**(11), 66–73 (2009)
8. Klinkowski, M., Walkowiak, K.: Routing and spectrum assignment inspectrum sliced elastic optical path network. IEEE Commun. Lett. **15**(8), 884–886 (2011)
9. Pan, Q., Green, R.J.: Bit-error-rate performance of lightwave hybrid AM/OFDM systems with comparison with AM/QAM systems in the presence of clipping impulse noise. IEEE Photon. Technol. Lett. **8**, 278–280 (1996)
10. Patel, A.N., Ji, P.N., Jue, J.P., Wang, T.: Routing, wavelength assignment, and spectrum allocation in transparent flexible optical WDM (FWDM) networks. Optical Switch. Netw. **9**(3), 191–204 (2012)
11. Rouskas, G.N.: Routing and wavelength assignment in optical WDM networks. Encycloped. Telecommun. **11**(2), 259–272 (2003). Wiley
12. Shieh, W.: Ofdm for flexible high-speed optical networks. J. Lightwave Technol. **29**(10), 1560–1577 (2011)

13. Shirazipourazad, S., Zhou, C., Derakhshandeh, Z., Sen, A.: On routing and spectrum allocation in spectrum-sliced optical networks. In: INFOCOM, pp. 385–389 (2013)
14. Talebi, S., Alam, F., Katib, I., Khamis, M., Khalifah, R., Rouskas, G.N.: Spectrum management techniques for elastic optical networks: a survey. Optical Switch. Netw. **13**, 34–48 (2014)
15. Talebi, S., Bampis, E., Lucarelli, G., Katib, I., Rouskas, G.N.: Spectrum assignment in optical networks: a multiprocessor scheduling perspective. J. Optical Commun. Netw. **6**(8), 754–763 (2014)
16. Talebi, S., Bampis, E., Lucarelli, G., Katib, I., Rouskas, G.N.: On routing and spectrum assignment in rings. J. Lightwave Technol. **33**(1), 151–160 (2015)
17. Wang, Y., Cao, X., Pan, Y.: A study of the routing and spectrum allocation in spectrum-sliced elastic optical path networks. In: INFOCOM, pp. 1503–1511 (2011)

Mixed Connectivity of Random Graphs

Ran Gu[1], Yongtang Shi[2], and Neng Fan[3]([✉])

[1] College of Science, Hohai University,
Nanjing 210098, Jiangsu, People's Republic of China
rangu@hhu.edu.cn
[2] Center for Combinatorics and LPMC, Nankai University,
Tianjin 300071, People's Republic of China
shi@nankai.edu.cn
[3] Department of Systems and Industrial Engineering,
University of Arizona, Tucson, AZ 85721, USA
nfan@email.arizona.edu

Abstract. For positive integers k and λ, a graph G is (k, λ)-connected if it satisfies the following two conditions: (1) $|V(G)| \geq k+1$, and (2) for any subset $S \subseteq V(G)$ and any subset $L \subseteq E(G)$ with $\lambda|S| + |L| < k\lambda$, $G - (S \cup L)$ is connected. For positive integers k and ℓ, a graph G with $|V(G)| \geq k + \ell + 1$ is said to be (k, ℓ)-mixed-connected if for any subset $S \subseteq V(G)$ and any subset $L \subseteq E(G)$ with $|S| \leq k, |L| \leq \ell$ and $|S| + |L| < k + \ell$, $G - (S \cup L)$ is connected. In this paper, we investigate the (k, λ)-connectivity and (k, ℓ)-mixed-connectivity of random graphs, and generalize the results of Erdős and Rényi (1959), and Stepanov (1970). Furthermore, our argument can show that in the random graph process $\tilde{G} = (G_t)_0^N$, $N = \binom{n}{2}$, the hitting times of minimum degree at least $k\lambda$ and of G_t being (k, λ)-connected coincide with high probability, and also the hitting times of minimum degree at least $k + \ell$ and of G_t being (k, ℓ)-mixed-connected coincide with high probability. These results are analogous to the work of Bollobás and Thomassen (1986) on classic connectivity.

Keywords: Connectivity · Edge-connectivity · Random graph · Threshold function · Hitting time

1 Introduction

All graphs in this paper are undirected, finite and simple. Additionally, we make this assumption: removal of a vertex in graph implies the removal of all its incident edges. A graph G is k-connected if $G - S$ is connected for any vertex subset S with $|S| < k$, and a graph G is ℓ-edge-connected if $G - L$ is connected for any edge subset L with $|L| < \ell$. The connectivity $\kappa(G)$ of graph G is the largest k for which the graph is k-connected. Similarly, the edge-connectivity $\lambda(G)$ of graph G is the largest ℓ for which the graph is ℓ edge-connected. There are many generalizations of connectivity and edge-connectivity, and we refer to [1, 10].

© Springer International Publishing AG 2017
X. Gao et al. (Eds.): COCOA 2017, Part I, LNCS 10627, pp. 133–140, 2017.
https://doi.org/10.1007/978-3-319-71150-8_13

In 2000, Kaneko and Ota [12] introduced the (k, λ)-*connectivity*. For a given graph G, let x and y be two distinct vertices. A pair (x, y) of vertices is said to be (k, λ)-*connected* in G if for any subset $S \subseteq V(G) - \{x, y\}$ and any subset $L \subseteq E(G)$ with $\lambda|S| + |L| < k\lambda$, the vertices x and y belong to the same component of $G - S - L$. Formally, a graph G is (k, λ)-*connected* if it satisfies the following conditions:

(i) $|V(G)| \geq k + 1$;
(ii) for any subset $S \subseteq V(G)$ and any subset $L \subseteq E(G)$ with $\lambda|S| + |L| < k\lambda$, $G - S - L$ is connected.

The well known Menger's Theorem characterizes the relationship of graph connectivity and the minimum number of disjoint paths between any pair of vertices. For example, the pair (x, y) of vertices is k-connected if and only if there are k internally disjoint (or vertex-disjoint) paths between x and y. Let (x, y)-k-fan be a union of k internally disjoint paths, and [6,12] showed that the pair (x, y) is (k, λ)-connected in G if and only if G contains λ edge-disjoint (x, y)-k-fans. To this end, the (k, λ)-connectivity can be considered as an extension of the classical connectivity and the edge-connectivity, considering the conclusion of Menger's theorem. More specifically, $(k, 1)$-connected graphs are k-connected graphs, and $(1, \lambda)$-connected graphs are λ-edge-connected graphs. In fact, we will show that the concept of $(k, 1)$-connected is equivalent to k-connected, and $(1, \lambda)$-connected is equivalent to λ-edge-connected.

As another generation of connectivity and the edge-connectivity was proposed by Beineke and Harary [3] in early 1960s. To avoid confusion with the (k, λ)-connectivity defined above, we here call this type of connectivity as (k, λ)-*mixed-connectivity*. Two distinct vertices x, y are said to be (k, ℓ)-*mixed-connected* (k, ℓ are two positive integers), if for any subset $S \subseteq V(G) - \{x, y\}$ and any subset $L \subseteq E(G)$ with $|S| + |L| < k + \ell$, the vertices x and y belong to the same component of $G - S - L$. Similarly, as generation of the conclusions in Menger's Theorem for (k, λ)-connectivity, [3] claimed to prove that a pair (x, y) is (k, ℓ)-mixed-connected if there are $(k + \ell)$ edge-disjoint paths of which k paths are vertex-disjoint. However, Mader [13] pointed out a gap in their proof. Recently, Sadeghi and Fan [14] modified the conclusion (by changing to $k + 1$ vertex-disjoint paths instead of k), and then proved it.

The two generations, (k, λ)-connectivity and (k, ℓ)-mixed-connectivity, consider both connectivity and edge-connectivity, and can be applied for vulnerability analysis for network design. As pointed out in [14], the survivable networks, with robustness against both vertex and edge failures, require the concepts of mixed connectivity.

In this paper, we investigate this two concepts of connectivity in the setting of random graphs. The Erdős-Rényi random graph model $G(n, p)$ consists of all graphs with n vertices in which the edges are chosen independently and with probability p. We say an event $\mathscr{A}$ happens *with high probability (w.h.p.)* if the probability that it happens approaches 1 as $n \to \infty$, i.e., $Pr[\mathscr{A}] = 1 - o_n(1)$. We will always assume that n is the variable that tends to infinity. Let G and H be two graphs on n vertices. A property P is said to be *monotone increasing* if whenever $G \subseteq H$ and G satisfies P, then H also satisfies P. For a graph property P, a function $p(n)$ is called a *threshold function* of P if:

- for every $r(n)$ with $r(n)/p(n) \to \infty$, $G(n, r(n))$ w.h.p. satisfies P; and
- for every $r'(n)$ with $r'(n)/p(n) \to 0$, $G(n, r'(n))$ w.h.p. does not satisfy P.

Furthermore, $p(n)$ is called a *sharp threshold function* of P if for any constants $0 < c < 1$ and $C > 1$, such that:

- for every $r(n) \geq C \cdot p(n)$, $G(n, r(n))$ w.h.p. satisfies P; and
- for every $r'(n) \leq c \cdot p(n)$, $G(n, r'(n))$ w.h.p. does not satisfy P.

In the extensive study of the properties of random graphs, many researchers observed that there are threshold functions for various natural graph properties. It is well known that all non-trivial monotone increasing graph properties have threshold functions (see [5] and [9]). In one of the first papers on random graphs, Erdős and Rényi [7] showed that $m = n \log n/2$ is a sharp threshold for connectivity in $G(n, m)$. Later, Stepanov [15] established a sharp threshold of connectivity for $G(n, p)$. For more results on this topic, we refer to Erdős-Rényi [8] and Ivchenko [11]. Especially, Bollobás and Thomassen [4] proved that for almost every graph process, the hitting time of the graph having the connectivity $\kappa(G)$ at least k is equal to the hitting time of the graph having the minimum degree at least k. Their result builds the bridge between the connectivity and the minimum degree.

In this paper, we extend these results for threshold functions and hitting times to (k, λ)-connectivity and (k, ℓ)-mixed-connectivity. First, we will generalize the result of Erdős and Rényi [7] and Stepanov [15], and provide the threshold functions for (k, λ)-connectivity and (k, ℓ)-mixed-connectivity of random graphs, respectively.

For (k, λ)-connectivity, we obtain that

Theorem 1. *For any two positive integers k and λ, if $p = \{\log n + k\lambda \log \log n - \omega(n)\}/n$, then w.h.p. $G(n, p)$ is (k, λ)-connected, if $p = \{\log n + (k\lambda - 1) \log \log n - \omega(n)\}/n$, then w.h.p. $G(n, p)$ is not (k, λ)-connected, where $\omega(n) \to \infty$ and $\omega(n) = o(\log \log n)$.*

Considering the definition of sharp threshold functions, the following result is an immediate consequence of Theorem 1.

Theorem 2. *For any two positive integers k and λ,*

$$p = \{\log n + k\lambda \log \log n - \omega(n)\}/n$$

is a sharp threshold function for the property that $G(n, p)$ is (k, λ)-connected, where $\omega(n) \to \infty$ and $\omega(n) = o(\log \log n)$.

For the mixed (k, ℓ)-connectivity, we obtain the following results.

Theorem 3. *For any two positive integers k and ℓ, if $p = \{\log n + (k + \ell) \log \log n - \omega(n)\}/n$, then w.h.p. $G(n, p)$ is (k, ℓ)-mixed-connected, if $p = \{\log n + (k + \ell - 1) \log \log n - \omega(n)\}/n$, then w.h.p. $G(n, p)$ is not (k, ℓ)-mixed-connected, where $\omega(n) \to \infty$ and $\omega(n) = o(\log \log n)$.*

Theorem 4. *For any two positive integers k and ℓ,*

$$p = \{\log n + (k + \ell) \log \log n - \omega(n)\}/n$$

is a sharp threshold function for the property that $G(n, p)$ is w.h.p. (k, ℓ)-mixed-connected, where $\omega(n) \to \infty$ and $\omega(n) = o(\log \log n)$.

In fact, we will prove something even stronger than the results above. A *random graph process* on $V = \{1, 2, \cdots, n\}$, or simply a *graph process*, is a Markov chain $\tilde{G} = (G_t)_0^N$, $N = \binom{n}{2}$, which starts with the empty graph on n vertices at time $t = 0$ and where at each step one edge is added, chosen uniformly at random from those not already present in the graph, until at time N we have a complete graph. We call G_t the state of a graph process $\tilde{G} = (G_t)_0^N$ at time t. For a monotone graph property P, the time τ at which P appears is the *hitting time* of P:

$$\tau = \tau_P = \tau_p(\tilde{G}) = \min\{t \ge 0 \colon G_t \text{ has property } P\}.$$

Consider the graph properties D_t, $F_{k,\lambda}$ and $R_{k,\ell}$ given by

$$D_t = \{G \colon \delta(G) \ge t\},$$

$$F_{k,\lambda} = \{G \colon G \text{ is } (k, \lambda)\text{-connected}\},$$

$$R_{k,\ell} = \{G \colon G \text{ is } (k, \ell)\text{-mixed-connected}\},$$

we prove the following results.

Theorem 5. *Given $k \in \mathbb{N}$, in the random graph process $\tilde{G} = (G_t)_0^N$, $N = \binom{n}{2}$, with high probability $\tau_{D_{k\lambda}} = \tau_{F_{k,\lambda}}$.*

Theorem 6. *Given $k \in \mathbb{N}$, in the random graph process $\tilde{G} = (G_t)_0^N$, $N = \binom{n}{2}$, with high probability $\tau_{D_{k+\ell}} = \tau_{R_{k,\ell}}$.*

The remainder of this paper is organized as follows. The results for (k, λ)-connectivity will be shown in Sect. 2, while the results for (k, ℓ)-mixed-connectivity are presented in Sect. 3.

2 (k, λ)-connectivity

We will use the following theorem proved by Ivchenko [11] to give proofs of our results.

Theorem 7 [11]. *If $p \le \{\log n + k \log \log n\}/n$ for some fixed k, then w.h.p. we have that*

$$\kappa(G(n, p)) = \lambda(G(n, p)) = \delta(G(n, p)).$$

Now we consider the (k, λ)-connectivity of random graphs. We shall prove Theorem 1 first.

2.1   Proof of Theorem 1

Let $p_1 = \{\log n + k\lambda \log \log n - \omega(n)\}/n$, $p_2 = \{\log n + (k\lambda - 1)\log\log n - \omega(n)\}/n$, where $\omega(n) \to \infty$ and $\omega(n) = o(\log\log n)$. When considering the (k,λ)-connectivity of a graph G, we only need to check the connectivity of $G - S - L$ such that $\lambda|S| + |L| < k\lambda$, where $S \subseteq V(G)$ and $L \subseteq E(G)$. Suppose that $|S| = i$, then it suffices to consider the case that $0 \le i \le k - 1$ and $|L|$ satisfying that $|L| < k\lambda - i\lambda = (k-i)\lambda$.

Note that if G is $(s+1)$-connected, i.e., $G - X$ is connected for any vertex subset X with $|X| \le s$, then we have that $G - D$ is connected for any edge subset D with $|D| \le s$. Furthermore, for any vertex subset X and edge subset D with $|S| + |D| \le s$, $G - X - D$ is still connected. It is known that the minimum degree $\delta(G(n, p_1))$ of $G(n, p_1)$ is w.h.p. equal to $k\lambda$ (see [2]). Combining with Theorem 7, we get that $G(n, p_1)$ is w.h.p. $k\lambda$-connected, so the new graph obtained by deleting any $k\lambda - 1$ vertices from $G(n, p_1)$ remains connected. Hence, for any vertex subset S and edge subset L with $|S| = i$ and $|L| < (k-i)\lambda$, $G(n, p_1) - S - L$ is w.h.p. connected, where $0 \le i \le k - 1$. Therefore, $G(n, p_1)$ is w.h.p. (k,λ)-connected.

For the second part of Theorem 1, since the minimum degree of $G(n, p_2)$ is w.h.p. equal to $k\lambda - 1$, if we let L be the set of edges incident to a vertex with minimum degree $k\lambda - 1$ in $G(n, p_2)$, then $G(n, p_2) - L$ is disconnected. Notice that $|L| = k\lambda - 1 < k\lambda$, we have that w.h.p. $G(n, p_2)$ is not (k,λ)-connected. ∎

Let $p = \{\log n + k\lambda \log \log n - \omega(n)\}/n$, where $\omega(n) \to \infty$ and $\omega(n) = o(\log\log n)$. From Theorem 1 and the monotonicity of (k,λ)-connectivity, it is easy to get that for every constant c_1 with $c_1 > 1$, $G(n, c_1 p)$ is w.h.p. (k,λ)-connected. And, for every constant c_2 with $0 < c_2 < 1$, we have that $c_2 p < \{\log n + (k\lambda - 1)\log\log n - \omega(n)\}/n$ for sufficiently large n. By the second part of Theorem 1, we get that $G(n, c_2 p)$ is w.h.p. not (k,λ)-connected. Thus, Theorem 2 follows.

From the definition of (k,λ)-connected graphs, we can obtain that $(k,1)$-connected graphs are k-connected graphs, and $(1,\lambda)$-connected graphs are λ-edge-connected graphs. Let G be a $(k,1)$-connected graph, for any vertex subset S and edge subset L such that $|S| + |L| < k$, we have that $G - S - L$ is connected. In particular, letting L be an empty set, then $G - S$ is connected for every vertex subset S with $|S| < k$. That implies G is k-connected. Similarly, if G is $(1,\lambda)$-connected, then for any vertex subset S and edge subset L with $\lambda|S| + |L| < \lambda$, $G - S - L$ is connected. Note that $\lambda|S| + |L| < \lambda$ holds if and only if $|S| = 0$ and $|L| < \lambda$. Thus, we have that $G - L$ is connected for any edge subset L with $|L| < \lambda$. Therefore, G is λ-edge-connected.

Moreover, a k-connected graph G is $(k,1)$-connected. As we showed in the proof of Theorem 1, for any k-connected graph G, $G - S - L$ is connected for any vertex subset S and edge set L with $|S| + |L| < k$, which implies that G is $(k,1)$-connected. Similarly, a λ-edge-connected graph G is $(1,\lambda)$-connected. If any vertex subset S and edge set L satisfies that $\lambda|S| + |L| < \lambda$, then S must be an empty set and $|L| < \lambda$. So $G - S - L$ is connected. Thus, if G is λ-edge-connected, then for any vertex subset S and edge subset L with $\lambda|S| + |L| < \lambda$, $G - S - L$ is connected. So we can obtain the following fact.

Corollary 1. *For a graph G with more than k vertices, G is $(k, 1)$-connected if and only if G is k-connected, and G is $(1, \lambda)$-connected if and only if G is λ-edge-connected.*

2.2   Proof of Theorem 5

Bollobás and Thomason [4] presented a fascinating result on the hitting time relation between the connectivity and the minimum degree. We will use it to prove our results.

Theorem 8 [4]. *For every function $k = k(n)$, $1 \leq k \leq n - 1$, in the random graph process $\tilde{G} = (G_t)_0^N$, $N = \binom{n}{2}$, w.h.p.*

$$\tau(\kappa(G) \geq k) = \tau(\delta(G) \geq k).$$

If G is (k, λ)-connected, then $\delta(G) \geq k\lambda$. Since otherwise, if $\delta(G) < k\lambda$, letting L be the edge subset consisting of the edges incident to a vertex with minimum degree in G, we have that $|L| < k\lambda$, and $G - L$ is disconnected, which contradicts to the assumption that G is (k, λ)-connected. So we have that w.h.p.

$$\tau_{D_{k\lambda}} \leq \tau_{F_{k,\lambda}}. \tag{1}$$

Let $t = \tau_{D_{k\lambda}}$. From Theorem 8, we get that w.h.p. $\kappa(G_t) \geq k\lambda$, so for any vertex subset S with $|S| < k\lambda$, $G_t - S$ is connected. Hence we know that for any vertex subset S and edge subset L with $|S| = i$ and $|L| < (k - i)\lambda$, w.h.p. $G_t - S - L$ is still connected, where $0 \leq i \leq k - 1$. By the definition of (k, λ)-connectivity, we know that G_t has been (k, λ)-connected already, which implies that w.h.p.

$$\tau_{D_{k\lambda}} \geq \tau_{F_{k,\lambda}}. \tag{2}$$

By (1) and (2), we obtain that w.h.p.

$$\tau_{D_{k\lambda}} = \tau_{F_{k,\lambda}}.$$

∎

Remark 1. For $\lambda = 1$, we have that $F_{k,\lambda} = F_{k,1} = \{G: G \text{ is } k\text{-connected }\}$ from Corollary 1, thus Theorem 5 is just the same with Theorem 8 given by Bollobás and Thomason. Hence Theorem 5 is a generalization of that proved by Bollobás and Thomason.

3 (k, ℓ)-mixed-connectivity

For the (k, ℓ)-mixed-connectivity, Sadeghi and Fan [14] gave a necessary and sufficient condition as follows, which will be used to prove Theorem 3.

Theorem 9 [14]. *Let $n \geq k + \ell + 1$; and $k, \ell \geq 1$. A graph G of order n is (k, ℓ)-mixed-connected if and only if*

(i) G *is* $(k + 1)$-*connected and*
(ii) G *is* $(k + \ell)$-*edge-connected.*

3.1   Proof of Theorem 3

Let $p_1 = \{\log n + (k+\ell)\log\log n - \omega(n)\}/n$, $p_2 = \{\log n + (k+\ell-1)\log\log n - \omega(n)\}/n$, where $\omega(n) \to \infty$ and $\omega(n) = o(\log\log n)$. From Theorem 7 and the fact that w.h.p. $\delta(G(n,p_1)) = k+\ell$, we have that w.h.p.

$$\kappa(G(n,p_1)) = \lambda(G(n,p_1)) = \delta(G(n,p_1)) = k+\ell.$$

Namely, $G(n,p_1)$ is w.h.p. $(k+\ell)$-connected and $(k+\ell)$-edge-connected. Thus, it is clear that $G(n,p_1)$ is w.h.p. $(k+1)$-connected and $(k+\ell)$-edge-connected. By Theorem 9, it follows that $G(n,p_1)$ is w.h.p. (k,ℓ)-mixed-connected.

For the second part of Theorem 3, since $\delta(G(n,p_2))$ is w.h.p. equal to $k+\ell-1$. Combining with Theorem 7, w.h.p. we have that $\lambda(G(n,p_2)) = \delta(G(n,p_2)) = k+\ell-1 < k+\ell$. From Theorem 9, we obtain that w.h.p. $G(n,p_2)$ is not (k,ℓ)-mixed-connected.

■

Let $p = \{\log n + (k+\ell)\log\log n - \omega(n)\}/n$, where $\omega(n) \to \infty$ and $\omega(n) = o(\log\log n)$. For any constant $c_1 \geq 1$, from Theorem 3 and the monotonicity of (k,ℓ)-mixed-connectivity, we know that $G(n,c_1 p)$ is w.h.p. (k,ℓ)-mixed-connected. On the other hand, for any constant $0 < c_2 < 1$, since $c_2 p < \{\log n + (k+\ell-1)\log\log n - \omega(n)\}/n$ for sufficiently large n, we have that w.h.p. $G(n,c_2 p)$ is not (k,ℓ)-mixed-connected by Theorem 3 and the monotonicity of (k,ℓ)-mixed-connectivity. Thus we obtain that Theorem 4 holds.

3.2   Proof of Theorem 6

From Theorem 8, we know that

$$\tau_{D_{k+\ell}} = \tau(\kappa(G) \geq k+\ell).$$

When $\kappa(G) \geq k+\ell$, we have that $\lambda(G) \geq \kappa(G) \geq k+\ell$. Thus, we obtain that G is (k,ℓ)-mixed-connected by Theorem 9. So, w.h.p.

$$\tau_{R_{k,\ell}} \leq \tau_{D_{k+\ell}}. \tag{3}$$

Let $t = \tau_{R_{k,\ell}}$, we have that w.h.p. G_t is $(k+\ell)$-edge-connected from Theorem 9. So it must hold that $\delta(G_t) \geq k+\ell$. Since otherwise, let L be the set of edges incident to a vertex with minimum degree in G_t, then $|L| \leq k+\ell-1$ and $G_t - L$ is disconnected, a contradiction to the fact that $\lambda(G_t) \geq k+\ell$. That means the minimum degree of G_t has already been at least $k+\ell$. Hence, w.h.p.

$$\tau_{R_{k,\ell}} \geq \tau_{D_{k+\ell}}. \tag{4}$$

Therefore, by (3) and (4), we have w.h.p.

$$\tau_{R_{k,\ell}} = \tau_{D_{k+\ell}}.$$

■

Acknowledgement. R. Gu was partially supported by Natural Science Foundation of Jiangsu Province (No. BK20170860), National Natural Science Foundation of China, and Fundamental Research Funds for the Central Universities (No. 2016B14214). Y. Shi was partially supported by the Natural Science Foundation of Tianjin (No. 17JCQNJC00300) and the National Natural Science Foundation of China.

References

1. Boesch, F.T., Chen, S.: A generalization of line connectivity and optimally invulnerable graphs. SIAM J. Appl. Math. **34**, 657–665 (1978)
2. Bollobás, B.: Random Graphs. Cambridge University Press, Cambridge (2001)
3. Beineke, L.W., Harary, F.: The connectivity function of a graph. Mathematika **14**, 197–202 (1967)
4. Bollobás, B., Thomason, A.: Random graphs of small order. Annals Discr. Math. **7**, 35–38 (1986)
5. Bollobás, B., Thomason, A.: Threshold functions. Combinatorica **7**, 35–38 (1986)
6. Egawa, Y., Kaneko, A., Matsumoto, M.: A mixed version of Menger's theorem. Combinatorica **11**, 71–74 (1991)
7. Erdős, P., Rényi, A.: On random graphs. Publ. Math. Debrecen **6**, 290–297 (1959)
8. Erdős, P., Rényi, A.: On the strength of connectedness of a random graph. Acta Math. Hung. **12**(1), 261–267 (1961)
9. Friedgut, E., Kalai, G.: Every monotone graph property has a sharp threshold. Proc. Amer. Math. Soc. **124**, 2993–3002 (1996)
10. Hennayake, K., Lai, H.-J., Li, D., Mao, J.: Minimally (k, k)-edge-connected graphs. J. Graph Theor. **44**, 116–131 (2003)
11. Ivchenko, G.I.: The strength of connectivity of a random graph. Theor. Probab. Applics **18**, 396–403 (1973)
12. Kaneko, A., Ota, K.: On minimally (n, λ)-connected graphs. J. Combin. Theor. Ser. B **80**, 156–171 (2000)
13. Mader, W.: Connectivity and edge-connectivity in finite graphs. In: Bollobás, B., (ed.): Surveys in Combinatorics proceedings of the Seventh British Combinatorial Conference, London Mathematical Society Lecture Note Series, vol. 38. Cambridge University Press, Cambridge, pp. 66–95 (1979)
14. Sadeghi, E., Fan, N.: On the survivable network design problem with mixed connectivity requirements (2015). http://www.optimization-online.org/DB_HTML/2015/10/5144.html
15. Stepanov, V.E.: On the probability of connectedness of a random graph $G_m(t)$. Theor. Probab. Applics **15**, 55–67 (1970)

Conflict-Free Connection Numbers
of Line Graphs

Bo Deng[1,2], Wenjing Li[1], Xueliang Li[1,2(✉)], Yaping Mao[2], and Haixing Zhao[2]

[1] Center for Combinatorics and LPMC, Nankai University,
Tianjin 300071, China
dengbo450@163.com, liwenjing610@mail.nankai.edu.cn, lxl@nankai.edu.cn
[2] School of Mathematics and Statistics, Qinghai Normal University, Xining 810008,
Qinghai, China
maoyaping@ymail.com, h.x.zhao@163.com

Abstract. A path in an edge-colored graph is called *conflict-free* if it contains a color that is used by exactly one of its edges. An edge-colored graph G is *conflict-free connected* if for any two distinct vertices of G, there is a conflict-free path connecting them. For a connected graph G, the *conflict-free connection number* of G, denoted by $cfc(G)$, is defined as the minimum number of colors that are required to make G conflict-free connected. In this paper, we investigate the conflict-free connection numbers of connected claw-free graphs, especially line graphs. We use $L(G)$ to denote the line graph of a graph G. In general, the *k-iterated line graph* of a graph G, denoted by $L^k(G)$, is the line graph of the graph $L^{k-1}(G)$, where $k \geq 2$ is a positive integer. We first show that for an arbitrary connected graph G, there exists a positive integer k such that $cfc(L^k(G)) \leq 2$. Secondly, we get the exact value of the conflict-free connection number of a connected claw-free graph, especially a connected line graph. Thirdly, we prove that for an arbitrary connected graph G and an arbitrary positive integer k, we always have $cfc(L^{k+1}(G)) \leq cfc(L^k(G))$, with only the exception that G is isomorphic to a star of order at least 5 and $k = 1$. Finally, we obtain the exact values of $cfc(L^k(G))$, and use them as an efficient tool to get the smallest nonnegative integer k_0 such that $cfc(L^{k_0}(G)) = 2$.

Keywords: Conflict-free connection number · Claw-free graphs · Line graphs · K-iterated line graphs

1 Introduction

All graphs considered in this paper are simple, finite, and undirected. We follow the terminology and notation of Bondy and Murty in [3] for those not defined here. For a connected graph G, let $V(G), E(G), \kappa(G)$ and $\lambda(G)$ denote the vertex

Supported by NSFC No. 11371205, 11531011 and 11701311, and NSFQH No. 2017-ZJ-790.

X. Gao et al. (Eds.): COCOA 2017, Part I, LNCS 10627, pp. 141–151, 2017.
https://doi.org/10.1007/978-3-319-71150-8_14

set, the edge set, the vertex-connectivity and the edge-connectivity of G, respectively. Throughout this paper, we use P_n, C_n and K_n to denote a path, a cycle and a complete graph of order n, respectively. And we call G a *star* of order $r+1$, denoted by $K_{1,r}$, which is a complete bipartite graph with the bipartition $V(G) = X \cup Y$ satisfying $|X| = 1$ and $|Y| = r$.

Let G be a nontrivial connected graph with an *edge-coloring* $c : E(G) \to \{0, 1, \ldots, t\}$, $t \in \mathbb{N}$, where adjacent edges may be colored with the same color. When adjacent edges of G receive different colors by c, the edge-coloring c is called *proper*. The *chromatic index* of G, denoted by $\chi'(G)$, is the minimum number of colors needed in a proper coloring of G. A path in G is called *a rainbow path* if no two edges of the path are colored with the same color. The graph G is called *rainbow connected* if for any two distinct vertices of G, there is a rainbow path connecting them. For a connected graph G, the *rainbow connection number* of G, denoted by $rc(G)$, is defined as the minimum number of colors that are needed to make G rainbow connected. These concepts were first introduced by Chartrand et al. in [5] and have been well-studied since then. For further details, we refer the reader to a book [11] and a survey paper [10].

Motivated by the rainbow connection coloring and proper coloring in graphs, Andrews et al. [1] and Borozan et al. [4] proposed the concept of proper-path coloring. Let G be a nontrivial connected graph with an edge-coloring. A path in G is called *a proper path* if no two adjacent edges of the path are colored with the same color. The graph G is called *proper connected* if for any two distinct vertices of G, there is a proper path connecting them. The *proper connection number* of G, denoted by $pc(G)$, is defined as the minimum number of colors that are needed to make G proper connected. For more details, we refer to a dynamic survey [9].

Inspired by the above mentioned two connection colorings and conflict-free colorings of graphs and hypergraphs [12], Czap et al. [7] recently introduced the concept of the conflict-free connection number of a nontrivial connected graph. Let G be a nontrivial connected graph with an edge-coloring c. A path in G is called *conflict-free* if it contains a color that is used by exactly one of its edges. The graph G is *conflict-free connected (with respect to the edge-coloring c)* if for any two distinct vertices of G, there is a conflict-free path connecting them. In this case, the edge-coloring c is called a *conflict-free connection coloring* (*CFC-coloring* for short). For a connected graph G, the *conflict-free connection number* of G, denoted by $cfc(G)$, is defined as the minimum number of colors that are required to make G conflict-free connected. For the graph with a single vertex or without any vertex, we assume the value of its conflict-free connection number equal to 0.

The conflict-free connection of graphs has the following application background. In a communication network between wireless signal towers, it is fundamental that the network is connected. We can assign signal connection paths between signal towers which may have other signal towers as intermediaries while requiring a large enough number of frequencies for this communication. If a same frequency appears more than once on a connection path between two towers A and B, then we cannot use this frequency to communicate between A and B along this path since mutual interference occurs, i.e., conflict happens. So we need a connection path between the two towers A and B on which there is a frequency that appears exactly once, and in this way we can use this unique frequency to communicate between A and B along this path without any mutual interference, i.e., conflict-free. Therefore, our goal is to allocate the minimum number of frequencies such that any two signal towers will be able to connect in this communication network without mutual interference, i.e., conflict-free. This situation can be modeled by a graph. Suppose that we assign a vertex to each signal tower, an edge between two vertices if the corresponding signal towers are directly connected by a signal and assign a color to each edge based on the assigned frequency used for the communication. Then, the minimum number of frequencies needed to assign the connections between towers so that there is always a connection path to communicate between each pair of towers with the property that there is a frequency that appears exactly once on this path, i.e., we can use this unique frequency to communicate between the pair of towers along this path without mutual interference (conflict-free), is precisely the conflict-free connection number of the corresponding graph.

The following observations are immediate.

Proposition 1. *Let G be a connected graph on $n \geq 2$ vertices. Then we have*

(i) $cfc(G) = 1$ if and only if G is complete;
(ii) $cfc(G) \geq 2$ if G is noncomplete;
(iii) $cfc(G) \leq n - 1$.

In [7], Czap et al. first gave the exact value of the conflict-free connection number for a path on n edges.

Theorem 1 ([7]). $cfc(P_n) = \lceil \log_2 n \rceil$.

Then they investigated the graphs with conflict-free connection number 2. If the number of components of G increases after removing an edge e from G, then e is called a cut-edge of G. Let $C(G)$ be the subgraph of a graph G induced by the set of cut-edges of G. A *linear forest* in a graph G is a subgraph each component of which is a path.

Theorem 2 ([7]). *If G is a noncomplete 2-connected graph, then $cfc(G) = 2$.*

Theorem 3 ([7]). *If G is a connected graph with at least 3 vertices and $C(G)$ is a linear forest whose each component is of order 2, then $cfc(G) = 2$.*

In fact, we can weaken the condition of Theorem 2, and get that the same result holds for 2-edge-connected graphs, whose proof is similar to that of Theorem 3 in [7]. For completeness, we give its proof here. Before we proceed to the result and its proof, we need the following lemmas which are useful in our proof, and can be found in [7].

Lemma 1 ([7]). *Let u, v be two distinct vertices and let $e = xy$ be an edge of a 2-connected graph G. Then there is a $u - v$ path in G containing the edge e.*

A *block* of a graph G is a maximal connected subgraph of G without cut-vertices. A connected graph with no cut-vertex therefore has just one block, namely the graph itself. An edge is a block if and only if it is a cut-edge. A block consisting of an edge is called trivial. Note that any nontrivial block is 2-connected.

Lemma 2 ([7]). *Let G be a connected graph. Then from its every nontrivial block an edge can be chosen so that the set of all such chosen edges forms a matching.*

Theorem 4. *Let G be a noncomplete 2-edge-connected graph. Then $cfc(G) = 2$.*

Proof. If G is a noncomplete 2-connected graph, then we are done. So we only consider the case that G has at least one cut-vertex. Note that G has a block decomposition with each block having at least 3 vertices, that is, each block is nontrivial. By Lemma 2, we choose from each block one edge so that all chosen edges create a matching S. Next we color the edges from S with color 2 and all remaining edges of G with color 1.

Now we prove this coloring makes G conflict-free connected, that is, for any two distinct vertices x and y, we need to find a conflict-free $x - y$ path.

Case 1. Let x and y belong to the same block B. Then by Lemma 1, there is an $x - y$ path, in B, containing the edge of B colored with color 2. Clearly, this $x - y$ path is conflict-free.

Case 2. Let x and y be in different blocks. Consider a shortest $x - y$ path in G. This path goes through blocks, say $B_1, B_2, \ldots, B_r$, $r \geq 2$, in this order, where $x \in V(B_1)$ and $y \in V(B_r)$. Let v_i be a common vertex of blocks B_i and B_{i+1}, $1 \leq i \leq r - 1$. Set $y = v_r$. Clearly, $x \neq v_1$. We choose an $x - v_1$ path in B_1 going through the edge assigned 2, and then a $v_i - v_{i+1}$ path in B_{i+1} omitting the edge colored with 2 in B_i for $1 \leq i \leq r - 1$. Obviously, the concatenation of the above r paths is an conflict-free $x - y$ path.

For a general graph G with connectivity 1, the authors of [7] gave the bounds on $cfc(G)$. Let G be a connected graph and $h(G) = max\{cfc(K) : K$ is a component of $C(G)\}$. In fact, $h(G) = 0$ if G is 2-edge-connected. So we restate that theorem as follows.

Theorem 5 ([7]). *If G is a connected graph with at least one cut-edge, then $h(G) \leq cfc(G) \leq h(G) + 1$. Moreover, these bounds are tight.*

Line graphs form one of the most important graph classes, and there have been a lot of results on line graphs, see [8]. In this paper we also deal with line graphs. Recall that the *line graph* of a graph G is the graph $L(G)$ whose vertex set $V(L(G)) = E(G)$ and two vertices e_1, e_2 of $L(G)$ are adjacent if and only if they are adjacent in G. The *iterated line graph* of a graph G, denoted by $L^2(G)$, is the line graph of the graph $L(G)$. In general, the *k-iterated line graph* of a graph G, denoted by $L^k(G)$, is the line graph of the graph $L^{k-1}(G)$, where $k \geq 2$ is a positive integer. We call a graph *claw-free* if it does not contain a claw $K_{1,3}$ as its induced subgraph. Notice that a line graph is claw-free; see [2] or [8].

This paper is organized as follows: In Sect. 2, we give some properties concerning the line graphs, and based on them, we show that for an arbitrary connected graph G, there exists a positive integer k such that $cfc(L^k(G)) \leq 2$. In Sect. 3, we start with the investigation of one special family of graphs, and then classify the graphs among them with $cfc(G) = h(G) + 1$. Using this result, we first get the exact value of the conflict-free connection number of a connected claw-free graph. As a corollary, for a connected line graph G, we obtain the value of $cfc(G)$. Then, we prove that for an arbitrary connected graph G and an arbitrary positive integer k, we always have $cfc(L^{k+1}(G)) \leq cfc(L^k(G))$, with only the exception that G is isomorphic to a star of order at least 5 and $k = 1$. Finally, we obtain the exact values of $cfc(L^k(G))$, and use them as an efficient tool to get the smallest nonnegative integer k_0 such that $cfc(L^{k_0}(G)) = 2$.

2   Dynamic Behavior of the Line Graph Operator

If one component $\mathcal{C}$ of $C(G)$ is either a cut-edge or a path of order at least 3 whose internal vertices are all of degree 2 in G, then we call $\mathcal{C}$ a *cut-path* of G.

Lemma 3. *For a connected claw-free graph G, each component of $C(G)$ is a cut-path of G.*

Proof. Firstly, $C(G)$ is a linear forest. Otherwise, there exists a vertex $v \in C(G)$ whose degree is larger than 2 in $C(G)$. Then v and three neighbors of v in $C(G)$ induce a $K_{1,3}$ in G, contradicting the condition that G is claw-free. Secondly, with a similar reason, if one component of $C(G)$ has at least 3 vertices, then all of its internal vertices must be of degree 2 in G. So, each component of $C(G)$ must be a cut-path of G.

Since a line graph is claw-free, Lemma 3 is valid for line graphs.

Corollary 1. *For a connected line graph G, every component of $C(G)$ is a cut-path of G.*

In 1969, Chartrand and Stewart [6] showed that $\kappa(L(G)) \geq \lambda(G)$, if $\lambda(G) \geq 2$. So, the following result is obvious.

Lemma 4. *The line graph of a 2-edge-connected graph is 2-connected.*

Now, we examine the dynamic behavior of the line graph operator, and get our main result of this section.

Theorem 6. *For any connected graph G, there exists a positive integer k such that $cfc(L^k(G)) \leq 2$.*

Proof. If G is a 2-edge-connected graph, then by Proposition 1, Theorem 2 and Lemma 4, we obtain $cfc(L(G)) \leq 2$. In this case, we set $k = 1$. In the following, we concentrate on the graphs having at least one cut-edge.

Let $\mathcal{P}$ be a set of paths in $C(G)$ who have at least one internal vertex and whose internal vertices are all of degree 2 in G. If $\mathcal{P} = \emptyset$, then we show that $L(G)$ is 2-edge-connected. Suppose otherwise, if there is a cut-edge e_1e_2 in $L(G)$, then there is a path of length 2 in G whose internal vertex is of degree 2, which is a contradiction. Thus, by Proposition 1 and Theorem 4, we have $cfc(L(G)) \leq 2$. Then we also set $k = 1$ in this case.

If $\mathcal{P} \neq \emptyset$, let p be the length of a longest path among $\mathcal{P}$. Notice that the cut-paths of $L^{i+1}(G)$ are the same as the cut-paths of $L^i(G)$, shortened by one edge (and no new cut-paths can appear). Then, by Corollary 1, each component of $C(L^i(G))$ must be a cut-path of $L^i(G)$ for $1 \leq i \leq p$. Since $L(P_j) = P_{j-1}$ for any positive integer $j \geq 1$, each component of $C(L^{p-1}(G))$ is of order 2. By Theorem 3, we have $cfc(L^{p-1}(G)) = 2$. Thus, we set $k = p - 1$ in this case.

The proof is thus complete.

3   The Values $cfc(L^k(G))$ of Iterated Line Graphs

In this section, we first investigate the connected graphs G having at least one cut-edge and each component of $C(G)$ is a cut-path of G. Among them, we characterize the graphs G satisfying $cfc(G) = h(G)$, and the graphs G satisfying $cfc(G) = h(G) + 1$. Let G be a connected graph of order n. If $n = 2$, $G \cong P_2$, and hence $cfc(G) = h(G) = 1$. In the following, we assume $n \geq 3$. If $h(G) = 1$, then by Theorem 3, we always have $cfc(G) = 2 = h(G) + 1$. So we only need to discuss the case of $h(G) \geq 2$.

Theorem 7. *Let G be a connected graph having at least one cut-edge, and $C(G)$ be its linear forest whose each component is a cut-path of G and $h(G) \geq 2$. Then $cfc(G) = h(G) + 1$ if and only if there are at least two components of $C(G)$ whose conflict-free connection numbers attain $h(G)$; and $cfc(G) = h(G)$ if and only if there is only one component of $C(G)$ whose conflict-free connection number attains $h(G)$.*

Proof. We first consider the case that there are at least two components of $C(G)$ whose conflict-free connection numbers attain $h(G)$, say $\mathcal{C}_1$ and $\mathcal{C}_2$. Consider the two vertices $v_1 \in V(\mathcal{C}_1)$ and $v_2 \in V(\mathcal{C}_2)$ such that the distance $d(v_1, v_2)$ between v_1 and v_2 is maximum. Assume that there exists a CFC-coloring c of G with $h(G)$ colors. Since any $v_1 - v_2$ path in G contains all the edges of $\mathcal{C}_1$ and $\mathcal{C}_2$, there is no conflict-free path connecting v_1 and v_2. Consequently, $h(G) < cfc(G)$. Together with Theorem 5, we have $cfc(G) = h(G) + 1$ in this case.

Next, we assume that there is only one component of $C(G)$ whose conflict-free connection number is $h(G)$, say $\mathcal{C}_0$. Now we give an edge-coloring of G. First, we color $\mathcal{C}_0$ with $h(G)$ colors, say $1, 2, \ldots, h(G)$, just like the coloring of a path stated in Theorem 1 of [7]. Let e_0 be the edge colored with color $h(G)$ in $\mathcal{C}_0$. Similarly, we color all the other components K of $C(G)$ with the colors from $\{1, \ldots, h(G) - 1\}$. Note that only e_0 is assigned $h(G)$ among all the edges of $C(G)$.

Then according to Lemma 2, we choose in any nontrivial block of G an edge so that all chosen edges form a matching S. We color the edges from S with color $h(G)$, and the remaining edges with color 1.

In the following we have to show that for any two distinct vertices x and y, there is a conflict-free $x - y$ path. If the vertices x and y are from the same component of $C(G)$, then such a path exists according to Theorem 1. If they are in the same nontrivial block, then by Lemma 1, there is an $x - y$ path going through the edge assigned $h(G)$. If none of the above situations appears, then x and y are either from distinct components of $C(G)$, or distinct nontrivial blocks, or one is from a component of $C(G)$ and the other from a nontrivial block.

Consider a shortest $x - y$ path P in G. Let $v_1, \ldots, v_{r-1}$ be all cut-vertices of G contained in P, in this order. Set $x = v_0$ and $y = v_r$. The path P goes through blocks $B_1, B_2 \ldots, B_r$ indicated by the vertices v_0 and v_1, v_1 and v_2, $\ldots$, v_{r-1} and v_r, respectively. At least one of the blocks is nontrivial. If P must go through the edge e_0, then in each block $B_i, 1 \leq i \leq r$, we choose a monochromatic $v_{i-1} - v_i$ path. The path concatenated of the above monochromatic paths is a desired one, since $h(G)$ only appears once. Otherwise, we consider the first nontrivial block $B_i, i \in \{1, \ldots, r\}$. In it, we choose a conflict-free $v_{i-1} - v_i$ path going through the edge of B_i colored with $h(G)$. Then in the remaining blocks $B_j, j \in \{1, \ldots, r\} \setminus \{i\}$, we choose a monochromatic $v_{j-1} - v_j$ path. The searched conflict-free $x - y$ path is then concatenated of these above paths. The resulting $x - y$ path contains only one edge assigned $h(G)$. Combining the fact $cfc(G) \geq h(G)$, we have $cfc(G) = h(G)$ in this case.

Therefore, from above, it is easy to see that there does not exist the case simultaneously satisfying $cfc(G) = h(G) + 1$ and there is only one component of $C(G)$ whose conflict-free connection number attains $h(G)$.

In contrast, if $cfc(G) = h(G)$, there is only one component of $C(G)$ whose conflict-free connection number attains $h(G)$. Otherwise, $cfc(G) = h(G) + 1$.

The result thus follows.

As a by product, we can immediately get the value of the conflict-free connection number of a connected claw-free graph. Before it, we state a structure theorem concerning a connected claw-free graph. Notice that a complete graph is claw-free. Recall that for a connected claw-free graph G, each component of $C(G)$ is a cut-path of G. Let $p(G)$, or simply p, be the length of a longest cut-path of G.

Theorem 8. *Let G be a connected claw-free graph. Then G must belong to one of the following four cases:*

i) *G is complete;*
ii) *G is noncomplete and 2-edge-connected;*
iii) *$C(G)$ has at least two components K satisfying $cfc(K) = \lceil \log_2(p + 1) \rceil$;*
iv) *$C(G)$ has only one component K satisfying $cfc(K) = \lceil \log_2(p + 1) \rceil$.*

Proof. There are two cases according to whether G has a cut-edge or not. If G has no cut-edge, we can distinguish two subcases according to whether G is complete or not. If G has a cut-edge, then we distinguish two subcases according to whether $C(G)$ has only one component K satisfying $cfc(K) = \lceil \log_2(p+1) \rceil$ or not. Thus, a connected claw-free graph G must be in one of the above four subcases.

According to Lemma 3, Theorems 3, 7 and 8, we get the following result.

Theorem 9. *Let G be a connected claw-free graph of order $n \geq 2$. Then we have*

i) $cfc(G) = 1$ *if G is complete;*
ii) $cfc(G) = 2$ *if G is noncomplete and 2-edge-connected, or $p = 1$ and $n \geq 3$;*
iii) $cfc(G) = \lceil \log_2(p+1) \rceil + 1$, *if $C(G)$ has at least two components K satisfying $cfc(K) = \lceil \log_2(p+1) \rceil$; otherwise, $cfc(G) = \lceil \log_2(p+1) \rceil$, where $p \geq 2$.*

Since line graphs are claw-free, from Theorems 8 and 9 we immediately get the following result.

Corollary 2. *Let G be a connected line graph of order $n \geq 2$. Then we have*

i) $cfc(G) = 1$ *if G is complete;*
ii) $cfc(G) = 2$ *if G is noncomplete and 2-edge-connected, or $p = 1$ and $n \geq 3$;*
iii) $cfc(G) = \lceil \log_2(p+1) \rceil + 1$, *if $C(G)$ has at least two components K satisfying $cfc(K) = \lceil \log_2(p+1) \rceil$; otherwise, $cfc(G) = \lceil \log_2(p+1) \rceil$, where $p \geq 2$.*

Next, for a connected graph G and a positive integer k, we compare $cfc(L^{k+1}(G))$ and $cfc(L^k(G))$. For almost all cases, we find that $cfc(L^{k+1}(G)) \leq cfc(L^k(G))$. However, note that if G is a complete graph of order $n \geq 4$, then $L(G)$ is noncomplete, since there exist two nonadjacent edges in G. In this case, we have $cfc(L(G)) \geq 2 > 1 = cfc(G)$. So we first characterize the connected graphs whose line graphs are complete graphs.

Lemma 5. *The line graph $L(G)$ of a connected graph G is complete if and only if G is isomorphic to a star or K_3.*

Proof. If G is isomorphic to a star or K_3, then obviously $L(G)$ is complete.

Conversely, suppose $L(G)$ is complete. From Whitney isomorphism theorem of line graphs (see [8]), i.e., two graphs H and H' have isomorphic line graphs if and only if H and H' are isomorphic, or one of them is isomorphic to the claw $K_{1,3}$ and the other is isomorphic to the triangle K_3, we immediately get that G is isomorphic to a star or K_3.

By Lemma 5 we get the following result.

Theorem 10. *Let G be a connected graph which is not isomorphic to a star of order at least 5, and k be an arbitrary positive integer. Then we have $cfc(L^{k+1}(G)) \leq cfc(L^k(G))$.*

Proof. To the contrary, we suppose that there exists a positive integer k_0 such that $cfc(L^{k_0+1}(G)) > cfc(L^{k_0}(G))$. We first claim that $L^{k_0+1}(G)$ has at least one cut-edge. Otherwise, by Proposition 1 and Theorem 4, we have $cfc(L^{k_0+1}(G)) \leq 2$. If $L^{k_0}(G)$ is complete, then it follows from Lemma 5 that $L^{k_0}(G) \cong C_3$. Then $L^{k_0+1}(G)$ is also complete, implying $cfc(L^{k_0+1}(G)) = cfc(L^{k_0}(G)) = 1$. If $L^{k_0}(G)$ is noncomplete, then by Proposition 1, we have $cfc(L^{k_0}(G)) \geq 2$; clearly, $cfc(L^{k_0+1}(G)) \leq cfc(L^{k_0}(G))$ in this case. In both cases, we have $cfc(L^{k_0+1}(G)) \leq cfc(L^{k_0}(G))$, a contradiction.

From Corollary 1, it follows that for a positive integer i, each component of $C(L^i(G))$ is a cut-path of $L^i(G)$. Let p_i be the length of a largest path of $C(L^i(G))$. Then we have $p_{i+1} = p_i - 1$, meaning $h(L^{i+1}(G)) \leq h(L^i(G))$. Set $h(L^{k_0}(G)) = q$. Since $L^{k_0+1}(G)$ has a cut-edge, we deduce that $q \geq 2$. And we know $h(L^{k_0+1}(G)) = q - 1$ or $h(L^{k_0+1}(G)) = q$. If $h(L^{k_0+1}(G)) = q - 1$, by Theorem 5, we have $q - 1 \leq cfc(L^{k_0+1}(G)) \leq q$. For the same reason, $q \leq cfc(L^{k_0}(G)) \leq q+1$. Thus, it makes a contradiction to the supposition that $cfc(L^{k_0+1}(G)) > cfc(L^{k_0}(G))$.

Then we have $h(L^{k_0+1}(G)) = q$, $cfc(L^{k_0+1}(G)) = q+1$ and $cfc(L^{k_0}(G)) = q$. By Theorem 7, there are at least two components of $C(L^{k_0+1}(G))$ whose conflict-free connection numbers are q, and there is only one component of $C(L^{k_0}(G))$ whose conflict-free connection number is q. Since every cut-path of $L^{k_0+1}(G)$ corresponds to a cut-path of $L^{k_0}(G)$, a cut-path of $L^{k_0+1}(G)$ is shorter than its corresponding cut-path of $L^{k_0}(G)$. So there is at most one component of $C(L^{k_0+1}(G))$ whose conflict-free connection number is q, a contradiction. Thus, we have $cfc(L^{k+1}(G)) \leq cfc(L^k(G))$ for any positive integer k.

If G is a star of order at least 5, then $L^i(G)$ $(i \geq 2)$ are noncomplete and 2-connected. The following result is easily obtained according to Theorem 2.

Theorem 11. *Let G be isomorphic to a star of order at least 5, and $k \geq 2$ be a positive integer. Then we have $cfc(L^{k+1}(G)) = cfc(L^k(G))$.*

Combining the above two theorems, we get a main result of this section.

Theorem 12. *For an arbitrary connected graph G and an arbitrary positive integer k, we always have $cfc(L^{k+1}(G)) \leq cfc(L^k(G))$, with only the exception that G is isomorphic to a star of order at least 5 and $k = 1$.*

From Theorem 6, we know the existence of a positive integer k such that $cfc(L^k(G)) \leq 2$. From Proposition 1 we know that only complete graphs have the cfc-value equal to 1. So, the iterated line graph $L^k(G)$ of a connected graph G has a cfc-value 1 if and only if G is complete for $k = 0$ from Proposition 1, or G is isomorphic to a star of order at least 3 for $k = 1$ from Lemma 5, or G is K_3 for all $k \geq 1$, or G is a path of order $n \geq 4$ for $k = n - 2$. Next, we want to find the smallest nonnegative integer k_0 such that $cfc(L^{k_0}(G)) = 2$. Let k an arbitrary nonnegative integer. Based on Proposition 1, Theorems 1 through 4, Lemmas 4 and 5, we begin with the investigation of the exact value of $cfc(L^k(G))$ when G is a path, a complete graph, a star, or a noncomplete 2-edge-connected graph.

Lemma 6. *Let $n \geq 2$ be a positive integer. Then $cfc(L^k(P_n)) = \lceil \log_2(n-k) \rceil$ if $k < n-1$; otherwise, $cfc(L^k(P_n)) = 0$.*

Lemma 7. *Let G be a complete graph of order $n \geq 3$. Then $cfc(L^k(G)) = 1$ for any nonnegative integer k if $n = 3$; $cfc(G) = 1$ and $cfc(L^k(G)) = 2$ for any positive integer k if $n \geq 4$.*

Lemma 8. *Let G be a star of order $n \geq 4$. Then $cfc(G) = n-1$; $cfc(L(G)) = 1$; for a positive integer $k \geq 2$, $cfc(L^k(G)) = 1$ if $n = 4$, $cfc(L^k(G)) = 2$ if $n \geq 5$.*

Lemma 9. *Let G be a noncomplete 2-edge-connected graph of order $n \geq 4$. Then $cfc(L^k(G)) = 2$ for a nonnegative integer k.*

Let $\mathcal{G} = \{G \mid G$ is a connected graph of order $n \geq 4$, G has a cut-edge, G is not a path or a star$\}$. Except for the above four kinds of graphs in Lemmas 6 through 9, we know little about the exact values of the conflict-free connection numbers of other connected graphs, even for a general tree. So for a graph $G \in \mathcal{G}$, it is difficult to give the value of $cfc(L^k(G))$ when $k = 0$. However, based on Corollaries 1 and 2, we can give the value of $cfc(L^k(G))$ when $k \geq 1$. Set p_0 be the length of a longest cut-path of $L(G)$, and let $p_0 = 0$ if $L(G)$ is 2-edge-connected.

Lemma 10. *Let $G \in \mathcal{G}$ and let k be an arbitrary positive integer. Then we have*

i) *$cfc(L^k(G)) = 2$ always holds if $p_0 \leq 1$ or there is only one component K of $C(L(G))$ satisfying $cfc(K) = h(L(G)) = 2$;*

ii) *otherwise, for $k \leq p_0 - 1$, $cfc(L^k(G)) = \lceil \log_2(p_0 - k + 2) \rceil$ if there is only one component K of $C(L^k(G))$ satisfying $cfc(K) = \lceil \log_2(p_0 - k + 2) \rceil$, and $cfc(L^k(G)) = \lceil \log_2(p_0 - k + 2) \rceil + 1$ if there are at least two components K of $C(L^k(G))$ satisfying $cfc(K) = \lceil \log_2(p_0 - k + 2) \rceil$; for $k > p_0 - 1$, $cfc(L^k(G)) = 2$ always holds.*

From Lemmas 6 through 10, we can easily get the smallest nonnegative integer k_0 such that $cfc(L^{k_0}(G)) = 2$.

Theorem 13. *Let G be a connected graph and k_0 be the smallest nonnegative integer such that $cfc(L^{k_0}(G)) = 2$. Then we have*

i) *for $G \in \{K_2, K_3, K_{1,3}\}$, k_0 does not exist;*

ii) *for a path of order 3, $k_0 = 0$; for a path of order $n \geq 4$, $k_0 = n - 4$;*

iii) *for a complete graph of order at least 4, $k_0 = 1$;*

iv) *for a star of order at least 5, $k_0 = 2$;*

v) *for a noncomplete 2-edge-connected graph, $k_0 = 0$;*

vi) *for a graph $G \in \mathcal{G}$, $k_0 = 0$ if $cfc(G) = 2$; $k_0 = 1$ if $C(L(G)) = \emptyset$ or $C(L(G))$ is a linear forest whose each component is of order 2 or there is only one component K of $C(L(G))$ satisfying $cfc(K) = h(L(G)) = 2$; otherwise, $k_0 = p_0 - 2$ if there is only one path of length p_0 in $C(L(G))$ and there is no path of length $p_0 - 1$ in $C(L(G))$ with $p_0 \geq 4$, $k_0 = p_0 - 1$ if there is only one path of length p_0 in $C(L(G))$ and there is a path of length $p_0 - 1$ in $C(L(G))$ with $p_0 \geq 3$, $k_0 = p_0$ if there are at least two paths of length p_0 in $C(L(G))$ with $p_0 \geq 2$.*

Proof. Obviously, we can get i) through v) from Lemmas 6 through 9.

For vi), if $cfc(G) = 2$, then we have $k_0 = 0$. If $C(L(G)) = \emptyset$, then we have $L(G)$ is 2-edge-connected, and hence, $cfc(L(G)) = 2$ by Theorem 4; if $C(L(G))$ is a linear forest whose each component is of order 2, then from Theorem 3 it follows that $cfc(L(G)) = 2$; if there is only one component K of $C(L(G))$ satisfying $cfc(K) = h(L(G)) = 2$, then it follows from Corollary 2 that $cfc(L(G)) = 2$. Thus, in the above three cases, we obtain $k_0 = 1$. In the following, we consider the case that both $cfc(G) \geq 3$ and $cfc(L(G)) \geq 3$. First, we give a fact that the largest integer ℓ such that $cfc(P_\ell) = 2$ is 4. Let G_0 be a connected graph, then from Corollary 2, we have $cfc(G_0) = 3$ if there is a component K satisfying $cfc(K) = h(G_0) = 3$ or there are at least two components K satisfying $cfc(K) = h(G_0) = 2$. However, $cfc(G_0) = 2$ if there is a path of length 3 and there is no path of length 2 in $C(G_0)$, or if there is a path of length 2 and there is a path of length 1 in $C(G_0)$, or if there are at least two components each of which is of order 2 in $C(G_0)$. Correspondingly, we get our results.

Acknowledgement. The authors would like to thank the reviewers for helpful suggestions and comments.

References

1. Andrews, E., Laforge, E., Lumduanhom, C., Zhang, P.: On proper-path colorings in graphs. J. Combin. Math. Combin. Comput. **97**, 189–207 (2016)
2. Beineke, L.W.: Characterizations of derived graphs. J. Combin. Theor. **9**(2), 129–135 (1970)
3. Bondy, J.A., Murty, U.S.R.: Graph Theory Applications. The Macmillan Press, London (1976)
4. Borozan, V., Fujita, S., Gerek, A., Magnant, C., Manoussakis, Y., Montero, L., Tuza, Z.S.: Proper connection of graphs. Discrete Math. **312**, 2550–2560 (2012)
5. Chartrand, G., Johns, G.L., McKeon, K.A., Zhang, P.: Rainbow connection in graphs. Math. Bohem. **133**(1), 85–98 (2008)
6. Chartrand, G., Stewart, M.J.: The connectivity of line-graphs. Math. Ann. **182**, 170–174 (1969)
7. Czap, J., Jendrol', S., Valiska, J.: Conflict-free connections of graphs, discussions mathematics graph theory. In: press
8. Hemminger, R.L., Beineke, L.W.: Line graphs and line digraphs. In: Beineke, L.W., Wilson, R.J. (eds.) Selected Topics in Graph Theory, pp. 271–305. Academic Press Inc., London (1978)
9. Li, X., Magnant, C.: Properly colored notions of connectivity - a dynamic survey, Theor. Appl. Graphs, 0(1), Art. 2 (2015)
10. Li, X., Shi, Y., Sun, Y.: Rainbow connections of graphs: a survey. Graphs Combin. **29**, 1–38 (2013)
11. Li, X., Sun, Y.: Rainbow Connections of Graphs. SpringerBriefs in Mathematics. Springer, New York (2012)
12. Pach, J., Tardos, G.: Conflict-free colourings of graphs and hypergraphs. Comb. Probab. Comput. **18**, 819–834 (2009)

The Coloring Reconfiguration Problem on Specific Graph Classes

Tatsuhiko Hatanaka$^{(\boxtimes)}$, Takehiro Ito, and Xiao Zhou

Graduate School of Information Sciences, Tohoku University,
Aoba-yama 6-6-05, Sendai 980-8579, Japan
{hatanaka,takehiro,zhou}@ecei.tohoku.ac.jp

Abstract. We study the problem of transforming one (vertex) k-coloring of a graph into another one by changing only one vertex color assignment at a time, while at all times maintaining a k-coloring, where k denotes the number of colors. This decision problem is known to be PSPACE-complete even for bipartite graphs and any fixed constant $k \geq 4$. In this paper, we study the problem from the viewpoint of graph classes. We first show that the problem remains PSPACE-complete for chordal graphs even if the number of colors is a fixed constant. We then demonstrate that, even when the number of colors is a part of input, the problem is solvable in polynomial time for several graph classes, such as split graphs and trivially perfect graphs.

1 Introduction

Recently, *reconfiguration problems* [13] have been intensively studied in the field of theoretical computer science. These problems model several "dynamic" situations where we wish to find a step-by-step transformation between two feasible solutions of a combinatorial (search) problem such that all intermediate results are also feasible and each step conforms to a fixed reconfiguration rule, that is, an adjacency relation defined on feasible solutions of the original search problem. This framework has been applied to several well-studied combinatorial problems, including SATISFIABILITY, INDEPENDENT SET, VERTEX COVER, DOMINATING SET, and so on. (See, e.g., a survey [12] and references in [9].)

1.1 Our Problem

In this paper, we study the reconfiguration problem for (vertex) colorings in a graph, called the COLORING RECONFIGURATION problem, which was introduced by Bonsma and Cereceda [3].

Let $C = \{c_1, c_2, \ldots, c_k\}$ be the set of k colors. Throughout the paper, k denotes the number of colors in C. A (proper) k-*coloring* of a graph $G = (V, E)$

This work is partially supported by JST CREST Grant Number JPMJCR1402, and by JSPS KAKENHI Grant Numbers JP16J02175, JP16K00003, and JP16K00004, Japan.

© Springer International Publishing AG 2017
X. Gao et al. (Eds.): COCOA 2017, Part I, LNCS 10627, pp. 152–162, 2017.
https://doi.org/10.1007/978-3-319-71150-8_15

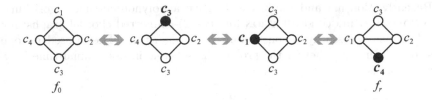

Fig. 1. A reconfiguration sequence between two 4-colorings f_0 and f_r of G.

is a mapping $f : V \to C$ such that $f(v) \neq f(w)$ for every edge $vw \in E$. Figure 1 illustrates four 4-colorings of the same graph G; the color assigned to each vertex is attached to the vertex.

Suppose that we are given two k-colorings f_0 and f_r of a graph G (e.g., the leftmost and rightmost ones in Fig. 1), and we are asked whether we can transform one into the other via k-colorings of G such that each differs from the previous one in only one vertex color assignment. This decision problem is called the COLORING RECONFIGURATION problem. For the particular instance of Fig. 1, the answer is "yes" as illustrated in the figure, where the vertex whose color assignment was changed from the previous one is depicted by a black circle. We emphatically write k-COLORING RECONFIGURATION when the number k of colors is fixed, that is, k is not a part of input.

1.2   Known and Related Results

COLORING RECONFIGURATION is one of the most well-studied reconfiguration problems from various viewpoints [1–8,11,14,16,17], including the parameterized complexity [4,14], (in)tractability with respect to graph classes [3,5,16], generalized variants such as the list coloring variant [3,11,16], the H-coloring variant [17] and the circular coloring variant [6].

Bonsma and Cereceda [3] proved that k-COLORING RECONFIGURATION is PSPACE-complete even for (i) bipartite graphs and any fixed $k \geq 4$, (ii) planar graphs and any fixed $4 \leq k \leq 6$, and (iii) bipartite planar graphs and $k = 4$. On the other hand, Cereceda et al. [8] gave a polynomial-time algorithm to solve COLORING RECONFIGURATION for any graph and $k \leq 3$. Thus, the complexity status of COLORING RECONFIGURATION is analyzed sharply with respect to k.

Because the problem remains PSPACE-complete even for very restricted instances, some sufficient conditions have been proposed so that any pair of k-colorings of a graph has a desired transformation [1,3,7]; in other words, if a given instance satisfies one of sufficient conditions, then it is a yes-instance (but, the opposite direction does not necessarily hold.) Bonsma and Cereceda [3] proved that if k is at least the degeneracy of a graph G plus two, then there is a desired transformation between any pair of k-colorings of G. Bonamy et al. [1] gave some sufficient condition with respect to graph structures: for example, chordal graphs and chordal bipartite graphs satisfy their sufficient condition.

Recently, Bonsma and Paulusma [5] gave a polynomial-time algorithm to solve COLORING RECONFIGURATION for $(k-2)$-connected chordal graphs; note that k is not necessarily a constant in their algorithm. They posed an open question which asks whether the problem is solvable in polynomial time for all chordal graphs.

1.3   Our Contribution

In this paper, we study COLORING RECONFIGURATION from the viewpoint of graph classes. More specifically, we first show that k-COLORING RECONFIG-URATION remains PSPACE-complete for chordal graphs; note that k is some fixed constant. Therefore, we answer the open question posed by Bonsma and Paulusma [5]. We then demonstrate that COLORING RECONFIGURATION is solvable in polynomial time for several graph classes, even when k is a part of input; such graph classes include 2-degenerate graphs, split graphs, and trivially perfect graphs.

2   Preliminaries

In this section, we define some basic terms and notation.

Let $G = (V, E)$ be a graph with vertex set V and edge set E; we sometimes denote by $V(G)$ and $E(G)$ the vertex set and the edge set of G, respectively. For a vertex v in G, we denote by $N(G, v)$ and $\deg(G, v)$ the neighborhood $\{w \in V \mid vw \in E\}$ and the degree $|N(G, v)|$ of v in G, respectively. We denote by $\omega(G)$ the size of a maximum clique in G.

2.1   LIST COLORING RECONFIGURATION

In this subsection, we formally define COLORING RECONFIGURATION. Because we sometimes use the notion of list colorings, we define it as a special case of LIST COLORING RECONFIGURATION as follows.

In list coloring, each vertex $v \in V(G)$ of a graph G has a set $L(v) \subseteq C = \{c_1, c_2, \ldots, c_k\}$ of colors, called the *list of* v; we sometimes call the list assignment $L : V \to 2^C$ itself a *list*. Then, a k-coloring f of G is called an *L-coloring* of G if $f(v) \in L(v)$ holds for every vertex $v \in V(G)$. Thus, a k-coloring of G is an L-coloring of G when $L(v) = C$ holds for every vertex v in G, and hence L-coloring is a generalization of k-coloring.

For two L-colorings f and f' of a graph G, a *reconfiguration sequence* between f and f' is a sequence $\langle f_p, f_{p+1}, \ldots, f_q \rangle$ of L-colorings of G such that $f_p = f$, $f_q = f'$, and $|\{v \in V(G): f_{i-1}(v) \neq f_i(v)\}| = 1$ holds for each $i \in \{p+1, p+2, \ldots, q\}$. Note that any reconfiguration sequence is *reversible*, that is, $\langle f_q, f_{q-1}, \ldots, f_p \rangle$ is a reconfiguration sequence between f' and f. We say that two L-colorings f and f' are *reconfigurable* if there is a reconfiguration sequence between them. Then, the LIST COLORING RECONFIGURATION problem is defined as follows:

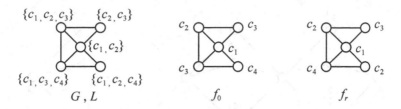

Fig. 2. Example for frozen vertices: The upper three vertices are frozen on f_0 and f_r because they form a clique of size three, and their lists contain only three colors in total.

> **Input:** A graph G, a list L, two L-colorings f_0 and f_r of G
> **Question:** Determine whether f_0 and f_r are reconfigurable or not.

Note that LIST COLORING RECONFIGURATION is a decision problem, and hence does not require the specification of an actual reconfiguration sequence.

We denote by a 4-tuple (G, L, f_0, f_r) an instance of LIST COLORING RECONFIGURATION. COLORING RECONFIGURATION is indeed LIST COLORING RECONFIGURATION when restricted to the case where $L(v) = C$ holds for every vertex v in an input graph G. We thus simply denote by a 4-tuple (G, k, f_0, f_r) an instance of COLORING RECONFIGURATION, and by a triple (G, f_0, f_r) an instance of k-COLORING RECONFIGURATION; recall that k is fixed in the latter case.

2.2 Frozen Vertices

In this subsection, we introduce the notion of "frozen vertices."

Let f be an L-coloring of a graph G with a list L. Then, a vertex $v \in V(G)$ is said to be *frozen on* f if $f'(v) = f(v)$ holds for every L-coloring f' of G which is reconfigurable from f. Therefore, v cannot be recolored in any reconfiguration sequence. Thus, (G, L, f_0, f_r) is a no-instance if $f_0(v) \neq f_r(v)$ holds for at least one frozen vertex v on f_0 or f_r. By the definition, a frozen vertex v on an L-coloring f stays frozen on any L-coloring which is reconfigurable from f.

Generally speaking, it is not easy to characterize such frozen vertices for a given L-coloring. However, there is a simple sufficient condition for which a vertex is frozen, as follows. (See Fig. 2 as an example of Observation 1.)

Observation 1. *Let G be a graph with a list L, and assume that G contains a clique V_Q of size q. If $|\bigcup_{v \in V_Q} L(v)| = q$, then all vertices $v \in V_Q$ are frozen on any L-coloring of G.*

3   PSPACE-Completeness

A graph is *chordal* if it contains no induced cycle of length at least four. In this section, we prove the following theorem.

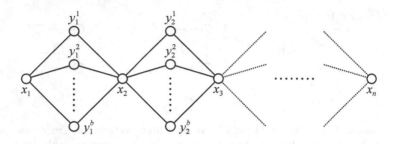

Fig. 3. Graph H.

Theorem 1. *There exists a fixed constant k' such that k-COLORING RECONFIG-URATION is PSPACE-complete for chordal graphs and every $k \geq k'$.*

It is known that k-COLORING RECONFIGURATION belongs to PSPACE [3]. Therefore, as a proof of Theorem 1, we show that there exists a fixed constant k' such that k-COLORING RECONFIGURATION is PSPACE-hard for chordal graphs and any $k \geq k'$, by giving a polynomial-time reduction from LIST COLORING RECONFIGURATION [16].

3.1   LIST COLORING RECONFIGURATION

Wrochna [16] proved that there exist two constants b and m such that LIST COLORING RECONFIGURATION remains PSPACE-complete even when an input instance (H, L, g_0, g_r) satisfies the following conditions (see also Fig. 3):

(a) $H = (X \cup Y, E)$ is a bipartite graph with bipartition X and Y such that $X = \{x_1, x_2, \ldots, x_n\}$, $Y = \{y_i^j \mid 1 \leq i \leq n-1, \ 1 \leq j \leq b\}$, and $E = \{x_i y_i^j, \ y_i^j x_{i+1} \mid 1 \leq i \leq n-1, \ 1 \leq j \leq b\}$;
(b) the list $L(v)$ of each vertex $v \in V(H)$ is a subset of the color set $C_1 \cup C_2$ such that $C_1 \cap C_2 = \emptyset$ and $|C_1| = |C_2| = m$;
(c) $L(x_i) = C_1$ if i is odd, $L(x_i) = C_2$ otherwise; and
(d) $L(y) \subseteq C_1 \cup C_2$ for all $y \in Y$.

The graph H above can be modified to an interval graph (and hence a chordal graph) H' by adding an edge $x_i x_{i+1}$ for each $i \in \{1, 2, \ldots, n-1\}$. This modification does not affect the existence and the reconfigurability of L-colorings, because any two vertices x_i and x_{i+1} joined by the new edge have distinct lists C_1 and C_2. We note in passing that this modification gives the following theorem. For an integer $d \geq 0$, a graph G is *d-degenerate* if every subgraph H of G has at least one vertex v such that $\deg(H, v) \leq d$.

Theorem 2. LIST COLORING RECONFIGURATION *is PSPACE-complete for 2-degenerate interval graphs.*

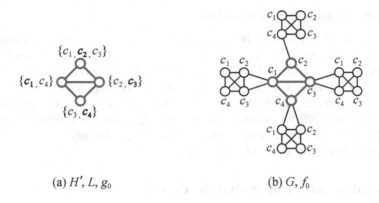

(a) H', L, g_0 (b) G, f_0

Fig. 4. (a) A graph H', a list L and an L-coloring g_0, and (b) a constructed graph G and k-coloring f_0.

3.2   Reduction

We then construct an instance (G, f_0, f_r) of k-COLORING RECONFIGURATION from the instance (H', L, g_0, g_r) above of LIST COLORING RECONFIGURATION, as follows.

Let $k \geq k' = |C_1 \cup C_2| = |\bigcup_{u \in V(H')} L(u)| = 2m$. For each vertex $u \in V(H')$, we introduce a complete graph W_u with k vertices, which is called a *frozen clique gadget*. (See Fig. 4 as an example, where $k = 4$.) The vertices in W_u are labeled as $w_1^u, w_2^u, \ldots, w_k^u$, and each vertex w_i^u corresponds to the color c_i for each $i \in \{1, 2, \ldots, k\}$. We denote by W the set of all vertices in frozen clique gadgets, that is, $W = \bigcup_{u \in V(H')} V(W_u)$.

We next add an edge between $u \in V(H')$ and $w_i^u \in V(W_u)$ if and only if $L(u)$ does *not* contain color c_i. The constructed graph G is chordal, because the addition of frozen clique gadgets does not produce any induced cycle with length at least four.

Finally, we define f_0 and f_r, as follows:

$$f_0(v) = \begin{cases} c_i & \text{if } v = w_i^u \in V(W_u) \text{ for some } u \in V(H'); \\ g_0(v) & \text{otherwise,} \end{cases}$$

and

$$f_r(v) = \begin{cases} c_i & \text{if } v = w_i^u \in V(W_u) \text{ for some } u \in V(H'); \\ g_r(v) & \text{otherwise.} \end{cases}$$

Therefore, we have $f_0(v) = f_r(v)$ for all vertices $v \in W$. From the construction, we note that both f_0 and f_r are proper k-colorings of G.

This completes our construction of the corresponding instance (G, f_0, f_r) of k-COLORING RECONFIGURATION. This construction can be done in polynomial time.

3.3 Correctness of the Reduction

We note that all vertices in W are frozen on both f_0 and f_r, because each frozen clique gadget W_u is a clique in G of size $|V(W_u)| = k$. Therefore, we can recolor vertices only in $V(H') = V(G) \setminus W$. In addition, we can use colors only in $L(u)$ for each vertex $u \in V(H')$; recall the construction with noting that $f_0(v) = f_r(v)$ for all vertices $v \in W$. Thus, (H', L, g_0, g_r) is a yes-instance of LIST COLORING RECONFIGURATION if and only if the corresponding instance (G, f_0, f_r) of k-COLORING RECONFIGURATION is a yes-instance.

This completes our proof of Theorem 1.   □

4 Polynomial-Time Solvable Cases

In this section, we demonstrate that COLORING RECONFIGURATION can be solved in polynomial time for some graph classes, even when the number k of colors is a part of input.

We start with noting the polynomial-time solvability for 2-degenerate graphs, which can be obtained straightforwardly by combining two known results. The class of 2-degenerate graphs properly contains graphs with treewidth at most two, and hence trees, cacti, outerplanar graphs, and series-parallel graphs.

Theorem 3. COLORING RECONFIGURATION *can be solved in* $O(nm)$ *time for* 2-*degenerate graphs, where* n *and* m *are the numbers of vertices and edges in an input graph, respectively.*

Proof. Let (G, k, f_0, f_r) be an instance for COLORING RECONFIGURATION. Bonsma and Cereceda [3, Theorem 11] proved that it is a yes-instance if G is d-degenerate and $k \geq d + 2$. Therefore, for 2-degenerate graphs, the answer is always yes if $k \geq 2 + 2 = 4$. On the other hand, Cereceda et al. [8, Theorem 1] gave an $O(nm)$-time algorithm to solve COLORING RECONFIGURATION for any graph if $k \leq 3$. Thus, the theorem follows.   □

In contrast to the polynomial-time solvability for 2-degenerate graphs even when k is a part of input, the reduction given by Bonsma and Cereceda [3, Theorem 3] indeed shows the following theorem.

Theorem 4 ([3]). 4-COLORING RECONFIGURATION *is PSPACE-complete for* 3-*degenerate planar graphs.*

4.1 Split Graphs

In this and next subsections, we consider split graphs and trivially perfect graphs, respectively, both of which are subclasses of chordal graphs. The following sufficient condition for yes-instances on chordal graphs will play an important role in those subsections.

Lemma 1 ([1]). *Let* (G, k, f_0, f_r) *be an instance of* COLORING RECONFIGURATION *such that* G *is a chordal graph. If* $\omega(G) \leq k - 1$, *then it is a yes-instance.*

In this subsection, we consider split graphs. A graph is *split* if its vertex set can be partitioned into a clique and an independent set.

Theorem 5. COLORING RECONFIGURATION *can be solved in linear time for split graphs.*

Proof. We give such a linear-time algorithm for split graphs. Let $\mathcal{I} = (G, k, f_0, f_r)$ be a given instance of COLORING RECONFIGURATION such that G is split. We first obtain a partition of $V(G)$ into a clique V_Q and an independent set V_I such that V_Q has the maximum size $\omega(G)$. Such a partition can be obtained in linear time [10]. Because f_0 and f_r are proper k-colorings of G, we have $|V_Q| = \omega(G) \leq k$. Therefore, there are two cases to consider.

Case 1: $|V_Q| < k$.

In this case, $|V_Q| = \omega(G) \leq k - 1$ holds. Since G is split and hence is a chordal graph, Lemma 1 implies that $\mathcal{I}$ is a yes-instance.

Case 2: $|V_Q| = k$.

In this case, every vertex in V_Q is frozen on f_0 and f_r. Thus, $\mathcal{I}$ is a no-instance if there exists a vertex $u \in V_Q$ such that $f_0(u) \neq f_r(u)$. Otherwise, because V_I is an independent set and both f_0 and f_r are proper k-colorings of G, we can directly recolor each vertex $w \in V_I$ from $f_0(w)$ to $f_r(w)$; $\mathcal{I}$ is a yes-instance.

We finally estimate the running time of our algorithm. We can obtain desired subsets V_Q and V_I in linear time [10]. Then, the algorithm simply checks if $|V_Q| < k$, and if $f_0(u) = f_r(u)$ holds for every vertex $u \in V_Q$. Therefore, our algorithm runs in linear time. $\square$

4.2   Trivially Perfect Graphs

In this subsection, we consider trivially perfect graphs. The class of trivially perfect graphs has many characterizations. We here give its recursive definition. For two graphs $G_1 = (V_1, E_1)$ and $G_2 = (V_2, E_2)$, their *union* $G_1 \cup G_2$ is the graph such that $V(G_1 \cup G_2) = V_1 \cup V_2$ and $E(G_1 \cup G_2) = E_1 \cup E_2$, while their *join* $G_1 \vee G_2$ is the graph such that $V(G_1 \vee G_2) = V_1 \cup V_2$ and $E(G_1 \vee G_2) = E_1 \cup E_2 \cup \{vw : v \in V_1, w \in V_2\}$. Then, a *trivially perfect graph* can be recursively defined, as follows:

(1) a graph consisting of a single vertex is a trivially perfect graph;
(2) if G_1 and G_2 are trivially perfect graphs, then their union $G_1 \cup G_2$ is a trivially perfect graph; and
(3) if G_1 and G_2 are trivially perfect graphs such that G_2 consists of a single vertex u, then their join $G_1 \vee G_2$ is a trivially perfect graph.

Notice that, by the join operation (3) above, the single vertex u in G_2 becomes a universal vertex in $G_1 \vee G_2$.

Theorem 6. COLORING RECONFIGURATION *can be solved in linear time for trivially perfect graphs.*

Proof. We give such a linear-time algorithm for trivially perfect graphs. Since any trivially perfect graph G is a cograph, we can represent G by a binary tree, called a cotree, which can be naturally obtained from the recursive definition of trivially perfect graphs: a *cotree* $T = (V_T, E_T)$ of a trivially perfect graph G is a binary tree such that each leaf of T corresponds to a single vertex in G, and each internal node of T has exactly two children and is labeled with either union ∪ or join ∨; notice that, for each join node in T, one of the two children must be a leaf of T. Such a cotree of G can be constructed in linear time [15]. Each node $i \in V_T$ corresponds to a subgraph G_i of G which is induced by all vertices corresponding to the leaves of T that are the descendants of i in T. Clearly, $G = G_0$ for the root 0 of T.

We note that the maximum clique sizes $\omega(G_i)$ for all $i \in V_T$ can be computed in linear time, by a bottom-up computation according to the cotree T, as follows:

$$\omega(G_i) = \begin{cases} 1 & \text{if } i \text{ is a leaf of } T; \\ \max\{\omega(G_x), \omega(G_y)\} & \text{if } i \text{ is a union node with children } x \text{ and } y; \\ \omega(G_x) + 1 & \text{if } i \text{ is a join node with children } x \text{ and } y \\ & \quad \text{such that } y \text{ is a leaf of } T. \end{cases}$$

Therefore, we assume without loss of generality that we are given a trivially perfect graph G together with its cotree $T = (V_T, E_T)$ such that the maximum clique size $\omega(G_i)$ is associated to each node $i \in V_T$.

Let $\mathcal{I} = (G, k, f_0, f_r)$ be a given instance of COLORING RECONFIGURATION such that G is a trivially perfect graph. For each node $i \in V_T$ and a k-coloring f of G, we denote by f^i the k-coloring of the subgraph G_i such that $f^i(v) = f(v)$ holds for every $v \in V(G_i)$. We propose the following algorithm to solve the problem, and will prove its correctness.

Input: An instance $\mathcal{I} = (G, k, f_0, f_r)$ of COLORING RECONFIGURATION such that G is a trivially perfect graph

Output: yes/no as the answer to $\mathcal{I}$

Step 1. If $|V(G)| = 1$ or $\omega(G) < k$, then return yes.

Step 2. In this step, G has more than one vertex, and hence the root of the cotree T is either a union node or a join node. Let x and y be two children of the root of T. Then, we execute either (a) or (b):

    Case (a): The root is a union node.
        Return yes if both (G_x, k, f_0^x, f_r^x) and (G_y, k, f_0^y, f_r^y) are yes-instances; otherwise return no.

    Case (b): The root is a join node.
        Assume that G_y consists of a single vertex u. Return no if $f_0(u) \neq f_r(u)$; otherwise return the answer to $(G_x, k - 1, f_0^x, f_r^x)$.

We first verify the correctness of Step 1. If $|V(G)| = 1$, then we can directly recolor the vertex w in G from $f_0(w)$ to $f_r(w)$; thus, $\mathcal{I}$ is a yes-instance. If $\omega(G) < k$, then Lemma 1 yields that $\mathcal{I}$ is a yes-instance because G is a trivially perfect graph and hence is a chordal graph. Thus, Step 1 correctly returns yes.

We then verify the correctness of Step 2(a). This step is executed when the root of T is a union node. Then, there is no edge between G_x and G_y. Therefore, it suffices to solve each of (G_x, k, f_0^x, f_r^x) and (G_y, k, f_0^y, f_r^y), and combine their answers. Thus, Step 2(a) works correctly.

We finally verify the correctness of Step 2(b). This step is executed when the root of T is a join node. In addition, $\omega(G) = k$ holds because it is executed after Step 1. Since $u \in V(G_y)$ becomes a universal vertex in $G = G_x \vee G_y$, it is contained in any maximum clique in G. Since $\omega(G) = k$ holds, u is frozen on f_0 and f_r. Thus, if $f_0(u) \neq f_r(u)$, then $\mathcal{I}$ is a no-instance. Otherwise no vertex in $V(G) \setminus \{u\}$ can use the color $f_0(u) = f_r(u)$ in any reconfiguration sequence, because u is a universal vertex in G and is frozen on f_0 and f_r. Therefore, (G, k, f_0, f_r) is a yes-instance if and only if $(G_x, k-1, f_0^x, f_r^x)$ is a yes-instance. Thus, Step 2(b) works correctly.

Although the algorithm above is written as a recursive function, it can be implemented so as to run in linear time, as follows: we first traverse the cotree T of a given (whole) trivially perfect graph G from the root to leaves, and assign the sub-instance (G_i, k, f_0^i, f_r^i) to each node $i \in V_T$; we then solve the sub-instances from leaves to the root of T by combining their children's answers.

This completes our proof of Theorem 6.   □

5 Conclusions

In this paper, we have studied COLORING RECONFIGURATION from the viewpoint of graph classes. We first proved that k-COLORING RECONFIGURATION is PSPACE-complete for chordal graphs; this answers the open question posed by Bonsma and Paulusma [5]. We then demonstrated that COLORING RECONFIGURATION is solvable in polynomial time for several graph classes, even when k is a part of input; such graph classes include 2-degenerate graphs, split graphs, and trivially perfect graphs.

One interesting open question is whether COLORING RECONFIGURATION is solvable in polynomial time for interval graphs or not. We note that LIST COLORING RECONFIGURATION is PSPACE-complete for 2-degenerate interval graphs, as mentioned in Theorem 2.

References

1. Bonamy, M., Johnson, M., Lignos, I., Patel, V., Paulusma, D.: Reconfiguration graphs for vertex colourings of chordal and chordal bipartite graphs. J. Comb. Optim. **27**, 132–143 (2014)
2. Bonamy, M., Bousquet, N.: Recoloring bounded tree width graphs. Electron. Notes Discr. Math. **44**, 257–262 (2013)
3. Bonsma, P., Cereceda, L.: Finding paths between graph colourings: PSPACE-completeness and superpolynomial distances. Theor. Comput. Sci. **410**, 5215–5226 (2009)

4. Bonsma, P., Mouawad, A.E., Nishimura, N., Raman, V.: The complexity of bounded length graph recoloring and CSP reconfiguration. In: Cygan, M., Heggernes, P. (eds.) IPEC 2014. LNCS, vol. 8894, pp. 110–121. Springer, Cham (2014). https://doi.org/10.1007/978-3-319-13524-3_10

5. Bonsma, P., Paulusma, D.: Using contracted solution graphs for solving reconfiguration problems. In: Proceedings of MFCS 2016, LIPIcs 58, pp. 20:1–20:15 (2016)

6. Brewster, R.C., McGuinness, S., Moore, B., Noel, J.A.: A dichotomy theorem for circular colouring reconfiguration. Theor. Comput. Sci. **639**, 1–13 (2016)

7. Cereceda, L.: Mixing Graph Colourings. Ph.D. Thesis, London School of Economics and Political Science (2007)

8. Cereceda, L., van den Heuvel, J., Johnson, M.: Finding paths between 3-colorings. J. Graph Theory **67**, 69–82 (2011)

9. Demaine, E.D., Demaine, M.L., Fox-Epstein, E., Hoang, D.A., Ito, T., Ono, H., Otachi, Y., Uehara, R., Yamada, T.: Linear-time algorithm for sliding tokens on trees. Theor. Comput. Sci. **600**, 132–142 (2015)

10. Hammer, P.L., Simeone, B.: The splittance of a graph. Combinatorica **1**, 275–284 (1981)

11. Hatanaka, T., Ito, T., Zhou, X.: The list coloring reconfiguration problem for bounded pathwidth graphs. IEICE Trans. Fundam. Electron. Commun. Comput. Sci. **E98-A**, 1168–1178 (2015)

12. van den Heuvel, J.: The complexity of change. Surveys in Combinatorics 2013, London Mathematical Society Lecture Notes Series 409 (2013)

13. Ito, T., Demaine, E.D., Harvey, N.J.A., Papadimitriou, C.H., Sideri, M., Uehara, R., Uno, Y.: On the complexity of reconfiguration problems. Theor. Comput. Sci. **412**, 1054–1065 (2011)

14. Johnson, M., Kratsch, D., Kratsch, S., Patel, V., Paulusma, D.: Finding shortest paths between graph colourings. Algorithmica **75**, 295–321 (2016)

15. McConnell, R.M., Spinrad, J.P.: Linear-time modular decomposition of directed graphs. Discr. Appl. Math. **145**, 198–209 (2005)

16. Wrochna, M.: Reconfiguration in bounded bandwidth and treedepth (2014). arXiv:1405.0847

17. Wrochna, M.: Homomorphism reconfiguration via homotopy. In: Proceedings of STACS 2015, LIPIcs 30, pp. 730–742 (2015)

Combinatorial Optimization

Minimizing Total Completion Time of Batch Scheduling with Nonidentical Job Sizes

Rongqi Li, Zhiyi Tan$^{(\boxtimes)}$, and Qianyu Zhu

Department of Mathematics, Zhejiang University,
Hangzhou 310027, People's Republic of China
tanzy@zju.edu.cn

Abstract. This paper concerns the problem of scheduling jobs with unit processing time and nonidentical sizes on single or parallel identical batch machines. The objective is to minimize the total completion time of all jobs. We show that the worst-case ratio of the algorithm based on the bin-packing algorithm First Fit Increasing (FFI) lies in the interval $[\frac{109}{82}, \frac{2+\sqrt{2}}{2}] \approx [1.3293, 1.7071]$ for the single machine case, and is no more than $\frac{6+\sqrt{2}}{4} \approx 1.8536$ for the parallel machines case.

1 Introduction

In this paper, we study the problem of scheduling jobs with nonidentical sizes on single or parallel identical batch machines [22]. We are given a non-empty set of jobs $\mathcal{J} = \{J_1, J_2, \ldots, J_n\}$. For $j = 1, \ldots, n$, the processing time and size of J_j is p_j and s_j, respectively. There are $m \geq 1$ machines $M_1, M_2, \ldots, M_m$ with the same capacity B. Each machine can simultaneously process a number of jobs as a batch as long as the total size of jobs in the batch is no greater than B. The processing time of a batch is the maximum of the processing times of jobs contained in the batch. The *cardinality* of a batch is the number of jobs contained in the batch. W. l. o. g., we will assume that $B = 1$ and $s_j \leq 1$ for all j. All jobs are available at time 0 and no preemption is allowed. The objective is to minimize the total completion time of all jobs.

Given a schedule σ^S of $\mathcal{J}$, denote by $TC^S(\mathcal{J})$ the total completion time of all jobs in σ^S. We will write it simply TC^S when no confusion can arise. Let $\sigma^{\mathcal{A}}$ and σ^* be the schedule produced by algorithm $\mathcal{A}$ and the optimal schedule, respectively. The *worst-case ratio* of algorithm $\mathcal{A}$ is then defined as

$$\inf\left\{\alpha | TC^{\mathcal{A}}(\mathcal{J}) \leq \alpha TC^*(\mathcal{J}) \text{ for all } \mathcal{J}\right\}.$$

The worst-case ratio of an algorithm for problems with other objectives can be defined accordingly.

Research on batch scheduling problems is motivated by burn-in operations in semi-conductor manufacturing, and dates back to the 1980's [13]. According

Supported by the National Natural Science Foundation of China (11671356, 11271324, 11471286).

X. Gao et al. (Eds.): COCOA 2017, Part I, LNCS 10627, pp. 165–179, 2017.
https://doi.org/10.1007/978-3-319-71150-8_16

to the type of capacity constraint of a batch, batch scheduling problems can be classified into three types: *unbounded batch model*, where the capacity of a batch is infinity; *bounded batch model with identical job sizes*; and *bounded batch model with nonidentical job sizes*. Problems in the last category are clearly the most difficult ones and do have particular features. We will first briefly survey some results on the classical batch scheduling problems belong to the first two types. More results on other objectives and more complex paradigms such as nonidentical release times can be found in [4,19], and references therein.

Most papers have been devoted to makespan (the maximum completion time of all jobs) minimization. If the batch has an unbounded capacity, combining all the jobs into a single batch is optimal for both the single and parallel machine cases. For bounded batch model with identical job sizes, the problem is still polynomially solvable for the single machine case. If there is more than one machine, the problem becomes NP-hard, but it still admits a *Polynomial Time Approximation Scheme* (PTAS) [17]. Problems with the objective of minimizing the total completion time are much more difficult. If the batch has an unbounded capacity, Brucker et al. [4] designed a polynomial time algorithm via a dynamic programming approach for the single machine case. They also remarked that such techniques can be applied to the parallel machine cases, and lead to a pseudopolynomial time dynamic programming algorithm when the number of machines is a given number. For bounded batch model with identical job sizes, the capacity constraint of a batch can be interpreted as at most b jobs can be packed into a batch. If b is a part of input, the complexity status is still open despite the intense research conducted. In the absence of complexity results, Hochbaum and Landy [12] presented an approximation algorithm with worst-case ratio 2 for the single machine case. Later, Deng et al. [6] and Li et al. [16] designed a PTAS for the single and parallel machines case, respectively. If b is a fixed number, the problem can be solved in $O(n^{b(b-1)})$ time for the single machine case [4]. The time complexity was improved to $O(n^{6b})$ by Poon and Yu [18].

Batch scheduling problems with nonidentical job sizes have a close relationship with the one-dimensional bin-packing problem, where a sequence of items with sizes between 0 and 1 are required to be packed into a minimum number of unit capacity bins. Almost all algorithms for batch scheduling consist of two phases, *batching* and *scheduling*. The former refers to a grouping of the jobs into batches such that the total size of all jobs in a batch does not exceed the capacity. The latter determines on which machine and in what order the batches are scheduled. Obviously, the batching phase is essentially a one-dimensional bin-packing problem, and any algorithm of the latter can be used as a procedure for the batching phase of a batch scheduling algorithm.

Let $\mathcal{P}$ denote a bin-packing algorithm. We will use $C^{\mathcal{P}}$ to denote the number of bins employed when $\mathcal{P}$ is applied, and C^* to denote the number of bins employed for a packing which uses a minimal number of bins. *First Fit* (FF) and *First Fit Decreasing* (FFD) are two of the most fundamental algorithms for one-dimensional bin-packing. Algorithm FF always packs the next unpacked item of the list (according to increasing subscript) into the first opened bin that has enough room to accommodate it. If no opened bin is suitable for this, a new bin is opened and the item is packed there. Algorithm FFD first sorts the items in non-increasing order of their sizes and then calls FF. In Johnson's pioneering work [14], it was proved that $C^{FFD} \leq \frac{11}{9}C^* + 4$. The additive term was gradually reduced, and eventually Dósa proved that the tight value is $\frac{6}{9}$ [7]. Simchi-Levi [20] proved that $C^{FFD} \leq \frac{3}{2}C^*$, and no polynomial-time algorithm can have a smaller worst-case ratio unless $P = NP$ [11]. For algorithm FF, Ullman [21] proved that $C^{FF} \leq \frac{17}{10}C^* + 3$. Simchi-Levi [20] proved that the worst-case ratio of FF is at most $\frac{7}{4}$. This upper bound was improved to $\frac{12}{7}$ in [3,23] independently. Finally, Dósa and Sgall [8] proved that the worst-case ratio of FF is exactly $\frac{17}{10}$.

A less popular algorithm for one-dimensional bin-packing is *First Fit Increasing* (FFI). Algorithm FFI first sorts the items in non-decreasing order of their sizes and then calls FF. Obviously, the worst-case ratio of FFI would not be larger than that of FF. In fact, nor is it smaller due to the instance given in [15]. Therefore, the presorting step of FFI seems to be redundant. However, as we will see later in our paper, FFI is superior to FFD to be adopted as a batching procedure for problem with objective of minimizing the total completion time.

So far theoretical results on batch scheduling problems with nonidentical job sizes were concentrated on makespan minimization. For the single machine case, there is no polynomial-time algorithm with worst-case ratio smaller than $\frac{3}{2}$ unless $P = NP$, since it contains the one-dimensional bin-packing problem as a special case. The current best worst-case ratio is $\frac{5}{3}$, which can be achieved by adopting a recently proposed algorithm of one-dimensional bin-packing [2] in the batching phase. The proof of the worst-case ratio follows directly from the general results established in [9]. For the case of parallel machines, Dosa et al. [9] proved that there does not exist any polynomial-time algorithm with worst-case ratio better than 2 unless $P = NP$, even if all jobs have unit processing time. They also proposed an algorithm with worst-case ratio arbitrarily close to 2. To the authors' knowledge, no approximation algorithm with performance guarantee is known for problems with the objective of minimizing the total completion time. In [5], the authors claimed to obtain an algorithm with worst-case ratio 2 for the parallel machines case. Unfortunately, the actual objective of the problem they have studied is minimizing the total completion time of all *machines*. Furthermore, serious errors exist in their proof, and the worst-case ratio of their algorithm is in fact unbound for any one of the three objectives mentioned above.

Given the considerable difficulties of the problem, we study a special case that all jobs have the same processing time. W. l. o. g., we will assume that $p_j = 1$ for all j. Then the processing time of any batch is also 1, and thus the scheduling phase becomes trivial. For the makespan minimization problem, if we use a bin-packing algorithm $\mathcal{P}$ as a procedure in the batching phase, then a schedule with makespan of $\lceil \frac{C^{\mathcal{P}}}{m} \rceil$ can be easily obtained. If the worst-case ratio of $\mathcal{P}$ is at most α for the classical one-dimensional bin-packing problem, then we can obtain an algorithm for the batch scheduling with worst-case ratio of at most $\alpha + \frac{m-1}{m}$ since

$$\frac{\lceil \frac{C^{\mathcal{P}}}{m} \rceil}{\lceil \frac{C^*}{m} \rceil} \leq \frac{\frac{C^{\mathcal{P}}}{m} + \frac{m-1}{m}}{\lceil \frac{C^*}{m} \rceil} \leq \frac{\frac{C^{\mathcal{P}}}{m}}{\frac{C^*}{m}} + \frac{\frac{m-1}{m}}{1} \leq \alpha + \frac{m-1}{m}.$$

If $\alpha \leq 2$, then $C^{\mathcal{P}} \leq 2m$ when $C^* \leq m$. We can further tighten the bound to $\max\{2, \alpha + \frac{m-1}{2m}\}$. Specifically, if we can pack jobs into a minimum number of batches, an optimal schedule for the batch scheduling problem can be obtained without difficulty.

The situation is similar but a little more complicated for problems with objective of minimizing the total completion time. Given a packing of jobs into batches, an optimal schedule can be obtained by ensuring that the batch with larger cardinality is completed no later than the batch with smaller cardinality. Therefore, any rational algorithm for batch scheduling is completely determined by the algorithm of bin-packing problem used in the batching phase. Nevertheless, it is still unknown yet which packing is the best one since no bin-packing variant ever studied has the same objective as minimizing the total completion time.

A somewhat similar variant is bin-packing with general costs [1]. It differs with classical one-dimensional bin-packing problem in that the objective is minimizing the total cost of all bins, where the cost of a bin is a concave and monotone function of the number of items assigned to it. Anily et al. [1] showed that there is no constant worst-case ratio for either FFD or FF, while FFI has a worst-case ratio of no more than 1.75. Later, Epstein and Levin [10] designed an *Asymptotic Fully Polynomial Time Approximation Scheme* (AFPTAS). Though the objective of minimizing the total completion time of all jobs can also be interpreted as total cost of all batches, the definitions of the cost of a batch is different in two problems of bin-packing with general costs and batch scheduling. The cost of a batch in the batch scheduling problem is determined not only its cardinality, but also its completion time. The cost of two batches with the same cardinality is identical in the former problem, but must be different in the latter problem.

In this paper, we give a first attempt to provide a worst-case performance analysis of algorithms for scheduling jobs with unit processing time and nonidentical sizes on single or parallel identical batch machines. For the single machine case, we show the worst-case ratio of algorithms based on FFD and FF are unbounded, and the worst-case ratio of the algorithm based on FFI is no more than $\frac{2+\sqrt{2}}{2}$. For the parallel machine case, we introduce an universal algorithm $RR - \mathcal{A}$ which uses a bin-packing algorithm $\mathcal{A}$ in the batching phase. The worst-case ratio of $RR - \mathcal{A}$ can be estimated by the worst-case ratio of $\mathcal{A}$. As a result, the worst-case ratio of the algorithm based on $RR - FFI$ is no more than $\frac{6+\sqrt{2}}{4}$ for any number of machines.

The structure of the paper is as follows. Following the introductory section, Sect. 2 presents two important technical lemmas. Sections 3 and 4 give an outline of single and parallel machines, respectively. Due to page limitations, the proofs of Lemma 2 and Theorem 4 regarding parallel machines are omitted.

2 Technical Preliminaries

In this section, we present two technical lemmas that will play an essential role when proving the worst-case ratio of the algorithm based on FFI. Let $\{x_i\}_{i=1}^n$, $\{y_i\}_{i=1}^n$ be two nonnegative integer series. The series $\{x_i\}$ begins with $p(\geq 2)$ positive items, and ends by $n-p$ items of 0. The series $\{y_i\}$ begins with $q(\geq 2)$ positive items, and ends by $n-q$ items of 0. Among the q positive items, there are r items greater than 1 and $q-r$ items of 1. Let

$$k = \max\left\{l\Big| \sum_{i=1}^{l} y_i + l - 1 < n\right\}. \tag{1}$$

It follows that

$$\sum_{i=1}^{k} y_i + k - 1 < n \tag{2}$$

and

$$\sum_{i=1}^{k+1} y_i + k \geq n. \tag{3}$$

For series $\{x_i'\}_{i=1}^n$ and $\{y_i'\}_{i=1}^n$, the corresponding values are denoted p', q', r' and k', respectively. Define

$$\alpha(x_i, y_i) = \frac{\sum_{i=1}^{q} iy_i}{\sum_{i=1}^{p} ix_i}.$$

Lemma 1. *If series $\{x_i\}_{i=1}^n$ and $\{y_i\}_{i=1}^n$ satisfy the following conditions:*

(C1) $\sum_{i=1}^{p} x_i = \sum_{i=1}^{q} y_i = n$.
(C2) $y_1 \geq y_2 \geq \cdots \geq y_q$.
(C3) For all $1 \leq l \leq q$, $\sum_{i=1}^{l} x_i \leq \sum_{i=1}^{l} y_i + (l-1)$.
(C4) $p \geq q - r$.

then $\alpha(x_i, y_i) \leq \frac{2+\sqrt{2}}{2}$.

Proof. Suppose that series $\{x_i\}$ and $\{y_i\}$ satisfy (C1)–(C4). If $p > q$, define a new series $\{x_i'\}_{i=1}^n$ such that

$$x_i' = \begin{cases} x_i & 1 \leq i \leq q-1, \\ \sum_{l=q}^{p} x_l & i = q, \\ 0 & q+1 \leq i \leq n. \end{cases}$$

Clearly, $p' = q \geq q - r$, and $\sum_{i=1}^{p'} x_i' = \sum_{i=1}^{p} x_i = n$. Thus $\{x_i'\}$ and $\{y_i\}$ satisfy (C1), (C2) and (C4). Since $\{x_i\}$ and $\{y_i\}$ satisfy (C3), we have

$$\sum_{i=1}^{l} x_i' = \sum_{i=1}^{l} x_i \leq \sum_{i=1}^{l} y_i + (l-1), \quad l = 1, \ldots, q-1,$$

and

$$\sum_{i=1}^{q} x_i' = n = \sum_{i=1}^{q} y_i \le \sum_{i=1}^{q} y_i + (q-1).$$

Hence, $\{x_i'\}$ and $\{y_i\}$ also satisfy (C3). Note that

$$\sum_{i=1}^{p'} i x_i' = \sum_{i=1}^{q} i x_i' = \sum_{i=1}^{q-1} i x_i' + q x_q' = \sum_{i=1}^{q-1} i x_i + q \sum_{i=q}^{p} x_i < \sum_{i=1}^{q-1} i x_i + \sum_{i=q}^{p} i x_i = \sum_{i=1}^{p} i x_i.$$

Therefore,

$$\alpha(x_i, y_i) = \frac{\sum_{i=1}^{q} i y_i}{\sum_{i=1}^{p} i x_i} \le \frac{\sum_{i=1}^{q} i y_i}{\sum_{i=1}^{p'} i x_i'} = \alpha(x_i', y_i)$$

It follows that to prove the lemma, it suffices to assume that $p \le q$. Under such circumstance, we have $\sum_{i=1}^{q} x_i = \sum_{i=1}^{p} x_i = \sum_{i=1}^{q} y_i$ by (C1). Hence,

$$\sum_{l=1}^{q} \left(\sum_{i=1}^{l} x_i - \sum_{i=1}^{l} y_i \right) = \sum_{l=1}^{q} \sum_{i=1}^{l} (x_i - y_i) = \sum_{i=1}^{q} \sum_{l=i}^{q} (x_i - y_i)$$

$$= \sum_{i=1}^{q} (q - i + 1)(x_i - y_i)$$

$$= (q+1) \sum_{i=1}^{q} (x_i - y_i) - \sum_{i=1}^{q} i (x_i - y_i)$$

$$= \sum_{i=1}^{q} i y_i - \sum_{i=1}^{p} i x_i. \qquad (4)$$

Note that as long as $p \le q$, (4) remains true without (C3) and (C4). The rest of the proof will be divided into two parts according to the value of y_k.

First assume that $y_k = 1$. Then

$$r \le k - 1. \qquad (5)$$

By (2), we have

$$\sum_{i=1}^{k} y_i + k - 1 < n = \sum_{i=1}^{q} y_i = \sum_{i=1}^{k} y_i + \sum_{i=k+1}^{q} y_i = \sum_{i=1}^{k} y_i + (q - k).$$

Hence,

$$q > 2k - 1 \ge 2r + 1.$$

Recall that $\{x_i\}$ and $\{y_i\}$ satisfy (C3). For $1 \le l \le r$,

$$\sum_{i=1}^{l} x_i \le \sum_{i=1}^{l} y_i + (l - 1).$$

For $r + 1 \leq l \leq q - r$, by (C4),

$$\sum_{i=1}^{l} x_i = n - \sum_{i=l+1}^{p} x_i \leq n - (p - l) \leq n - (q - r - l)$$

$$= \sum_{i=1}^{q} y_i - (q - r - l) = \sum_{i=1}^{l} y_i + \sum_{i=l+1}^{q} y_i - (q - r - l)$$

$$= \sum_{i=1}^{l} y_i + (q - l) - (q - r - l) = \sum_{i=1}^{l} y_i + r.$$

For $q - r + 1 \leq l \leq q$,

$$\sum_{i=1}^{l} x_i \leq n = \sum_{i=1}^{q} y_i = \sum_{i=1}^{l} y_i + \sum_{i=l+1}^{q} y_i = \sum_{i=1}^{l} y_i + (q - l).$$

Hence,

$$\sum_{l=1}^{q} \left(\sum_{i=1}^{l} x_i - \sum_{i=1}^{l} y_i \right) \leq \sum_{l=1}^{r} (l - 1) + \sum_{l=r+1}^{q-r} r + \sum_{l=q-r+1}^{q} (q - l)$$

$$= \frac{r(r-1)}{2} + r(q - 2r) + \frac{r(r-1)}{2} = (q - r - 1)r. \quad (6)$$

On the other hand,

$$\sum_{i=1}^{q} iy_i = \sum_{i=1}^{r} iy_i + \sum_{i=r+1}^{q} iy_i \geq 2 \sum_{i=1}^{r} i + \sum_{i=r+1}^{q} i = \sum_{i=1}^{r} i + \sum_{i=1}^{q} i$$

$$= \frac{r(r+1)}{2} + \frac{q(q+1)}{2} > \frac{q^2 + r^2}{2} = \frac{1}{2}((q - r)^2 + 2r^2) + r(q - r)$$

$$\geq \sqrt{2}r(q - r) + r(q - r) = (1 + \sqrt{2})r(q - r). \quad (7)$$

From (4), (6) and (7), it is easy to get

$$(2 + \sqrt{2}) \sum_{i=1}^{p} ix_i - 2 \sum_{i=1}^{q} iy_i = \sqrt{2} \sum_{i=1}^{q} iy_i - (2 + \sqrt{2}) \left(\sum_{i=1}^{q} iy_i - \sum_{i=1}^{p} ix_i \right)$$

$$= \sqrt{2} \sum_{i=1}^{q} iy_i - (2 + \sqrt{2}) \sum_{l=1}^{q} \left(\sum_{i=1}^{l} x_i - \sum_{i=1}^{l} y_i \right)$$

$$\geq \sqrt{2}(1 + \sqrt{2})r(q - r) - (2 + \sqrt{2})(q - r - 1)r \geq 0.$$

Hence,

$$\alpha(x_i, y_i) = \frac{\sum_{i=1}^{q} iy_i}{\sum_{i=1}^{p} ix_i} \leq \frac{2 + \sqrt{2}}{2}.$$

We now turn to the case of $y_k \geq 2$. By (C1), $\sum_{i=1}^q y_i + q - 1 = n + q - 1 \geq n$, Thus $k \leq q - 1$ by the definition of k. Construct a new series $\{y_i'\}_{i=1}^n$ with $q' = n + k + 1 - \sum_{i=1}^{k+1} y_i$ as follows:

$$y_i' = \begin{cases} y_i & 1 \leq i \leq k + 1, \\ 1 & k + 2 \leq i \leq q', \\ 0 & q' + 1 \leq i \leq n. \end{cases}$$

Clearly, $y_1' \geq y_2' \geq \cdots \geq y_{q'}'$, and

$$\sum_{i=1}^{q'} y_i' = \sum_{i=1}^{k+1} y_i' + \sum_{i=k+2}^{q'} y_i' = \sum_{i=1}^{k+1} y_i + (q' - k - 1) = n. \tag{8}$$

By (3) and the definition of q', we have

$$\sum_{i=1}^{k+1} y_i + k \geq n = \sum_{i=1}^{k+1} y_i + q' - (k+1).$$

Thus $q' \leq 2k + 1$. Since $y_i = y_i'$ for $1 \leq i \leq k + 1$ and $y_i \geq 1 \geq y_i'$ for $k + 2 \leq i \leq q$, $\sum_{i=1}^q y_i' \leq \sum_{i=1}^q y_i = n$, and thus $q' \geq q$. Combining above two inequalities regarding q' with $k \leq q - 1$, we have

$$k + 1 \leq q' \leq 2k + 1. \tag{9}$$

Since $\{x_i\}$ and $\{y_i\}$ satisfy (C3),

$$\sum_{i=1}^l x_i \leq \sum_{i=1}^l y_i + (l-1) = \sum_{i=1}^l y_i' + (l-1), \quad l = 1, \ldots, k,$$

and

$$\sum_{i=1}^l x_i \leq n = \sum_{i=1}^{q'} y_i' = \sum_{i=1}^l y_i' + \sum_{i=l+1}^{q'} y_i' = \sum_{i=1}^l y_i' + (q' - l), \quad l = k+1, \ldots, q'.$$

Hence,

$$\sum_{l=1}^{q'} \left(\sum_{i=1}^l x_i - \sum_{i=1}^l y_i' \right) \leq \sum_{l=1}^k (l-1) + \sum_{l=k+1}^{q'} (q' - l)$$

$$= \frac{k(k-1)}{2} + \frac{(q'-k)(q'-k-1)}{2}. \tag{10}$$

Define

$$g_1(\gamma) = 2 \left((k+1)(k+2) + \frac{\gamma(\gamma+1)}{2} \right) - 5 \left(\frac{k(k-1)}{2} + \frac{(\gamma-k)(\gamma-k-1)}{2} \right)$$

and

$$g_2(\gamma) = 2\left(\frac{k(k+1)}{2} + \frac{\gamma(\gamma+1)}{2}\right) - 5\left(\frac{k(k-1)}{2} + \frac{(\gamma-k)(\gamma-k-1)}{2}\right)$$

Both are quadratic functions with negative quadratic coefficients. We further distinguish two subcases according to the value of y'_{k+1}.

If $y'_{k+1} \geq 3$, then

$$\sum_{i=1}^{q'} iy'_i = \sum_{i=1}^{k+1} iy'_i + \sum_{i=k+2}^{q'} iy'_i \geq y'_{k+1}\sum_{i=1}^{k+1} i + \sum_{i=k+2}^{q'} i \geq 3\sum_{i=1}^{k+1} i + \sum_{i=k+2}^{q'} i$$

$$= 2\sum_{i=1}^{k+1} i + \sum_{i=1}^{q'} i = (k+1)(k+2) + \frac{q'(q'+1)}{2} \tag{11}$$

Therefore, by (4), (9), (10) and (11), we have

$$5\sum_{i=1}^{p} ix_i - 3\sum_{i=1}^{q'} iy'_i = 2\sum_{i=1}^{q'} iy'_i - 5\left(\sum_{i=1}^{q'} iy'_i - \sum_{i=1}^{p} ix_i\right)$$

$$= 2\sum_{i=1}^{q'} iy'_i - 5\sum_{l=1}^{q'}\left(\sum_{i=1}^{l} x_i - \sum_{i=1}^{l} y'_i\right) \geq g_1(q')$$

$$\geq \min\{g_1(2k+1), g_1(k+1)\}$$

$$= \min\left\{k^2 + 12k + 6, \frac{k^2 + 23k + 12}{2}\right\} \geq 0. \tag{12}$$

If $y'_{k+1} \leq 2$, by (2) and (8),

$$\sum_{i=1}^{k} y'_i + k - 1 = \sum_{i=1}^{k} y_i + k - 1 < n = \sum_{i=1}^{q'} y'_i = \sum_{i=1}^{k} y'_i + y'_{k+1} + \sum_{i=k+2}^{q'} y'_i$$

$$\leq \sum_{i=1}^{k} y'_i + 2 + (q' - k - 1).$$

Hence, $q' > 2k - 2$. Combining the above inequality with (9), we have

$$2k - 1 \leq q' \leq 2k + 1. \tag{13}$$

Recall that $y'_k = y_k \geq 2$. We have

$$\sum_{i=1}^{q'} iy'_i = \sum_{i=1}^{k} iy'_i + \sum_{i=k+1}^{q'} iy'_i \geq y'_k\sum_{i=1}^{k} i + \sum_{i=k+1}^{q'} i \geq 2\sum_{i=1}^{k} i + \sum_{i=k+1}^{q'} i$$

$$= \sum_{i=1}^{k} i + \sum_{i=1}^{q'} i = \frac{k(k+1)}{2} + \frac{q'(q'+1)}{2} \tag{14}$$

Therefore, by (4), (10), (13) and (14) we have

$$5\sum_{i=1}^{p} ix_i - 3\sum_{i=1}^{q'} iy_i' = 2\sum_{i=1}^{q'} iy_i' - 5\left(\sum_{i=1}^{q'} iy_i' - \sum_{i=1}^{p} ix_i\right)$$

$$= 2\sum_{i=1}^{q'} iy_i' - 5\sum_{l=1}^{l}\left(\sum_{i=1}^{l} x_i - \sum_{i=1}^{l} y_i'\right) \geq g_2(q')$$

$$\geq \min\{g_2(2k+1), g_2(2k-1)\}$$

$$= \min\{7k+2, 9k-5\} \geq 0. \tag{15}$$

By the definition of q', we have

$$\sum_{i=k+2}^{q}(y_i - 1) = \sum_{i=k+2}^{q} y_i - (q-k-1) = n - \sum_{i=1}^{k+1} y_i - (q-k-1) = q' - q.$$

Hence,

$$\sum_{i=1}^{q} iy_i = \sum_{i=1}^{k+1} iy_i + \sum_{i=k+2}^{q} iy_i \leq \sum_{i=1}^{k+1} iy_i + \sum_{i=k+2}^{q} i + q(q'-q)$$

$$\leq \sum_{i=1}^{k+1} iy_i + \sum_{i=k+2}^{q} i + \sum_{i=q+1}^{q'} i = \sum_{i=1}^{q'} iy_i'.$$

Therefore, by (12) and (15),

$$\alpha(x_i, y_i) = \frac{\sum_{i=1}^{q} iy_i}{\sum_{i=1}^{p} ix_i} \leq \frac{\sum_{i=1}^{q'} iy_i'}{\sum_{i=1}^{p} ix_i} \leq \frac{5}{3} \leq \frac{2+\sqrt{2}}{2}.$$

The proof of the lemma is thus completed.   □

Given m is an integer. Define

$$\beta(x_i, y_i) = \frac{\sum_{i=1}^{q} \lceil\frac{i}{m}\rceil y_i}{\sum_{i=1}^{p} \lceil\frac{i}{m}\rceil x_i}.$$

Lemma 2. *If series $\{x_i\}_{i=1}^{n}$ and $\{y_i\}_{i=1}^{n}$ satisfy the following conditions:*

(C0) $\alpha(x_i, y_i) \leq 2$,
(C1) $\sum_{i=1}^{p} x_i = \sum_{i=1}^{q} y_i = n$,
(C2) $y_1 \geq y_2 \geq \cdots \geq y_q$,
(C2') $x_1 \geq x_2 \geq \cdots \geq x_p$,

then $\beta(x_i, y_i) \leq \frac{m+1}{2m}\alpha(x_i, y_i) + \frac{m-1}{m}$, and the bound is tight.

3   Single Machine

In this section, we present complexity results and approximation algorithms for the single machine batch scheduling problem of minimizing the total completion time with non-identical job sizes and unit processing times.

Theorem 1. *The single machine batch scheduling problem of minimizing the total completion time with non-identical job sizes and unit processing times is NP-hard.*

The theorem can be proved by using the reduction from the Equal Cardinality Partition [11], and the detail is omitted here. Due to the NP-hardness of the problem, we turn our attention to approximation algorithms. As we have pointed out before, algorithms for batch scheduling with unit processing times are completely determined by the bin-packing procedure used in the batching phase. Therefore, we name a algorithm after the bin-packing procedure being involved. We begin with FFD, the most commonly used algorithm for the classical bin-packing and its many variants due to its simplicity and good performance. Unfortunately, FFD has very poor performance for problems with objective of minimizing the total completion time.

Theorem 2. *For single machine batch scheduling problem of minimizing the total completion time with non-identical job sizes and unit processing times, the worst-case ratio of FFD is unbounded, even if the sizes of all jobs lie in $\left[0, \frac{1}{l}\right]$, where l is an arbitrary integer.*

Proof. For any integer $l > 0$, let $N > l + 1$ be also an integer. Consider a job set $\mathcal{J}$ that consists of N^2 *small* jobs of size $\frac{1}{N^2}$ and lN *large* jobs of size $\frac{N-1}{lN}$. Note that $0 < \frac{1}{N^2} < \frac{N-1}{lN} < \frac{1}{l}$ and $(l+1)\frac{N-1}{lN} = \frac{lN+N-l-1}{lN} > 1$. Clearly, FFD packs every l large jobs and N small jobs into a batch. Thus the schedule has totally N batches, with each batch has a cardinality of $N + l$. Hence,

$$TC^{FFD}(\mathcal{J}) = \sum_{i=1}^{N} i(N + l) = \frac{N(N+1)(N+l)}{2}.$$

On the other hand, FFI packs all small jobs into a single batch, and every l large jobs into a batch. The schedule has totally $N + 1$ batches, with one batch of cardinality of N^2 and N batches of cardinality of l. Hence,

$$TC^{FFI}(\mathcal{J}) = N^2 + \sum_{i=2}^{N+1} il = N^2 + \frac{lN(N+3)}{2}.$$

Therefore,

$$\frac{TC^{FFD}(\mathcal{J})}{TC^*(\mathcal{J})} \geq \frac{TC^{FFD}(\mathcal{J})}{TC^{FFI}(\mathcal{J})} = \frac{\frac{N(N+1)(N+l)}{2}}{N^2 + \frac{lN(N+3)}{2}} \to \infty (N \to \infty). \qquad \square$$

It is not difficult to see that the packing given by FFD in fact uses a minimum number of batches. Therefore, use better algorithms of the classical one-dimensional bin-packing problem is of no use. This is owing to the fundamental difference between two objectives. On the other hand, we will see below that FFI does have a better performance than FFD.

Denote by $\mathcal{B}^S$ the set of batches in schedule σ^S. For ease of narration, we add *dummy* batches of cardinality of 0 such that $\mathcal{B}^S$ contains exactly n batches. All dummy batches do not contain any jobs, and the objective will remain unchanged. For batch $B_i^S \in \mathcal{B}^S$, $i = 1,\ldots,n$, the cardinality of B_i^S will be denoted by $|B_i^S|$. Reindex the batches in non-increasing order of their cardinalities, i.e., $|B_1^S| \geq |B_2^S| \geq \cdots \geq |B_n^S|$. Write $x_i = |B_i^*|$ and $y_i = |B_i^{FFI}|$ for any $1 \leq i \leq n$. In the following lemma, we will prove that series $\{x_i\}_{i=1}^n$ and $\{y_i\}_{i=1}^n$ do satisfy conditions (C1)-(C4).

Lemma 3. *Series $\{x_i\}_{i=1}^n$ and $\{y_i\}_{i=1}^n$ satisfy conditions (C1)-(C4).*

Proof. Note that (C1) trivially holds. Let z_i, $i = 1,\ldots,q$ be the number of jobs packed in the first i batches of $\mathcal{B}^{FFI}$, i.e., $z_i = \sum_{l=1}^i y_l$. Set $z_0 = 0$. By the description of the algorithm, for any $1 \leq i \leq q$, B_i^{FFI} consists of y_i jobs of $J_{z_{i-1}+1}, J_{z_{i-1}+2}, \ldots, J_{z_i}$, and J_{z_i+1} can not be packed into B_i^{FFI} due to limit of space. Thus

$$\sum_{j=z_{i-1}+1}^{z_i} s_j \leq 1 < \sum_{j=z_{i-1}+1}^{z_i+1} s_j. \tag{16}$$

Since FFI packs the jobs in non-decreasing order of their sizes, we assume that $s_1 \leq s_2 \leq \ldots \leq s_n$. Hence, we have

$$\sum_{j=z_i+1}^{z_i+y_i+1} s_j \geq \sum_{j=z_{i-1}+1}^{z_{i-1}+y_i+1} s_j = \sum_{j=z_{i-1}+1}^{z_i+1} s_j > 1 \geq \sum_{j=z_i+1}^{z_i+1} s_j.$$

Hence, $y_{i+1} = z_{i+1} - z_i < (z_i + y_i + 1) - z_i = y_i + 1$, which is equivalent to $y_{i+1} \leq y_i$. (C2) is valid.

To prove (C3), it is sufficient to prove $\sum_{i=1}^l x_i < \sum_{i=1}^l y_i + l$ as both $\{x_i\}_{i=1}^n$ and $\{y_i\}_{i=1}^n$ are integer series. If $z_l + l > n$, then

$$\sum_{i=1}^l x_i \leq n < z_l + l = \sum_{i=1}^l y_i + l.$$

Otherwise, by (16) and $s_1 \leq s_2 \leq \ldots \leq s_n$,

$$\sum_{j=1}^{z_l+l} s_j = \sum_{j=1}^{z_l} s_j + \sum_{j=z_l+1}^{z_l+l} s_j \geq \sum_{j=1}^l \sum_{j=z_{i-1}+1}^{z_i} s_j + \sum_{i=1}^l s_{z_i+1} = \sum_{i=1}^l \sum_{j=z_{i-1}+1}^{z_i+1} s_j > l.$$

It indicates that the smallest $z_l + l$ jobs has a total size greater than l. Thus, these jobs can not be packed into l batches in any schedule. Therefore,

$$\sum_{i=1}^l x_i < z_l + l = \sum_{i=1}^l y_i + l,$$

which is the desired result.

If $q - r \leq 1$, $p \geq q - r$ trivially holds. If there are at least two batches of $\mathcal{B}^{FFI}$ of cardinality of 1, then any two jobs belong to these batches has a total size of greater than 1. Clearly, any two of them can not be packed into the same batch in any schedule. (C4) is also valid. □

Theorem 3. *For single machine batch scheduling problem of minimizing the total completion time with non-identical job sizes and unit processing times, the worst-case ratio of FFI is at most $\frac{2+\sqrt{2}}{2}$, and at least $\frac{109}{82}$.*

Proof. The upper bound of the worst-case ratio is a direct corollary of Lemmas 1 and 3. To prove the lower bound, consider a job set $\mathcal{J}$ given in Table 1, where K and L are integers, ϵ is a sufficient small positive number. The schedule σ^{FFI} and another feasible schedule σ^A are illustrated in Table 2. Thus,

$$TC^{FFI}(\mathcal{J}) = 4 + 4 + \sum_{i=3}^{K+2} 2i + \sum_{i=K+3}^{3K+5} i + \sum_{i=3K+6}^{3K+5+L} i$$

$$= 4 + 4 + \frac{2K(K+5)}{2} + \frac{(2K+3)(4K+8)}{2} + \frac{L(6K+11+L)}{2}$$

$$= 5K^2 + 19K + 20 + \frac{L(6K+11+L)}{2}.$$

and

$$TC^*(\mathcal{J}) \leq 3 + \sum_{i=2}^{3} 3i + \sum_{i=4}^{2K+3} 2i + \sum_{i=2K+4}^{2K+3+L} i$$

$$= 3 + (6+9) + \frac{4K(2K+7)}{2} + \frac{L(4K+7+L)}{2}$$

$$= 4K^2 + 14K + 18 + \frac{L(4K+7+L)}{2}.$$

Hence,

$$\frac{TC^{FFI}(\mathcal{J})}{TC^*(\mathcal{J})} \geq \frac{10K^2 + 38K + 40 + L(6K+11+L)}{8K^2 + 28K + 36 + L(4K+7+L)}.$$

The right term achieve its maximum $\frac{109}{82} \approx 1.329$ at $K = 5$ and $L = 8$. □

Table 1. Job set $\mathcal{J}$

Number of jobs	Size of jobs	Index of jobs
1	0.1	J_1
2	0.2	J_2, J_3
2	$0.3 - \epsilon$	J_4, J_5
1	$0.4 - \epsilon$	J_6
$2K$	$0.5 - \epsilon$	$J_7, \cdots, J_{2K+6}$
$2K + 3$	$0.5 + \epsilon$	$J_{2K+7}, \cdots, J_{4K+9}$
L	1	$J_{4K+10}, \cdots, J_{4K+9+L}$

Table 2. Schedules σ^{FFI} and σ^A

σ^{FFI}				σ^A			
Number of batches	Cardinality of a batch	Index of batches	Index of Jobs in batches	Number of batches	Cardinality of a batch	Index of batches	Index of Jobs in batches
1	4	B_1	J_1, J_2, J_3, J_4	1	3	B_1	J_1, J_6, J_{2K+7}
1	2	B_2	J_5, J_6	2	3	B_2	J_2, J_4, J_{2K+8}
K	2	B_3	J_7, J_8			B_3	$J_3, J_5.J_{2K+9}$
		...	...	$2K$	2	B_4	J_7, J_{2K+10}
		B_{K+2}	J_{2K+5}, J_{2K+6}			B_5	J_8, J_{2K+11}
$2K+3$	1	B_{K+3}	J_{2K+7}			...	...
		...	...			B_{2K+3}	J_{2K+6}, J_{4K+9}
		B_{3K+5}	J_{4K+9}	L	1	B_{2K+4}	J_{4K+10}
L	1	B_{3K+6}	J_{4K+10}			B_{2K+5}	J_{4K+11}
		...	...			...	...
		B_{3K+5+L}	J_{4K+9+L}			B_{2K+3+L}	J_{4K+9+L}

4 Parallel Machines

In this section, we generalize our discussion to parallel machines. We introduce a universal algorithm $RR - \mathcal{A}$, which uses a bin-packing algorithm $\mathcal{A}$ in the batching phase, and then schedules all the batches on m parallel identical machines in a round-robin way. More specifically, add dummy batches of cardinality of 0 such that there are exactly nm batches. For any $1 \leq t \leq m$ and $1 \leq l \leq n$, schedule $B^{\mathcal{A}}_{t+(l-1)m}$ as the lth batch on M_t. It is obvious that the schedule is well-defined. The worst-case ratio of $RR - \mathcal{A}$ can be derived based on the worst-case ratio of $\mathcal{A}$ as follows.

Theorem 4. *For the batch scheduling problem of minimizing the total completion time with non-identical job sizes and unit processing times, if $\mathcal{A}$ is an algorithm for the single machine case with worst-case ratio of at most $\alpha \leq 2$, then $RR - \mathcal{A}$ is an algorithm for the m parallel machines case with worst-case ratio of at most $\frac{m+1}{2m}\alpha + \frac{m-1}{m}$.*

By Theorems 3 and 4, we have the following corollary.

Corollary 1. *For batch scheduling problem on m parallel machines of minimizing the total completion time with non-identical job sizes and unit processing times, the worst-case ratio of $RR - FFI$ is at most $\frac{6+\sqrt{2}}{4} - \frac{2-\sqrt{2}}{4m}$.*

References

1. Anily, S., Bramel, J., Simchi-Levi, D.: Worst-case analysis of heuristics for the bin packing problem with general cost structures. Oper. Res. **42**(2), 287–298 (1994)
2. Balogh, J., Békési, J., Dósa, G., Sgall, J., van Stee, R.: The optimal absolute ratio for online bin packing. In: Proceedings of the 26th Annual ACM-SIAM Symposium on Discrete Algorithms, pp. 1425–1438 (2015)

3. Boyar, J., Dósa, G., Epstein, L.: On the absolute approximation ratio for First Fit and related results. Discrete Appl. Math. **160**, 1914–1923 (2012)
4. Brucker, P., Gladky, A., Hoogeveen, H., Kovalyov, M.Y., Potts, C.N., Tautenhahn, T., van de Velde, S.L.: Scheduling a batching machine. J. Sched. **1**, 31–54 (1998)
5. Cheng, B., Yang, S., Hu, X., Chen, B.: Minimizing makespan and total completion time for parallel batch processing machines with non-identical job sizes. Appl. Math. Model. **36**(7), 3161–3167 (2012)
6. Deng, X., Feng, H., Li, G., Liu, G.: A PTAS for minimizing total completion time of bounded batch scheduling. Int. J. Found. Comput. Sci. **13**, 817–827 (2002)
7. Dósa, G.: The tight bound of First Fit decreasing bin-packing algorithm is $FFD(L) \leq \frac{11}{9}OPT(L) + \frac{6}{9}$. In: Chen, B., Paterson, M., Zhang, G. (eds.) ESCAPE 2007. LNCS, vol. 4614, pp. 1–11. Springer, Heidelberg (2007). https://doi.org/10.1007/978-3-540-74450-4_1
8. Dósa, G., Sgall, J.: First Fit bin packing: a tight analysis. In: Proceedings of the 30th Symposium on Theoretical Aspects of Computer Science, pp. 538–549 (2013)
9. Dósa, G., Tan, Z.Y., Tuza, Z., Yan, Y., Lányi, C.S.: Improved bounds for batch scheduling with nonidentical job sizes. Naval Res. Logistics **61**, 351–358 (2014)
10. Epstein, L., Levin, A.: Bin packing with general cost structures. Math. Program. **132**(1), 355–391 (2012)
11. Garey, M.R., Johnson, D.S.: Computers and Intractability: A Guide to the Theory of NP-Completeness. Freeman, New York (1978)
12. Hochbaum, D.S., Landy, D.: Scheduling semiconductor burn-in operations to minimize total flowtime. Oper. Res. **45**(6), 874–8859 (1997)
13. Ikura, Y., Gimple, M.: Scheduling algorithms for a single batching processing machine. Oper. Res. Letters **5**, 61–65 (1986)
14. Johnson, D. S.: Near-optimal bin packing algorithms. Doctoral Thesis, MIT (1973)
15. Johnson, D.S., Demers, A., Ullman, J.D., Garey, M.R., Graham, R.L.: Worst-case performance bounds for simple one-dimensional packing algorithms. SIAM J. Comput. **3**, 299–325 (1974)
16. Li, S., Li, G., Qi, X.: Minimizing total weighted completion time on identical parallel batch machines. Int. J. Found. Comput. Sci. **17**(6), 1441–1453 (2006)
17. Li, S., Li, G., Zhang, S.: Minimizing makespan with release times on identical parallel batching machines. Discrete Appl. Math. **148**, 127–134 (2005)
18. Poon, C.K., Yu, W.: On minimizing total completion time in batch machine scheduling. Int. J. Found. Comput. Sci. **15**(4), 593–607 (2004)
19. Potts, C.N., Kovalyov, M.Y.: Scheduling with batching: a review. Eur. J. Oper. Res. **120**, 228–249 (2000)
20. Simchi-Levi, D.: New worst-case results for the bin-packing problem. Naval Res. Logistics **41**, 579–585 (1994)
21. Ullman, J.D.: The performance of a memory allocation algorithm. Technical report 100, Princeton University (1971)
22. Uzsoy, R.: A single batch processing machine with non-identical job sizes. Int. J. Prod. Res. **32**, 1615–1635 (1994)
23. Xia, B., Tan, Z.Y.: Tighter bounds of the First Fit algorithm for the bin-packing problem. Discrete Appl. Math. **158**, 1668–1675 (2010)

New Insights for Power Edge Set Problem

Benoit Darties[1], Annie Chateau[2]($^{\boxtimes}$), Rodolphe Giroudeau[2],
and Mathias Weller[2]

[1] Le2i FRE2005, CNRS, Arts et Métiers, Univ. Bourgogne Franche-Comté,
Dijon, France
benoit.darties@u-bourgogne.fr
[2] LIRMM - CNRS UMR 5506, Montpellier, France
{chateau,giroudeau,weller}@lirmm.fr

Abstract. We study the computational complexity of POWER EDGE
SET (PES), for restricted graph classes with degree bounded by three
(bipartite graph, Unit disk graphs and Grid graphs). This problem is
devoted to the monitoring of an electric network. The aim is to minimize
the number of edge-allocated PMUs in a network such that all vertices
are monitored according to two spreading rules. We improve known complexity results using an L-reduction. We also derive some lowers bounds
according to classic complexity hypothesis ($\mathcal{P} \neq \mathcal{NP}, \mathcal{UGC}, \mathcal{ETH}$).

1 Introduction

Motivation. In an high voltage electrical network, synchrophasors are time-
synchronized numbers that represent both the magnitude and phase angle of
the sine waves on network links. A Phasor Measurement Unit (PMU) is an
expensive measuring device used to continuously collect the voltage and phase
angle of a single electrical (sub-)station as well as the links and incident stations
connected to it. Using among others Ohm's and Kirchhoff's Laws, if one station
is monitored and all but one of its neighbors are too, then the unmonitored
station becomes monitored. The problem of minimizing the number of PMUs
to place on stations for complete network monitoring and known as POWER
DOMINATING SET [16] is an important challenge for operators and has gained a
considerable attention over the past decade. We consider in this work a recent
variant of the problem [15], called POWER EDGE SET (PES), placing PMUs on
the network links rather than the stations.

Model. We model the electrical network with an undirected graph $G = (V, E)$,
and will use the functional notation $V(G)$ and $E(G)$ to refer to respectively the
set of vertices (electrical stations) of a graph G, and its set of edges (direct links).
Let $n = |V|$ and $m = |E|$. We note $N_G(v)$ the set of neighbors of $v \in V$ in G and
$d_G(v) = |N_G(v)|$ its degree in G. Further, we let $N_G[v] := N_G(v) \cup \{v\}$ denote
the *closed* neighborhood of v in G. The problem POWER EDGE SET can be seen
as a problem of color propagation on G with colors 0 (white) and 1 (black),
respectively designating the states *not monitored* and *monitored* of a vertex.

© Springer International Publishing AG 2017
X. Gao et al. (Eds.): COCOA 2017, Part I, LNCS 10627, pp. 180–194, 2017.
https://doi.org/10.1007/978-3-319-71150-8_17

Let $c(v)$ be the color assigned to vertex v (abusing notation, we abbreviate $\bigcup_{v \in X} c(v) =: c(X)$). Before placing the PMUs, we have $c(V) = 0$. Given a set $E' \subseteq E$ of edges on which to place PMUs, colors propagate according to the following rules:

Rule R_1: if $(u,v) \in E'$, then $c(u) = c(v) = 1$
Rule R_2: for u, u' with $c(u) = 1$, $u' \in N_G(u)$ and $c(v) = 1$ for all $v \in N_G(u) \backslash \{u'\}$, then $c(u') = 1$ (u' is the only uncoloured neighbor of a colored vertex u, then u' is colored by *color propagation* from u).

The objective of POWER EDGE SET is to find a smallest set of edges $E' \subseteq E$ on which to place the PMUs such that $c(V) = \{1\}$ after exhaustive application of RULE R_1 and RULE R_2. We call such a set a *power edge set* of G (see Fig. 1 for a guided example of RULE R_1 and RULE R_2 on a simple graph, leading to an optimal solution with two PMUs).

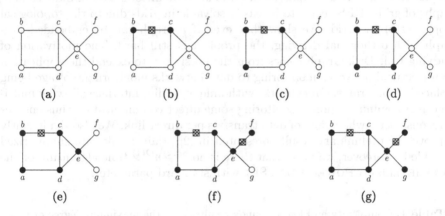

(a)   (b)   (c)   (d)

(e)   (f)   (g)

Fig. 1. PMU propagation: before any placement, all vertices are white (**a**). A PMU on $\{b,c\}$ induce $c(b) = c(c) = 1$ (black) by RULE R_1 (**b**). By applying RULE R_2 on b, we obtain $c(a) = 1$ (**c**). Then RULE R_2 on a induces $c(d) = 1$ (**d**), and RULE R_2 on c or d induces $c(e) = 1$ (**e**). A second PMU is required to complete the coloration, i.e. on $\{e,f\}$ to obtain $c(f) = 1$ by RULE R_1 (**f**). Finally, RULE R_2 on e induces $c(g) = 1$ (**g**).

Related Work. The Problem POWER DOMINATING SET is $\mathcal{NP}$-complete in general graphs [6]. A large literature is devoted to this problem, describing a large range of approaches, either exact such as integer linear programming [5] or branch-and-cut [13], or heuristic, such as greedy algorithm [8]. Transversely, many studies have been led on interesting classes of graphs. For instance, the problem is polynomial on Grids [3], but is $\mathcal{NP}$-complete in Unit Disk Graphs, and in Disk Graphs [14].

The problem of assigning a minimum number of PMUs on the links to monitor the whole network, POWER EDGE SET, is known to be $\mathcal{NP}$-hard in the general case. The authors in [15] propose the first complexity result and a lower bound of a value $1.12 - \epsilon$ with $\epsilon > 0$ based on an E-reduction from VERTEX COVER. They also propose a linear-time algorithm on trees by performing a polynomial reduction to PATH COVER. Moreover, the authors in [11] develop an exact method, a linear program with binary variables, indexed on the necessary iterations using propagation RULE R_1 and RULE R_2, extended to a linear program in mixed variables, with the goal of being efficient in practice.

Recently we proposed in [2] some preliminary complexity results according to variant topologies (planar, bipartite,...). An interesting open question stems from the assumption that power lines run in a bounded degree, at least in few planes or surfaces of low genus.

Our Contribution. In this work, we address this question, developing hardness results on (bipartite) planar graphs with bounded degree, covering both approximation and parameterized complexity. This negative result is extended to Grid Graphs and Unit Disk Graphs. Grid Graphs are defined as vertex-induced subgraphs of grids. PES seems to be easily solvable in grids due to the topological properties of the grid. We show that even if a graph can be embedded in a graph with orthogonal drawing, the problem is still hard. The motivation of studying Unit Disk Graph comes from the possibility to extend the application from electrical networks monitoring to data networks monitoring, voltage being replaced by bits rates. On networks with unique (radio) interface allowing multiple direct communications, monitoring some direct communication thus ensures to reconstruct a whole map of data transfers per direct link. We also significantly improve the preliminary results presented in [2]. Main results are summarized in Table 1. Moreover, we prove that there is no $2^{o(k)}n^{O(1)}$ time algorithm for the PARAMETERIZED POWER EDGE SET with standard parameter.

Table 1. Complexity and lower bounds results. Δ is the maximum degree of G.

Topology	Complexity	Lower bound ($\mathcal{P} \neq \mathcal{NP}, \mathcal{ETH}, \mathcal{UGC}$)
Bipartite planar, $\Delta \leq 3$	$\mathcal{NP}$-C (Theorem 1)	$\rho \geq 1 + \frac{(1-(2+O_\delta(1))\frac{\log\delta}{\log\log\delta}}{1+2\delta}$ under $\mathcal{UGC}$ (Theorem 3) $\nexists 2^{o(n^{\frac{1}{4}})}$ under $\mathcal{ETH}$ (Corollary 2)
Grid graph $\Delta \leq 3$	$\mathcal{NP}$-C (Theorem 2)	
Unit disk graph	$\mathcal{NP}$-C (Corollary 1)	

Organization of the Paper. In Sect. 2, we present some properties simplifying the presentation of the problem and the discussion on considered instances. Section 3 describes results on complexity issues, whereas Sect. 4 explores approximation questions for restricted classes of graphs, under classical hypotheses.

2   Preliminaries

First we present some observations and transformations concerning parts of optimal solutions to PES on a graph G.

We represent a total order $<$ of vertices of a graph G, by a sequence $(v_1, v_2, \ldots)$ such that v_i occurs before v_j in the sequence if and only if $v_i < v_j$.

Definition 1 (Valid total order $<$ in PES). *Given a total order $<$ of vertices of a graph G, $<$ is valid for any $S \subseteq E(G)$ if, for each $v \in V(G)$, there is an edge incident with v in S or there is some $u \in N_G(v)$ with $N_G[u] \leq v$.*

Since we consider PES as a color propagation problem, a valid order describes how vertices are consecutively colored under the propagation process of S in G.

Claim 1. *PES is stable under the operation of contracting an edge $\{x, y\}$ with $d_G(x) = d_G(y) = 2$ or with $d_G(x) = 2$ and $d_G(y) = 1$.*

Proof. Let $G = (V, E)$ an instance of PES.

Case 1: let x, y be two adjacent vertices of G with $d_G(x) = d_G(y) = 2$. Let x' and y' be respectively the unique other neighbour of x and y. Let $S \subseteq E$ is the set of edges corresponding to an optimal placement of PMUs on edges. If x and y are monitored because there is a PMU on $\{x, y\}$ and propagate the monitoring to x' and y', then moving this PMU on $\{x, x'\}$ does not affect the monitoring of x, x', y, y'. Thus we can suppose w.l.o.g. that $\{x, y\}$ is free from PMU. There are then two possibilities to propagate the monitoring: from x to y and then to y', or from y to x, and then to x'. In both cases, contracting $\{x, y\}$ does not affect the propagation from the contracted node to, respectively, y' or x'.

Case 2: let x, y be two adjacent vertices of G with $d_G(x) = 2$ and $d_G(y) = 1$. Let x' be other neighbor of x. If $\{x, y\} \in S$, one can move the PMU from $\{x, y\}$ to $\{x, x'\}$ and color y by application of RULE R_2 on x. As x and y have no other neighbor to color, one can contract x and y. $\qquad\square$

Claim 2. *PES is stable under the operation of splitting a vertex of degree two, i.e. the replacement of the vertex v with $N(v) = \{x, y\}$ by adjacent vertices v^1 and v^2 and the replacement of edges $\{x, v\}, \{y, v\}$ by edges $\{x, v^1\}, \{y, v^2\}$.*

Proof. By reciprocity of Claim 1. $\qquad\square$

Definition 2 (Ribbon). *Let G be a graph, let C be a cycle. We say that C is a ribbon iff all but exactly one vertex v of C have degree two in G and we call v the knot of C. Figure 2 presents an example of a ribbon.*

Observation 1. *Let G be a graph, let C be a ribbon with knot v and let e be an edge of C. Then, there is an optimal power edge set S for G with $S \cap E(C) = \{e\}$.*

Proof. Suppose that no PMU is placed on the edges of C. Then, even if $c(v) = 1$, none of the neighbors of v in C can become colored and, thus, v cannot propagate on any of them. If one PMU is placed on e, we obtain $c(V(C)) = \{1\}$ by consecutive propagation of vertices of degree two. $\qquad\square$

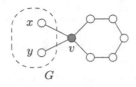

Fig. 2. A ribbon with knot v.

3   Complexity Results

We present here new complexity and lower bounds results for PES according to the degree of the graph. It is clear that graph with degree bounded by two, a trivial polynomial-time algorithm exists. It is sufficient to put one PMU on an edge by connected component. Thus, using RULE R_2 all vertices are colored.

On the other way, we show that PES remains $\mathcal{NP}$-complete even if G is a Grid Graph with bounded degree at most three (by definition planar and bipartite graph may be representable as Unit Disk Graph). From our actual knowledge, this is the most restricted class of graphs on which POWER EDGE SET is $\mathcal{NP}$-complete.

3.1   Hardness on Bipartite Planar Graphs of Degree Three

To prove these results, we use a reduction from 3-REGULAR PLANAR VERTEX COVER defined as follows:

3-REGULAR PLANAR VERTEX COVER (3-RPVC)
Input: a 3-regular planar graph $G = (V, E)$ and some $k \in \mathbb{N}$.
Question: Is there a size-k set $S \subseteq V$ covering E, i.e. $\forall_{e \in E} \ e \cap S \neq \varnothing$?

Construction 1. *First, we introduce the gadget graph H_v presented in Fig. 3.*
Let $G = (V, E)$ be 3-regular planar graph. Given a vertex $v \in V$ of degree three with $N_G(v) = \{x, y, z\}$, the gadget H_v is composed of

1. *An inner planar sub-graph (represented by the triangle (v_1, v_7, v_{12}) in Fig. 3), composed of five faces $(v_1, v_{31}, v_3, v_{13}, v_{14}, v_{10}, v_{29}, v_{12}), (v_3, v_4, v_{16}, v_{15}, v_{14}, v_{13}), (v_4, v_5, v_6, v_{18}, v_{17}, v_{16}), (v_6, v_7, v_8, v_{18}),$ and $(v_8, v_9, v_{10}, v_{14}, \ldots, v_{18}),$*
2. *An outer circle: vertices $(v_{19}, v_{20}, v_{11}, v_{22} \ldots, v_{28}),$*
3. *Three paths $(v_1, v_2, v_{11}), (v_7, v_{21}, v_{30}, v_{22})$ and (v_{12}, v_{32}, v_{26}) connecting the planar-subgraph to the outer circle at three points,*
4. *A set of three border-vertices v_i for each $i \in N_G(v)$, connected to the external circle with v_{19}, resp. with v_{25}, resp. with a path (v_{28}, v_{33}).*

Therefore, the gadget graph H_v contains 36 vertices and 43 edges.
From any 3-regular planar graph G, we construct a planar bipartite graph G' as follows:

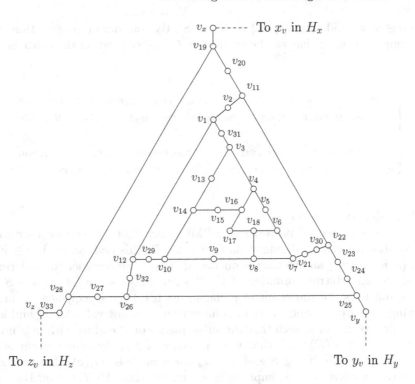

Fig. 3. Gadget H_v for a vertex v with neighbors x, y and z.

1. for each $v \in V(G)$, we add H_v to G',
2. for each $\{u,v\} \in E$, we add a connecting path $\{u_v, t_{uv}, v_u\}$ to G', linking the gadgets H_u and H_v

Remark 1. By construction H_v is a bipartite planar graph with v_x, v_y, v_z in the same sets: all the distances in H_v between all pairs of these vertices are even. Gadgets graphs are connected together by path of length two only, using the same planar scheme from G. As a result, clearly G' is bipartite, planar and with degree at most three.

Theorem 1. POWER EDGE SET *remains $\mathcal{NP}$-complete in planar bipartite graphs of degree at most three.*

Proof. Let $G = (V, E)$ be 3-regular planar graph, and G' the graph obtained by Construction 1 using G as input. We show that 3-RPVC has a solution of size k on G iff PES has a solution of size $n + k$ on G'.

For clarity, $\forall \{u, v\} \in E(G)$, we contract path (u_v, t_{uv}, v_u) into an edge (u_v, v_u) in G' using Claim 1. Any solution to PES on G' remains unchanged, as t_{uv} were introduced to guarantee only the bipartite property on G'.

$\boxed{\Rightarrow}$ Let S be a vertex cover of size k on G. We build a solution S' to PES on G' as follows: for each $v \in V(G)$, we add edge $\{v_{14}, v_{15}\}$ of H_v to S' and for each $v \in S$, we add edge $\{v_4, v_5\}$ of H_v to S'. By considering propagation rules, it is simple to verify that all the vertices of G' are colored, in the order defined below:

$$
<_v := \begin{cases}
v_4, v_5, v_{14}, v_{15}, v_6, v_{16}, v_{17}, v_{18}, v_3, v_7, v_8, v_9, v_{10}, v_{13}, v_{29}, v_{31}, v_1, v_{12}, v_2, v_{11}, \\
v_{21}, v_{30}, v_{22}, v_{32}, v_{26}, v_{20}, v_{19}, v_{23}, v_{24}, v_{25}, v_{27}, v_{28}, v_{33}, v_x, v_y, v_z \text{ if } v \in S \\
\\
v_{14}, v_{15}, v_x, v_y, v_z, v_{16}, v_{19}, v_{25}, v_{33}, v_{28}, v_{20}, v_{11}, v_{27}, v_{26}, v_{24}, v_{23}, v_{22}, v_{30}, \\
v_{21}, v_7, v_{32}, v_{12}, v_2, v_1, v_{31}, v_3, v_{29}, v_{10}, v_9, v_8, v_{18}, v_6, v_{17}, v_4, v_5, v_{13} \text{ if } v \notin S
\end{cases}
\tag{1}
$$

Note that $|S'| = n + k$.

$\Leftarrow$ Assume that G' possess $n + k$ PMUs such that all vertices are colored. We construct a valid ordering $<$ on G' for S'. To this end, for each $v \in V(G)$ with (x, y, z) being an arbitrary sequence of $N_G(v)$, we consider the ordering 1. Let $<^*$ be an arbitrary ordering of $V(G)$ such that $u <^* v$ for all $u \in S$ and $v \notin S$ and let $<$ be the result of replacing each v by the sequence $<_v$ in this ordering. Towards a contradiction, assume that $<$ is not valid for S' and let w be the first vertex of $<$ such that the subsequence of $<$ ending with w is invalid for S'. Let $v \in V(G)$ such that w is a vertex of H_v. By construction of $<_v$, this is only possible if $v \notin S$ and $w = v_x$ for some $x \in N_G(v)$. However, since S is a vertex cover, $x \in S$, implying $x <^* v$ and, thus, $V(H_x) < w$. But then, $N_{G'}[x_v] \le v_x$ for some $x_v \in V(H_x)$), contradicting that the subsequence of S' ending with w is invalid.

Claim 3. S' contains an edge incident with a vertex from $\{v_4-v_6, v_8, v_9, v_{13}-v_{18}\}$ for all $v \in V(G)$.

Proof. Let $C_v = \{v_4-v_6, v_8, v_9, v_{13}-v_{18}\}$. For each $v \in V(G)$, vertices set $V(S_v) = \{v_3, v_7, v_{10}\}$ is a vertex separator in G'. Obviously $G'[C_v]$ is a connected component if v_3, v_7 and v_{10} are removed from G'. All vertices from $V(S_v)$ have two pairwise distinct neighbors in C_v. Thus it is impossible to produce a valid order on G' unless S contains at least one edge incident to a node in C_v. $\square$

Claim 4. $S' \subseteq \bigcup_{v \in V(G)} E(H_v)$.

Proof. Towards a contradiction, assume that S' contains $\{v_x, x_v\}$ for some $\{x, v\} \in E(G)$. Then, we can swap $\{v_x, x_v\}$ and the edges in $S' \cap E(H_v)$ for $\{v_4, v_5\}$ and $\{v_{14}, v_{15}\}$ in S', allowing us to start $<$ with
$(v_4, v_5, v_{14}, v_{15}, v_6, v_{16}, v_{17}, v_{18}, v_3, v_7, v_8, v_9, v_{10}, v_{13}, v_{29}, v_{31}, v_1, v_{12}, v_2,$
$v_{11}, v_{21}, v_{30}, v_{22}, v_{32}, v_{26}, v_{20}, v_{19}, v_{23}, v_{24}, v_{25}, v_{27}, v_{28}, v_{33}, v_x, v_y, v_z, x_v)$
assuming $N_G(v) = \{x, y, z\}$. Thus v_x and x_v precede all $w \notin V(H_v)$ and the new order is still valid in G'. $\square$

Claim 5. Let $v \in V(G)$ with $|S' \cap E(H_v)| = 1$, let $x \in N_G(v)$ and let $w \in \{v_1 - v_{33}\}$ such that w is not incident with an edge of S'. Then, $v_x < w$.

Claim 6. *Let* $\{x, v\} \in E(G)$. *Then* $|S' \cap E(H_x)| > 1$ *or* $|S' \cap E(H_v)| > 1$.

Claim 6 implies that $\{v \mid |S' \cap E(H_v)| > 1\}$ is a vertex cover of G and, by Claim 3, its size is at most $|S'| - n = k$.

Thus, for any solution S' on PES on G', each gadget I_v contains either one or two PMUs. By noting S the set of all vertices v such that I_v contains two PMUs, we have proved that S is a vertex cover on G. □

3.2 Extension of Hardness Result to Grid Graphs of Degree Three and Unit Disk Graph

In this section, we extend the results of Theorem 1 to Grid Graphs with holes of maximum degree three using a slight modification of gadget H_v and known results on embedding a 3-regular planar graph in a grid.

Construction 2. *Let* $G = (V, E)$ *be a 3-regular planar graph. We introduce the gadget graph* I_v *presented in Fig. 4, and constructed from* H_v *(Fig. 3).*

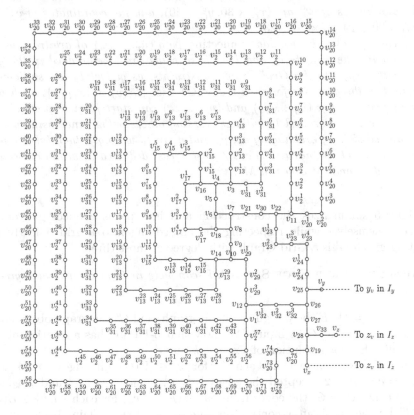

Fig. 4. Illustration of a Grid Graph using in Construction 2.

For a vertex v of degree two with $N(v) = \{x, y\}$, let $str(v, i)$ be the operation consisting to stretch v i^{th} times, i.e. replacing v with a path $(v^1, v^2, \ldots, v^i)$ and the edges $\{v, x\}, \{v, y\}$ by the edges $\{v^1, x\}, \{v^i, y\}$. Using the reciprocal of Claim 1, this operation has no incidence on any solution to PES.

Given a vertex $v \in V(G)$ with $N_G(v) = \{x, y, z\}$ and the gadget H_v presented in Construction 1, we copy H_v into I_v we apply $str(v_1, 13)$, $str(v_2, 43)$, $str(v_{11}, 3)$, $str(v_{15}, 15)$, $str(v_{17}, 29)$, $str(v_{20}, 75)$, $str(v_{23}, 4)$, $str(v_{24}, 2)$, $str(v_{29}, 57)$, $str(v_{32}, 3)$ on I_v.

Figure 4 presents an embedding of I_v into a 2-dimensional grid with dimensions 20×24. By construction I_v is a Grid Graph of degree at most three. We use this embedding in the following.

From any 3-regular planar graph G, we construct a Grid Graph G' with maximum degree three as follows:

1. *First we perform an embedding of G in a grid M_1 with orthogonal drawing, i.e. using [12]. This embedding always exists for 3-regular planar graphs. Let (x_G^v, y_G^v) be the coordinates of $v \in V(G)$ in the grid.*
2. *For each $v \in V(G)$, we add a gadget I_v, and embed it in a grid M_2 so that the coordinates of v_{20}^{56} are $(x_G^v \times 30, y_G^v \times 40)$, and the coordinates of v_{20}^{33} are $(x_G^v \times 30 + 19, y_G^v \times 40 + 23)$.*
3. *For each $\{u, v\} \in E$, we add connecting path $(u_v, \ldots, v_u)$ of arbitrary length $l_{u,v}$ linking the gadgets I_u and I_v. These paths can be embedded in the grid G_2 with respect to the Grid Graph constraints, as gadget I_v are sufficiently distant in the grid G_2: for each $v \in V(G)$ with $N_G(v) = \{x, y, z\}$, each path among $(v_x, \ldots, x_v)$, $(v_y, \ldots, y_v)$ and $(v_z, \ldots, z_v)$ can start from respectively v_x, v_y, and v_z, turn around I_v without crossing others paths (uncrossing path is always possible by exchanging v_x, v_y and v_z in I_v w.l.o.g.), then takes one of the four directions north / south / west / east deducted from the embedding of v and it incident edges in M_1.*

Figure 5 summarizes the construction and how paths connect gadgets I_v between themselves while respecting the embedding into a grid. Clearly the resulting graph G' is a Grid Graph with degree at most three.

Theorem 2. POWER EDGE SET *is $\mathcal{NP}$-complete in Grid Graphs of degree at most three.*

Proof. Let $G = (V, E)$ be 3-regular planar graph, and G' the graph obtained by Construction 2 using G as input. We show that 3-RPVC has a solution of size k on G iff PES has a solution of size $n + k$ on G'.

From G', we contract in each I_v for $v \in V(G)$ all edges $\{x, y\}$ with $d_{G'}(x) = d_{G'}(y) = 2$ excepted the edges $\{v_{23}^4, v_{24}^1\}$ and $\{v_{30}, v_{21}\}$ and $\{v_{33}, v_z\}$. We also contract in G' the edges from paths $(u_v, \ldots, v_u)$ for each $\{u, v\} \in E(G)$. Let G'' be the graph after contracting operations. Using Claim 1, clearly any optimal solution S' on G' leads to an optimal solution on G'' with same cost, and reciprocally.

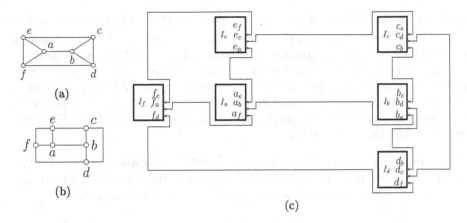

Fig. 5. Construction of a Grid Graph G' from a 3-regular planar graph G. From G **(a)** we produce an embedding into a grid **(b)**. Using one gadget I_v per vertex $v \in V(G)$, we produce a Grid Graph G' and connect the gadgets with paths turning around a gadget to get a direction to the other gadget. **(c)** presents a macroscopic view of G' and the paths linking these gadgets.

Finally, let us note that G'' is identical to the graph constructed in Construction 1 and used in the proof of Theorem 1, and there is an immediate extension of the result.   □

As Grid Graphs are also a Unit Disk Graph, the following corollary is immediate:

Corollary 1. POWER EDGE SET *remains* $\mathcal{NP}$*-complete for Unit Disk Graphs.*

4 Lower Bounds

4.1 Lower Bounds for Exact and $\mathcal{FPT}$ Algorithms

In this section, we propose some negative results for POWER EDGE SET concerning the existence of subexponential-time algorithms under $\mathcal{ETH}$ [7], and $\mathcal{FPT}$ Algorithms. We introduce PARAMETERIZED POWER EDGE SET, the parameterized version of POWER EDGE SET:

> PARAMETERIZED POWER EDGE SET (PPES)
> **Input:** Given $G = (V, E)$, a graph and $k \in \mathbb{N}^*$
> **Question:** Does it exist a set $E' \subseteq E$ defining a function $c : V \to \mathbb{N}$, with
> $c(u) = c(v) = 1 \; \forall \{u, v\} \in E'$ (by RULE R_2) of size k such that after
> exhaustive use of RULE R_2, we have $c(u) = 1 \; \forall u \in V$?
> **Parameter:** k

Since the polynomial-time transformation given in the proof of Theorem 1 (by making a slight modification of the gadget in the absence of regularity) is quadratic in the number of vertices, and since PLANAR VERTEX COVER does not admit a $2^{o(\sqrt{n})}$ algorithm there is no hope to find a $2^{o(n^{\frac{1}{4}})}$ time algorithm for POWER EDGE SET in presence of bipartite planar graph bounded by three.

Moreover, since $k \leq n$, a $2^{o(k)}n^c$ time algorithm directly implies a $2^{o(n)}$ time algorithm for PLANAR VERTEX COVER. However, we know VERTEX COVER is as hard to approximate in regular graphs [4], and does not admit an algorithm with running time $2^{o(n)}$ even in planar case unless $\mathcal{ETH}$ fails ([9]). Therefore, we obtain the following results.

Corollary 2. *Assuming $\mathcal{ETH}$, there is no $2^{o(n^{\frac{1}{4}})}$ time algorithm for* POWER EDGE SET, *and there is no $2^{o(k)}n^{O(1)}$ time algorithm for* PARAMETERIZED POWER EDGE SET *(even in bipartite planar of bounded at most three).*

4.2 Non-approximability Results According to Complexity Hypothesis

In this section, we derive new lower bounds based on the *L-reduction*.

First recall the definition of *L-reduction* between two difficult problems Π and Π', described by [10]. This reduction consists of polynomial-time computable functions f and g such that, for each instance x of Π, $f(x)$ is an instance of Π' and for each feasible solution y' for $f(x)$, $g(y')$ is a feasible solution for x. Moreover there are constants $\alpha_1, \alpha_2 > 0$ such that:

1. $OPT_{\Pi'}(f(x)) \leq \alpha_1 OPT_{\Pi}(x)$ and
2. $|val_{\Pi}(g(y')) - OPT_{\Pi}(x)| \leq \alpha_2 |val_{\Pi'}(y') - OPT_{\Pi'}(f(x))|$.

Construction 3. *Let $G = (V, E)$ be a graph of maximum degree δ. From G we construct an instance G' of PES as follows:*

First we introduce the gadget graph J_x for every vertex $x \in V(G)$:

let $(y_1, y_2, \ldots, y_{d_x})$ be an arbitrary sequence of $N_G(x)$, with $d_x = |N_G(x)|$.

- *We add a chain $(x_1, x_2, x_{y_1}^1, x_{y_1}^2, x_{y_1}^3, x_{y_1}^4, x_{y_1}^5, x_{y_2}^1, x_{y_2}^2, x_{y_2}^3, x_{y_2}^4, x_{y_2}^5, \ldots, x_{y_{d_x}-1}^1, x_{y_{d_x}-1}^2, x_{y_{d_x}-1}^3, x_{y_{d_x}-1}^4, x_{y_{d_x}-1}^5, x_{y_{d_x}}^1)$ to J_x.*
- *For every $y_i \in N_G(x)$ we add a ribbon with knot node $x_{y_i}^r$, three nodes $x_{y_i}^6$, $x_{y_i}^7$, $x_{y_i}^\ell$, and the edges $(x_{y_i}^7, x_{y_i}^\ell), (x_{y_i}^7, x_{y_i}^r), (x_{y_i}^6, x_{y_i}^1)$ and $(x_{y_i}^r, x_{y_i}^6)$.*
- *For every y_i in the sequence excepted the last element y_{d_x}, we add the edge $(x_{y_i}^4, x_{y_i}^7)$.*

Figure 6 presents a construction of J_x for a 9-neighbors node x and the neighbor sequence $(y_1, y_2, \ldots, y_9)$.

Note that J_v is a bipartite graph with maximum degree 3.

We add all the gadgets $J_v | v \in V(G)$ into G', and we link two gadgets J_x and J_y with an edge (x_y^ℓ, y_x^ℓ) if and only if $\{x, y\} \in E(G)$.

Using Claims 1 and 2, we can stretch or contract edges (x_y^ℓ, y_x^ℓ) to ensure that G' remains bipartite for clarity in the rest of the paper we restrict the construction to G' without stretching or contracting edges, and consider that G' is implicitly bipartite.

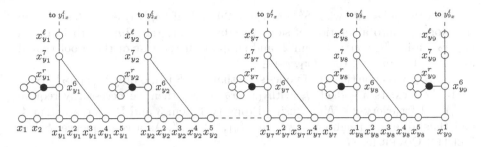

Fig. 6. An example of J_x-gadget for a node x with $d_G(x) = 9$ and a sequence of neighbors $(y_1, y_2, \ldots, y_9)$. Here blacks nodes represent knots nodes from ribbons.

Property 1. The gadget J_x has the following properties:

1. By construction, we have $2m$ ribbons in G'.
2. If we place one PMU on the edge (x_1, x_2) and a PMU per ribbon, then the exhaustive application of RULE R_2 colors all the nodes of J_x.
3. If we place a PMU per ribbon, and have $c(x_y^\ell) = 1$ for every $y \in N(x)$ (i.e. after application of RULE R_2 on each node y_x^ℓ from each J_y), then the exhaustive application of RULE R_2 colors all the nodes of J_x.
4. However the coloring remains incomplete if there exists at least one neighbor y_i with $c(x_{y_i}^\ell) = 0$.

Theorem 3. *For every sufficient large integer δ, POWER EDGE SET is inapproximable to less than $1 + \frac{(1 - (2 + O_\delta(1)) \frac{\log \delta}{\log \log \delta}}{1 + 2\delta}$ under $\mathcal{UGC}$, even for bipartite graphs of maximum degree 3.*

Proof. We use a reduction to VERTEX COVER on simple graph of bounded degree δ. Under $\mathcal{UGC}$, this problem cannot be approximated to less than $(2 - (2 + O_\delta(1)) \frac{\log \log \delta}{\log \delta})$ [1].

Given an instance $I = (G, k)$ of VERTEX COVER such that each vertex of G as degree at most δ, an instance $I' = (G', k + 2m)$ of PES is constructed: let G' be the bipartite graph returned by Construction 3 with G as input.

$\boxed{\Rightarrow}$ Let $S \subseteq V$ be an optimal k-size vertex cover on G. Then one can construct a $(k + 2m)$ solution to PES by placing one PMU on each of the $2m$ ribbons, and a PMU on the edge (x_1, x_2) if and only if $x \in S$. Using Property 1 item 2, all the nodes from $J_x | x \in S$ are colored. Applying RULE R_2 on every $x_y^\ell, \forall x \in S, \forall y \in N(x)$ gives $c(y_x^\ell) = 1$. As S is a solution to vertex cover, then we have $c(y_x^\ell) = 1 \forall (x, y) \in E$. Using Property 1 item 2, all the nodes are colored.

$\boxed{\Leftarrow}$ Let S' be an optimal $(k + 2m)$-size solution to I'. By construction, exactly $2m$ PMUs require to be set on each of the $2m$ ribbons, using Observation Construction 1. If two or more other PMUs are on the same gadget J_x, then the solution cannot be optimal, as a consequence of Property 1 item 2, and we have $S' \cap E(J_x) = 2m$ or $2m + 1$.

In every gadget J_x with $S' \cap E(J_x) = 2m + 1$, if $(x_1, x_2) \notin S'$, then we can transform S' into a solution of same cost by moving the $(2m + 1)$th PMU on (x_1, x_2), using Property 1 item 2, and suppose in the rest of the proof that if $S' \cap E(J_x) = 2m + 1$, then $(x_1, x_2) \in S'$.

Let $S \subseteq V(G)$ be the set of nodes such that $v \in S$ if and only if $\{v_1, v_2\} \cap S' \neq \varnothing$. If there exists a gadget J_x with $(x_1, x_2) \notin S'$, then we necessarily require $c(x_y^\ell) = 1$ for every $y \in N(x)$, using Property 1 item 3 and Property 1 item 4, and thus for every neighbor y we have $(y_1, y_2) \in S'$. As a consequence, S is a VERTEX COVER for G.

To show that the above constitutes an L-reduction, let f be a function transforming any instance I of VERTEX COVER into an instance I' of PES as above, let S' be any feasible solution for I', and let g be the function that transforms S' into a solution S'' that contains exactly $d_G(v)$ or $d_G(v) + 1$ edges per gadget J_v, and then outputs the set of vertices v for which S'' assigns $d_G(v) + 1$ PMUs to J_v. First, the above argument shows that $g(S')$ is a feasible solution for VERTEX COVER. Second, by construction,

$$OPT(I') = OPT(I) + 2m \tag{2}$$

While S is a solution to VERTEX COVER in I, we have $m \leq \sum_{v \in S} d_G(v)$, and $d_G(v) \leq \delta | v \in S$. We obtain $m \leq \sum_{v \in S} \delta$ and $2m \leq 2\delta.OPT(I)$ as $|S| = OPT(I)$.

We then obtain $OPT(I') \leq (1 + 2\delta) \cdot OPT(I)$. Third, by construction of g, we have

$$val(g(S')) \leq val(S') - 2m \leq val(S') - OPT(I') + OPT(I) \tag{3}$$

Thus, we constructed an L-reduction with $\alpha_1 = 1 + 2\delta$, $\alpha_2 = 1$.

$$val(g(S')) \leq val(S') - 2m \leq val(S') - OPT(I') + OPT(I) \tag{4}$$

Assuming $\mathcal{UGC}$, VERTEX COVER is hard to approximate to a factor of $(2 - (2 + O_\delta(1))\frac{\log \log \delta}{\log \delta})$ [1] and, thus:

$$
\begin{aligned}
val(S') &\geq val(g(S')) + OPT(I') - OPT(I) \\
&\geq (2 - (2 + O_\delta(1))\frac{\log \log \delta}{\log \delta}) \cdot OPT(I) + OPT(I') - OPT(I) \\
&\geq (1 - (2 + O_\delta(1))\frac{\log \log \delta}{\log \delta}) \cdot \frac{OPT(I')}{(1 + 2\delta)} + OPT(I') \\
&\geq (1 + \frac{(1 - (2 + O_\delta(1))\frac{\log \log \delta}{\log \delta})}{(1 + 2\delta)})OPT(I'),
\end{aligned}
$$

then we obtain the desired result. $\square$

5   Conclusion and Perspectives

In this article, we focus on determinating the demarcation line between polynomial and hardness cases according to the degree of the considered graph. We presented several new complexity results and some lowers bounds according to classical complexity hypothesis ($P \neq NP$, ETH, UGC) which improved existing results. In particular, we exhibit the most constrained classes of graphs known so far on which POWER EDGE SET is NP-hard. We consider the following perspectives to this work. First, it would be interesting to explore some particular class of graphs to understand in what extent the regularity of the graph affects the complexity of the problem. Actually, we focus our work on the study of two sub-classes of planar graphs, that is to say cactus graphs and outplanar graphs. From preliminary results, we conjecture that the problem may be polynomial on cactus graphs. On the other side, the problem is still open on outplanar graphs, but may be 2-approximable.

Another interesting question concerns the development of some efficient FPT algorithms with standard parameter, treewidth, structural parameter, or the number of steps of RULE R_2. We also plan to study how special patterns and minors, may influence the complexity of the problem, as well as its FPT tractability and approximability. Since approximability is complicated in the general case, it would also be interesting to design exact methods, like branch-and-bound exploration or meta-heuristics, hopefully sufficiently efficient on real networks.

References

1. Austrin, P., Khot, S., Safra, M.: Inapproximability of vertex cover and independent set in bounded degree graphs. In: Proceedings of the Annual IEEE Conference on Computational Complexity, pp. 74–80 (2009)
2. Darties, B., Chateau, A., Giroudeau, R., Weller, M.: Improved complexity for power edge set problem. In: 28th International Workshop on Combinatorial Algorithms (2017)
3. Dorfling, M., Henning, M.: A note on power domination in grid graphs. Discr. Appl. Math. **154**(6), 1023–1027 (2006)
4. Feige, U.: Vertex cover is hardest to approximate on regular graphs. Technical report MCS03-15, the Weizmann Institute (2003)
5. Gou, B.: Generalized integer linear programming formulation for optimal PMU placement. IEEE Trans. Power Syst. **23**(3), 1099–1104 (2008)
6. Haynes, T.W., Hedetniemi, S.M., Hedetniemi, S.T., Henning, M.A.: Domination in graphs applied to electric power networks. SIAM J. Discr. Math. **15**(4), 519–529 (2002)
7. Impagliazzo, R., Paturi, R.: On the complexity of k-SAT. J. Comput. Syst. Sci. **62**(2), 367–375 (2001)
8. Li, Y., Thai, M.T., Wang, F., Yi, ℭ., Wan, P., Du, D.: On greedy construction of connected dominating sets in wireless networks. Wirel. Commun. Mobile Comput. **5**(8), 927–932 (2005)

9. Lokshtanov, D., Marx, D., Saurabh, S.: Lower bounds based on the exponential time hypothesis. Bull. EATCS **105**, 41–72 (2011)

10. Papadimitriou, C., Yannakakis, M.: Optimization, approximation, and complexity classes. J. Comput. Syst. Sci. **43**(3), 425–440 (1991)

11. Poirion, P., Toubaline, S., D'Ambrosio, C., Liberti, L.: The power edge set problem. Networks **68**(2), 104–120 (2016)

12. Radwan, A.: A new algorithm for orthogonal drawings of 3-planar graphs. Int. J. Appl. Math. **6**(3), 301–317 (2001)

13. Simonetti, L., Salles da Cunha, A., Lucena, A.: The minimum connected dominating set problem: formulation, valid inequalities and a branch-and-cut algorithm. In: Pahl, J., Reiners, T., Voß, S. (eds.) INOC 2011. LNCS, vol. 6701, pp. 162–169. Springer, Heidelberg (2011). https://doi.org/10.1007/978-3-642-21527-8_21

14. Thai, M.T., Du, D.Z.: Connected dominating sets in disk graphs with bidirectional links. IEEE Commun. Lett. **10**(3), 138–140 (2006)

15. Toubaline, S., Poirion, P.-L., D'Ambrosio, C., Liberti, L.: Observing the state of a smart grid using bilevel programming. In: Lu, Z., Kim, D., Wu, W., Li, W., Du, D.-Z. (eds.) COCOA 2015. LNCS, vol. 9486, pp. 364–376. Springer, Cham (2015). https://doi.org/10.1007/978-3-319-26626-8_27

16. Yuill, W., Edwards, A., Chowdhury, S., Chowdhury, S.P.: Optimal PMU placement: a comprehensive literature review. In: 2011 IEEE Power and Energy Society General Meeting, pp. 1–8, July 2011

Extended Spanning Star Forest Problems

Kaveh Khoshkhah[1], Mehdi Khosravian Ghadikolaei[2], Jérôme Monnot[2(✉)],
and Dirk Oliver Theis[1]

[1] Institute of Computer Science, University of Tartu, Tartu, Estonia
{kavehkho,dotheis}@ut.ee
[2] Université Paris-Dauphine, PSL Research University, CNRS UMR [7243]
LAMSADE, 75016 Paris, France
m.khosravian@gmail.com, jerome.monnot@dauphine.fr

Abstract. We continue the investigation proposed in [COCOA 2016, Weller, Chateau, Giroudeau, König and Pollet "On Residual Approximation in Solution Extension Problems"] about the study of extended problems. In this context, a partial feasible solution is given in advance and the goal is to extend this partial solution. In this paper, we focus on the edge-weighted spanning star forest problem for both minimization and maximization versions. The goal here is to find a vertex partition of an edge-weighted complete graph into disjoint non-trivial stars and the value of a solution is given by the sum of the edge-weights of the stars. We propose **NP**-hardness, parameterized complexity, positive and negative approximation results.

1 Introduction

In [2], a *diversity problem* with application in the automobile industry is introduced. Here, each node corresponds to a cable configuration u (cable with a set of active option connections) and the cost $w_{u,v}$ between nodes u and v means that the cable configuration v will be supplied by the cable configuration u. In this application, a decision maker wishes to produce a set of cable configurations of minimum global cost under the constraint, that only configurations in the form of node-disjoint directed trees of depth at most 1 are feasible. In other words, the feasible solutions of this problem are directed spanning star forests; also called spanning star branchings. Several generalizations of this problem known as *Carpooling* problems have been studied in the literature (see the survey on ride-sharing [1]). One of them, called *Maximum Carpool Matching* models an automatic service to match commuting trips between passengers and drivers [11,13]. In this case, new constraints are added: each star has an upper bound on its size. These bounds may be different for each potential driver and they represent the capacity of the car according to the number of passengers each user

K. Khoshkhah, D.O. Theis supported by the Estonian Research Council (PUT Exploratory Grant #620); M. Khosravian Ghadikolaei supported by a *DORA Plus* scholarship of the Archimedes Foundation (funded by the European Regional Development Fund).

X. Gao et al. (Eds.): COCOA 2017, Part I, LNCS 10627, pp. 195–209, 2017.
https://doi.org/10.1007/978-3-319-71150-8_18

can drive if she was selected as a driver. Sometimes it is convenient to enforce some option connections in scheduling plans for the diversity problem or pre-select some drivers with passengers together in the carpool Matching problem (enforcing them may be desirable for reasons outside of the scope of the model itself). This corresponds to requiring some arcs to belong to any feasible collection of spanning star branchings. Motivated by scheduling or control networking applications [7,9,17,18], this type of problem, which consists of extending partial solutions, has recently drawn attention of the research community.

The undirected maximization version of the *spanning star forest* problem also has several motivations in bioinformatics [6,12,14]. Subproblems involving extending partial solutions can be used as a subroutine to design approximation algorithms for the master problem. In particular, He and Liang in [12] define and solve the *complementary partial dominating set* problem.

1.1   Graph Terminology and Definitions

Throughout this paper, we consider edge-weighed undirected graphs $G = (V, E)$ on $n = |V|$ vertices and $m = |E|$ edges without isolated vertices. Each edge $e = uv \in E$ between vertices u and v is weighted by a non-negative weight $w(e) \geq 0$; K_n denotes the *complete graph* on n vertices; a *split graph* $G = (L \cup R, E)$ is an undirected graph where the vertex set $L \cup R$ is decomposable into a clique L and a independent set R. The *degree* $d_G(v)$ of vertex $v \in V$ in G is the number of edges incident to v. A *star* $S \subset E$ of a graph $G = (V, E)$ is a tree of G where at most one vertex has a degree greater than 1, or, equivalently, it is isomorphic to $K_{1,\ell}$ for some $\ell \geq 0$. The vertices of degree 1 (except the center when $\ell \leq 1$) are called *leafs* of the star while the remaining vertex is called *center* of the star. A ℓ-star is a star of ℓ leafs; when $\ell = 0$, the star is called *trivial* and it is reduced to a single vertex (the center). A *spanning star forest* $\mathcal{S} = \{S_1, \ldots, S_r\} \subseteq E$ of G is a spanning forest into stars, that is, each S_i is a non trivial star, $V(S_i) \cap V(S_j) = \emptyset$ and $\cup_{i=1}^p V(S_i) = V$. Hence in this paper, a spanning star forest of a graph G is a collection of node disjoint non trivial stars (without isolated vertices, i.e., $K_{1,0}$) that covers all vertices of G. A *matching* $M \subseteq E$ is a subset of pairwise non-adjacent edges. A matching M of G is *perfect* if all vertices of G are covered by M. A *claw* is a $K_{1,3}$. In this paper, we will consider the following two problems:

SPANNING STAR FOREST PROBLEM
Input: A weighted complete graph (K_n, w) on n vertices where $K_n = (V(K_n), E(K_n))$ and $w(e) \geq 0$ for $e \in E(K_n)$.
Solution: Spanning star forest $\mathcal{S} = \{S_1, \ldots, S_p\} \subseteq E$.
Output: Optimizing $w(\mathcal{S}) = \sum_{e \in \mathcal{S}} w(e) = \sum_{i=1}^p \sum_{e \in S_i} w(e)$.

The extended version of the SPANNING STAR FOREST PROBLEM, called EXTENDED SPANNING STAR FOREST PROBLEM consists of extending a given packing of stars into a spanning star forest. Formally, we have:

> **EXTENDED SPANNING STAR FOREST PROBLEM**
> **Input:** A weighted complete graph (K_n, w) on n vertices and a packing of stars $\mathcal{U} = \{U_1, \ldots, U_r\}$ where $K_n = (V(K_n), E(K_n))$ and $w(e) \geq 0$ for $e \in E(K_n)$.
> **Solution:** Spanning star forest $\mathcal{S} = \{S_1, \ldots, S_p\} \subseteq E$ containing $\mathcal{U}$.
> **Output:** Optimizing $w(\mathcal{S}) = \sum_{e \in \mathcal{S}} w(e) = \sum_{i=1}^{p} \sum_{e \in S_i} w(e)$.

The first problem is the restriction of the second to $\mathcal{U} = \emptyset$. Given an instance $I = (K_n, w)$ of SPANNING STAR FOREST PROBLEM (resp., $I = (K_n, w, \mathcal{U})$ of EXTENDED SPANNING STAR FOREST PROBLEM), $\mathcal{S}^* = \{S_1^*, \ldots, S_{\ell^*}^*\}$ will be an optimal solution with stars S_i^* and value $w(\mathcal{S}^*) = opt_{SSF}(I)$. The $\{0, 1\}$-SPANNING STAR FOREST PROBLEM is the restriction to binary weights $w(e) \in \{0, 1\}$. For each of these problems, two possible goals will be considered in this paper: maximization and minimization. Hence, the MAX SPANNING STAR FOREST PROBLEM consists of finding a spanning star forest $\mathcal{S} = \{S_1, \ldots, S_r\} \subseteq E$ maximizing its weight. For instance, in MAX $\{0, 1\}$-SPANNING STAR FOREST PROBLEM the size of a spanning star forest is the number of leaves in all its components. The goal in this case is to find the maximum-size spanning star forest of a given graph (induced by the edges of unit weight). From now, we assume the forced set $\mathcal{U} = M_{\mathcal{U}} \cup S_{\mathcal{U}}$ is decomposed into a matching $M_{\mathcal{U}} = \{p_i q_i : i = 1, \ldots, k'\}$ of k' edges and a set $S_{\mathcal{U}} = \{F_1, \ldots, F_k\}$ of k vertex-disjoint stars with at least two leafs (c_i will be the center of the ith star, and $C = \{c_1, \ldots, c_k\}$ is the set of centers). An illustration of these definitions is depicted in Fig. 1.

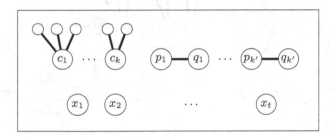

Fig. 1. Bold Edges corresponds to forced edges of $\mathcal{U}$. Sets $S_{\mathcal{U}}$ and $M_{\mathcal{U}}$ are indicated to the left and to the right of the figure respectively.

H-Extended Procedure. In several parts of this article, we will consider the weighted graph $I' = (H, w_H)$ built from an instance $I = (K_n, w, \mathcal{U})$ of the MIN EXTENDED SPANNING STAR FOREST PROBLEM where $\mathcal{U}$ is a packing of stars. $H = (V_H, E_H)$ is a complete weighted split graph defined as follows:

- $V_H = (V(K_n) \setminus V(\mathcal{U})) \cup (R \cup C)$ where $R = \{r_1, \ldots, r_{k'}\}$ is a new set (see Fig. 1 for definition of C, k and k');
- E_H is the set of edges of a complete split graph where the left side is a complete graph on $V(K_n) \setminus V(\mathcal{U})$, the right side is an independent set on $R \cup C$ and we have a complete bipartite graph between them;

$$\bullet \ w_H(uv) = \begin{cases} w(uv) & \text{if } u, v \notin R \cup C, \ u \neq v \\ w(uv) & \text{if } u \notin R \cup C, \ v \in C \\ \min\{w(up_i); w(uq_i)\} & \text{if } u \notin R \cup C, \ v = r_i. \end{cases}$$

Figure 2 gives an illustration of the construction. The *H-extended procedure* transforms any subset $F \subseteq E(H)$ into a subset $F' \subseteq E(K_n)$ by adding $\mathcal{U}$ and replacing any edge $xr_i \in F$ by edge xp_i if $w(xr_i) = w(xp_i)$, else by xq_i. Obviously, these two constructions (H and H-extension procedure) are done in polynomial-time. Figure 3 propose an example of the H-extended procedure.

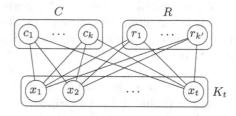

Fig. 2. Illustration of the construction of the split graph H.

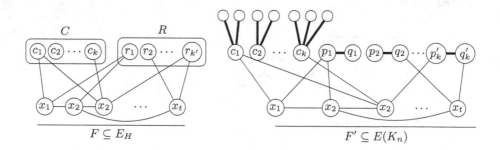

Fig. 3. H-extended procedure. Bold edges are in $\mathcal{U}$.

Cost Function Variants. In this paper, we will consider variants of the problem according to the cost function w: One variant assumes that w is any non-negative integer weight function; another that w satisfies the *c-relaxed triangle inequality*. Mainly consider that the c-relaxed triangle inequality might be satisfied outside the subgraph induced by $V(\mathcal{U})$, i.e., inside $V \setminus V(\mathcal{U})$ because the structure of feasible solutions are strongly constrained by subset $\mathcal{U}$.

Definition 1 (c-relaxed triangle inequality). *For a fixed $c > 1/2$, a weight function w on K_n satisfies the c-relaxed triangle inequality, if:*

$$\forall x, y, z \in V(K_n), \quad w(x,y) \leq c\left(w(x,z) + w(z,y)\right) \tag{1}$$

The case $c = 1$ is usually called in the literature *triangle inequality* while for $c \in (1/2; 1)$ it is called *sharpened triangle inequality*. Note that the extreme case $c = 1/2$ becomes trivial since all edges must have the same weight. A detailed motivation of the study of the *Traveling Salesman* problem satisfying sharpened triangle inequalities is given in [4]. In the context of extended problems, Definition 1 leads to a new definition called the *Extended c-relaxed triangle inequality*.

Definition 2 (Extended c-relaxed triangle inequality). *For a fixed $c \geq 1$, a weight function w on K_n satisfies the extended c-relaxed triangle inequality, if:*

(i) $w(e) = 0$ for $e \in \mathcal{U}$;
(ii) for all $\{x, y, z\} \nsubseteq V(\mathcal{U})$, w satisfies the c-relaxed triangle inequality.

Condition (*i*) of Definition 2 refers to the discussion in [18] which argues that the "residue" part of a feasible solution S, i.e., the part given in $S \setminus \mathcal{U}$, is the most important of the valuation. Another consequence of conditions (*i*) and (*ii*) concerns the valuation of w restricted to the subgraph induced by $V(\mathcal{U})$ (except for edges of $\mathcal{U}$): this function does not satisfy any specified property. The main reason is that they could never contribute in any spanning star forest containing $\mathcal{U}$. Finally, the reason for assuming $c \geq 1$ is that condition (*ii*) implies $\max\{w(xz); w(yz)\} \leq c \min\{w(xz); w(yz)\}$ when $xy \in \mathcal{U}$ and $z \notin V(\mathcal{U})$.

1.2   Related Work

The maximization version of SPANNING STAR FOREST PROBLEM, called here MAX STAR FOREST PROBLEM, in general graphs has been investigated intensively in recent years for unweighted graphs. Usually, the input is an undirected graph (weighted or not) and trivial stars are allowed as part of a feasible spanning star forests. In [14], an **APX**-hardness proof with explicit inapproximability bound is proposed, together with a polynomial-time algorithms for trees. A combinatorial 0.6-approximation algorithm which mainly solves the *dominating set* problem is presented as well while better algorithms with approximation ratio 0.71 and 0.803 are given respectively in [6] and [3]. In contrast, for edge weighted graphs with non-negative weights, few results are proposed in the literature. As indicated above, *trivial stars* (i.e., isolated vertices) are allowed because we want to maximize the total weight of the stars. This requirement is equivalent to find a *packing star forest* (i.e. a collection of vertex disjoint stars): a 0.5-approximation is given in [14] (which is the best ratio obtained so far) and polynomial-time algorithms for special classes of graphs such as trees and cactus graphs are presented in [14, 15]. Negative approximation results are presented in [5, 6, 14]. For any $\varepsilon > 0$, the unweighted version (or equivalently the MAX $\{0, 1\}$-SPANNING STAR FOREST PROBLEM) is hard to approximate within a factor of $\frac{545}{546} + \varepsilon$ unless $\mathbf{P} = \mathbf{NP}$ [14]. The edge-weighted version is **NP**-hard to approximate within $\frac{10}{11} + \varepsilon$ [5]. For the MAXIMUM CARPOOL MATCHING PROBLEM, a 0.33-approximation algorithm and a 0.5-approximation algorithm for both the general problem and the unweighted variant are given in [13]. To the best of our knowledge, the minimization version

MIN SPANNING STAR FOREST PROBLEM and the extended versions of both problems have not been studied in the literature.

As indicated in introduction, extending a partial solution into a feasible solution has been studied from a computational complexity for INDEPENDENT DOMINATING SET[1], CONFERENCE PROGRAMS and COLORATION in [7,9,17] respectively. Dealing with approximation algorithms with performance guarantee of **NP**-hard of optimization problems, results on extension problems are given in [7,18] for several problems including VERTEX COVER, CONNECTED VERTEX COVER FEEDBACK VERTEX SET, STEINER TREE, MAX LEAF and BIN PACKING. For algorithms finding an optimal solution, it often does not matter whether we optimize the weight of whole solution S or the weight of the *residue part* $S \setminus \mathcal{U}$. However, in the context of approximation algorithms, this difference may produce important modifications as for BIN PACKING. It is the main reason explaining the works given in [18] where the authors define and propose the approximation classes **FRAPX** and **RAPX** capturing approximability of the residue. The *residue approximation* of a solution S is the approximation of $w(S) - w(\mathcal{U})$. In the conclusion of their paper [18], the authors elaborate that obtaining parameterized complexity results [8] of extension problems parameterized by $|\mathcal{U}|$ leads to challenging open problems.

1.3 Organization and Contribution

We prove the following results in this paper. In Sect. 2 the minimization version is studied. A dichotomy result of the computational complexity is presented depending on parameter c of the (extended) c-relaxed triangle inequality (Theorems 1 and 2). Then, a parameterized complexity showing that this version is **FPT** is given. Positive and negative approximation results conclude this section. In Sect. 4, the maximization extended version is studied. We prove that, compared to the unextended version, the same positive approximation result is reached, while we strengthen the negative approximation result to hold even for the MAX EXTENDED $\{0,1\}$-SPANNING STAR FOREST PROBLEM. Table 1 summarizes the results obtained in the paper.

Table 1. The results given in this paper.

	w-general	c-relaxed	Extended c-relaxed
Min Extended Spanning Star Forest	NP-hard inapproximable at all in FPT parameterized by $|\mathcal{U}|$	NP-hard $c > 1$ polynomial $\frac{1}{2} \leq c \leq 1$ in RAPX with c apx-ratio inapproximable with $\frac{7+c}{8} - \epsilon$	NP-hard $c > 1$ polynomial $c = 1$ in RAPX with $\frac{c+1}{2}$ apx-ratio inapproximable with $\frac{7+c}{8} - \epsilon$
Max Extended Spanning Star Forest	in RAPX with $\frac{1}{2}$ apx-ratio inapproximable within $\frac{7}{8} + \epsilon$ even for $w(e) \in \{0,1\}$	–	–

---

[1] In this case, it is also required that some vertices are forbidden.

2   Spanning Star Forest Problem: Minimization Case

We start with the unextended version.

Proposition 1. MIN SPANNING STAR FOREST PROBLEM *is polynomial-time solvable.*

We now prove that the extended version of MIN SPANNING STAR FOREST PROBLEM can be much harder. Actually, we will give a dichotomy result depending on parameter c of the (extended) c-relaxed triangle inequality.

Theorem 1. MIN EXTENDED SPANNING STAR FOREST PROBLEM *is **NP**-hard for both c- and extended c-relaxed triangle inequality when $c > 1$.*

Démonstration. Let $c > 1$ be a constant. For both cases, we propose a reduction from SAT to the MIN EXTENDED SPANNING STAR FOREST PROBLEM where the weight function satisfies both conditions. SAT is an **NP**-complete problem [10] which consists of deciding if an instance $I = (\mathcal{C}, \mathcal{X})$ of SAT is satisfiable. Here, $\mathcal{C} = \{c_1, \ldots, c_m\}$ and $\mathcal{X} = \{x_1, \ldots, x_n\}$ are the set of clauses and variables respectively; a variable x_i which appear negatively will be denoted $\neg x_i$. From $I = (\mathcal{C}, \mathcal{X})$, we build an instance $I' = (K_{2n+m}, w, \mathcal{U})$ of MIN EXTENDED SPANNING STAR FOREST PROBLEM as follows:

- $V(K_{2n+m}) = V(\mathcal{C}) \cup V(\mathcal{X})$ where $V(\mathcal{C}) = \{v_j : c_j \in \mathcal{C}\}$ and $V(\mathcal{X}) = \{v_i^0, v_i^1 : i = 1, \ldots, n\}$,
- $\mathcal{U} = \{v_i^0 v_i^1 : x_i \in \mathcal{X}\}$ and let $M = \{v_j v_i^1 : x_i \in c_j\} \cup \{v_j v_i^0 : \neg x_i \in c_j\}$.

It is clear that I' is built in polynomial-time. The weight function w is defined by, $\forall xy \in E(K_{2n+m})$,

$$
w(xy) = \begin{cases}
0 & \text{if } xy \in \mathcal{U}, \\
1 & \text{if } xy \in M, \\
c & \text{if } xy \notin M, \ x \in V(\mathcal{C}), \ y \in V(\mathcal{X}), \\
2c & \text{otherwise.}
\end{cases}
$$

We can easily verify that w satisfies the extended c-relaxed (and c-relaxed) triangle inequality. We claim that I is satisfiable if and only if $opt_{SSF}(I') \leq m$.

Suppose that I is satisfiable and let T be a truth assignment of I. For each clause c_j let $x_{f(j)}$ be a variable satisfying it in T; we build a spanning star forest $\mathcal{S}$ containing $\mathcal{U}$ such that $w(\mathcal{S}) = \sum_{e \in S} w(e) = m$ as follows: $\mathcal{S} = \{v_{f(j)}^1 v_j : T(x_i) = \texttt{true}\} \cup \{v_{f(j)}^0 v_j : T(x_i) = \texttt{false}\} \cup \mathcal{U}$.

Conversely, assume that $\mathcal{S}^*$ is a spanning star forest containing $\mathcal{U}$ with $w(\mathcal{S}^*) = opt_{SSF}(I') \leq m$. Since $\mathcal{U}$ is a matching of size n, and by construction of the weights, if $\mathcal{S}^*$ contains ℓ edges of weights $2c$, then $w(\mathcal{S}^*) \geq 2c\ell + (m - 2\ell) = m - 2\ell(c - 1)$ because these ℓ edges cover at most 2ℓ vertices of $V(\mathcal{C})$ and the weight of any other edge is at least 1 (recall $c > 1$). Hence, we deduce $\ell = 0$. Now, if $\mathcal{S}^*$ contains ℓ' edges of weight $c > 1$, then these ℓ' edges cover exactly

ℓ' vertices of $V(\mathcal{C})$ and $w(\mathcal{S}^*) \geq \ell'c + (m - \ell') > m$, contradiction. Hence, $\mathcal{S}^*$ only contains unit weights. We can build a truth assignment T as follows: if $v_j v_i^1 \in \mathcal{S}^*$, then $T(x_j) = \texttt{true}$ and $T(x_i) = \texttt{false}$ otherwise. ∎

Corollary 1. *The* MIN EXTENDED SPANNING STAR FOREST PROBLEM *for general weight function w is not approximable at all unless* **P=NP**.

Theorem 2. MIN EXTENDED SPANNING STAR FOREST PROBLEM *is solvable in polynomial-time for the c-relaxed triangle inequality when $1/2 \leq c \leq 1$ and the extended c-relaxed triangle inequality when $c = 1$.*

Démonstration. We only deal with the c-relaxed triangle inequality case, because the other case is simpler. Let c be a constant with $1/2 \leq c \leq 1$. We solve MIN EXTENDED SPANNING STAR FOREST PROBLEM for the c-relaxed triangle inequality via the help of the MIN WEIGHTED LOWER-UPPER-COVER PROBLEM. This latter problem consists, given a edge-weighted graph (G, w) where $G = (V, E)$ and two non-negative functions a, b from V such that $\forall v \in V$, $0 \leq a(v) \leq b(v) \leq d_G(v)$, of finding a subset $M \subseteq E$ such that the subgraph $G_M = (V, M)$ induced by M satisfies $a(v) \leq d_{G_M}(v) \leq b(v)$ (such a solution will be called a *lower-upper-cover*) and minimizing its weight $w(M) = \sum_{e \in M} w(e)$ among all such solutions (if any). Given $I = (G, w)$ instance of the MIN WEIGHTED LOWER-UPPER-COVER PROBLEM $opt_{LUC}(I)$ denotes the value of an optimal solution. The MIN WEIGHTED LOWER-UPPER-COVER PROBLEM is known to be solvable in polynomial-time (Theorem 35.2 Chap. 35 of Volume A in [16]).

Let $I = (K_n, w, \mathcal{U})$ be an instance of the MIN EXTENDED SPANNING STAR FOREST PROBLEM where w satisfies the c-relaxed triangle inequality and $\mathcal{U}$ is a packing of stars. From I, we build the instance $I' = (H, w_H)$ described Fig. 2 of Subsect. 1.1. Moreover, we consider two functions a, b of the MIN WEIGHTED LOWER-UPPER-COVER PROBLEM as follows: if $v \in V_H \setminus (R \cup C)$, then $a(v) = 1$ and $b(v) = 2$. Otherwise, $v \in R \cup C$ and $a(v) = 0$ and $b(v) = 1$. Figure 4 proposes an illustration of the construction.

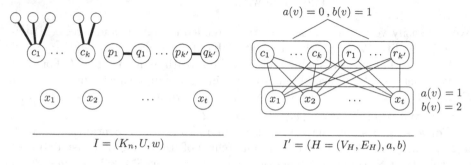

Fig. 4. An instance I of the MIN WEIGHTED LOWER-UPPER-COVER PROBLEM is shown on the right hand. Bold edges are in $\mathcal{U}$.

By construction of I', an optimal lower-upper-cover with parameters a, b is $\{P_4, C_3\}$-free and then is an extended spanning star forest of I. Hence,

$$opt_{LUC}(I') \geq opt_{SSF}(I) \tag{2}$$

Conversely, let $\mathcal{S}^*$ be an optimal extended spanning star forest of I. The next property allow us to focus on spanning star forest claw $\mathcal{U}$-free where a claw $F = K_{1,3}$ is not $\mathcal{U}$-free if at least two edges of the claw F do not belongs to $\mathcal{U}$, i.e., $|F \cap \mathcal{U}| \geq 2$. Where a claw $F = K_{1,3}$ is $\mathcal{U}$-free if at most one edge of the claw F belongs to $\mathcal{U}$, i.e.., $|F \cap \mathcal{U}| \leq 1$.

Property 1. There is an optimal extended spanning star forest of I which is claw $\mathcal{U}$-free.

Démonstration. Let $\mathcal{S}$ be an optimal extended spanning star forest. Assume $\mathcal{S}$ is not claw $\mathcal{U}$-free and let $S = \{uv_i : i = 1, 2, 3\}$ be a claw not $\mathcal{U}$-free with $uv_i \notin \mathcal{U}$ for $i = 1, 2$. In particular, vertices v_1 and v_2 are not adjacent to $\mathcal{U}$; hence, $\mathcal{S}^* = (\mathcal{S} \setminus S) \cup \{v_1 v_2, uv_3\}$ is an extended spanning star forest with $w(\mathcal{S}^*) \leq w(\mathcal{S})$. By repeating this process, we get the expected result. Note that if $c < 1$, all optimal extended spanning star forests are indeed claw $\mathcal{U}$-free.

Hence, we can assume that $\mathcal{S}^*$ is claw $\mathcal{U}$-free, and then it is a lower-upper-cover with parameters a, b of I':

$$opt_{SSF}(I) = w(\mathcal{S}^*) \geq w_H(\mathcal{S}^*) \geq opt_{LUC}(I') \tag{3}$$

Inequalities (2) and (3) give the expected result.   ∎

We end this subsection by giving a parameterized complexity result depending on the number of forced edges.

Theorem 3. Min Extended Spanning star forest problem *parameterized by* $|\mathcal{U}|$ *is* **FPT** *and under* **ETH**, Min Extended Spanning star forest problem *cannot be solved in time* $O^*(2^{s|\mathcal{U}|})$ *for some* $s > 0$.

Démonstration. Let $I = (K_n, w, \mathcal{U})$ be an instance of the Min Extended Spanning star forest problem where we recall that $\mathcal{U} = M_\mathcal{U} \cup S_\mathcal{U}$ with $M_\mathcal{U} = \{p_i q_i : i = 1, \ldots, k'\}$ and $S_\mathcal{U} = \{F_1, \ldots, F_k\}$. The set of centers is $C = \{c_1, \ldots, c_k\}$ where c_i is the center of star F_i. As in Theorem 2, we solve several instances I_J of the Min Weighted Lower-upper-cover problem for each set $J \subseteq \{1, \ldots, k'\}$. At the end, we return the solution minimizing $w(\mathcal{S}_J) = opt_{LUC}(I_J)$ among all possible sets J, that is $\mathcal{S} = \mathrm{argmin}_J opt_{LUC}(I_J)$ where $opt_{LUC}(I_J)$ is the optimal value of the Min Weighted Lower-upper-cover problem on instance I_J.

Let $I = (K_n, w, \mathcal{U})$ be an instance of the Min Extended Spanning star forest problem where $\mathcal{U}$ is a packing of stars. From I and a set $J \subseteq \{1, \ldots, k'\}$, we built an instance $I_J = (H_J, w)$ where $H_J = (V_{H_J}, E_{H_J})$ is a complete subgraph of K_n and two functions a_J, b_J of the Min Weighted Lower-upper-cover problem as follows: $V_{H_J} = (V(K_n) \setminus V(\mathcal{U})) \cup$

$(\{p_j : j \in J\} \cup \{q_j : j \in \{1, \ldots, k'\} \setminus J\} \cup C)$. Finally, if $v \in V(K_n) \setminus V(\mathcal{U})$, then $a_J(v) = 1$ and $b_J(v) = d_{H_J}(v)$. Otherwise, for $v \in \{p_j : j \in J\} \cup \{q_j : j \in \{1, \ldots, k'\} \setminus J\} \cup C$, $a_J(v) = 0$ and $b_J(v) = d_{H_J}(v)$. Let $\mathcal{S}_J$ be an optimal solution of the MIN WEIGHTED LOWER-UPPER-COVER PROBLEM on (I_J, a_J, b_J) Clearly, $\mathcal{S}_J$ is a spanning star forest on I and by construction there exists J^* such that $w(\mathcal{S}_{J^*}) = w(\mathcal{S}^*) = opt_{SSF}(I)$. The complexity of the whole algorithm is $O^*(2^{|M_\mathcal{U}|}) = O^*(2^{|\mathcal{U}|})$ and then MIN EXTENDED SPANNING STAR FOREST PROBLEM is **FPT**.

The second part of the proof is a direct consequence of Corollary 1 for the MIN EXTENDED $\{0,1\}$-SPANNING STAR FOREST PROBLEM and use the *Exponential Time Hypothesis* (**ETH** in short): $\exists s > 0$ such that 3-CNF-Sat with n variables cannot be solved in time $O^*(2^{sn})$ [8]. ∎

3 Approximation Results

From Corollary 1 and Theorem 2, we focus on the approximation of the MIN EXTENDED SPANNING STAR FOREST PROBLEM for both c- and extended c-relaxed triangle inequality. Hence, let $c > 1$ be a fixed constant. In algorithm APPROX 1 we use optimal solution of the MIN WEIGHTED LOWER-UPPER-COVER PROBLEM as subroutine. This latter problem consists, given a edge-weighted graph (G, w) (not necessarily complete) where $G = (V, E)$ and two non-negative functions a, b from V such that $\forall v \in V$, $0 \leq a(v) \leq b(v) \leq d_G(v)$, of finding a subset $M \subseteq E$ such that the subgraph $G_M = (V, M)$ induced by M satisfies $a(v) \leq d_{G_M}(v) \leq b(v)$ (such a solution will be called a *lower-upper-cover*) and minimizing its weight $w(M) = \sum_{e \in M} w(e)$ among all such solutions (if any). The MIN WEIGHTED LOWER-UPPER-COVER PROBLEM is known to be solvable in polynomial-time (Theorem 35.2 Chap. 35 of Volume A in [16]).

Algorithm 1. Approx 1

Input: $I = (K_n, w, \mathcal{U})$ where $\mathcal{U}$) is a packing of forced stars.
Output: A spanning star forest S of I containing $\mathcal{U}$).

1 Build instance $I' = (H, w_H)$ described in Fig. 2 where $H = (V(K_n) \setminus V(\mathcal{U})), R \cup C, E_H)$ is a split complete graph;
2 Find an optimal solution $S^* \subseteq E_H$ of the MIN WEIGHTED LOWER-UPPER-COVER PROBLEM on (I', a_1, b_1) with $a_1(v) = 1$ and $b_1(v) = d_H(v)$ if $v \in V_H \setminus (R \cup C)$, $a_1(v) = 0$ and $b_1(v) = 1$ for $v \in R$ and $a_1(v) = 0$ and $b_1(v) = d_H(v)$ for $v \in C$;
3 Convert S^* into S using the H-extended procedure (see Subsect. 1.1);
4 Return S.

Theorem 4. APPROX 1 *is a c-approximation of* MIN EXTENDED SPANNING STAR FOREST PROBLEM *for both c- and extended c-relaxed triangle inequality.*

Corollary 2. MIN EXTENDED SPANNING STAR FOREST PROBLEM *is in* **RAPX** *for both c- and extended c-relaxed triangle inequality.*

Algorithm 2. Approx 2

Input: $I = (K_n, w, \mathcal{U})$ where $\mathcal{U}$ is a packing of forced stars.
Output: A spanning star forest S of I containing $\mathcal{U}$.

1   Build instance $I' = (H, w_H)$ described in Figure 2 where
   $H = (V(K_n) \setminus V(\mathcal{U}), R \cup C, E_H)$ is a split complete graph;
2   Find an optimal solution $S^* \subseteq E_H$ of the MIN WEIGHTED LOWER-UPPER-COVER
   PROBLEM on (I', a_2, b_2) with $a_2(v) = 1$ and $b'(v) = d_H(v)$ if $v \in V_H \setminus (R \cup C)$ and
   $a_2(v) = 0$ and $b_2(v) = d_H(v)$ for $v \in R \cup C$;
3   Convert S^* into S using the H-extended procedure;
4   **for** *(each connected component F_i of S with L_i as leafs such that $p_i q_i \in F_i$ and*
   $L_i \cap \{p_i, q_i\} = \emptyset$) **do**
5 | Build two stars $S_i^1 = \{p_i x : x \in (L_i \cup \{q_i\})\}$ and $S_i^2 = \{q_i x : x \in (L_i \cup \{p_i\})\}$;
6 | **if** $w(S_i^1) \leq w(S_i^2)$ **then** $S \leftarrow (S \setminus F_i) \cup S_i^1$;
7 | **else** $S \leftarrow (S \setminus F_i) \cup S_i^2$;
8   Return $S \leftarrow S$.

By construction, each connected component F_i of S with $p_i q_i \in F_i$ and $L_i \cap \{p_i, q_i\} = \emptyset$ has a diameter equals to 3 (some leafs are connected to p_i while the others leafs are connected to q_i). The other connected components are stars. Hence, S is a spanning star forest of I. Figure 5 proposes an illustration of the construction of stars S_i^1 and S_i^2.

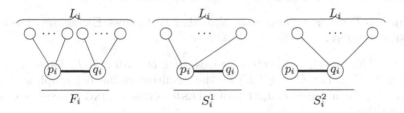

Fig. 5. Illustration of construction of stars S_i^1 and S_i^2 from F_i. Bold edges are in $\mathcal{U}$.

Theorem 5. APPROX 2 *is a $\frac{c+1}{2}$-approximation of* MIN EXTENDED SPANNING STAR FOREST PROBLEM *for extended c-relaxed triangle inequality.*

Theorem 6. *For any $\epsilon > 0$ it is* **NP**-*hard to approximate and residue approximate the* MIN EXTENDED SPANNING STAR FOREST PROBLEM *within $\frac{7+c}{8} - \epsilon$ for both c and extended c-relaxed triangle inequalities.*

4   Spanning Star Forest Problem: Maximization Case

In this section, we study the maximization case when the weight function w is general, but non-negative and the graph is complete. Usually (see Subsect. 1.2), the SPANNING STAR FOREST PROBLEM is defined in general graphs (i.e., not necessarily complete), and allowing trivial stars. This assumption is not restrictive because by completing the graph by weights 0, the two problems become equivalent. Moreover, by replacing the weights of required edges $\mathcal{U}$ by an large enough value, then MAX SPANNING STAR FOREST PROBLEM and EXTENDED SPANNING STAR FOREST PROBLEM are completely equivalent from a computational complexity point of view. However, these modifications affect the approximability of the problem. Hence, here we are interested in the hardest case which corresponds to $w(e) = 0$ for $\forall e \in \mathcal{U}$. This means that the obtained results will be valid for the *residual approximation* [18]. Recall that $\mathcal{U} = \{U_1, \ldots, U_r\} = M_{\mathcal{U}} \cup S_{\mathcal{U}}$ where $r = k + k'$, $M_{\mathcal{U}} = \{e_i : i = 1, \ldots, k'\}$ is a matching of k' edges and $S_{\mathcal{U}} = \{F_1, \ldots, F_k\}$ is a set of k vertex-disjoint stars with at least two leafs each. The set of centers is $C = \{c_1, \ldots, c_k\}$ and L_i are the leafs of F_i.

We study an intermediary problem called here EXTENDED DISJOINT SPANNING FOREST because it will provide an upper bound of our problem:

EXTENDED DISJOINT SPANNING FOREST

Input: A weighted connected graph (G, w) and a packing of non trivial stars $\mathcal{U} = \{U_1, \ldots, U_r\}$.

Solution: Spanning forest $\mathcal{S} = \{S_1, \ldots, S_r\} \subseteq E$ of G such that $U_i \subseteq S_i$.

Output: Maximizing $w(\mathcal{S}) = \sum_{e \in \mathcal{S}} w(e) = \sum_{i=1}^{r} \sum_{e \in S_i} w(e)$.

Solving EXTENDED DISJOINT SPANNING FOREST is polynomial and use the same arguments that solving *maximum weighted spanning tree*.

Lemma 1. *There is a linear-time algorithm that solves* EXTENDED DISJOINT SPANNING FOREST.

From $I = (K_n, w, \mathcal{U})$, we delete all edges $xy \notin \mathcal{U}$ with $x \in L_i$ for some $i \leq k$ and $y \in V(K_n)$. Let $G = (V, E)$ be the resulting connected graph and $I' = (G, w, \mathcal{U})$ be the instance of EXTENDED DISJOINT SPANNING FOREST. Consider the Algorithm 3.

Let us formally explain how solutions are built during Step 5. Here, $U_i \subset S_i^*$; first we root subtree S_i^* at the center of U_i (if $U_i = \{p_i q_i\}$, we root S_i^* at p_i). Then, we construct a first partial solution which consider edges of $S_i^* \setminus U_i$ with odd levels and another partial solution with even levels. At the end of this Step 5. we add edges of U_i for both partial solutions. Figure 6 propose an illustration on the construction of the two spanning star forests (containing trivial stars at this stage) S_i^1 and S_i^2 of the induced subgraph (V, S_i^*) according to the structure of U_i.

Theorem 7. APPROX 3 *is a* $\frac{1}{2}$-*approximation of* MAX EXTENDED SPANNING STAR FOREST PROBLEM.

Setting $w(e) = 0$ for $e \in \mathcal{U}$ leads to the following corollary.

Corollary 3. MAX EXTENDED SPANNING STAR FOREST PROBLEM *is in* **RAPX**.

Theorem 8. *For any $\epsilon > 0$ it is **NP**-hard to approximate and residue approximate* MAX EXTENDED SPANNING STAR FOREST PROBLEM *within* $\frac{7}{8} + \epsilon$.

Algorithm 3. Approx 3

Input: $I = (K_n, w, \mathcal{U})$ where $\mathcal{U}$ is a packing of forced stars.
Output: A spanning star forest $\mathcal{S}$ of I containing $\mathcal{U}$.

1 Build instance $I' = (G, w, \mathcal{U})$ of EXTENDED DISJOINT SPANNING FOREST;
2 Find an optimal solution $\mathcal{S}_1^* = \{S_1^*, \ldots, S_r^*\}$ such that $U_i \subseteq S_i^*$ of EXTENDED DISJOINT SPANNING FOREST;
3 **for** *(each subtree S_i^*)* **do**
4   **if** $S_i^* = U_i$ **then** $S \leftarrow S \cup S_i^*$;
5   **else**
6   Split S_i^* into two spanning star forest (with possibly trivial stars) S_i^1 and S_i^2 such that $S_i^1 \cap S_i^2 = U_i$ and $S_i^1 \cup S_i^2 = S_i^*$ by dividing subtree S_i^* into alternating levels (even and odd from center of U_i);
7   **if** $w(S_i^1) \geq w(S_i^2)$ **then** $S \leftarrow S \cup S_i^1$;
8   **else** $S \leftarrow S \cup S_i^2$;

9 Complete S into a spanning star forest by connecting each isolated vertex to some center;
10 Return $\mathcal{S} \leftarrow S$.

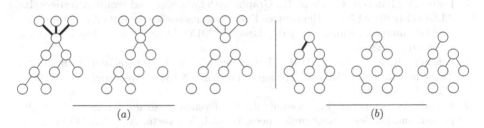

(a)   (b)

Fig. 6. Construction of solutions S_i^1 and S_i^2 depending on S_i^* contains $p_i q_i$ (case (b)) or not (case (a)); bold edges are in U_i. For each case both solutions S_i^1 and S_i^2 are indicated (at this stage, trivial stars are allowed).

5 Conclusion

In this article, we have studied two MAX and MIN EXTENDED SPANNING STAR FOREST PROBLEMS. We considered different types of weight function w for edges of input graphs in MIN EXTENDED SPANNING STAR FOREST PROBLEM. We have shown for general-w, the problem is not approximable at all but for c-relaxed and extended c-relaxed triangle, it is in **RAPX**. Moreover, we have shown the MIN EXTENDED SPANNING STAR FOREST PROBLEM parameterized by the cardinality of $\mathcal{U}$ is in **FPT**. Furthermore, we proved MAX EXTENDED SPANNING STAR FOREST PROBLEM is in **RAPX** for general-w. It would be interesting to study parameterized complexity of the maximizing version on the future.

References

1. Agatz, N.A.H., Erera, A.L., Savelsbergh, M.W.P., Wang, X.: Optimization for dynamic ride-sharing: a review. Eur. J. Oper. Res. **223**(2), 295–303 (2012)
2. Agra, A., Cardoso, D., Cerfeira, O., Rocha, E.: A spanning star forest model for the diversity problem in automobile industry. In: ECCO XVIII, Minsk (2005)
3. Athanassopoulos, S., Caragiannis, I., Kaklamanis, C., Papaioannou, E.: Energy-efficient communication in multi-interface wireless networks. In: Královič, R., Niwiński, D. (eds.) MFCS 2009. LNCS, vol. 5734, pp. 102–111. Springer, Heidelberg (2009). https://doi.org/10.1007/978-3-642-03816-7_10
4. Böckenhauer, H.-J., Hromkovic, J., Klasing, R., Seibert, S., Unger, W.: Approximation algorithms for the TSP with sharpened triangle inequality. Inf. Process. Lett. **75**(3), 133–138 (2000)
5. Chakrabarty, D., Goel, G.: On the approximability of budgeted allocations and improved lower bounds for submodular welfare maximization and GAP. SIAM J. Comput. **39**(6), 2189–2211 (2010)
6. Chen, N., Engelberg, R., Nguyen, C.T., Raghavendra, P., Rudra, A., Singh, G.: Improved approximation algorithms for the spanning star forest problem. Algorithmica **65**(3), 498–516 (2013)
7. Delbot, F., Laforest, C., Phan, R.: Graphs with forbidden and required vertices. In: ALGOTEL 2015-17emes Rencontres Francophones sur les Aspects Algorithmiques des Télécommunications Jun 2015, Beaune (2015). https://hal.archives-ouvertes.fr/hal-01148233
8. Downey, R.G., Fellows, M.R.: Fundamentals of Parameterized Complexity. Texts in Computer Science. Springer, London (2013). https://doi.org/10.1007/978-1-4471-5559-1
9. Fotakis, D., Gourvès, L., Monnot, J.: Conference program design with single-peaked and single-crossing preferences. In: Cai, Y., Vetta, A. (eds.) WINE 2016. LNCS, vol. 10123, pp. 221–235. Springer, Heidelberg (2016). https://doi.org/10.1007/978-3-662-54110-4_16
10. Garey, M.R., Johnson, D.S.: Computers and Intractability: A Guide to the Theory of NP-Completeness. W. H. Freeman & Co., New York (1979)
11. Hartman, I.B.-A., Keren, D., Dbai, A.A., Cohen, E., Knapen, L., Yasar, A.-U.-H., Janssens, D.: Theory and practice in large carpooling problems. In: Shakshuki, E.M., Yasar, A.-U.-H. (eds.) Proceedings of the 5th International Conference on Ambient Systems, Networks and Technologies (ANT 2014), The 4th International Conference on Sustainable Energy Information Technology (SEIT-2014), Hasselt, 2–5 June 2014, vol. 32. Procedia Computer Science, pp. 339–347. Elsevier (2014)
12. He, J., Liang, H.: Improved approximation for spanning star forest in dense graphs. J. Comb. Optim. **25**(2), 255–264 (2013)
13. Kutiel, G.: Approximation algorithms for the maximum carpool matching problem. In: Weil, P. (ed.) CSR 2017. LNCS, vol. 10304, pp. 206–216. Springer, Cham (2017). https://doi.org/10.1007/978-3-319-58747-9_19
14. Nguyen, C.T., Shen, J., Hou, M., Sheng, L., Miller, W., Zhang, L.: Approximating the spanning star forest problem and its application to genomic sequence alignment. SIAM J. Comput. **38**(3), 946–962 (2008)
15. Nguyen, V.H.: The maximum weight spanning star forest problem on cactus graphs. Discrete Math. Algorithms Appl. **7**(2), 1550018 (2015)
16. Schrijver, A.: Combinatorial Optimization: Polyhedra and Efficiency. Springer, Heidelberg (2003)

17. Tuza, Z.: Graph colorings with local constraints - a survey. Discussiones Mathematicae Graph Theory **17**(2), 161–228 (1997)
18. Weller, M., Chateau, A., Giroudeau, R., König, J.-C., Pollet, V.: On residual approximation in solution extension problems. In: Chan, T.-H.H., Li, M., Wang, L. (eds.) COCOA 2016. LNCS, vol. 10043, pp. 463–476. Springer, Cham (2016). https://doi.org/10.1007/978-3-319-48749-6_34

Faster and Enhanced Inclusion-Minimal Cograph Completion

Christophe Crespelle[1](✉), Daniel Lokshtanov[2], Thi Ha Duong Phan[3], and Eric Thierry[1]

[1] University of Lyon, UCB Lyon 1, ENS de Lyon, CNRS, Inria, LIP UMR 5668, 15 Parvis René Descartes, 69342 Lyon, France
christophe.crespelle@inria.fr, eric.thierry@ens-lyon.fr
[2] Department of Informatics, University of Bergen, 5020 Bergen, Norway
daniello@ii.uib.no
[3] Institute of Mathematics, Vietnam Academy of Science and Technology, 18 Hoang Quoc Viet, Hanoi, Vietnam
phanhaduong@math.ac.vn

Abstract. We design two incremental algorithms for computing an inclusion-minimal completion of an arbitrary graph into a cograph. The first one is able to do so while providing an additional property which is crucial in practice to obtain inclusion-minimal completions using as few edges as possible: it is able to compute a minimum-cardinality completion of the neighbourhood of the new vertex introduced at each incremental step. It runs in $O(n + m')$ time, where m' is the number of edges in the completed graph. This matches the complexity of the algorithm in [24] and positively answers one of their open questions. Our second algorithm improves the complexity of inclusion-minimal completion to $O(n + m \log^2 n)$ when the additional property above is not required. Moreover, we prove that many very sparse graphs, having only $O(n)$ edges, require $\Omega(n^2)$ edges in any of their cograph completions. For these graphs, which include many of those encountered in applications, the improvement we obtain on the complexity scales as $O(n/ \log^2 n)$.

1 Introduction

We consider the problem of completion of an arbitrary graph into a *cograph*, i.e. a graph with no induced path on 4 vertices. This is a particular case of *graph modification problem*, in which one wants to perform elementary modifications to an input graph, typically adding and removing edges and vertices, in order to obtain a graph belonging to a given target class of graphs, which satisfies some additional property compared to the input. Ideally, one would like to do

This work was partially funded by the PICS program of CNRS.

This work was partially funded by the Vietnam National Foundation for Science and Technology Development (NAFOSTED) under the grant number 101.99-2016.16 and by the Vietnam Institute for Advanced Study in Mathematics (VIASM).

X. Gao et al. (Eds.): COCOA 2017, Part I, LNCS 10627, pp. 210–224, 2017.
https://doi.org/10.1007/978-3-319-71150-8_19

so by performing a minimum number of elementary modifications. This is a fundamental problem in graph algorithms, which determines how far is a given graph to satisfy a property.

Here, we consider the modification problem called *completion*, where only one operation is allowed: adding an edge. In this case, the quantity to be minimised, called the *cost* of the completion, is the number of edges added, which are called *fill edges*. The particular case of completion problems has been shown very useful in computer science and other disciplines such as archaeology [22], molecular biology [3] and genomics [14].

Unfortunately, finding the minimum number of edges to be added in a completion problem is NP-hard for most of the target classes of interest (see, e.g., the thesis of Mancini [25] for further discussion and references). To deal with this difficulty of computation, the domain has developed a number of approaches. This includes, approximation, restricted input, parameterization and inclusion-minimal completions. In the latter approach, one does not ask for a completion having the minimum number of fill edges but only ask for a set of fill edges which is minimal for inclusion, i.e. which does not contain any proper subset of fill edges that also results in a graph in the target class. This is the approach we follow here. In addition to the case of cographs [24], it has been followed for many other graph classes, including chordal graphs [17], interval graphs [10,26], proper interval graphs [28], split graphs [18], comparability graphs [16] and permutation graphs [9].

The rationale behind the inclusion-minimal approach is that minimum-cardinality completions are in particular inclusion-minimal. Therefore, if one is able to sample[1] efficiently the space of inclusion-minimal completions, one can compute several of them, pick the one of minimum cost and hope to get a value close to the optimal one. One of the reason of the success of inclusion-minimal completion algorithms is that this heuristic approach was shown to perform quite well in practice [2]. The second reason of this success, which is a key point for the approach, is that it is usually possible to design algorithms of low complexity for the inclusion-minimal relaxation of completion problems.

Modification problems into the class of cographs have already received a great amount of attention [15,19,23,24], as well as modification problems into some of its subclasses, such as *quasi-threshold graphs* [4] and *threshold graphs* [12]. One reason for this is that cographs are among the most widely studied graph classes. They have been discovered independently in many contexts [6] and they are known to admit very efficient algorithms for problems that are hard in general. Moreover, very recently, cograph modification was shown a powerful approach to solve problems arising in complex networks analysis, e.g. community detection [21] and inference of phylogenomics [19]. This growing need for treating real-world datasets, whose size is often huge, asks for more efficient algorithms

---

[1] Usually, minimal completion algorithms are not fully deterministic. There are some choices to be made arbitrarily along the algorithm and different choices lead to different minimal completions.

both with regard to the running time and with regard to the quality (number of modifications) of the solution returned by the algorithm.

Our results. Our main contribution is to design two algorithms for inclusion-minimal cograph completion. The first one (Sect. 4) is an improvement of the incremental algorithm in [24]. It runs in the same $O(n + m')$ complexity, where m' is the number of edges in the completed graph, and is in addition able to select one minimum-cardinality completion of the neighbourhood of the new incoming vertex at each incremental step of the algorithm, which is an open question in [24] (Question 3 in the conclusion) which we positively answer here. It must be clear that this does not guarantee that the completion computed at the end of the algorithm has minimum cardinality but this feature is highly desirable in practice to obtain completions using as few fill edges as possible.

When this additional feature is not required, our second algorithm (Sect. 5) solves the inclusion-minimal problem in $O(n + m \log^2 n)$ time, which only depends on the size of the input. Furthermore, we prove that many sparse graphs, namely those having mean degree fixed to a constant, require $\Omega(n^2)$ edges in any of their cograph completions. This result is worth of interest in itself and implies that, for such graphs, which have only $O(n)$ edges, the improvement of the complexity we obtain with our second algorithm is quite significant : a factor $n/\log^2 n$.

2   Preliminaries

All graphs considered here are finite, undirected, simple and loopless. In the following, G is a graph, V (or $V(G)$) is its vertex set and E (or $E(G)$) is its edge set. We use the notation $G = (V, E)$, $n = |V|$ stands for the cardinality of V and $m = |E|$ for the cardinality of E. An edge between vertices x and y will be arbitrarily denoted by xy or yx. The neighbourhood of x is denoted by $N(x)$ (or $N_G(x)$) and for a subset $X \subseteq V$, we define $N(X) = (\bigcup_{x \in X} N(x)) \setminus X$. The subgraph of G induced by some $X \subseteq V$ is denoted by $G[X]$.

For a rooted tree T and a node $u \in T$, we denote $parent(u)$, $\mathcal{C}(u)$, $Anc(u)$ and $Desc(u)$ the *parent* and the set of *children*, *ancestors* and *descendants* of u respectively, using the usual terminology and with u belonging to $Anc(u)$ and $Desc(u)$. The *lowest common ancestor* of two nodes u and v, denoted $lca(u, v)$, is the lowest node in T which is an ancestor of both u and v. The subtree of T rooted at u, denoted T_u, is the tree induced by node u and all its descendants in T. We use two other notions of subtree, which we call *upper tree* and *extracted tree*. The upper tree of a subset of nodes S of T is the tree, denoted T_S^{up}, induced by the set $Anc(S)$ of all the ancestors of the nodes of S, i.e. $Anc(S) = \bigcup_{s \in S} Anc(s)$. The tree extracted from S in T, denoted T_S^{xtr}, is defined as the tree whose set of nodes is S and whose parent relationship is the transitive reduction of the ancestor relationship in T. More explicitly, for $u, v \in S$, u is the parent of v in T_S^{xtr} iff u is an ancestor of v in T and there exist no node $v' \in S$ such that v' is a strict ancestor of v and a strict descendant of u in T.

Cographs. One of their simpler definitions is that they are the graphs that do not admit the P_4 (path on 4 vertices) as an induced subgraph. This shows that

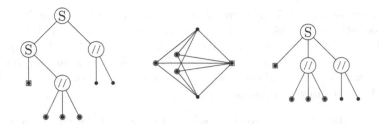

Fig. 1. Example of a labelled construction tree (left), the cograph it represents (centre), and the associated cotree (right). Some vertices are decorated in order to ease the reading.

the class is *hereditary*, i.e., an induced subgraph of a cograph is also a cograph. Equivalently, they are the graphs obtained from a single vertex under the closure of the *parallel* composition and the *series* composition. The parallel composition of two graphs $G_1 = (V_1, E_1)$ and $G_2 = (V_2, E_2)$ is their disjoint union, i.e., the graph $G_{par} = (V_1 \cup V_2, E_1 \cup E_2)$. The series composition of G_1 and G_2 is their disjoint union plus all possible edges between vertices of G_1 and vertices of G_2, i.e., the graph $G_{ser} = (V_1 \cup V_2, E_1 \cup E_2 \cup \{xy \mid x \in V_1, y \in V_2\})$. These operations can naturally be extended to an arbitrary finite number of graphs.

This gives a nice representation of a cograph G by a tree whose leaves are the vertices of G and whose internal nodes (non-leaf nodes) are labelled $//$, for parallel, or S, for series, corresponding to the operations used in the construction of G. It is always possible to find such a labelled tree T representing G such that every internal node has at least two children, no two parallel nodes are adjacent in T and no two series nodes are adjacent. This tree T is unique [6] and is called the *cotree* of G, see example in Fig. 1. Note that the subtree T_u rooted at some node u of cotree T also defines a cograph, denoted G_u, whose set of vertices is the set of leaves of T_u, denoted $V(u)$ in the following. The adjacencies between vertices of a cograph can easily be read on its cotree, in the following way.

Remark 1. *Two vertices x and y of a cograph G having cotree T are adjacent iff the lowest common ancestor u of leaves x and y in T is a series node. Otherwise, if u is a parallel node, x and y are not adjacent.*

The incremental approach. Our approach for computing a minimal cograph completion of an arbitrary graph G is incremental, in the sense that we consider the vertices of G one by one, in an arbitrary order $(x_1, \ldots, x_n)$, and at step i we compute a minimal cograph completion H_i of $G_i = G[\{x_1, \ldots, x_i\}]$ from a minimal cograph completion H_{i-1} of G_{i-1}, by adding only edges incident to x_i. This is possible thanks to the following observation that is general to all hereditary graph classes that are also stable by addition of a universal vertex, which holds in particular for cographs.

Lemma 1 (see e.g. [26]). *Let $G = (V, E)$ be an arbitrary graph and let H be a minimal cograph completion of G. Consider a new vertex $x \notin V$ adjacent to an* *arbitrary subset $N(x) \subseteq V$ of vertices and denote $G' = G + x$ and $H' = H + x$ the graphs obtained by adding x to G and H respectively. Then, there exists a subset $M \subseteq V \setminus N(x)$ of vertices such that $H'' = (V, E(H') \cup \{xy \mid y \in M\})$ is a cograph. Moreover, for any such set M which is minimal for inclusion, H'' is an inclusion-minimal cograph completion of G'. We call such completions (minimal) constrained completions of $G + x$.*

For any subset $S \subseteq V$ of vertices, we say that we *fill* S in H'' if we make all the vertices of $S \setminus N(x)$ adjacent to x in the completion H'' of $G + x$. The edges added in a completion are called *fill edges* and the *cost* of the completion is its number of fill edges.

The new problem. From now on, we consider the following problem, with slightly modified notations. $G = (V, E)$ is a cograph, and $G + x$ is the graph obtained by adding to G a new vertex x adjacent to some arbitrary subset $N(x)$ of vertices of G. Both our algorithms take as input the cotree of G and the neighbourhood $N(x)$ of the new vertex x. They compute the set $N'(x) \supseteq N(x)$ of neighbours of x in some minimal constrained cograph completion H of $G + x$, i.e. obtained by adding only edges incident to x (cf. Lemma 1). Then, the cotree of G is updated under the insertion of x with neighbourhood $N'(x)$, in order to obtain the cotree of H which will serve as input in the next incremental step.

We now introduce some definitions and characterisations we use in the following.

Definition 1 (Full, hollow, mixed). *Let G be a cograph and let x be a vertex to be inserted in G with neighbourhood $N(x) \subseteq V(G)$. A subset $S \subseteq V(G)$ is* full *if $S \subseteq N(x)$,* hollow *if $S \cap N(x) = \varnothing$ and* mixed *if S is neither full nor hollow. When S is full or hollow, we say that S is* uniform.

We use these notions for nodes u of the cotree as well, referring to their associated set of vertices $V(u)$. We denote $\mathcal{C}_{nh}(u)$ the subset of non-hollow children of a node u.

Theorem 1 below gives a characterisation of the neighbourhood of a new vertex x so that $G + x$ is a cograph.

Theorem 1 ([7,8]). *(Cf. Fig. 2) Let G be a cograph with cotree T and let x be a vertex to be inserted in G with neighbourhood $N(x) \subseteq V(G)$. If the root of T is mixed, then $G + x$ is a cograph iff there exists a mixed node u of T such that:*

1. *all children of u are uniform and*
2. *for all vertices $y \in V(G) \setminus V(u)$, $y \in N(x)$ iff $lca(y, u)$ is a series node.*

Moreover, when such a node u exists, it is unique and it is called the insertion node.

Remark 2. *In all the rest of the article, we do not consider the case where the new vertex x is adjacent to none of the vertices of G or to all of them. Therefore, the root of the cotree T of G is always mixed wrt. x.*

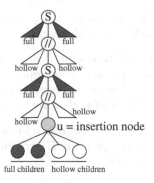

Fig. 2. Illustration of Theorem 1: characterisation of the neighbourhood of a new vertex x so that $G + x$ is a cograph. The nodes and triangles in black (resp. white) correspond to the parts of the tree that are full wrt. x (resp. hollow wrt. x). The insertion node u, which is mixed, appears in grey colour.

The reason for this is that the case where the root is uniform is straightforward: the only minimal completion of $G + x$ adds an empty set of edges and the update of cotree T is very simple. By definition, inserting x in G with its neighbourhood $N'(x)$ in some constrained cograph completion H of $G + x$ results in a cograph, namely H. Therefore, to any such completion H we can associate one insertion node which is uniquely defined, from Theorem 1 and from the restriction stated in Remark 2.

Definition 2. *Let G be a cograph with cotree T and let x be a vertex to be inserted in G. A node u of T is called a* completion-minimal insertion node *iff there exists a minimal constrained completion H of $G + x$ such that u is the insertion node associated to H.*

From now and until the end of the article, G is a cograph, T is its cotree, x is a vertex to be inserted in G and we consider only constrained cograph completions of $G + x$. We therefore omit to systematically precise it.

3 Characterisation of Minimal Constrained Completions

The goal of this section is to give necessary and sufficient conditions for a node u of T to be a completion-minimal insertion node. From Theorem 1, the subtrees attached to the parallel strict ancestors of the insertion node u must be hollow. As we can modify the neighbourhood of x only by adding edges, it follows that if u is the insertion node of some completion, then u is *eligible*, as defined below.

Definition 3 (eligible). *A node u of T is* eligible *iff for all the strict ancestors v of u that are parallel nodes, all the children of v distinct from its unique child $u' \in \mathcal{C}(v) \cap Anc(u)$ are hollow.*

When a node u is eligible, there is a natural way to obtain a completion of the neighbourhood of x, which we call the completion anchored at u.

Definition 4 (Completion anchored at u). *Let u be an eligible node of T. The completion anchored at u is the one obtained by making x adjacent to all the vertices of $V(G) \setminus V(u)$ whose lowest common ancestor with u is a series node and by filling all the children of u that are non-hollow.*

The completion anchored at some eligible node u may not be minimal but, on the other hand, all minimal completions H are completions anchored at some eligible node u, namely the insertion node of H.

Lemma 2. *For any completion-minimal insertion node u of T, there exists a unique minimal completion H of $G+x$ such that u is the insertion node associated to H and this unique completion is the completion anchored at u.*

Sketch of proof. Consider a minimal completion H that has u as insertion node (there exists one by definition). Note that the set of edges added between x and the vertices of $V(G) \setminus V(u)$ is necessarily the same for all such completions: this is the set given in the definition of the completion anchored at u. Moreover, from Theorem 1, H must make each child of u either full or hollow. Consequently, the children of u that were non-hollow before completion must be full after completion. Since letting the children of u that were hollow before completion remain hollow results in a valid completion, it follows that, by minimality, this is what H does. Thus, H is the completion anchored at u. □

To characterise completion-minimal insertion nodes, we will use the notion of *forced nodes*. Their main property (see Lemma 3 below) is that they are full in any completion of $G + x$.

Definition 5 (Completion-forced). *Let G be a cograph with cotree T and let x be a vertex to be inserted in G. A* completion-forced *(or simply* forced*) node u is inductively defined as a node satisfying at least one of the three following conditions:*

1. *u is full, or*
2. *u is a parallel node with all its children non-hollow, or*
3. *u is a series node with all its children completion-forced.*

Lemma 3. *Let u be a completion-forced node of T. Then, u is filled in all the completions of $G + x$.*

Sketch of proof. It can be proven by induction on $|V(u)|$. The only interesting case is when u is parallel. Since all the children of u are non-hollow, it follows that no node of $T_u \setminus \{u\}$ is eligible. Moreover, u itself cannot be the insertion node of some completion since, in this completion, u should have at least one hollow child, which is impossible as all the children of u are already non-hollow before completion. As a consequence, the insertion node v of any completion H is necessarily out of T_u. And since u is not hollow, Theorem 1 implies that u is full in H. □

The next remark directly follows from Theorem 1 and Lemma 2.

Remark 3. *The insertion node u of any minimal completion of $G + x$ has at least one hollow child and at least one non-hollow child. Therefore, u is non-hollow and non-completion-forced.*

We now characterise the nodes u that contain some minimal-insertion node in their subtree T_u (including u itself). In our algorithms, we will use this characterisation to decide whether we have to explore the subtree of a given node.

Lemma 4. *For any node u of T, T_u contains some completion-minimal insertion node iff u is eligible, non-hollow and non-completion-forced.*

Sketch of proof. A minimal insertion node is eligible and, from Remark 3, non-hollow and non-completion forced. So are all its ancestors, proving that the conditions of the Lemma are necessary for T_u to contain a minimal insertion node. To show that they are sufficient, consider a node $v \in T_u$ that satisfies these three conditions and that is lower possible in T_u. By definition, the children of v are non-eligible or hollow or forced. One can show that, whether v is parallel or series, in both cases, v must have at least one hollow child and at least one non-hollow child. Therefore, the completion H' anchored at v leaves v mixed, and so does a minimal completion H included in H'. Then, H also leaves u mixed, which, from Theorem 1, is true only for completions whose insertion node is in T_u.   □

Lemma 5 below gives additional conditions for u itself to be an insertion node.

Lemma 5. *A node u of T is a completion-minimal insertion node iff u is eligible, non-hollow and non-completion-forced and u satisfies in addition one of the two following conditions:*

1. u is a series node and u has at least one hollow child, or

2. u is a parallel node and u has no eligible non-completion-forced child.

Sketch of proof. To show that the conditions of the lemma are sufficient, we have to show that the completion H anchored at u is minimal (cf. Lemma 2). Because of Condition 1 when u is series and because u is non-completion forced when u is parallel, in both cases, H leaves some child v of u hollow. Since u is not hollow, the completions H' whose insertion node u' is out of T_u must fill u, and so v. It follows that such completions H' are not strictly included in H. The same holds if $u' = u$, since in this case $H' = H$ from Lemma 2. Then, the only possibility for H' to be strictly included in H is that its insertion node u' is a strict descendant of u. But, in that case, if u is a parallel node, u does not satisfy Condition 2. Consequently, u must be a series node and therefore v is not hollow in H', which shows that H' is not included in H. Thus, H is minimal.

Conversely, if u is a minimal insertion node, then Lemma 4 gives that u is eligible, non-hollow and non-completion-forced. Moreover, from Remark 3, Condition 1 is satisfied. Finally, if u is parallel and does not satisfy Condition 2, then u has a child v which is eligible non-hollow and non-completion forced. By Lemma 4, T_v contains a completion-minimal insertion node. Then, the completion H' anchored at v is included in the completion anchored at u, which implies that u is not a completion-minimal insertion node. By contraposition, if u is a parallel node, then u satisfies Condition 2. □

4   An $O(n + m')$ algorithm with incremental minimum

In this section, we design an incremental algorithm whose overall time complexity is $O(n + m')$, where m' is the number of edges in the output completed cograph. We concentrate on one incremental step, whose input is the cotree T of some cograph G (the completion computed so far) and a new vertex x together with the list of its neighbours $N(x) \subseteq V(G)$. Each node $u \in T$ stores its number $|\mathcal{C}(u)|$ of children and the number $|V(u)|$ of leaves in T_u. One incremental step takes time $O(d')$, where d' is the degree of x in the completion of $G + x$ computed by the algorithm. Within this complexity, our algorithm scans all the minimal completions of the neighbourhood of x and select one of minimum cardinality. Our description is in two steps.

First step. For each non-hollow node u of T we determine: (i) the list of its non-hollow children $\mathcal{C}_{nh}(u)$, (ii) the number of neighbours of x in $V(u)$ and (iii) whether it is completion forced or not. To this purpose, we perform two bottom-up searches of T from the leaves of T that are in $N(x)$ up until the root of T. Note that each of these searches discovers exactly the set $\mathcal{NH}(T)$ of non-hollow nodes of T (for which we show later that their number is $O(d')$). In the first search, we label each node encountered as non-hollow, we build the list of its non-hollow children and count them. In the second search, for each non-hollow node u we determine the rest of its information, that is items (ii) and (iii) above.

For the leaves of T in $N(x)$, it is straightforward to get this information. Then, the bottom-up search starts in an asynchronous manner: as soon as a node determines its information, it forwards its to its parent. When a node has received the information from all its non-hollow children (we determined their number in the first search), it can easily determine its own information and the process goes on, until the root of T has determined its information.

Second step. We search the set of all non-hollow, eligible and non-completion-forced nodes of T. For each of them, we determine whether it is a minimal insertion node and, in the positive, we compute the number of edges to be added in its associated minimal completion. Then, at the end of the search we select the completion of minimum cardinality.

Since, all the ancestors of a non-hollow eligible non-completion-forced node also satisfy these three properties, it follows that the part of T we have to search is a connected subset of nodes containing the root. Then, our search starts by determining whether the root is non-completion-forced. In the negative, we are done: there exists one unique minimal completion of $G + x$ which is obtained by adding all missing edges between x and the vertices of G. Otherwise, if the root is non-completion-forced (it is always eligible and non-hollow, cf. Remark 2), we start our search. For all the non-hollow children of the current node (we built their list in the first step), we check whether they are eligible and non-completion-forced and search, in a depth-first manner, the subtrees of those for which the test is positive (cf. Lemma 4).

During this depth-first search, we compute for each node u encountered the number of edges, denoted $cost-above(u)$, to be added between x and the vertices of $V(G) \setminus V(u)$ in the completion anchored at u. It can be determined as follows: if the parent v of u is a parallel node, then $cost - above(u) = cost - above(v)$; otherwise, if the parent v of u is a series node, then $cost - above(u) = cost - above(v) + |V(v) \setminus N(x)| - |V(u) \setminus N(x)|$. We also determine whether u is a minimal insertion node by testing whether it satisfies Condition 1 or 2 of Lemma 5. Importantly for the complexity, this can be done by scanning only the list of its non-hollow children, and by using the information collected in the first step. If u is a minimal insertion node, then we determine the number of edges $cost(u)$ to be added in the completion anchored at u as $cost(u) = cost - above(u) + \sum_{v \in C_{nh}(u)} |V(v) \setminus N(x)|$.

From Lemma 5, minimal insertion nodes are non-hollow, eligible and non-completion-forced. Therefore, our search discovers all of them and returns one that achieves the minimum cost among all completions of the neighbourhood of x. Finally, we need to update the cotree T for the next incremental step of the algorithm, as explained below.

Complexity. The key of the $O(d')$ time complexity is that we search and manipulate only the set $\mathcal{NH}(T)$ of non-hollow nodes of T. For each of them u, we need to scan the list of its non-hollow children $C_{nh}(u)$ and to perform a constant number of tests and operations that all take $O(1)$ time (thanks to the information collected in the first step). Thus, the execution of the two steps takes $O(|\mathcal{NH}(T)|)$ time, which is also $O(d')$ as shown in [24]. Indeed, one can observe that during completion, all non-hollow nodes are filled (see Definition 4), except the ancestors of the insertion node u, but their number is also $O(d')$.

When, the insertion node u has been determined, the completed neighbourhood $N'(x)$ of x can be computed in extension by a search of the part of T that is filled, which takes $O(d')$ time. Then, the cotree of the completion H of $G + x$ is obtained from the cotree of G (as depicted in Fig. 3) in the same time complexity thanks to the algorithm of [7]. Overall, one incremental step takes $O(d')$ time and the whole running time of the algorithm is $O(n + m')$.

5   An O(n + m log²n) algorithm

Even though it is linear in the number of edges in the output cograph, the $O(n + m')$ complexity achieved by the algorithm in [24] and the one we presented in Sect. 4 is not necessarily optimal, as the output cograph can actually be represented in $O(n)$ space using its cotree. We then design a refined version of the inclusion-minimal completion algorithm that runs in $O(n + m \log^2 n)$ time, when no additional condition is required on the completion output at each incremental step. This improvement is further motivated by the fact that, as we show below, there exist graphs having only $O(n)$ edges and which require $\Omega(n^2)$ edges in any of their cograph completions. For such graphs, the new complexity we achieve also writes $O(n \log^2 n)$ (since $m = O(n)$) and constitutes a significant improvement over the $O(n^2)$ complexity of the previous algorithm (since $m' = \Omega(n^2)$).

Worst-case minimum-cardinality completion of very sparse graphs. Our proof is based on vertex expander graphs (see [20] for a survey on the topic), which require $\Omega(n^2)$ edges in any of their cograph completions, as stated by Theorem 2 below.

Definition 6. *A graph G is a c-expander if, for every vertex subset $S \subseteq V(G)$ with $|S| \leq \frac{|V(G)|}{2}$ we have $|N(S)| \geq c \cdot |S|$.*

Theorem 2. *Let $c > 0$ be a real number and G a c-expander. For any cograph completion H of G, $|E(H)| \geq \Omega(c^2.n^2)$.*

Sketch of proof. Cographs are known to be also distance hereditary graphs, and therefore totally decomposable by split decomposition, or equivalently of rank-width 1. It implies (for example from a result of [27]) that H contains a split $(S, V \setminus S)$ (meaning that the graph induced by the edges crossing the bipartition $(S, V \setminus S)$ is a complete bipartite graph) such that $n/3 \leq |S| \leq n/2$.

Since G is a c-expander, then $|N(S)| \geq c.|S|$. One can show that the expansion property of G also implies that $|N(V \setminus S)| \geq c.|S|/3$. Consequently, in H, the complete bipartite graph induced by the edges crossing the split $(S, V \setminus S)$ has at least $c.|S|/3$ vertices in S and at least $c.|S|$ in $V \setminus S$. Thus, it contains at least $c^2.|S|^2/3 \geq c^2.n^2/27$ edges.   □

There exist deterministic constructions of very sparse graphs that are c-expanders, see for example the construction of 3-regular c-expanders by Alon and Boppana [1], for some fixed c. Such graphs have only $O(n)$ edges but, from Theorem 2, require $\Omega(n^2)$ edges in any of their cograph completions. More generally, it is part of the folklore that, for any constant $a > 1$, there exist $c > 0$ and $p > 0$ such that, for any $n \in \mathbb{N}$ sufficiently large, the proportion of graphs on n vertices and $a.n$ edges that are c-expanders is at least p. This means that many graphs of fixed mean degree have the vertex expansion property and therefore require $\Omega(n^2)$ edges in any cograph completion. Motivated by this frequent worst-case for the $O(n+m')$ complexity, we now describe an $O(n+m \log^2 n)$-time algorithm for inclusion-minimal cograph completion of arbitrary graphs.

Data structure. We store two distinct copies of the cotree T of G. The first one is a basic data structure in which each node u stores its number of children $|\mathcal{C}(u)|$ and a bidirectional couple of pointers to the corresponding node in the second copy of T, so that we can move from one copy to the other one in $O(1)$ time. In addition, in the first copy of T, each node u stores a copy of the list of its children using the *order data structure* of [11]. It allows to determine which of two given children of u precedes the other one in the list, in $O(1)$ time. It also supports two update operations, delete and insert, that respectively remove and insert an element in the list (just after a specified element), in $O(1)$ time as well.

The second copy of T is stored using the dynamic data structure developed in [29]. This data structure allows to answer two kinds of query: lowest-common-Ancestor?, which provides $lca(u, v)$ for two given nodes u, v, and next-step-to-descendant?, which given a node u of T and one of its strict descendants v, provides the child of u which is an ancestor of v. These two queries are treated in $O(\log n)$ worst-case time. To be precise, the latter query is obtained as a combination of three other basic operations provided by [29] that we do not use here, namely root?, evert and parent?. This data structure is dynamic, meaning that it supports update operations on the structure of the tree (which is actually a forest, as it is allowed to be disconnected). Operation cut removes the edge between one given node and its parent and link makes the root of one tree of the forest become the child of a given node in another tree. These update operations also have $O(\log n)$ worst-case time complexity.

Algorithm. Our algorithm determines the set W of the eligible non-hollow non-completion-forced nodes that are minimal for the ancestor relationship (i.e. none of their descendants satisfies the considered property), and arbitrarily picks one of them to be the insertion node of the minimal completion returned at this incremental step. Indeed, since nodes of W satisfy the conditions of Lemma 4 and none of their children does, it follows that nodes of W are completion-minimal insertion nodes. In order to get the improved $O(n + m \log^2 n)$ complexity, we avoid to completely search the upper tree $T^{up}_{N(x)}$ to determine W. Instead, we use a limited number of lowest-common-Ancestor? queries.

Clearly, if a parallel node u of T is the lca of two leaves in $N(x)$ then $T_u \setminus \{u\}$ contains no eligible node. Let P_{max} be the set of parallel common ancestors of vertices of $N(x)$ that are maximal for the ancestor relationship and let us denote $W' = P_{max} \cup N_{out}$, where N_{out} is the set of vertices of $N(x)$ that are not descendant of any node of P_{max}, i.e. $N_{out} = N(x) \setminus \bigcup_{p \in P_{max}} V(p)$. Note that all the nodes $w' \in W'$ are eligible, and so are their ancestors. It follows that the set W we aim at determining is the set of the lowest non-completion-forced nodes in the upper tree $T^{up}_{W'}$.

In order to compute the set W, we start by computing the tree $\widetilde{T} = T^{xtr}_{N(x) \cup A_x}$ extracted from (see Sect. 2) the leaves that belong to $N(x)$ and the set A_x of their lowest common ancestors, i.e. nodes u such that $u = lca(l_1, l_2)$ for some leaves $l_1, l_2 \in N(x)$. Then, we search $\widetilde{T}$ to find its parallel nodes P_{max} that are maximal for the ancestor relationship and thereby obtain the set W'. Finally, for each node $w' \in W'$ we determine its lowest non-completion-forced ancestor $nfa(w')$ in T and we keep only the $nfa(w')$'s that are minimal for the ancestor relationship: this is the set W. It is worth noting from the beginning that since $\widetilde{T}$ has exactly d leaves and since all its internal nodes have degree at least 2, then the size of $\widetilde{T}$ is $O(d)$.

We now show how to compute $\widetilde{T}$ in $O(d \log^2 n)$ time. To this purpose, we sort the neighbours of x according to a special order of the vertices of the cograph G called a *factorising permutation* [5], which is the order π in which the vertices of G (which are the leaves of T) are encountered when performing a depth-first search of T. One can determine whether a vertex y_1 is before or after a vertex y_2 in π as follows: (1) find $u = lca(y_1, y_2)$ and the two children u_1 and u_2 of u that are respectively the ancestor of y_1 and y_2, and (2) determine whether u_1 is before or after u_2 in the list of children of u. Operation (1) can be implemented by one `lowest-common-ancestor?` query and two `next-step-to-descendant?` queries, which takes $O(\log n)$ time. Operation (2) can be executed in $O(1)$ time using the order data structure of [11]. Therefore, since one comparison takes $O(\log n)$ time, the neighbours of x can be sorted according to π in $O(d \log d \log n) = O(d \log^2 n)$ time.

The benefit of doing so is that $\widetilde{T}$ can be built efficiently by considering the neighbours of x one by one in the order $x_1, x_2, \ldots, x_d$ in which they appear in π (we say from left to right). At each step, we build the tree T_i extracted from $\{x_1, \ldots, x_i\}$ and their lowest common ancestors, then, at the end $T_d = \widetilde{T}$. When inserting x_{i+1} in T_i, at most one new internal node v_{i+1} is created in T_{i+1}. Because we consider the x_i's in the order they appear in π, we have necessarily $v_{i+1} = lca_T(x_i, x_{i+1})$ and v_{i+1} must be inserted in the rightmost branch of T_i, or it already appears in this branch if $v_{i+1} \in T_i$. Consequently, we climb up the rightmost branch of T_i, starting from the father of x_i, and for each node v encountered we determine whether v_{i+1} is above v by computing $lca(v, v_{i+1})$. The total number of lca queries needed to build the whole tree $\widetilde{T}$ is proportional to its size, since every time we pass above a node v on the rightmost branch, v leaves this branch for ever and will then never participate again to any lca query (cf. [13]). Since the size of $\widetilde{T}$ is $O(d)$ and each query takes $O(\log n)$ time, building $\widetilde{T}$ from the sorted list of neighbours of x takes $O(d \log n)$ time.

Once $\widetilde{T}$ is built, a simple search starting from its root determines the set P_{max} of its parallel nodes that are maximal for the ancestor relationship, and we cut off from $\widetilde{T}$ all the subtrees rooted at the children of nodes in P_{max}. The leaves of the resulting tree are precisely the nodes of W'. As $\widetilde{T}$ has size $O(d)$, this step takes $O(d)$ time. Then, for each $w' \in W'$, we determine its lowest non-completion-forced ancestor $nfa(w')$ in T. From the definition of P_{max}, the lowest parallel ancestor of w' is non-completion-forced. Then, $nfa(w')$ cannot be higher in T than the grand-parent of w'. It follows that we have to check the non-completion-forced condition only for w' and its parent, which can be done, for each of them u, in $O(|\mathcal{C}_{nh}(u)|)$ time. Then, we remove the $nfa(w')$'s that are not minimal for the ancestor relationship to obtain the set W, this takes $O(d)$ time, and we arbitrarily pick one node w in W. The minimal completion of the neighbourhood of x returned is the one anchored at w and the total complexity of one incremental step is $O(d + d \log n + d \log^2 n) = O(d \log^2 n)$.

Updating the data structure. After the insertion node w has been determined, the cotree T must be modified as shown in Fig. 3, and the data structure of [29] must be updated accordingly. The key for preserving the complexity is to perform operations involving only the non-hollow children of w. After the insertion of x, w is split into two new nodes w_h and w_{nh} in T', which are parent of respectively the hollow children of w and the non-hollow children of w.

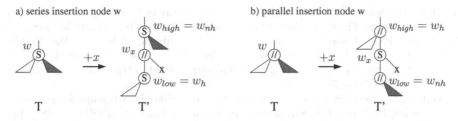

a) series insertion node w   b) parallel insertion node w

Fig. 3. Modification of the cotree under the insertion of x at insertion node w. The triangles in black (resp. white) correspond to the parts of the tree that are filled (resp. that remain hollow) in the completion anchored at w.

To form these two nodes, we cut from w its non-hollow children to obtain w_h, still linked to the hollow children, and we link the non-hollow children to a new node w_{nh}. This takes $O(d \log n)$ as it requires $O(d)$ cut and link operations. The rest of the operations to build T' are less sensitive and, for lack of space, we do not describe them.

As a conclusion, the complexity of one incremental step of the algorithm is $O(d \log^2 n)$ and overall, the complexity of the whole algorithm is $O(n + m \log^2 n)$.

References

1. Alon, N.: Eigenvalues and expanders. Combinatorica **6**(2), 83–96 (1986)
2. Berry, A., Heggernes, P., Simonet, G.: The minimum degree heuristic and the minimal triangulation process. In: Bodlaender, H.L. (ed.) WG 2003. LNCS, vol. 2880, pp. 58–70. Springer, Heidelberg (2003). https://doi.org/10.1007/978-3-540-39890-5_6
3. Bodlaender, H., Downey, R., Fellows, M., Hallett, M., Wareham, H.: Parameterized complexity analysis in computational biology. Comput. Appl. Biosci. **11**, 49–57 (1995)
4. Brandes, U., Hamann, M., Strasser, B., Wagner, D.: Fast Quasi-threshold editing. In: Bansal, N., Finocchi, I. (eds.) ESA 2015. LNCS, vol. 9294, pp. 251–262. Springer, Heidelberg (2015). https://doi.org/10.1007/978-3-662-48350-3_22
5. Capelle, C., Habib, M., de Montgolfier, F.: Graph decompositions and factorizing permutations. Discrete Math. Theoret. Comput. Sci. **5**(1), 55–70 (2002)
6. Corneil, D., Lerchs, H., Burlingham, L.S.: Complement reducible graphs. Discrete Appl. Math. **3**(3), 163–174 (1981)
7. Corneil, D., Perl, Y., Stewart, L.: A linear time recognition algorithm for cographs. SIAM J. Comput. **14**(4), 926–934 (1985)
8. Crespelle, C., Paul, C.: Fully dynamic recognition algorithm and certificate for directed cographs. Discrete Appl. Math. **154**(12), 1722–1741 (2006)
9. Crespelle, C., Perez, A., Todinca, I.: An $\mathcal{O}(n^2)$ time algorithm for the minimal permutation completion problem. In: Mayr, E.W. (ed.) WG 2015. LNCS, vol. 9224, pp. 103–115. Springer, Heidelberg (2016). https://doi.org/10.1007/978-3-662-53174-7_8
10. Crespelle, C., Todinca, I.: An $O(n^2)$-time algorithm for the minimal interval completion problem. Theor. Comput. Sci. **494**, 75–85 (2013)

11. Dietz, P., Sleator, D.: Two algorithms for maintaining order in a list. In: 19th ACM Symposium on Theory of Computing (STOC 1987), pp. 365–372. ACM (1987)
12. Drange, P.G., Dregi, M.S., Lokshtanov, D., Sullivan, B.D.: On the threshold of intractability. In: Bansal, N., Finocchi, I. (eds.) ESA 2015. LNCS, vol. 9294, pp. 411–423. Springer, Heidelberg (2015). https://doi.org/10.1007/978-3-662-48350-3_35
13. Gabow, H., Bentley, J., Tarjan, R.: Scaling and related techniques for geometry problems. In: 16th ACM Symposium on Theory of Computing (STOC 1984), pp. 135–143. ACM (1984)
14. Goldberg, P., Golumbic, M., Kaplan, H., Shamir, R.: Four strikes against physical mapping of DNA. J. Comput. Biol. **2**, 139–152 (1995)
15. Guillemot, S., Havet, F., Paul, C., Perez, A.: On the (non-)existence of polynomial kernels for P_l-free edge modification problems. Algorithmica **65**(4), 900–926 (2012)
16. Heggernes, P., Mancini, F., Papadopoulos, C.: Minimal comparability completions of arbitrary graphs. Discrete Appl. Math. **156**(5), 705–718 (2008)
17. Heggernes, P., Telle, J.A., Villanger, Y.: Computing minimal triangulations in time $O(n^{\alpha \log n}) = o(n^{2.376})$. SIAM J. Discrete Math. **19**(4), 900–913 (2005)
18. Heggernes, P., Mancini, F.: Minimal split completions. Discrete Appl. Math. **157**(12), 2659–2669 (2009)
19. Hellmuth, M., Wieseke, N., Lechner, M., Lenhof, H.P., Middendorf, M., Stadler, P.F.: Phylogenomics with paralogs. PNAS **112**(7), 2058–2063 (2015)
20. Hoory, S., Linial, N., Wigderson, A.: Expander graphs and their applications. Bull. Am. Math. Society **43**(4), 439–561 (2006)
21. Jia, S., Gao, L., Gao, Y., Nastos, J., Wang, Y., Zhang, X., Wang, H.: Defining and identifying cograph communities in complex networks. New J. Phys. **17**(1), 013044 (2015)
22. Kendall, D.: Incidence matrices, interval graphs, and seriation in archeology. Pacific J. Math. **28**, 565–570 (1969)
23. Liu, Y., Wang, J., Guo, J., Chen, J.: Complexity and parameterized algorithms for cograph editing. Theoret. Comput. Sci. **461**, 45–54 (2012)
24. Lokshtanov, D., Mancini, F., Papadopoulos, C.: Characterizing and computing minimal cograph completions. Discrete Appl. Math. **158**(7), 755–764 (2010)
25. Mancini, F.: Graph modification problems related to graph classes. Ph.D. thesis, University of Bergen, Norway (2008)
26. Ohtsuki, T., Mori, H., Kashiwabara, T., Fujisawa, T.: On minimal augmentation of a graph to obtain an interval graph. J. Comput. Syst. Sci. **22**(1), 60–97 (1981)
27. Oum, S., Seymour, P.D.: Testing branch-width. J. Comb. Theory Ser. B **97**(3), 385–393 (2007)
28. Rapaport, I., Suchan, K., Todinca, I.: Minimal proper interval completions. Inf. Process. Lett. **106**(5), 195–202 (2008)
29. Sleator, D.D., Tarjan, R.E.: A data structure for dynamic trees. J. Comput. Syst. Sci. **26**(3), 362–391 (1983)

Structure of Towers and a New Proof of the Tight Cut Lemma

Nanao Kita[✉]

National Institute of Informatics, 2-1-2 Hitotsubashi, Chiyoda-ku,
Tokyo 101-8430, Japan
kita@nii.ac.jp

Abstract. In the first part of our study, we extend the theory of basilica canonical decomposition by introducing new concepts known as *towers* and *tower-sequences*. The basilica canonical decomposition is a recently proposed tool in matching theory that can be applied non-trivially even for general graphs with perfect matchings. When studying matchings, the structure of *alternating paths* frequently needs to be considered. We show how a graph is made up of towers and tower-sequences, and thus obtain the structure of alternating paths in terms of the basilica canonical decomposition. This result provides a strong tool for analyzing general graphs with perfect matchings.

The second part of our study is a new graph theoretic proof of the so-called *Tight Cut Lemma* derived from the first part of our study. To derive a characterization of the perfect matchings polytope, Edmonds, Lovász, and Pulleyblank introduced the Tight Cut Lemma as the most challenging aspect of their work. The Tight Cut Lemma in fact claims *bricks* as the fundamental building blocks that constitute a graph and can be referred to as a key result in this field. Although the Tight Cut Lemma itself is a purely graph theoretic statement, there was no known graph theoretic proof for decades until Szigeti provided such a proof using Frank-Szigeti's optimal ear decomposition theory.

By contrast, we provide a new proof using the extended theory of basilica canonical decomposition as the only preliminary result, and accordingly proposes a new strategy for studying bricks and tight cuts or matching theory in general. Our proof shows how the discussions on alternating paths construct the Tight Cut Lemma from first principles via the basilica canonical decomposition, even without using *barriers*, that is, the dual notion of matchings. The distinguishing features of our proof are that it is purely graph theoretic, purely matching (cardinality 1-matching) theoretic, and purely "primal" with respect to matchings.

1 Introduction

In this paper, we extend the theory of basilica canonical decomposition [8,10,11] by introducing *towers* and *tower-sequences* and then use this extension to provide

Supported by JSPS KAKENHI Grant Number 15J09683.

X. Gao et al. (Eds.): COCOA 2017, Part I, LNCS 10627, pp. 225–239, 2017.
https://doi.org/10.1007/978-3-319-71150-8_20

a new graph theoretic proof of the Tight Cut Lemma originally proposed and proved by Edmonds, Lovász, and Pulleyblank [5].

The basilica canonical decomposition [8,10,11] is a new tool in matching theory that has been recently proposed. As the term "canonical" conventionally means in the mathematical context, *canonical decompositions* are a standard tool to analyze graphs in matching theory [18]. Several types of canonical decompositions have been known, and each canonical decomposition partitions a graph into substructures in a way determined uniquely for the graph, and thus describes the structure of all maximum matchings in the graph using this partition. The classically known canonical decompositions are the *Gallai-Edmonds* [4,7], *Kotzig-Lovász* [13–16], and *Dulmage-Mendelsohn* [1–3]. However, they target particular classes of graphs or are too sparse to provide sufficient information; the Kotzig-Lovász and Dulmage-Mendelsohn decompositions are applicable to *factor-connected graphs* and bipartite graphs, respectively, whereas the Gallai-Edmonds decomposition is so sparse that any graph with perfect matchings falls into an irreducible case and cannot be decomposed nontrivially. Therefore, no known canonical decomposition could be used to analyze general graphs with perfect matchings, until Kita [8,10,11] introduced a new one: the *basilica canonical decomposition*. This decomposition is applicable nontrivially to general graphs with perfect matchings, and therefore provides a refinement of the Gallai-Edmonds decomposition.

The first contribution of this paper is an extension of the theory of basilica canonical decomposition by the introduction of new tools: *towers* and *tower-sequences*. A *tower* is a type of subgraph defined using the basilica canonical decomposition. In studying matchings, the structure of *alternating paths*[1] in a given graph often needs to be considered. The structure of alternating paths in a single tower is rather easy to observe [8,9,12], and a graph with perfect matchings can be viewed as being constructed by "gluing" towers together. Hence, we introduce a notion of a *tower-sequence* to capture the interrelationship between towers, that is, how towers are "glued" to each other in a graph. Thus, we are able to analyze alternating paths in an entire graph. This provides the structure of alternating paths in terms of the basilica canonical decomposition, and therefore can be a strong tool for studying matchings. This contribution is applicable to general graphs with perfect matchings.

The second contribution of this paper is a new proof of the Tight Cut Lemma derived from the first contribution. Edmonds et al. [5] introduced the *Tight Cut Lemma* as a key result in their paper to characterize the perfect matching polytope. They stated that proving the Tight Cut Lemma was the most difficult part of their study.

Tight Cut Lemma. Any tight cut in a brick is trivial.

---

[1] Regarding the duality of the maximum matching problem, alternating paths are essential to the primal optimality and algorithms. Hence, we say that alternating paths have a "primal" nature regarding matchings, in contradistinction to barriers.

A graph is a *brick* if deleting any two vertices results in a connected graph with a perfect matching. A cut is *tight* if it shares exactly one edge with any perfect matching. A tight cut is *trivial* if it is a star cut.

The Tight Cut Lemma is in fact a characterization of *bricks* as the fundamental building blocks that constitute a graph in the polyhedral study of matchings. Provided a given graph has a non-trivial tight cut, we can decompose it into two smaller graphs by contracting each shore; this operation is the *tight cut decomposition*. Naddef [19] proved that the property of the perfect matching polytope of a graph is determined by the set of small graphs obtained by repeatedly applying the tight cut decomposition until it gets stuck. Therefore, the question of what type of graphs are the irreducible class of the tight cut decomposition is inevitable, and the answer is bricks, according to the Tight Cut Lemma. It is easily observed from known facts that if a graph is not a brick, then it has a non-trivial tight cut. Hence, the Tight Cut Lemma proves the essential part of the above characterization.

Since it was proposed, the Tight Cut Lemma has directed the study of the perfect matching polytope. From the Tight Cut Lemma, Edmonds et al. [5] derived the dimension of the perfect matching polytope using as a parameter the number of bricks that constitute a given graph. Consequently, they determine the minimal set of inequalities that defines the perfect matching polytope. Since the work of Edmonds et al. [5], the study of bricks and the consequential results on the perfect matching polytope (and lattice) have flourished; see Lovász [17] or the surveys [18, 20].

To contribute to the theoretical basis of combinatorial optimization, this paper focuses on how to derive the Tight Cut Lemma. When we wish to further develop a systematic theory, we need to understand clearly how it is organized, and this requires a close examination of the way in which important theorems can be derived. In recent decades, polyhedral combinatorics has succeeded using matchings as an archetypal problem [18, 20]; that is, matching theory is a branch with a special meaning that can affect the entire field of combinatorial optimization, and the Tight Cut Lemma is definitely one of its milestones. A good proof of the Tight Cut Lemma will provide us with a technique to handle bricks and tight cuts.

There is more than one proof of the Tight Cut Lemma. The original proof by Edmonds et al. [5] used a linear programming argument. As the Tight Cut Lemma itself is a purely graph theoretic statement, a purely graph theoretic proof has been awaited. Szigeti [22, 24] later provided a purely graph theoretic proof using Frank-Szigeti's *optimal ear decomposition theory* [6, 21, 23].

By contrast, we derive a new graph theoretic proof of the Tight Cut Lemma from the extended theory of basilica canonical decomposition. To prove the Tight Cut Lemma, we need to detect an "alternating circuit" in a brick, and this can be reduced to the task of detecting an alternating path with a certain condition in a graph with perfect matchings. As this graph does not fall into any of the classes targeted by classical canonical decompositions, the device of tower-sequences is effective for solving our problem.

Our poof method is novel in that we use the basilica canonical decomposition theory as the only preliminary result, and that our proof shows how the Tight Cut Lemma is constructed from first principles via the basilica canonical decomposition by combinatorial discussions on alternating paths; our proof does not even use *barriers*. In the following, we explain these in more detail. The basilica canonical decomposition is established from first principles by discussing the structure of alternating paths in graphs, and our results on towers and tower-sequences are derived from the basilica canonical decomposition by discussing the structure of alternating paths. Furthermore, our proof of the Tight Cut Lemma is derived from these results by discussing the structure of alternating paths. In this entire process, no other known results are used. Hence, our proof shows how discussions on alternating paths construct the Tight Cut Lemma from first principles.

Therefore, our proof has distinguishing features that it is *purely matching* (*cardinality 1-matching*) *theoretic* and *purely "primal"* with respect to matchings, in addition to that it is purely graph theoretic. Besides alternating paths, *barriers*, which are a combinatorial notion that corresponds to the dual of matchings, are frequently used in matching theory discussions[2]; barriers are defined by a min-max theorem called the *Berge formula*, in which cardinality maximum 1-matchings are the optimizers on one side of this formula, whereas barriers are the optimizers of the other side. Hence, regarding the study of matchings, discussions of alternating paths and barriers can be referred to as primal and dual methods, respectively. Each known proof of the Tight Cut Lemma uses barriers. In fact, Szigeti [22, 24] made use of barriers in his proof of the Tight Cut Lemma, in addition to Frank-Szigeti's optimal ear decomposition theory, which also uses barriers. The proof by Edmonds et al. [5] also uses barriers, as they uses Lovász's result [16]. By contrast, our proof does not use barriers directly or indirectly and is purely primal. Frank-Szigeti's optimal ear decomposition theory is established using T-join theory, therefore Szigeti's proof indirectly uses T-joins. By contrast, our proof uses results closed within the cardinality 1-matching theory; this seems reasonable considering the statement of the Tight Cut Lemma, which is also regarding cardinality 1-matchings.

These features indicate that our proof provides a new direction for how to proceed in the study of bricks and tight cuts, or matching theory.

The remainder of this paper is organized as follows. Section 2 presents preliminary definitions and results: Sect. 2.1 explains fundamental notation and definitions; Sect. 2.2 introduces the canonical decomposition given by Kita [10, 11]. Section 3 introduces new results of us; here, we further develop a device to analyze the structure of graphs with perfect matchings, which will be used in Sect. 4. Section 4 gives the new proof of the Tight Cut Lemma.

---

[2] We believe that almost all substantial results regarding cardinality 1-matchings use barriers. This may be problematic because, as Lovász and Plummer state [18], not so much is known about barriers and therefore the limitation of our knowledge about barriers can be the limitation of our ability to proceed in studying matchings.

2 Preliminaries

2.1 Notation and Definitions

For standard notation and definitions on sets and graphs, we mainly follow Schrijver [20]. In this section, we list those that are exceptional or non-standard. We denote the vertex set and the edge set of a graph G by $V(G)$ and $E(G)$. We sometimes refer to the vertex set of a graph G simply as G. We treat paths and circuits as graphs. Given a path P and two vertices x and y in $V(P)$, xPy denotes the connected subgraph of P, which is of course a path, that has the ends x and y. In referring to graphs obtained by usual graph theoretic operations such as contractions and taking subgraphs, we often identify items such as vertices and edges with the naturally corresponding items of old graphs.

Let G be a graph, and let $X \subseteq V(G)$. The subgraph of G induced by X is denoted by $G[X]$. The notation $G - X$ denotes the graph $G[V(G) \setminus X]$. The contraction of G by X is denoted by G/X. Let $\hat{G}$ be a supergraph of G, and let $F \subseteq E(\hat{G})$. The notation $G + F$ and $G - F$ denotes the graphs obtained by adding and by deleting F from G. Given another subgraph H of $\hat{G}$, the graph $G + H$ denotes the union of G and H. In referring to graphs obtained by these operations, we often identify their items such as vertices and edges with the naturally corresponding items of old graphs.

The set of neighbors of $X \subseteq V(G)$ in a graph G is denoted by $N_G(X)$; namely, $N_G(X) := \{u \in V(G) \setminus X : \exists v \in X \text{ s.t. } uv \in E(G)\}$. Given $X, Y \subseteq V(G)$, $E_G[X, Y]$ denotes the set of edges of G that join X and Y. We denote $E_G[X, V(G) \setminus X]$ by $\delta_G(X)$. We often omit the subscripts "G" in using this notation.

Given a graph, a *matching* is a set of edges in which any two are disjoint. A matching is a *perfect matching* if every vertex of the graph is adjacent to one of its edges. A graph is *factorizable* if it has a perfect matching. An edge of a factorizable graph is *allowed* if it is contained in a perfect matching. A graph G is *factor-critical* if it has only a single vertex or for, any $v \in V(G)$, $G - v$ is factorizable.

Given a set of edges M, a circuit C is *M-alternating* if $E(C) \cap M$ is a perfect matching of C. A path P with two ends x and y is *M-saturated* (resp. *M-exposed*) between x and y if $E(P) \cap M$ (resp. $E(P) \setminus M$) is a perfect matching of P. A path P with ends x and y is *M-forwarding* from x to y if $E(P) \cap M$ is a matching of P and, among the vertices in $V(P)$, only y is disjoint from the edges in $E(P) \cap M$. We define a trivial graph, i.e., a graph with a single vertex and no edges, as an M-forwarding path.

Given a set of vertices X, an M-exposed path is an *M-ear* relative to X if the ends are in X while the other vertices are disjoint from X; also, a circuit C is an *M-ear* relative to X if $V(C) \cap X = \{x\}$ holds and $C - x$ is an M-saturated path. In the first case, we say the M-ear is *proper*. Even in the second case, we call x an end of the M-ear for convenience. An M-ear is *trivial* if it consists of only a single edge. We say an M-ear *traverses* a set of vertices Y if it has a vertex other than the ends that is in Y.

2.2   Canonical Decomposition for General Factorizable Graphs

We now introduce the basilica canonical decomposition [10,11], which will be used in Sects. 3 and 4 as the only preliminary result to derive the Tight Cut Lemma. The principal results that constitute the theory of this canonical decomposition are Theorems 1, 2, and 3. In this section, unless otherwise stated, G denotes a factorizable graph.

Definition 1. *Let $\hat{M}$ be the union of all perfect matchings of G. A factor-component of G is the subgraph induced by $V(C)$, where C is a connected component of the subgraph of G determined by $\hat{M}$. The set of factor-components of G is denoted by $\mathcal{G}(G)$. That is to say, a factorizable graph consists of factor-components and edges joining distinct factor-components. A separating set of G is a set of vertices that is the union of the vertex sets of some factor-components of G. Note that if $X \subseteq V(G)$ is a separating set, then $\delta_G(X) \cap M = \emptyset$ for any perfect matching M of G.*

Definition 2. *Given $G_1, G_2 \in \mathcal{G}(G)$, we say $G_1 \lhd_G G_2$ if there is a separating set $X \subseteq V(G)$ such that $V(G_1) \cup V(G_2) \subseteq X$ holds and $G[X]/V(G_1)$ is a factor-critical graph. We sometimes denote $\lhd_G$ simply by $\lhd$.*

The next theorem is highly analogous to the known *Dulmage-Mendelsohn decomposition* for bipartite graphs [18], in that it describes a partial order over $\mathcal{G}(G)$:

Theorem 1 (Kita [10,11]). *In any factorizable graph, $\lhd$ is a partial order over $\mathcal{G}(G)$.*

Under Theorem 1, we denote the poset of $\lhd$ over $\mathcal{G}(G)$ by $\mathcal{O}(G)$. For $H \in \mathcal{G}(G)$, the set of upper bounds of H in $\mathcal{O}(G)$ is denoted by $\mathcal{U}_G^*(H)$. The union of vertex sets of all upper bounds of H is denoted by $U_G^*(H)$. We denote $\mathcal{U}_G^*(H) \setminus \{H\}$ by $\mathcal{U}_G(H)$ and $U_G^*(H) \setminus V(H)$ by $U_G(H)$. We sometimes write them by omitting the subscripts "G".

Definition 3. *Given $u, v \in V(G)$, we say $u \sim_G v$ if u and v are contained in the same factor-component and $G - u - v$ has no perfect matching.*

Theorem 2 (Kita [10,11]). *In any factorizable graph G, $\sim_G$ is an equivalence relation on $V(G)$. Each equivalence class is contained in the vertex set of a factor-component.*

Given $H \in \mathcal{G}(G)$, we denote by $\mathcal{P}_G(H)$ the family of equivalence classes of $\sim_G$ that are contained in $V(H)$. Note that $\mathcal{P}_G(H)$ gives a partition of $V(H)$. The structure given by Theorem 2 is called the *generalized Kotzig-Lovász partition* as it is a generalization of the results given by Kotzig [13–15] and Lovász [16].

Even though Theorems 1 and 2 were established independently, a natural relationship between the two is shown by the next theorem.

Theorem 3 (Kita [10,11]). *Let G be a factorizable graph, and let $H \in \mathcal{G}(G)$. Let K be a connected component of $G[U_G(H)]$. Then, there exists $S \in \mathcal{P}_G(H)$ with $N_G(K) \cap V(H) \subseteq S$.*

Intuitively, Theorem 3 states that each proper upper bound of a factor-component H is tagged with a single member from $\mathcal{P}_G(H)$. As a result of Theorem 3, the two structures given by Theorems 1 and 2 are unified naturally to produce a new canonical decomposition that enables us to analyze a factorizable graph as a building-like structure in which each factor-component serves as a floor and each equivalence class serves as a foundation.

As given in Theorem 3, for $H \in \mathcal{G}(G)$ and $S \in \mathcal{P}_G(H)$, we define $\mathcal{U}_G(S) \subseteq \mathcal{U}_G(H)$ as follows: $I \in \mathcal{U}_G(H)$ is in $\mathcal{U}_G(S)$ if the connected component K of $G[U_G(H)]$ with $V(I) \subseteq V(K)$ satisfies $N_G(K) \cap V(H) \subseteq S$. The union of vertex sets of factor-components in $\mathcal{U}_G(S)$ is denoted by $U_G^*(S)$. The sets $U_G^*(S) \setminus S$ and $U_G^*(H) \setminus U_G^*(S)$ are denoted by $U_G(S)$ and $^\top U_G(S)$, respectively. Note that the family $\{U^*(S) : S \in \mathcal{P}_G(H)\}$ (resp. $\{U(S) : S \in \mathcal{P}_G(H)\}$) gives a partition of $U^*(H)$ (resp. $U(H)$). We sometimes omit the subscript "G" if the meaning is apparent from the context.

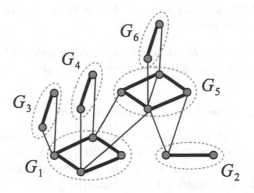

Fig. 1. A graph G and its factor-components $G_1, \ldots, G_6$: Thick edges indicate allowed edges.

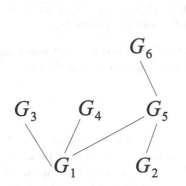

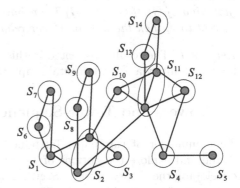

Fig. 2. The Hasse diagram of $(\mathcal{G}(G), \vartriangleleft)$

Fig. 3. The generalized Kotzig-Lovász partition of G.

Example 1. As for the graph G given in Fig. 1, the poset $\mathcal{O}(G)$ is described by the Hasse diagram in Fig. 2. The generalized Kotzig-Lovász partition $\mathcal{P}(G)$ is described in Fig. 3.

In the remainder of this section, we present some pertinent properties that will be used in later sections.

Lemma 1 (Kita [9,12]). *Let G be a factorizable graph and M be a perfect matching of G, and let $H \in \mathcal{G}(G)$. Let $S \in \mathcal{P}_G(H)$, and let $T \in \mathcal{P}_G(H)$ be such with $S \neq T$.*

(i) *For any $x \in U^*(S)$, there exists $y \in S$ such that there is an M-forwarding path from x to y whose vertices except for y are in $U(S)$.*

(ii) *For any $x \in S$ and any $y \in T$, there is an M-saturated path between x and y whose vertices are in $U^*(H) \setminus U(S) \setminus U(T)$.*

(iii) *For any $x \in S$ and any $y \in {}^\top U(S)$, there is an M-forwarding path from x to y whose vertices are in $U^*(H) \setminus U(S)$.*

(iv) *For any $x \in U^*(S)$ and any $y \in U^*(T)$ there is an M-saturated path between x and y whose vertices are in $U^*(H)$.*

Lemma 2 (Kita [10,11]). *Let G be a factorizable graph, and let M be a perfect matching of G. If there is an M-ear relative to $H_1 \in \mathcal{G}(G)$ and traversing $H_2 \in \mathcal{G}(G)$, then $H_1 \lhd H_2$ holds.*

From Lemma 2, the next lemma is easily derived.

Lemma 3 (Kita [10,11]). *Let G be a factorizable graph and M be a perfect matching of G. Let $x \in V(G)$, and let $H \in \mathcal{G}(G)$ be the factor-component that contains x. If there is an M-ear P relative to $\{x\}$, then the connected components of $P - E(H)$ are M-ears relative to H. Accordingly, if $I \in \mathcal{G}(G)$ has common vertices with P, then $H \lhd I$ holds.*

Lemma 4 (Kita [10,11]). *Let G be a factorizable graph, and M be a perfect matching of G. If $G_1 \in \mathcal{G}(G)$ is an immediate lower-bound of $G_2 \in \mathcal{G}(G)$ with respect to $\lhd$, then there is an M-ear relative to G_1 and traversing G_2.*

Remark 1. The results presented in this section are obtained without using any known results via a fundamental graph theoretic discussion on matchings.

3 Towers and Tower-Sequences

The remainder of this paper introduces the new results. In this section, We define and explore the notions of *towers*, *arcs*, and *tower-sequences* by further developing the theory given in Sect. 2.2, and obtain a new device to capture the structure of factorizable graphs. The results in this section will be used in Sect. 4.

In this section, unless otherwise stated, let G be a factorizable graph and M be a perfect matching. The set of minimal element in the poset $\mathcal{O}(G)$ is denoted by $\mathrm{Min}\mathcal{O}(G)$.

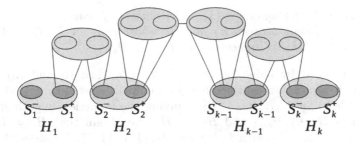

Fig. 4. An abstract image of a tower-sequence with $k = 4$

Definition 4. *Let $H \in \mathcal{G}(G)$. A* tower *over H is the subgraph $G[U^*(H)]$ and is denoted by $\mathcal{T}_G(H)$ or simply by $\mathcal{T}(H)$. Given $H_1, H_2 \in \mathcal{G}(G)$ such that neither $H_1 \lhd H_2$ nor $H_2 \lhd H_1$ holds, we say $\mathcal{T}(H_1)$ and $\mathcal{T}(H_2)$ are* tower-adjacent *or* t-adjacent *if $U(H_1) \cap U(H_2) \neq \emptyset$ or $E[U^*(H_1), U^*(H_2)] \neq \emptyset$ holds. Here, for each $i, j \in \{1, 2\}$ with $i \neq j$, $S_i \in \mathcal{P}_G(H_i)$ is a* port *of this adjacency if $U(S_i) \cap U(H_j) \neq \emptyset$ or $E[U^*(S_i), U^*(H_j)] \neq \emptyset$ hold.*

Definition 5. *Let $H_1, H_2 \in \mathcal{G}(G)$ be two distinct factor-components. An M-exposed path P is an M-arc between H_1 and H_2 if the ends of P are in H_1 and H_2 whereas the internal vertices are disjoint from H_1 and H_2.*

The next lemma states that t-adjacency implies an M-arc with a certain property.

Lemma 5. *Let G be a factorizable graph, and M be a perfect matching of G. Let $H_1, H_2 \in \mathcal{G}(G)$ be such that neither $H_1 \lhd H_2$ nor $H_2 \lhd H_1$ hold. If $\mathcal{T}(H_1)$ and $\mathcal{T}(H_2)$ are t-adjacent, with ports $S_1 \in \mathcal{P}_G(H_1)$ and $S_2 \in \mathcal{P}_G(H_2)$, then there is an M-arc between H_1 and H_2, whose ends are in S_1 and S_2 whereas the internal vertices are contained in $U(S_1) \cup U(S_2)$.*

Proof. Let $uv \in E[U^*(S_1), U^*(S_2) \setminus U^*(S_1)]$, where $u \in U^*(S_1)$ and $v \in U^*(S_2) \setminus U^*(S_1)$. By Lemma 1 (i), there is an M-forwarding path P_2 from v to a vertex $w \in S_2$ with $V(P_2) \setminus \{w\} \subseteq U(S_2)$. Also, there is an M-forwarding path P_1 from u to a vertex $z \in S_2$ with $V(P_1) \setminus \{z\} \subseteq U(S_1)$. By Lemma 3, we have that $P_1 + uv + P_2$ is a desired M-arc. $\qquad\square$

Definition 6. *Let $H_1, \ldots, H_k \in \mathcal{G}(G)$, where $k \geq 1$. For each $i \in \{1, \ldots, k\}$, let $S_i^+, S_i^- \in \mathcal{P}_G(H_i)$ be such with $S_i^+ \neq S_i^-$. We say $H_1, \ldots, H_k$ is a* tower-sequence, *from H_1 to H_k, if $k = 1$ holds or if $k > 1$ holds and for each $i \in \{1, \ldots, k-1\}$, $\mathcal{T}(H_i)$ and $\mathcal{T}(H_{i+1})$ are t-adjacent with ports S_i^+ and S_{i+1}^-.*

See Fig. 4, which describes an abstract image of a tower-sequence.

Theorem 4. *Let G be a factorizable graph, and M be a perfect matching of G. Let $H_1, \ldots, H_k \in \mathrm{Min}\mathcal{O}(G)$, where $k > 1$, be a tower-sequence with ports $S_i^+, S_i^- \in \mathcal{P}_G(H_i)$ for $i \in \{1, \ldots, k\}$. Then,*

(i) $H_i \neq H_j$ holds for any $i, j \in \{1, \ldots, k\}$ with $i \neq j$, and

(ii) there is an M-arc between H_1 and H_k whose ends are in S_1^+ and S_k^- and which, if $k \geq 3$ holds, traverses each $H_2, \ldots, H_{k-1}$.

Proof. We proceed by induction on k. If $k = 2$, then (i) and (ii) hold by the definition of a tower-sequence and by Lemma 5. Let $k > 2$, and suppose (i) and (ii) hold for $1, \ldots, k-1$. By applying the induction hypothesis to the substructures $H_1, \ldots, H_{k-1}$ and $H_2, \ldots, H_k$, we obtain that $H_i \neq H_j$ holds for any $i, j \in \{1, \ldots, k\}$ with $i \neq j$ and $\{i, j\} \neq \{1, k\}$. Consider the subsequence $H_1, \ldots, H_{k-1}$. There is an M-arc $\hat{P}$ between H_1 and H_{k-1} that satisfies (ii). Let $\hat{s} \in S_1^+$ and $\hat{t} \in S_{k-1}^-$ be the ends of $\hat{P}$. By Lemma 5, there is an M-arc P between H_{k-1} and H_k, whose ends are $s \in S_{k-1}^+$ and $t \in S_k^-$, such that its vertices, except for s and t, are in $U(S_{k-1}^+) \cup U(S_k^-)$. By Lemma 1 (ii), there is an M-saturated path Q between $\hat{t}$ and s with $V(Q) \subseteq U^*(H_{k-1}) \setminus U(S_{k-1}^-) \setminus U(S_{k-1}^+)$. Let $\hat{Q} := P + Q$; then, $\hat{Q}$ is an M-forwarding path from $\hat{t}$ to t that traverses H_{k-1}. By Lemma 3, $\hat{P} + \hat{Q}$ is an M-exposed path that traverses $H_2, \ldots, H_{k-1}$. If $H_1 = H_k$ holds, then $\hat{P} + \hat{Q}$ is an M-ear relative to H_1. This is a contradiction by Lemma 2, because $H_2, \ldots, H_{k-1} \in \mathrm{Min}\mathcal{O}(G)$. Hence, we obtain $H_1 \neq H_k$, and so $H_1, \ldots, H_k$ are all mutually distinct. Accordingly, $\hat{P} + \hat{Q}$ is an M-arc satisfying the statement. $\square$

Definition 7. *A factor-component $H \in \mathrm{Min}\mathcal{O}(G)$ is a* border *of G if $\mathcal{T}(H)$ is t-adjacent with no other tower or if exactly one member S from $\mathcal{P}_G(H)$ can be a* port *by which $\mathcal{T}(H)$ is t-adjacent with other towers, i.e., $E[U^*(S), V(G) \setminus U^*(H)] \neq \emptyset$ and $E[U^*(T), V(G) \setminus U^*(H)] = \emptyset$ for any $T \in \mathcal{P}_G(H) \setminus \{S\}$ hold. Here, S is the* port *of the border H. We denote the set of borders of G by $\partial\mathcal{O}(G)$.*

Definition 8. *We say a tower-sequence $H_1, \ldots, H_k \in \mathrm{Min}\mathcal{O}(G)$, where $k \geq 1$, is* spanning *if H_1 and H_k are borders of G. An M-arc between $H \in \mathcal{G}(G)$ and $I \in \mathcal{G}(G)$ is* spanning *if H and I are borders of G.*

As Theorem 4 (i) states that no members are repeated in any tower-sequence, the next theorem follows:

Theorem 5. *Let G be a factorizable graph. For a tower-sequence $H_1, \ldots, H_k \in \mathrm{Min}\mathcal{O}(G)$, there is a spanning tower-sequence $I_1, \ldots, I_l \in \mathrm{Min}\mathcal{O}(G)$ with $l \geq k$ and $I_i = H_1, \ldots, I_{i+k} = H_k$ for some $i \in \{1, \ldots, k-l\}$.*

4   A New Proof of the Tight Cut Lemma

4.1   Shared Definitions, Assumptions, Lemmas

In Sect. 4, we introduce our new proof of the Tight Cut Lemma.

Formal Statement of the Tight Cut Lemma. Let $\hat{G}$ be a brick, and $\hat{S} \subseteq V(\hat{G})$ be such with $1 < |\hat{S}| < |V(\hat{G})| - 1$. Then, there is a perfect matching with more than one edge in $\delta_{\hat{G}}(\hat{S})$.

In the following, we prove the above. Let $\hat{M}$ be a perfect matching of G. If $|\delta_{\hat{G}}(\hat{S}) \cap \hat{M}| > 1$ holds, then we have nothing to do. Hence, in the following, we assume $|\delta_{\hat{G}}(\hat{S}) \cap \hat{M}| = 1$ and prove $\hat{S}$ is not a tight cut by finding an $\hat{S}$-fat perfect matching, i.e., a perfect matching with more than one edge in $\delta_{\hat{G}}(\hat{S})$.

Let $\hat{S}^c$ be $V(\hat{G}) \setminus \hat{S}$. Let $u \in \hat{S}$ and $v \in \hat{S}^c$ be such that $\delta_{\hat{G}}(\hat{S}) \cap \hat{M} = \{uv\}$. We denote $\hat{G} - u - v$ by G, $\hat{S} - u$ by S, $\hat{S}^c - v$ by S^c, and $\hat{M} - uv$ by M.

Note that G is connected and has a perfect matching M. Also, $\delta_G(S) \cap M = \emptyset$ holds in G. If S is not a separating set, then of course $\delta_{\hat{G}}(\hat{S})$ is not a tight cut in G, and we are done. So, in the following, we assume that S *is a separating set of G* and prove the Tight Cut Lemma for this case.

Without loss of generality, we also assume in the following that G *has a border whose vertices are contained in S*.

In the following, we present lemmas that will be used in Sects. 4.2 and 4.3 when we find a cut vertex in G. They are relatively easy to confirm.

Lemma 6. *Let x be a cut vertex of G, and let C be one of the connected components of $G - x$. Then, $N_{\hat{G}}(w) \cap V(C) \neq \emptyset$ holds for each $w \in \{u, v\}$.*

Lemma 7. *Let x be a cut vertex of G, and let C be one of the connected components of $G - x$. If $V(C) \cup \{x\}$ is a separating set of G, then, for each $w \in \{u, v\}$, there exists $y \in V(C) \cap N_{\hat{G}}(w)$ such that G has an M-saturated path between x and y.*

4.2 When There Exists a Factor-Component in $\mathrm{Min}\mathcal{O}(G)$ Whose Vertices are in S^c

Here in Sect. 4.2, we assume that $\mathrm{Min}\mathcal{O}(G)$ has a factor-component whose vertex set is contained in S^c, and prove the Tight Cut Lemma for this case, using mainly the results obtained in Sect. 3.

Lemma 8. *If G has a spanning M-arc with an edge in $E_G[S, S^c]$, then $\hat{G}$ has an $\hat{S}$-fatperfect matching.*

Proof. Let P be a spanning M-arc, between two borders H_1 and H_2. Let S_1 and S_2 be the ports of H_1 and H_2, respectively, and let $s_1 \in S_1$ and $s_2 \in S_2$ be the ends of P. Let $i \in \{1, 2\}$, and let $w_1 := v$ and $w_2 := u$. If $|S_i| = 1$, then by Lemma 6, there exists $x_i \in N_{\hat{G}}(w_i) \cap {}^\top U_G(S_i)$. Otherwise, by considering an M-saturated path between two vertices in S_i, we again obtain $w_i \in N_{\hat{G}}(v) \cap {}^\top U_G(S_i)$. By Lemma 1 (iii), there is an M-saturated path Q_i between x_i and s_i. Whether $x_1 \in S$ holds or not, Q_1 has an edge in $E_G[S, S^c]$. Hence, $Q_1 + P + Q_2 + x_2 u + uv + vx_1$ is an M-alternating circuit with more than one edges in $E_G[S, S^c]$. Thus an $\hat{S}$-fatmatching is obtained. $\square$

As Lemma 8 is obtained, we give the following two lemmas to find such a spanning M-arc. They treat the cases that are counterparts to each other.

Lemma 9. *If G also has a border whose vertices are in S^c, then there is a spanning M-arc that has an edge in $E_G[S, S^c]$.*

Proof. Define $\mathcal{H}_1 \subseteq \mathrm{Min}\mathcal{O}(G)$ (resp. $\mathcal{H}_2 \subseteq \mathrm{Min}\mathcal{O}(G)$) as follows: $H \in \mathrm{Min}\mathcal{O}(G)$ is in $\mathcal{H}_1$ (resp. $\mathcal{H}_2$) if there is a tower-sequence from a border whose vertex set is contained in S (resp. S^c) to H.

By Theorem 5, $\mathcal{H}_1 \cup \mathcal{H}_2 = \mathrm{Min}\mathcal{O}(G)$.

Claim 1. *There is a spanning tower-sequence from a border whose vertex set is contained in S to a border whose vertex set is contained in S^c.*

Proof. As G is connected, there exists $H \in \mathcal{H}_1 \cap \mathcal{H}_2$. By $H \in \mathcal{H}_1$, there is a tower-sequence from $H_1 \in \partial \mathcal{O}(G)$ to H with $V(H_1) \subseteq S$. Hence, by Theorem 5, there is a spanning tower-sequence $H_1, \ldots, H_k \in \mathrm{Min}\mathcal{O}(G)$ with $k \geq 2$ and $H = H_i$ for some $i \in \{1, \ldots, k\}$. If $V(H_k) \subseteq S^c$ holds, we are done. Thus, let $V(H_k) \subseteq S$. By $H \in \mathcal{H}_2$, there is a tower-sequence $I_1, \ldots, I_l \in \mathrm{Min}\mathcal{O}(G)$ with $l \geq 1$, $I_1 \in \partial \mathcal{O}(G)$, $V(I_1) \subseteq S^c$, and $I_l = H$. Either $H_1, \ldots, H_i = H = I_l, \ldots, I_1$ or $H_k, \ldots, H_i = H = I_l, \ldots, I_1$ forms a spanning tower-sequence, satisfying the statement of this claim.   □

By Theorem 4 and Claim 1, we obtain a desired spanning M-arc.   □

Lemma 10. *Assume every border of G has the vertex set that is contained in S. If there exists a non-border element of $\mathrm{Min}\mathcal{O}(G)$ whose vertex set is contained in S^c, then there is a spanning M-arc that has some edges in $E_G[S, S^c]$.*

Proof. Let H be such a non-border element. Applying Theorems 4 (ii) and 5 to H, this lemma is proved.   □

By combining Lemmas 9 and 10 with Lemma 8, the Tight Cut Lemma is proved for the case of Sect. 4.2.

4.3   When Every Factor-Component in $\mathrm{Min}\mathcal{O}(G)$ has the Vertex Set Contained in S

Shared Assumptions and Lemmas. Here in Sect. 4.3, we assume that the vertex set of any factor-component in $\mathrm{Min}\mathcal{O}(G)$ is contained in S, which is namely the counterpart case to Sect. 4.2. Let $S_0 \subseteq S$ be the inclusion-wise maximal separating subset of S such that $\{H_1, \ldots, H_p\}$ is a lower-ideal of $\mathcal{O}(G - S)$, where $S_0 = V(H_1) \dot\cup \cdots \dot\cup V(H_p)$. Choose arbitrarily a connected component C of $G - S_0$. Note that $V(C)$ is a separating set in G and C is factorizable. The following two lemmas will be used in both of the succeeding case analyses. The first one is obtained rather easily from Lemma 4.

Lemma 11. *For each $H \in \mathrm{Min}\mathcal{O}(C)$, G has an M-ear, P_H, relative to S_0 and traversing H.*

Under Lemma 11, for each $H \in \mathrm{Min}\mathcal{O}(C)$, choose and fix arbitrarily an M-ear relative to S_0 and traversing H; in the rest of this paper, we denote it by P_H.

Lemma 12. *Let $y \in V(C)$, and let $H \in \mathrm{Min}\mathcal{O}(C)$ be such that $y \in U_C^*(H)$. Then, there is an M-forwarding path Q_H^y from y to x_H, one of the ends of the M-ear P_H, with $V(Q_H^y) \setminus \{x_H\} \subseteq V(C)$.*

Proof. Let $R := P_H[U_C^*(H)]$; then, R is an M-saturated path. Let z_1 and z_2 be its ends. By Lemma 1 (iv), for either z_1 or z_2, say, for z_1, there is an M-saturated path Q between y and z_1 whose vertices are in $U_C^*(H)$. By taking one of the connected components L of $P_H - E(R)$ appropriately, $Q + L$ forms a desired path.   □

After the above, in the rest of this paper, for each $H \in \text{Min}\mathcal{O}(C)$ and each $y \in U_C^*(H)$, let Q_H^y be the path as given in Lemma 12, and let x_H be the end of the M-ear P_H that is the end of Q_H^y.
Case with $|N_G(C) \cap S_0|=1$. Here, we assume that there exists $x_0 \in S_0$ with $N_G(C) \cap S_0 = \{x_0\}$, and prove the Tight Cut Lemma under this assumption. Of course, $x_H = x_0$ holds for each $H \in \text{Min}\mathcal{O}(C)$.

Lemma 13. *If $|N_G(C) \cap S_0| = 1$ holds, then $\hat{G}$ has an $\hat{S}$-fatperfect matching.*

Proof. According to Lemma 7, there exists $z \in N_{\hat{G}}(v)$ such that there is an M-saturated path R between x_0 and z with $V(R) \subseteq S_0$. By Lemma 6, there exists $y \in V(C) \cap N_{\hat{G}}(u)$. Let $H \in \text{Min}\mathcal{O}(C)$ be such with $y \in U_C^*(H)$, and take a path Q_H^y as given in Lemma 12. Let $K := R + zv + uv + vy + Q_H^y$. Note that K is an $\hat{M}$-alternating circuit of $\hat{G}$. Note $zv \in E_{\hat{G}}[\hat{S}, \hat{S}^c] \setminus \hat{M}$. Regardless of $y \in \hat{S}$ or $y \in \hat{S}^c$, $Q_H^y + uy$ has an edge in $E_{\hat{G}}[\hat{S}, \hat{S}^c] \setminus \hat{M}$. Hence, this lemma is proven.   □

From Lemma 13, the proof of the Tight Cut Lemma for the case analysis of Sect. 4.3 is completed.

Case with $|N_G(C) \cap S_0| > 1$. Here, we treat the counterpart to the previous case analysis; namely, we assume $|N_G(C) \cap S_0| > 1$. We use the next lemma as the main strategy to obtain a desired perfect matching:

Lemma 14. *If G has a proper M-ear relative to S_0 and traversing S^c, then $\hat{G}$ has an $\hat{S}$-fat perfect matching.*

Proof. Let P be such an M-ear, and let x and y be its ends. As $\hat{G}$ is a brick, there is an M-saturated path Q between x and y with $uv \in E(Q)$. By Lemma 3, $P + Q$ forms an M-alternating circuit. As this circuit has more than one edge in $\delta_{\hat{G}}(\hat{S})$, this lemma is proved.   □

As given in Lemma 14, we aim to find such a proper M-ear. If the M-ear P_H is proper for some $H \in \text{Min}\mathcal{O}(C)$, then Lemma 14 gives an $\hat{S}$-fatmatching of $\hat{G}$. Hence, in the rest of this proof, we assume that P_H *is not proper, having the unique end x_H.* The next two lemmas finds desired M-ears and so $\hat{S}$-fat perfect matchings under the assumptions that are counterparts to each other. The first one comes rather immediately from Lemma 12.

Lemma 15. *Let $H \in \text{Min}\mathcal{O}(C)$. If $N_G(U_C^*(H)) \cap S_0$ contains a vertex other than x_H, then $\hat{G}$ has an $\hat{S}$-fatmatching.*

Lemma 16. *If $N_G(U_C^*(H)) \cap S_0 = \{x_H\}$ holds for any $H \in \text{Min}\mathcal{O}(C)$, then $\hat{G}$ has an $\hat{S}$-fatperfect matching.*

Proof. As C is connected, there exist $H, I \in \mathrm{Min}\mathcal{O}(C)$ such that $\mathcal{T}_C(H)$ and $\mathcal{T}_C(I)$ are t-adjacent. Under Lemma 5, take an M-arc P between H and I. Let $x \in V(H)$ and $y \in V(I)$ be the ends of P. Then, by Lemmas 3, $Q_H^x + P + Q_I^y$ is a desired M-ear. □

By Lemmas 15 and 16, the last remaining case is proved. This completes the whole proof of the Tight Cut Lemma.

References

1. Dulmage, A.L., Mendelsohn, N.S.: Coverings of bipartite graphs. Can. J. Math. **10**, 517–534 (1958)
2. Dulmage, A.L., Mendelsohn, N.S.: A structure theory of bipartite graphs of finte exterior dimension. Trans. Roy. Soc. Can. Sect. **III**(53), 1–13 (1959)
3. Dulmage, A.L., Mendelsohn, N.S.: Two algorithms for bipartite graphs. J. Soc. Ind. Appl. Math. **11**(1), 183–194 (1963)
4. Edmonds, J.: Paths, trees and flowers. Can. J. Math. **17**, 449–467 (1965)
5. Edmonds, J., Lovász, L., Pulleyblank, W.R.: Brick decompositions and the matching rank of graphs. Combinatorica **2**(3), 247–274 (1982)
6. Frank, A.: Conservative weightings and ear-decompositions of graphs. Combinatorica **13**, 65–81 (1993)
7. Gallai, T.: Maximale systeme unabhängiger kanten. A Magyer Tudományos Akadémia: Intézetének Közleményei **8**, 401–413 (1964)
8. Kita, N.: A new canonical decomposition in matching theory, under review
9. Kita, N.: A canonical characterization of the family of barriers in general graphs. CoRR abs/1212.5960 (2012)
10. Kita, N.: A partially ordered structure and a generalization of the canonical partition for general graphs with perfect matchings. In: Chao, K.-M., Hsu, T., Lee, D.-T. (eds.) ISAAC 2012. LNCS, vol. 7676, pp. 85–94. Springer, Heidelberg (2012). https://doi.org/10.1007/978-3-642-35261-4_12
11. Kita, N.: A partially ordered structure and a generalization of the canonical partition for general graphs with perfect matchings. CoRR abs/1205.3 (2012)
12. Kita, N.: Disclosing barriers: a generalization of the canonical partition based on Lovász's formulation. In: Widmayer, P., Xu, Y., Zhu, B. (eds.) COCOA 2013. LNCS, vol. 8287, pp. 402–413. Springer, Cham (2013). https://doi.org/10.1007/978-3-319-03780-6_35
13. Kotzig, A.: Z teórie konečných grafov s lineárnym faktorom. I. Mat. časopis **9**(2), 73–91 (1959)
14. Kotzig, A.: Z teórie konečných grafov s lineárnym faktorom. II. Mat. časopis **9**(3), 136–159 (1959)
15. Kotzig, A.: Z teórie konečných grafov s lineárnym faktorom. III. Mat. časopis **10**(4), 205–215 (1960)
16. Lovász, L.: On the structure of factorizable graphs. Acta Math. Hungarica **23**(1–2), 179–195 (1972)
17. Lovász, L.: Matching structure and the matching lattice. J. Comb. Theory, Ser. B **43**(2), 187–222 (1987)
18. Lovász, L., Plummer, M.D.: Matching theory, vol. 367. American Mathematical Soc. (2009)

19. Naddef, D.: Rank of maximum matchings in a graph. Math. Program. **22**(1), 52–70 (1982)
20. Schrijver, A.: Combinatorial Optimization: Polyhedra and Efficiency, vol. 24. Springer Science & Business Media, Heidelberg (2003)
21. Szigeti, Z.: On a matroid defined by ear-decompositions of graphs. Combinatorica **16**(2), 233–241 (1996)
22. Szigeti, Z.: On optimal ear-decompositions of graphs. In: Cornuéjols, G., Burkard, R.E., Woeginger, G.J. (eds.) IPCO 1999. LNCS, vol. 1610, pp. 415–428. Springer, Heidelberg (1999). https://doi.org/10.1007/3-540-48777-8_31
23. Szigeti, Z.: On generalizations of matching-covered graphs. Eur. J. Comb. **22**(6), 865–877 (2001)
24. Szigeti, Z.: Perfect matchings versus odd cuts. Combinatorica **22**(4), 575–589 (2002)

On the Complexity of Detecting k-Length Negative Cost Cycles

Longkun Guo[1(✉)] and Peng Li[2(✉)]

[1] College of Mathematics and Computer Science, Fuzhou University, Fuzhou, China
`lkguo@fzu.edu.cn`
[2] Amazon Web Services, Amazon.com Inc., Seattle, WA, USA
`lipeng.net@gmail.com`

Abstract. Given a positive integer k and a directed graph G with a real number cost on each edge, the k-length negative cost cycle ($kLNCC$) problem that first emerged in deadlock avoidance in synchronized streaming computing network [14] is to determine whether G contains a negative cost cycle of at least k edges. The paper first shows a related problem of $kLNCC$, namely the fixed-point trail with k-length negative cost cycle ($FPTkLNCC$) problem which is to determine whether there exists a negative closed trail enrouting a given vertex as the fixed point and containing only cycles with at least k edges, is $\mathcal{NP}$-complete in multigraphs even when $k = 3$ by reducing from the $3SAT$ problem. Then as the main result, we prove the $\mathcal{NP}$-completeness of $kLNCC$ by giving a more sophisticated reduction from the 3 Occurrence 3-Satisfiability ($3O3SAT$) problem, a known $\mathcal{NP}$-complete special case of 3SAT in which a variable occurs at most three times. The complexity result is surprising, since polynomial time algorithms are known for both $2LNCC$ (essentially no restriction on the value of k) and the k-cycle problem with k being fixed which is to determine whether there exists a cycle of at least length k in a given graph. This closes the open problem proposed by Li et al. in [14,15] whether $kLNCC$ admits polynomial-time algorithms.

Keywords: k-length negative cost cycle · $\mathcal{NP}$-complete · 3 occurrence 3-satisfiability · 3-satisfiability

1 Introduction

We define the following *k-length negative cost cycle problem ($kLNCC$)*:

Definition 1. *Given a fixed integer k and a directed graph $G = (V, E)$, in which each edge $e \in E$ is with a cost $c(e) \to \mathbb{R}$ and a length $l(e) = 1$, $kLNCC$ is to determine whether there exists a cycle O with total length $l(O) = \sum_{e \in O} l(e) \geq k$ and total cost $c(O) < 0$.*

This research work is supported by Natural Science Foundation of China #61772005, #61300025 and Natural Science Foundation of Fujian Province #2017J01753.

X. Gao et al. (Eds.): COCOA 2017, Part I, LNCS 10627, pp. 240–250, 2017.
https://doi.org/10.1007/978-3-319-71150-8_21

The $kLNCC$ problem arises in deadlock avoidance for streaming computing, which is widely used in realtime analytics, machine learning, robotics, and computational biology, etc. A streaming computing system consists of networked nodes communicating through finite first-in first-out (FIFO) channels, and a data stream referring to data transmitted through a channel. In streaming computing, a compute node might need to synchronize different incoming data streams. If the synchronized streams have different rates (e.g. due to filtering [14]), the computing network might deadlock. Several deadlock avoidance algorithms have been proposed in [14], which rely on inserting heartbeat messages into data streams. When to insert those heartbeat messages, however, depends on the network topology and buffer size configurations. An open problem is deciding whether a given heartbeat message schedule can guarantee deadlock freedom, which raises the negative-cost cycle detection problem. Further, we are only interested in the negative-cost cycle with length at least $k \geq 3$, as deadlocks of few nodes can be easily eliminated. This raises the above $kLNCC$ problem. Besides, in many cases, we are also interested in whether a particular node in a streaming computing system is involved in a deadlock or not, which brings the *fixed-point trail with k-length negative-cost cycle problem* (*FPTkLNCC*):

Definition 2. *Given a graph $G = (V, E)$ in which each edge $e \in E$ is with a cost $c(e) \to \mathbb{R}$ and a length $l(e) = 1$, a fixed integer k, and a fixed point $p \in G$, the FPTkLNCC problem is to determine whether there exists a closed trail T containing p, such that $c(T) < 0$ and $l(O) = \sum_{e \in O} l(e) \geq k$ for every $O \subseteq T$ (with $c(O) < 0$), where a closed trail is a trail which begins and ends on the same vertex.*

Note that $c(O) < 0$ makes no difference for the above definition, as a negative cost closed trail must contain at least a negative cost cycle. Further, if G contains a vertex p as a fixed point such that the *FPTkLNCC* problem is feasible, then G must contain at least a cycle O with $c(O) < 0$ and $\sum_{e \in O} l(e) \geq k$, and *vice versa*. So if *FPTkLNCC* admits a polynomial-time algorithm, then $kLNCC$ is also polynomially solvable. That is because we can run the polynomial-time algorithm as a subroutine to verify whether G contains a vertex p wrt which *FPTkLNCC* is feasible, and then to verify whether $kLNCC$ is feasible. Conversely, if the $\mathcal{NP}$-completeness of $kLNCC$ is proven, it can be immediately concluded that *FPTkLNCC* is also $\mathcal{NP}$-complete.

Throughout this paper, by *walk* we mean an alternating sequence of vertices and connecting edges; by *trail* we mean a walk that does not pass over the same edge twice; by *path* we mean a *trail* that does not include any vertex twice; and by *cycle* we mean a path that begins and ends on the same vertex.

1.1 Related Works

The $kLNCC$ problem generalizes two well known problems: the negative cycle detection problem of determining whether there exist negative cycles in a given graph, and the k-cycle problem (or namely the long directed cycle problem [2]) of determining whether there exists a cycle with at least k edges. The former problem is known polynomially solvable via the Bellman-Ford algorithm [1,5] and is actually $kLNCC$ of $k = 2$. The latter problem, to determine whether a given graph contains a cycle O with $l(O) \geq k$ or not, is in fact $kLNCC$ when $c(e) = -1$ for every $e \in E$. The problem is shown fixed parameter tractable in [7], where an algorithm with a time complexity $k^{O(k)}n^{O(1)}$ is proposed. The runtime is then improved to $O(c^k n^{O(1)})$ for a constant $c > 0$ by using representative sets [4], and later to $6.75^k n^{O(1)}$ independently by [3,17]. Compared to the two results above, i.e. negative cycle detection and the k-cycle problem with k being fixed are both polynomially solvable, it is interesting that $3LNCC$ is $\mathcal{NP}$-complete because $3LNCC$ is exactly a combination of the two problems that belonging to $\mathcal{P}$.

Besides negative cycle detection and k-cycle, $kLNCC$ also generalizes the Hamiltonian cycle problem which, for a given graph, is to determine whether there exists a cycle containing all the vertices. Apparently, the Hamiltonian cycle problem is in fact $kLNCC$ when k equals n, the number of vertices in the graph. It is known that the Hamiltonian cycle problem is $\mathcal{NP}$-complete [9], so $kLNCC$ is also $\mathcal{NP}$-complete for general k.

Except for applications in deadlock avoidance for steaming computing and its self-own theoretical value, the hardness results over the $kLNCC$ problem can also shed some light on related problems whose currently existing algorithms are mainly based on computing bicriteria cost cycles. For example, the k-disjoint constrained shortest path (kCSP) problem, to compute k disjoint paths between a pair of specified vertices such that the total cost of the paths is minimized while the length sum is bounded by a given integer, was first proposed by Orda et al. [16] and known admits single factor approximation ratio $O(\ln n)$ [10,13] and bifactor ratio $O(1+\epsilon, 2+\epsilon)$ [11]. These approximation algorithms are mainly based on cycle cancellation method, so the hardness results of $kLNCC$ indicate the difficulty of further improving the approximation ratio by merely canceling bicriteria cost cycles.

1.2   Our Results

The main result of this paper is proving the $\mathcal{NP}$-completeness of $kLNCC$ in a simple directed graph. Since the proof is constructive and complicated, we will first accomplish an easier task of proving the $\mathcal{NP}$-completeness of the $FPTkLNCC$ problem in multigraphs by simply reducing from the 3-Satisfiability ($3SAT$) problem. A multigraph is a graph that allows multiple edges between two nodes.

Lemma 3. *For any fixed integer $k \geq 3$, FPTkLNCC is $\mathcal{NP}$-complete in multi-graphs.*

Then following a main idea similar to that of the proof of Lemma 3 while acquiring more sophisticated details, we will prove the $\mathcal{NP}$-completeness of $kLNCC$ (and hence also $FPTkLNCC$) in a simple directed graph.

Theorem 4. *For any fixed integer $k \geq 3$, kLNCC is $\mathcal{NP}$-complete in a simple graph.*

2   The $\mathcal{NP}$-completeness of *FPTkLNCC* in Multigraphs

In this section, we will prove Lemma 3 by reducing from *3SAT* that is known to be $\mathcal{NP}$-complete [8]. In an instance of *3SAT*, we are given n variables $\{x_1, \ldots, x_n\}$ and m clauses $\{C_1, \ldots, C_m\}$, where C_i is the *OR* of at most three *literals*, and each literal is an occurrence of the variable x_j or its negation. The *3SAT* problem is to determine whether there is an assignment satisfying all the m clauses.

For any given instance of *3SAT*, the key idea of our reduction is to construct a digraph G, such that G contains a negative cost closed *trail* with only cycles of length at least 3 and enrouting a given vertex iff the instance of *3SAT* is satisfiable. The construction is composed with the following three parts. First, for each variable x_i with a_i occurrences of x_i and b_i occurrences of $\overline{x}_i$ in the clauses, we construct a *lobe*[1] which contains two vertices, denoted as y_i and z_i, and $a_i + b_i$ edges of cost -1 between the two vertices, i.e. a_i copies of edge (y_i, z_i) and b_i copies of (z_i, y_i) (A lobe is depicted as in the dashed circles in Fig. 1). For briefness, we say an edge in the lobes is a *lobe-edge*. Then, for each clause C_j, add two vertices u_i and v_i, as well as edge (v_i, u_{i+1}), $1 \le i \le m - 1$, with cost 0 and edge (v_m, u_1) with cost $m - 1$. Last but not the least, for the relationship between the variables and the clauses, say variable x_j occurs in clause C_i, we add two edges with cost 0 to connect the lobes and the vertices of clauses:

- If C_i contains x_j, then add two edges (u_i, y_j), (z_j, v_i);
- If C_i contains $\overline{x}_j$, then add two edges (u_i, z_j), (y_j, v_i).

An example of the construction for a *3SAT* instance is depicted in Fig. 1.

Then since *FPTkLNCC* is clearly in $\mathcal{NP}$, the correctness of Lemma 3 can be immediately obtained from the following lemma:

Lemma 5. *An instance of 3SAT is satisfiable iff in its corresponding auxiliary graph G there exists a negative-cost closed trail containing u_1 but no length-2 cycles.*

Proof. Suppose there exists a negative-cost closed trail T, which contains u_1 but *NO* negative cost length-2 cycle. Then let τ be a true assignment for the *3SAT* instance according to T: if T goes through (y_i, z_i), then set $\tau(x_i) = true$; Otherwise, set $\tau(x_i) = false$. It remains to show such an assignment will satisfy all the clauses. Firstly, we show that the path $P = T \setminus \{(v_m, u_1)\}$ must go through all vertices of $\{v_i | i = 1, \ldots, m\}$. Since T contains u_1, T has to go through edge (v_m, u_1), as the edge is the only one entering u_1. Then because T is with negative cost and the cost of edge (v_m, u_1) is $m - 1$, P has to go through at least m edges within the n lobes, as only the edges of lobes has a negative cost -1. According to the construction of G, between two lobe-edges on P, there must exist at least an edge of $\{v_i, u_{i+1} | i = 1, \ldots, m - 1\}$, since v_i has only one out-going edge (v_i, u_{i+1}) while every edge leaving a lobe must enter a vertex of

---

[1] The term *lobe* was used to denote an unit of the auxiliary graph constructed for an instance of SAT, as in [12] and many others [6,19].

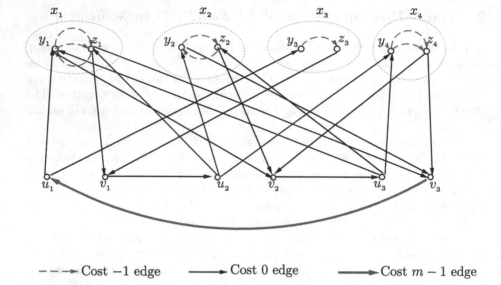

Fig. 1. The construction of G for an *3SAT* instance $(x_1 \vee x_3) \wedge (\overline{x}_1 \vee x_2 \vee x_4) \wedge (x_1 \vee \overline{x}_2 \vee x_4)$.

$\{v_i | i = 1, \ldots, m-1\}$. So $P = T \backslash \{v_m, u_1\}$ has to go through all the $m-1$ edges of $\{v_i, u_{i+1} | i = 1, \ldots, m-1\}$, and hence through all vertices of $\{v_i | i = 1, \ldots, m\}$. Secondly, assume v_j and $v_{j'}$ are two vertices of $\{v_i | i = 1, \ldots, m\}$, such that $P(v_j, v_{j'}) \cap \{v_i | i = 1, \ldots, m\} = \{v_j, v_{j'}\}$. Again, because there must be at least an edge of $\{v_i, u_{i+1} | i = 1, \ldots, m-1\}$ between two lobe-edges, there must be at least a lobe-edge appearing on $P(v_j, v_{j'})$, otherwise there will be at most $m-1$ lobe-edges on T. That is, there must be exactly a lobe edge, say (y_i, z_i), on $P(v_j, v_{j'})$. Then according to the construction of graph G, x_i appears in C_j, and hence $\tau(x_i) = true$ satisfies C_j. The case for (z_i, y_i) appears on $P(v_j, v_{j'})$ is similar. Therefore, the *3SAT* instance is feasible as it can be satisfied by τ.

Conversely, suppose the instance of *3SAT* is satisfiable, and a true assignment is $\tau : x \to \{true, false\}$. Then for clause C_k, there must exist a literal, say w_k, such that $\tau(w_k) = true$. If w_k is an occurrence of x_i, then set the corresponding subpath as $P_k = u_k - y_i - z_i - v_k$; otherwise set $P_k = u_k - z_i - y_i - v_k$. Then clearly, $P = \{P_k | k = 1, \ldots, m\} \cup \{(v_h, u_{h+1}) | h = 1, \ldots, m-1\}$ exactly composes a path from u_1 to v_m with a cost of $-m$, as it contains m lobe-edge and other edges of cost 0. So $T = P \cup \{v_m, u_1\}$ is a negative cost closed trail of length at least 3. Besides, since $\tau(x_i)$ must be either *true* or *false*, there exist no length-2 cycles on P. This completes the proof.   □

However, the above proof can not be immediately extended to prove the $\mathcal{NP}$-completeness of $kLNCC$, since there are two tricky obstacles. Firstly, in the above proof, there might exist negative cycles with length at least three but without going through u_1. Thus, in Lemma 5, containing u_1 is mandatory. Secondly,

Lemma 5 holds only for multigraphs as some of the lobes are already multigraphs. Thus, to extend the proof to $kLNCC$, we need to eliminate negative cycles bypassing u_1 and to transform the (multigraph) lobes to simple graphs.

3   The $\mathcal{NP}$-completeness Proof of $kLNCC$

In this section, to avoid the two obstacles as analyzed in the last section, we will prove Theorem 4 by reducing from the 3 occurrence $3SAT$ ($3O3SAT$) problem that is known $\mathcal{NP}$-complete [18]. Similar to $3SAT$, in an instance of $3O3SAT$ we are also given m clauses $\{C_1, \ldots, C_m\}$ and n variables $\{x_1, \ldots, x_n\}$, and the task is to determine whether there is an assignment satisfying all the m clauses. The only difference is, however, each variable x_i (including both literal x_i and $\overline{x}_i$) appears at most 3 times in all the m clauses. To simplify the reduction, we assume that the possible occurrences of a variable x in an instance of $3O3SAT$ fall in the following three cases:

- **Case 1:** All occurrences of x are all positive literal x;
- **Case 2:** The occurrences of x are exactly one positive literal and one negative literal;
- **Case 3:** The occurrences of x are exactly two positive literals and one negative literal.

The above assumption is without loss of generality. We note that there are still two other cases:

- **Case 4:** All occurrences of x are negative literals;
- **Case 5:** Exactly two occurrences of negative literals and one positive literal.

However, Cases 4 and 5 can be respectively reduced to Cases 1 and 3, by replacing occurrences of $\overline{x}$ and x respectively with y and $\overline{y}$. Therefore, we need only to consider $3O3SAT$ instances with variables satisfying Case 1–3.

The key idea of the proof is, for any given instance of $3O3SAT$, to construct a graph G, such that there exists a cycle O with $c(O) < 0$ and $l(O) \geq 3$ in G if and only if the instance is satisfiable. An important fact used in the construction is that every variable appears at most 3 times in a $3O3SAT$ instance. In the following, we will show how to construct G according to clauses, variables, and the relation between clauses and variables.

1. For each C_k:
   Add to G two vertices u_k and v_k, as well as edge (v_k, u_{k+1}), $1 \leq k \leq m-1$ with cost 0, and edge (v_m, u_1) with cost -1.
2. For each variable x_i, construct a lobe according to the occurrences of x_i and $\overline{x}_i$ (The construction a lobe is as depicted in Fig. 2):
   - **Case 1:** All occurrences of x_i in are positive literal x_i, such as x_4 in Fig. 3.
     For the jth occurrence of x_i, add a directed edge (y_i^j, z_i^j) and assign cost $-2m$ to it.

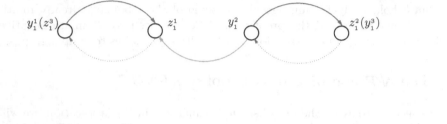

$$\longrightarrow \text{ Cost } -2m \text{ edge} \qquad \longrightarrow \text{ Cost } 0 \text{ edge} \qquad \cdots\cdots\triangleright \text{ Cost } \tfrac{1}{2m+2} \text{ edge}$$

Fig. 2. A lobe for x_1 with respect to an instance of 3O3SAT $(x_1 \vee x_3) \wedge (\overline{x}_1 \vee x_2 \vee x_4) \wedge (x_1 \vee \overline{x}_2 \vee x_4)$.

- **Case 2:** Exactly one occurrence for each of positive literal x_i and negation $\overline{x}_i$, such as x_2 in Fig. 3.
    - (a) Add two vertices $z_i^{j_2} = y_i^{j_1}$ and $y_i^{j_2} = z_i^{j_1}$, and connect them with directed edges $(y_i^{j_1}, z_i^{j_1})$ and $(y_i^{j_2}, z_i^{j_2})$.
    - (b) Assign edge $(y_i^{j_1}, z_i^{j_1})$ with cost $-2m$ and $(y_i^{j_2}, z_i^{j_2})$ with cost $\tfrac{1}{m+1}$;
- **Case 3:** Exactly 2 occurrences of x_i and one occurrence $\overline{x}_i$, such as x_1 in Fig. 3.
    - (a) For the two positive literals of x_i, say the j_1th and j_2th occurrence of x_i, $j_1 < j_2$, add four vertices $y_i^{j_1}, z_i^{j_1}, y_i^{j_2}, z_i^{j_2}$, and two directed edges $(y_i^{j_1}, z_i^{j_1}), (y_i^{j_2}, z_i^{j_2})$ connecting them with cost $-2m$;
    - (b) For the negation $\overline{x}_i$, say the j_3th occurrence, set $z_i^{j_3} = y_i^{j_1}$ and $y_i^{j_3} = z_i^{j_2}$, and add three directed edges $(y_i^{j_3}, y_i^{j_2}), (y_i^{j_2}, z_i^{j_1}), (z_i^{j_1}, z_i^{j_3})$ with costs $\tfrac{1}{2m+2}$, 0 and $\tfrac{1}{2m+2}$, respectively.
3. For the relation between the variables and the clauses, say C_k is the clause containing the jth occurrence of x_i, i.e. C_k is the jth clause x_i appears in, add directed edges (u_k, y_i^j) and (z_i^j, v_k). If the occurrence of x_i in C_k is a positive literal, assign the newly added edges with cost m; Otherwise, assign them with cost 0. Note that no edges will be added between lobes and u_k, v_k if x_i does not appear in C_k.

An example of the construction of G according to $F = (x_1 \vee x_3) \wedge (\overline{x}_1 \vee x_2 \vee x_4) \wedge (x_1 \vee \overline{x}_2 \vee x_4)$ is as depicted in Fig. 3.

As $kLNCC$ is apparently in $\mathcal{NP}$, it remains only to prove the following lemma to complete the proof of Theorem 4.

Lemma 6. *An instance of 3O3SAT is feasible iff the corresponding graph G contains a cycle O with $l(O) \geq 3$ and $c(O) < 0$.*

Let $U = \{u_i, v_i | i = 1, \dots, m\}$ be the set of vertices that correspond to the clauses. We first prove a proposition that if a path from u_h to v_l $(h \neq l)$ does not enroute any other $u \in U$, the cost of the path is at least m.

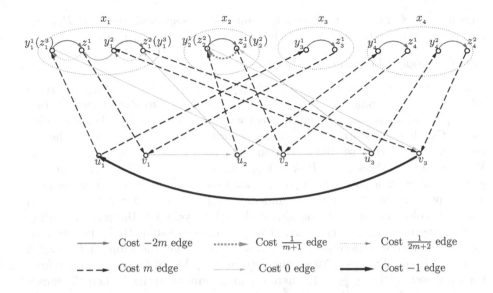

Fig. 3. The construction of G for an $3O3SAT$ instance $(x_1 \vee x_3) \wedge (\overline{x}_1 \vee x_2 \vee x_4) \wedge (x_1 \vee \overline{x}_2 \vee x_4)$.

Proposition 7. *Let $P(u, v)$ be a path from u to v. For any path $P(u_h, v_l)$ that satisfies $P(u_h, v_l) \cap U = \{u_h, v_l\}$, if $h \neq l$, then $c(p(u_h, v_l)) \geq m$.*

Proof. For every edge (y, z) with cost $-2m$, clearly there exists only one edge e_1 entering y, and only one edge e_2 leaving z. Furthermore, $e_1 = (u_{h'}, y)$ and $e_2 = (z, v_{h'})$. So $P(u_h, v_l)$, $h \neq l$, as in the proposition can not go through any cost $-2m$ edge. That is, $P(u_h, v_l)$, $h \neq l$, can only go through the non-negative cost edges. It remains to show $P(u_h, v_l)$, $h \neq l$, must go through at least one cost m edge.

Suppose $P(u_h, v_l)$, $h \neq l$ does not go through any cost m edges. Let $(u_h, y_i^{j_1})$, $(z_i^{j_2}, v_l) \in P(u_h, v_l)$ be the two edges leaving u_h and entering v_l, respectively. Then $y_i^{j_1}$ and $z_i^{j_2}$ must incident to two edges that corresponds to the negation of the two variables. Further, the two variable must be identical, since vertices of two distinct lobes will be separated by U, and hence for $P(y_i^{j_1}, z_i^{j_2}) \subset P(u_h, v_l)$, $P(y_i^{j_1}, z_i^{j_2}) \cap U \neq \emptyset$. This contradicts with $P(u_h, v_l) \cap U = \{u_h, v_l\}$.   □

Proposition 8. *In graph $G \setminus e(v_m, u_1)$, every path $P(u, v)$, $u, v \in U$, has a non-negative cost.*

Proof. Apparently, in $G \setminus e(v_m, u_1)$, every edge with negative cost is exactly with cost $-2m$. Let (y_i^j, z_i^j) be such an edge with cost $-2m$. From the structure of G, there exists exactly one edge entering y_i^j, and exactly one leaving z_i^j, each of which is with exactly cost m. So for every path $P(u, v)$, $u, v \in U$, if it contains edge (y_i^j, z_i^j), then it must also go through both the edge entering y_i^j and the edge leaving z_i^j. That is, the three edges must all present or all absent in $P(u, v)$, and contribute a total cost 0. Therefore, $c(P(u, v)) \geq 0$ must hold. □

Now the proof of Lemma 6 is as below:

Proof. Suppose that there exists an assignment $\tau : x \to \{true, false\}$ satisfying all the m clauses. Since $c(v_m, u_1) = -1$, we need only to show there exists a u_1v_m-path with cost smaller than 1 by construction such one path. For a satisfied clause C_k, there must exist a literal, say w_k with $\tau(w_k) = true$. If w_k is the jth occurrence of x_i, then set the corresponding subpath as $P_k = u_k - y_i^j - z_i^j - v_k$. We need only to show $P = \{P_k | k = 1, \ldots, m\} \cup \{(v_h, u_{h+1}) | h = 1, \ldots, m-1\}$ exactly composes a path from u_1 to v_m with cost smaller than 1. For the first, P is a path. Because τ is an assignment, $\tau(x_i) = true$ and $\tau(\overline{x}_i) = true$ can not both hold, and hence P contains no length-2 cycle. For the cost, according to the construction, if $\tau(w_k) = \tau(x_i) = true$ then the subpath P_k is with cost exactly equal to 0; otherwise, i.e. $\tau(w_k) = \tau(\overline{x}_i) = true$, the subpath P_k is with cost exactly equal to $\frac{1}{m+1}$. Meanwhile, $c(e(v_h, u_{h+1})) = 0$ for each h. Therefore, the total cost $c(P) \leq \frac{m}{m+1} < 1$, where the maximum is attained when all clauses are all satisfied by negative of the variables.

Conversely, assume that there exists a negative-cost cycle O, which contains *NO* negative cost length-2 cycle. According to Proposition 8, $e(v_m, u_1)$ must appears on O, so that $c(O) < 0$ can hold. Let $P = O \setminus e(v_m, u_1)$ and τ be a true assignment according to P: if P goes through literal $\overline{x}_i$, set $\tau(x_i) = false$ and set $\tau(x_i) = true$ otherwise. It remains to show such the assignment according to P satisfies all the clauses. To do this, we shall firstly show P will go through all the vertices of U in the order $u_1 \prec v_1 \prec \cdots \prec u_i \prec v_i \prec \cdots \prec u_m \prec v_m$; and secondly show that $P(u_h, v_h)$ has to go through exactly a subpath corresponding to a literal, say w, for which if $\tau(w) = true$, then C_h is satisfied. Then from the fact that P contains no negative cost length-2 cycle, τ is a feasible assignment satisfying all the clauses.

For the first, according to Proposition 7, if $P(u_i, v_j) \cap U = \{u_i, v_j\}$, then $j = i$ (i.e. $v_j = v_i$) must hold. Since otherwise, according to Proposition 7 $c(P(u_i, v_j)) \geq m$ must hold; while according to Proposition 8, the other parts of P is with $c(P(u_1, u_i)) \geq 0$ and $c(P(v_j, v_m)) \geq 0$. That is, $c(P) \geq m$. On the other hand, since $c(e(v_m, u_1)) = -1$ and $c(O) < 0$, we have $c(P) < 1$, a contradiction. Furthermore, since there exists only one edge leaving v_i, i.e. (v_i, u_{i+1}), P must go through every edge $e(v_i, u_{i+1})$ incrementally on i, i.e. in the order of $u_1 \prec v_1 \prec \cdots \prec u_i \prec v_i \prec \cdots \prec u_m \prec v_m$. For the second, according to the structure of G and $c(P) < 1$, $c(P(u_h, v_h)) \leq \frac{1}{m+1}$ must hold. Then $c(P(u_h, v_h))$ has to go through exactly a subpath corresponding to a literal. □

Note that, a simple undirected graph does not allow length-2 cycles. Anyhow, by replacing length-2 cycles with length-3 cycles in the above proof, i.e. replacing every edge (y_j, z_j) with two edges (y_j, w_j) and (w_j, z_j) of the same cost, and ignoring the direction of the edges, we immediately have the correctness of Corollary 9.

Corollary 9. *For any fixed integer $k \geq 4$, $kLNCC$ is $\mathcal{NP}$-complete in a simple undirected graph.*

4 Conclusion

In this paper, we have shown the $\mathcal{NP}$-completeness for both the k-length negative cost cycle ($kLNCC$) problem in a simple directed graph and the fixed-point trail with k-length negative cost cycle ($FPTkLNCC$) problem in a directed multigraph, which have wide applications in parallel computing, particularly in deadlock avoidance for streaming computing systems. Consequently, it can be concluded that $kLNCC$ is also $\mathcal{NP}$-complete in a simple *undirected* graph. In future, we will investigate approximation algorithms for the two problems.

Acknowledgment. The authors would like to thank the anonymous reviewers of COCOA 2017 for their insightful comments, which helped us improve the quality of the paper. Part of the work was completed when Peng Li was with Washington University in St. Louis as a PhD candidate.

References

1. Bellman, R.: On a routing problem. Technical report, DTIC Document (1956)
2. Cygan, M., Fomin, F.V., Kowalik, Ł., Lokshtanov, D., Marx, D., Pilipczuk, M., Pilipczuk, M., Saurabh, S.: Parameterized Algorithms. Springer, Cham (2015). https://doi.org/10.1007/978-3-319-21275-3
3. Fomin, F.V., Lokshtanov, D., Panolan, F., Saurabh, S.: Representative sets of product families. In: Schulz, A.S., Wagner, D. (eds.) ESA 2014. LNCS, vol. 8737, pp. 443–454. Springer, Heidelberg (2014). https://doi.org/10.1007/978-3-662-44777-2_37
4. Fomin, F.V., Lokshtanov, D., Saurabh, S.: Efficient computation of representative sets with applications in parameterized and exact algorithms. In: Proceedings of the Twenty-Fifth Annual ACM-SIAM Symposium on Discrete Algorithms, pp. 142–151. SIAM (2014)
5. Ford, L.R.: Network flow theory (1956)
6. Fortune, S., Hopcroft, J., Wyllie, J.: The directed subgraph homeomorphism problem. Theoret. Comput. Sci. **10**(2), 111–121 (1980)
7. Gabow, H.N., Nie, S.: Finding a long directed cycle. ACM Trans. Algorithms (TALG) **4**(1), 7 (2008)
8. Garey, M.R., Johnson, D.S.: Computers and intractability: a guide to the theory of NP-completeness (1979)
9. Garey, M.R., Johnson, D.S., Stockmeyer, L.: Some simplified NP-complete problems. In: Proceedings of the Sixth Annual ACM Symposium on Theory of Computing, pp. 47–63. ACM (1974)
10. Guo, L.: Efficient approximation algorithms for computing k disjoint constrained shortest paths. J. Comb. Optim. **32**(1), 144–158 (2016)
11. Guo, L., Liao, K., Shen, H., Li, P.: Efficient approximation algorithms for computing k disjoint restricted shortest paths. In: Proceedings of the 27th ACM on Symposium on Parallelism in Algorithms and Architectures, SPAA 2015, Portland, OR, USA, 13–15 June, pp. 62–64 (2015)

12. Guo, L., Shen, H.: On finding min-min disjoint paths. Algorithmica **66**(3), 641–653 (2013)
13. Guo, L., Shen, H., Liao, K.: Improved approximation algorithms for computing k disjoint paths subject to two constraints. J. Comb. Optim. **29**(1), 153–164 (2015)
14. Li, P., Agrawal, K., Buhler, J., Chamberlain, R.D.: Deadlock avoidance for streaming computations with filtering. In: Proceedings of the 22nd ACM Symposium on Parallelism in Algorithms and Architectures, pp. 243–252 (2010)
15. Li, P., Beard, J.C., Buhler, J.D.: Deadlock-free buffer configuration for stream computing. Int. J. High Perform. Comput. Appl. **31**(5), 441–450 (2017)
16. Orda, A., Sprintson, A.: Efficient algorithms for computing disjoint qos paths. In: INFOCOM 2004. Twenty-Third Annual Joint Conference of the IEEE Computer and Communications Societies, vol. 1. IEEE (2004)
17. Shachnai, H., Zehavi, M.: Faster computation of representative families for uniform matroids with applications. CoRR, abs/1402.3547 (2014)
18. Tovey, C.A.: A simplified NP-complete satisfiability problem. Discrete Appl. Math. **8**(1), 85–89 (1984)
19. Xu, D., Chen, Y., Xiong, Y., Qiao, C., He, X.: On the complexity of and algorithms for finding the shortest path with a disjoint counterpart. IEEE/ACM Trans. Networking **14**(1), 147–158 (2006)

A Refined Characteristic of Minimum Contingency Set for Conjunctive Query

Dongjing Miao$^{(\boxtimes)}$ and Zhipeng Cai

Department of Computer Science, Georgia State University, Atlanta, GA 30303, USA
dmiao1@student.gsu.edu

Abstract. Given a database instance d, a self join free conjunctive query q and its result $q(d)$, contingency set $\Gamma(q, d)$ is a set of tuples from d such that $q(d \setminus \Gamma)$ is *false* but $q(d)$ is *true* initially. Finding minimum contingency set $\Gamma_{\min}(q, d)$ is an important problem in database area. An important dichotomy for this problem was identified in the most recent result, Freire *et al.* showed that $\Gamma_{\min}(q_\triangle, d)$ is NP-complete if the input query includes a triad of form "R_{xy}, S_{yz}, T_{zx}" in a particular manner, PTime otherwise. However, we have two observations: (a) if two clauses have a common variable, then this database instance should be too complex, formally speaking, the visualization of its query result will not be of *planarity*, this requirement is too strict, (b) there is no limitation on the length of every circle in the visualization of the query result. This makes the previous theorem of dichotomy too weak. Therefore, the natural question is that, if the query result of input database is not of planarity or there is a fixed limitation on the length of every circle, is it $\Gamma_{\min}(q_\triangle, d)$ still NP-complete? To this end, we strengthen the hardness result, that $\Gamma_{\min}(q_\triangle, d)$ is still NP-complete, if the input database instance is of planarity, or the maximum length of every circle is limited. Our theorems also generalize the result of triangle edge deletion problem defined on general graph into directed graph, make a contribution to graph theory.

Keywords: Minimum contigency set · Database · Complexity

1 Introduction

The problem of making query result empty by tuple deletion from the corresponding source database can be stated as follows [1],

MINIMUM CONTINGENCY SET $\Gamma_{\min}(q, d)$

INPUT   Given a database d, a natural number k, and a boolean conjunctive query q, its corresponding query result $q(d)$ is *true* (say, not empty) initially.

OUTPUT   Yes, if there exists a contingency set $\Gamma \subseteq d$ of size k such that $q(d \setminus \Gamma)$ is *false* (say, empty).

This work is supported by the National Science Foundation (NSF) under grant NOs 1252292, 1741277 and 1704287.

To understand this problem, consider an example of the Minimum Contingency Set problem. A database instance d is given as follows (Fig. 1),

$$R: \begin{array}{cc} x & y \\ \hline 1 & a \\ 2 & a \\ 1 & b \\ \hline \end{array} \quad S: \begin{array}{cc} y & z \\ \hline a & \alpha \\ b & \alpha \\ b & \beta \\ b & \gamma \\ \hline \end{array} \quad T: \begin{array}{cc} z & x \\ \hline \alpha & 1 \\ \beta & 1 \\ \alpha & 2 \\ \hline \end{array} \quad q(d): \begin{array}{ccc} x & y & z \\ \hline 1 & a & \alpha \\ 1 & b & \alpha \\ 1 & b & \beta \\ 2 & a & \alpha \\ \hline \end{array}$$

$$q(d) :\!- R(x,y), S(y,z), Z(z,x)$$

$$|\Gamma_{\min}(q,d)| = 2 : \{(1,b),(a,\alpha)\}$$

$$R: \begin{array}{cc} x & y \\ \hline 1 & a \\ 2 & a \\ \hline \end{array} \quad S: \begin{array}{cc} y & z \\ \hline b & \alpha \\ b & \beta \\ b & \gamma \\ \hline \end{array} \quad T: \begin{array}{cc} z & x \\ \hline \alpha & 1 \\ \beta & 1 \\ \alpha & 2 \\ \hline \end{array} \quad q(d) : \emptyset, \text{ is } \textit{false}$$

Fig. 1. Example of contingency set corresponding to (q,d)

Here, if the given query defined as the shown in the figure, then for given instance (q,d), a contingency set could be $\Gamma(q,d) = \{(1,a),(2,a),(1,b)\}$, because its absence is able to make query becoming *false*. However, if to find a minimum contingency set, it must be $\Gamma(q,d) = \{(a,\alpha),(1,b)\}$ of size 2, which can not be less than 2.

Minimum Contingency Set problem is a fundamental problem of minimum side-effect deletion propagation, complexity of it and its related problem has been widely studied, previous works [1–4] on source side effect decision problem show some complexity results on the source side-effect problem on both data and combined complexity. Basically, Freire et al. show that for $\Gamma_{\min}(q,d)$ studied in this paper is PTime if q is a conjunctive query without structure of *triad*, NP-complete otherwise. They also extend the dichotomy condition '*triad*' into a more general one '*fd-induced triad*' for case with presence of functional dependencies. Besides research on view side effect, there are previous work on complexity results on the view side effect free problem [2–7]. On the data complexity of deletion propagation, Kimelfeld et al. [5] showed the dichotomy '*head domination*' for every conjunctive query without self join, deletion propagation is either APX-hard or solvable (in polynomial time) by the unidimensional algorithm. For functional dependency restricted version, it is radically different from the case without functional dependency, they also showed the corresponding dichotomy of '*fd-head domination*' [6]. For multiple or group deletion [7], they especially showed the trichotomy for group deletion a more general case including *level-k head domination* and so on; On the combined complexity of deletion propagation, Cong et al. [2,3] showed the variety results for different combination of relational algebraic operators. At the same time, Miao et al. [8] studied the functional dependency restricted version deletion propagation problem and showed the tractable and intractable results on both data and combined complexity aspects.

In this paper, we study the data complexity of problem $\Gamma_{min}(q_\triangle, d)$. *Data complexity* is different from combined complexity [9], it is the complexity measured in terms of the size of the database only. We consider two conditions on the input database instance: (a) the input database is more general that its query result is of planarity and any value occurs in at most 7 tuples in the result, (b) circles in the visualization of the query result has a length of no more than 3. We examine the impact of such two conditions on computational complexity of this problem and show they can still make problem $\Gamma_{min}(q_\triangle, d)$ hard, so the result of complexity is strengthened.

2 Preparation

We first given an introduction of some necessary terms and notations.

Database. A database schema is a finite set $\{R_1, \ldots, R_m\}$ of distinct relations. Each relation R_i has r_i attributes, say $\{A_1, \ldots, A_{r_i}\}$, where r_i is the arity of R_i. Each attribute A_j has a corresponding domain $dom(A_j)$ which is a set of valid values. A domain $dom(R_i)$ of a relation R_i is a set $dom(A_1) \times \cdots \times dom(A_{r_i})$. Any element of $dom(R_i)$ is called a fact. A database d can be written as $\{D; R_1, \ldots, R_m\}$, representing a schema over certain domain D, where D is a set $dom(R_1) \times \cdots \times dom(R_m)$.

Conjunctive query. By *datalog*-style notation, a boolean conjunctive query can be written as following

$$q :- R_{i_1}(\bar{x}_1), R_{i_2}(\bar{x}_2), \ldots, R_{i_k}(\bar{x}_k)$$

where each $\bar{x}_i$ has an arity of r_{i_1} consisting of constants and variables. Query result $q(d)$ is *true* if there exists facts $\{t_1, t_2, \ldots, t_k\}$ in d can be mapped to build-in variables $\bar{x}_1, \bar{x}_2, \ldots, \bar{x}_k$ consistently, say consistent with constants in each $\bar{x}_1$; Otherwise, $q(d)$ is *false*. Intuitively, the $\{t_1, t_2, \ldots, t_k\}$ is a witness such that q is true. From the perspective of relational algebra, conjunctive query written as a paradigm with combination of selection, projection and join operation equivalently. A special case in this paper is conjunctive query of form

$$q_\triangle :- R(x, y), S(y, z), T(z, x)$$

which is the same query as the example figure above.

Boolean queries. A boolean query q is a function mapping database d to $\{true, false\}$, it is *true* if query result of $q(d)$ is not empty, i.e., there exists at least a witness in d such that $t \models q$. We limit our study inside the first order query language, so that queries can be written by a certain fragment of the first order query language. We consider the most important query fragments, the conjunctive query mentioned above, because it is the key fragment in the dichotomy of the data complexity of Minimum Contingency Set.

2.1 Analysis of Previous Work

We next point out the limitation of an important previous result before introducing our results.

Triad. Freire *et al.* identified a trichotomy of conjunctive query with *triad* [1] where *triad* is a set of three atoms that are connected in a special way,

$$q_\triangle :- R(x, y), S(y, z), T(z, x)$$

Roughly speaking, if there is no triad existing in those rewritten queries in some simple way, Minimum Contingency Set problem for these queries is in PTime, otherwise it is in NP-complete. That is to say, triad makes queries hard, and this can be formally stated as the following theorem,

Theorem 1 (dichotomy of Minimum Contingency Set [1]). *Let q be a self join free conjunctive query and let q' be the result of making all dominated atoms exogenous. If q' has a triad, then $\Gamma_{min}(q, d)$ is NP-complete in terms of data complexity, otherwise it is in PTime.*

However, the input is not only the query, but also the input database instance, however, if we consider both input parameters, the instance built in the proof of theorem above is too strict, so that the condition is somehow too weak. Before we strengthen the theorem, we first revisit the instance built constructed in their proof.

Reduction in dichotomy of Minimum Contingency Set. The reduction is made from 3SAT to $\Gamma_{min}(q_\triangle, d)$. Then it immediately follows that $\Gamma_{min}(q_\triangle, d)$ is NP-complete. Given 3SAT instance ϕ with n variables x_i ($1 \leq i \leq n$) and m clauses C_j ($0 \leq j < m$), they build a $\Gamma_{min}(q_\triangle, d)$ instance (D_ϕ, k_ϕ) such that $q_\triangle(D_\phi)$ is *true*, then the goal is to guarantee

$$\phi \in 3SAT \Leftrightarrow (D_\phi, k_\phi) \in \Gamma_{min}(q_\triangle, d)$$

The basic idea of the proof is to build circles corresponding to variables made up of triangles satisfying the $q_\triangle$, and then connected those circles in a triangle manner to mimic those clauses, then guarantee that all results will be destroying by deleting only k_ϕ tuples if $\phi \in 3SAT$. Concretely, speaking in a way of visualization, insert triads of form $R(a, b)$, $S(b, c)$ and $T(c, a)$ into database D_ϕ, so that an triangle is formed by each witness (a, b, c) if $d_\phi \models q_\triangle$, and each tuple could be seen as an edge. To guarantee that a minimum contingency set for D_ϕ refers to a corresponding truth assignment of the variables x_i $(1 \le i \le n)$ in ϕ, so that the only way to remove all triangles depends on contingency set constructed for $q_\triangle$. Database instance D_ϕ contains one circular gadget g_i for each variable x_i. The circle consists of $12m$ edges, half of them labeled x_i and the other half labeled $\bar{x}_i$. Therefore, $12m$ triangles are formed by this and they can be minimally broken by choosing the $6m$ x_i edges or the $6m$ $\bar{x}_i$ edges. Any other way would require more edges removed. Therefore, set the instance parameter $k_\phi = 6mn$, such that there exists a contingency set of size k_ϕ in D_ϕ for $q_\triangle$ if and only if $\phi \in$ 3SAT. Based on this, a logically way to connect all the circles g_i is designed. Let clause $C_j = x_1 + \bar{x}_2 + x_3$, adding 3 tuples to connect circles of g_1, g_2 and g_3, this also form a triangle with edges labeled x_1, x_2 and x_3, specifically using edges (b^1_{4j+1}, b^2_{4j+1}) of g_1 and g_2, (c^2_{4j+1}, c^3_{4j+1}) of g_2 and g_3, and (a^3_{4j+2}, a^1_{4j+1}) of g_3 and g_1. In order to do this in D_ϕ, let a^1_{4j+1} in g_1 also be a^3_{4j+2} in g_3, say they are the same value from the domain of D_ϕ. This triangle is destroyed if and only if the chosen variable assignment satisfies C_j.

Here we consider if there are clauses $C_0 : (x_1 + \bar{x}_2 + x_3)$ and $C_1 : (x_1 + x_4 + x_5)$, then we will have a construction as Fig. 2.

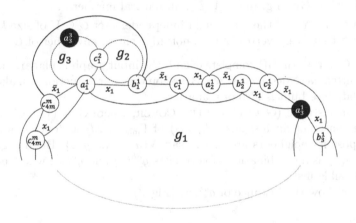

Fig. 2. Example for the simple reduction.

Now, we claim that the database instance got by such reduction will not be of *planarity*. According to the construction, we first build circles g_1, g_2 and g_3. To mimic C_0, we have that a^1_1 should be a^3_2 of g_3 and b^1_1 should be b^2_1 of g_2, also g_2 and g_3 should share a common vertex. Thus, due to the *"chain-like"* way of connections in g_1, in order to avoid breaking the planarity, g_2 and g_3 should locate inside the face of either (a^1_1, b^1_1, c^m_{4m}) or (a^1_1, b^1_1, c^1_1). However, no matter which face is chosen, when mimicking C_1, there must be a vertex, say a^1_3, shared by g_1 and g_3. Then, one can easily verify that, there is no way to do this while preserving the planarity.

So we have two observations: (a). If two clauses have a common variable, then query result of this database instance will not be of *planarity* and degree of vertex is unbounded, (b). There is no limitation on the length of every circle. These requirements are too strict, such database instance should be too complex. Therefore, a natural question is raised here: "*What if the input database is not of planarity or there is a fixed limitation on the length of every circle, is the problem* $\Gamma_{\min}(q_\triangle, d)$ *still* NP-complete?"

3   Results

To answer such questions, we provide the following results to strengthen the hardness result of $\Gamma_{\min}(q_\triangle, d)$.

Theorem 2. *Let q be an self join free conjunctive query, If q has a triad and input database is not of planarity, then finding $\Gamma_{\min}(q, d)$ is still* NP-complete.

Proof. To strengthen the hardness result for query with triad, construct a Karp-reduction from the NP-complete Vertex Cover problem to $\Gamma_{\min}(q_\triangle, d)$.

VERTEX COVER

INPUT   Given graph $G(V, E)$, an natural number k.

OUTPUT   Yes, if there exists an independent set $C \subseteq V$ of size k such that every vertex of C is not adjacent to any other of C.

Vertex Cover is still NP-complete even if the input graph is planar, and cubic, *i.e.*, very vertex has a degree of three. Based on this, our reduction can also ensure that the database built is also of planarity.

Given an instance (G, k) of VERTEX COVER, where $G = (V, E)$ is a planar graph, we construct an instance (d_G, k_G) of $\Gamma_{\min}(q_\triangle, d)$ as follows, for each u of V, suppose its neighbors are x, y, z, say $N(u) = \{x, y, z\}$. Build a gadget of planarity, $g(u)$ as a combination of two parts $g^{ab}(u)$ and $g^c(u)$, where each part has tow possible instances.

First, we show the instance of $g^{ab}(u)$ (Fig. 3),

	x	y
	1^u	2^u
	1^u	4^u
R	1^u	6^u
	ux	2^u
	uy	6^u

	y	z
	2^u	3^u
	4^u	3^u
S	4^u	5^u
	6^u	5^u
	2^u	$u(x)$
	6^u	$u(y)$

	z	x
	3^u	1^u
	5^u	1^u
T	$u(x)$	ux
	$u(x)$	1^u
	$u(y)$	1^u
	$u(y)$	uy

Fig. 3. Gadget for the instance of $g^{a,b}(u)$.

and its counterpart (Fig. 4),

<table>
<tr><th colspan="2">R</th><th colspan="2">S</th><th colspan="2">T</th></tr>
</table>

R	x	y
	2^u	1^u
	4^u	1^u
	2^u	ux
	6^u	1^u
	6^u	uy

S	y	z
	1^u	3^u
	1^u	5^u
	1^u	$u(x)$
	ux	$u(x)$
	1^u	$u(y)$
	uy	$u(y)$

T	z	x
	3^u	2^u
	3^u	4^u
	5^u	4^u
	5^u	6^u
	$u(x)$	2^u
	$u(y)$	6^u

Fig. 4. Gadget for the counterpart instance of $g^{a,b}(u)$.

To illustrate the intuition of the instance, we provide the visualization of part $g^{ab}(u)$ and its counterpart as following (Fig. 5),

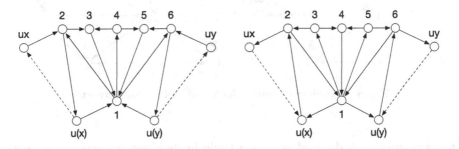

Fig. 5. Gadget visualization for $g^{a,b}(u)$ and its counterpart.

And we also should show the instance of second part $g^c(u)$ as following, the counterpart of it can be shown as following (Figs. 6 and 7),

R	x	y
	2^u	7^u
	uz	7^u

S	y	z
	7^u	6^u
	7^u	8^u
	7^u	$u(z)$

T	z	x
	6^u	2^u
	6^u	uz
	8^u	uz
	$u(z)$	uz

Fig. 6. Gadget for the instance of $g^c(u)$.

x	y
2^u	6^u
uz	6^u
uz	8^u
uz	$u(z)$

R

y	z
6^u	7^u
8^u	7^u
$u(z)$	7^u

S

z	x
7^u	2^u
7^u	uz

T

Fig. 7. Gadget for the counterpart instance of $g^c(u)$.

We also show a visualization of part $g^c(u)$ and its counterpart as following (Fig. 8),

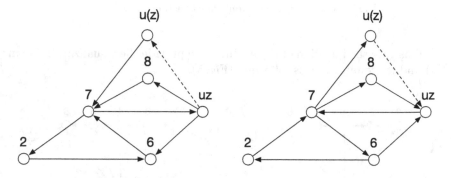

Fig. 8. Gadget visualization for $g^c(u)$ and its counterpart.

Observation: each directed edge is a tuple in database instance d, and each directed triangle is a tuple of $q_\triangle(d)$.

To build a complete gadget $g(u)$, merge g^{ab} and g^c with node 2^u and 6^u, equivalent to a union operation in d. Then, the gadget graph also has three "docking" edges, namely $(1, 2)$, $(1, 6)$, and $(6, 7)$, (or the opposite directions), which are used for connecting the edge gadgets. Note that for each edge (u, y), build a triangle $\triangle_{uy} : \{uy, u(y), y(u)\}$ showed as a dash lined triangle in the following Fig. 9.

The idea is similar to [10], the triangle is then attached to the vertex gadgets $g(u)$ and $g(y)$ as follows. For triangle $\triangle_{uy}$ and vertex gadget $g(u)$, add the edges $y(u), u(y)$ by inserting a tuple $y(u), u(y)$ into the next corresponding atom, where $(1, 6)$ is a docking edge that has not been used before. Vertex gadget $g(y)$ is attached analogously. By taking no account of direction consistency, since G is cubic, the three docking edges that each vertex gadget provides suffice and each docking edge is used. Planarity could be ensured when taking no account

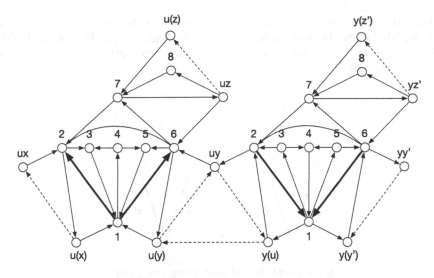

Fig. 9. Gadget connection visualization.

of direction consistency, by using the docking edges of u according to the relative order of neighbors of u given by an embedding of G into database. Since all gadgets are planar, this yields a planar graph when taking no account of directions.

The intuition of this construction is the following. Each edge (u, y) of the original graph G must have at least one of its endpoints in the vertex cover. Correspondingly, for each triangle $\triangle_{uy}$ at least one edge must be deleted. Consider the graph $\kappa(u)$ induced by the vertex set

$$\{1, 2, 3, 4, 5, 6, 7, 8\} \cup \{ux, u(x), uy, u(y), uz, u(z)\},$$

the minimum number of edge deletions to make $\kappa(u)$ triangle-free is six.

However, if one of the outer edges $(ux, u(x)), (uy, u(y)), (uz, u(z))$ is deleted, it is possible to delete the other two outer edges while only deleting seven edges. Note that this is the minimum number of edge deletions, equals to the size of minimum contingency set, to make $\kappa(u)$ triangle-free, say query result is $\emptyset$, under the constraint of having to use one of the outer edges. If we do so, all triangles in edge gadgets for edges incident to u are destroyed. Conversely, if there is a solution for the constructed instance of $\Gamma_{\min}(q_\triangle, d)$, there always is an optimal solution for $\Gamma_{\min}(q_\triangle, d)$ which does not contain the third edge $(u(y), y(u))$ and consequently activates $\kappa(u)$ or $\kappa(y)$, making the deletion of all the outer edges of one of these two graphs possible. There are at most k vertex gadgets corresponding to members of the vertex cover, then we set

$$k_d := 8k + 6(|V|k) = 6|V| + 2k$$

then, it follows immediately that $\Gamma_{\min}(q_\triangle, d)$ is NP-complete.

However, we can not ignore the consistency of edge direction in each common triangle. We denote the docking edge $(1, 2)$ as A-edge, $(1, 6)$ as B-edge, and $(7, 6)$ as C-edge one by one. A local view of connecting two vertices can be shown as following (Fig. 10),

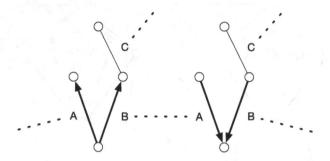

Fig. 10. Local view of connecting two gadgets.

Observation: inside every gadget g, the direction of B-edge is the same as that of A-edge, and for two adjacent gadgets g and g' connected by no C-edges, the directions of A- and B-edge of g are different from that of g'.

This brings a problem that the consistency of directed edges after connecting all these gadgets in a casual manner while keeping planarity. Consider such an odd circle as following (Fig. 11),

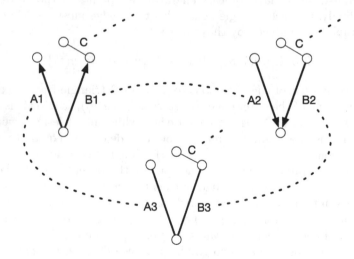

Fig. 11. Conflict resulting from odd circle of connecting gadgets.

Obviously, according to the observation mentioned above, there is a conflict of the directions of A3-edge and B3-edge. It is infeasible to implement this in

$\lceil_{\min}(q_\triangle, d)$ if do not change the manner of connection. However, a good news is that, an observation is that the C-edge of each gadget could be used to tolerate either, so that each C-edge provide a chance to solve a conflict. We can connect these gadgets consistently by a careful arrangement of the neighbor of every C-edge. To do this, we distribute the C-edges to resolve all the direction conflict greedily. That is, for each vertex $u \in V$ and at least two adjacent edges $(u, x_1), (u, x_2)$ of u are unlabeled, if there is an adjacent vertex, say x_i, whose C-edge is unused yet, label (u, x_i) by C-edge connection, repeat until the edge labeled graph G_d is obtained.

Pumping algorithm. To reset the graph to be planar, we design this algorithm to reset the layout of the gadgets corresponding to all vertices. The basic idea is based on an observation that, g^c of every gadget g can be set as either clockwise or anticlockwise direction, therefore, the gadget of the vertex adjacent to g^c part can be set up in any manner without resulting in a conflict.

C-edge Pumping Algorithm

INPUT -   cubic graph G
OUTPUT -   edge labeled graph G_d of directed consistency

1: **for** each vertex $u \in V$ and at least two adjacent edges
     $(u, x_1), (u, x_2)$ of u are unlabeled **do**
2:     **if** there is an adjacent vertex, say x_i, whose C-edge is unused
         yet **then**
3:         $label(u, x_i) \leftarrow x_i.c$
4: **return** the edge labeled graph G_d

We observe that pumping algorithm could deal with the arrangement of not only cubic graph but also non-planar 3-regular graph. An example solution on Peterson graph well known as a 3-regular non-planar graph as following, where the number is the order to distribute the C-edges to the corresponding position of three alternatives.
Here we claim that:

(a) the pumping algorithm distributes the C-edges in polynomial time $O(|d|)$;
(b) by means of clockwise rotating and vertical flipping, every gadget $g(u)$ will keep the planarity after rearrangement of connecting manner (In fact, g^{ab} and g^c could be two separated part of $g(u)$, then the planarity could never be violated);
(c) after C-edge rearrangement, there is no odd circle in which every edge is not connected by C-edge (Fig. 12).

Here the claim accomplishes this proof.   □

We also have the following results for limitation on the length of circle.

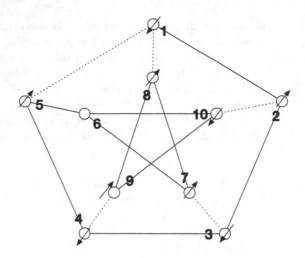

Fig. 12. Example of result on Peterson graph.

Theorem 3. *The problem* $\Gamma_{min}(q_\triangle, d)$ *is still* NP-complete, *if all the circles has a fixed length less than* 4.

This result follows by a simple reduction from the triangle edge deletion problem, which is still NP-complete even each circle is of length no more than 3, to $\Gamma_{min}(q_\triangle, d)$. We omit the detail here.

Next we show the upper bound of this problem is fixed-parametric tractable, and provide a kernelization algorithm.

Theorem 4. *The problem* $\Gamma_{min}(q_\triangle, d)$ *admits a problem kernel with vertices less than* $121k^2$ *tuples, which can be computed in* $O(kn\sqrt{n})$.

We here provide a kernelization algorithm, one for instances of planarity property, which produces a kernel consisting of $11k$ tuples. This kernelization for instance of planarity is based on idea for kernelizing triangle edge deletion [10]. We show the reductions in the following.

Reduction 1. *Remove all tuples that are not included in any witness of any query result* $t \in q_\triangle(d)$.

We here apply this simple data reduction, and it follows immediately. Then the second reduction rule is very powerful.

Reduction 2. *If a witness of some query result in* $q_\triangle(d)$ *contains only one tuple* t *contained in a witness of another different query result, delete* t, *and reduce* k *by 1; If a witness of some query result does not contain a tuple which is in a witness of another query result, then delete an arbitrary tuple of this witness, and reduce* k *by 1.*

The correctness of this reduction can be easily verified. One tuple of the three tuples r, s, t in a witness of any query result has to be removed. Then this reduction always chooses a tuple r, which covers all witness covered by s or t.

Let any two tuples $(u, a), (u, b) \in d$ be a dock of cell c iff $\{(u, c, a), (u, c, b)\} \subseteq q_{\triangle}(d)$ or $\{(a, u, c), (b, u, c)\} \subseteq q_{\triangle}(d)$ or $\{(c, a, u), (c, b, u)\} \subseteq q_{\triangle}(d)$. The cell contained in both edges of a dock is called the dock cell of c.

We have following property identifying a structure which holds in the database instance of planarity: *Let d be a database instance and S is the solution of $\Gamma_{\min}(q_{\triangle}, d)$, let cell set M including all the cells included in S, and for each cell $c \in M$, let $|S[u]|$ be the number of tuples in S including u. Then, let $|S[u]| \geq 2$. If then d is of planarity, then cell set C excluding M, say $C \setminus M$ should contain less than $2|S[u]| - 1$ cells with a dock $B \subseteq S$ with u as the dock cell.*

For simplicity, we first suppose each tuple t only occurs once in the database. Thus there are at most $2|S[u]|1$ neighbors of u, and they are contained in $C \setminus M$ and have a dock $B \subseteq S$ with u as the base vertex. As already shown, every vertex $c \in C \setminus M$ has a dock $B \subseteq S$, and this implies that the corresponding dock cell is in M. It follows that

$$|C \setminus M| \leq 2|S[u]| - 2 \leq 4|S| - 2|M|$$

We used the fact that the sum over all $|S[u]|$ counts every edge in S exactly two times. The cell set C is partitioned into M and $C \setminus M$, hence

$$|C| \leq |M| + |C \setminus M| \leq 4|S| - |M|.$$

Due to the planarity of database instance, then we have the Eulers formula implying that

$$|S| \leq 3|M| - 6.$$

Therefore we have

$$|M| \geq |S| + 2.$$

Using this, we obtain an upper bound

$$|C| \leq 4|S| - |S| - 2 \leq 11|S| - 2 \leq 11k$$

Then, for the general case, we know that each tuple t occurs at most three times in the database, and distributes in the three tables. Therefore, the final results of the kernel size should be

$$|d'| \leq (3|C|)^2 \leq 121k^2.$$

4   Conclusion

We studied the data complexity of Minimum Contingency Set problem for triad conjunctive query. If two clauses have a common variable, then this database instance should be too complex, formally speaking, the visualization of its query result will not be of *planarity*, this requirement is too strict. And there is no limitation on the length of every circle in the visualization of the query result. This makes the previous theorem of dichotomy weak. To answer such questions, we provide the following results to strengthen the hardness result, that the data complexity of $\Gamma_{min}(q_\triangle, d)$ is still NP-complete, if the input database instance is of planarity, or the maximum length of every circle is limited. Our theorems also generalize the result of triangle edge deletion problem defined on general graph into directed graph, make a contribution to graph theorem.

References

1. Freire, C., Gatterbauer, W., Immerman, N., Meliou, A.: The complexity of resilience and responsibility for self-join-free conjunctive queries. Proc. VLDB Endow. **9**(3), 180–191 (2015)
2. Cong, G., Fan, W., Geerts, F., Li, J., Luo, J.: On the complexity of view update analysis and its application to annotation propagation. IEEE Trans. Knowl. Data Eng. **24**(3), 506–519 (2012)
3. Cong, G., Fan, W., Geerts, F.: Annotation propagation revisited for key preserving views. In: Proceedings of the 15th ACM International Conference on Information and Knowledge Management, CIKM 2006, pp. 632–641. ACM, New York (2006)
4. Buneman, P., Khanna, S., Tan, W.C.: On propagation of deletions and annotations through views. In: Proceedings of the Twenty-First ACM SIGMOD-SIGACT-SIGART Symposium on Principles of Database Systems, PODS 2002, pp. 150–158. ACM, New York (2002)
5. Kimelfeld, B., Vondrák, J., Williams, R.: Maximizing conjunctive views in deletion propagation. ACM Trans. Database Syst. **37**(4), 24:1–24:37 (2012)
6. Kimelfeld, B.: A dichotomy in the complexity of deletion propagation with functional dependencies. In: Proceedings of the 31st Symposium on Principles of Database Systems, PODS 2012, pp. 191–202. ACM, New York (2012)
7. Kimelfeld, B., Vondrák, J., Woodruff, D.P.: Multi-tuple deletion propagation: approximations and complexity. Proc. VLDB Endow. **6**(13), 1558–1569 (2013)
8. Miao, D., Liu, X., Li, J.: On the complexity of sampling query feedback restricted database repair of functional dependency violations. Theoret. Comput. Sci. **609**, 594–605 (2016)
9. Vardi, M.Y.: The complexity of relational query languages (extended abstract). In: Proceedings of the Fourteenth Annual ACM Symposium on Theory of Computing, STOC 1982, pp. 137–146. ACM, New York (1982)
10. Brügmann, D., Komusiewicz, C., Moser, H.: On generating triangle-free graphs. Electron. Notes Discrete Math. **32**, 51–58 (2009)

Generalized Pyramidal Tours
for the Generalized Traveling Salesman Problem

Michael Khachay[1,2(✉)] and Katherine Neznakhina[1]

[1] Krasovsky Institute of Mathematics and Mechanics, Ural Federal University,
Ekaterinburg, Russia
mkhachay@imm.uran.ru, eneznakhina@yandex.ru
[2] Omsk State Technical University, Omsk, Russia

Abstract. In this paper, we introduce the notion of l-quasi-pyramidal and l-pseudo-pyramidal tours extending the classic notion of pyramidal tours to the case of the Generalized Traveling Salesman Problem (GTSP). We show that, for the instance of GTSP on n cities and k clusters with arbitrary weights, l-quasi-pyramidal and l-pseudo-pyramidal optimal tours can be found in time $O(4^l n^3)$ and $O(2^l k^{l+4} n^3)$, respectively. Consequently, we show that, in the most general setting, GTSP belongs to FPT for parametrizations induced by these special kinds of tours. Also, we describe a non-trivial polynomially solvable subclass of GTSP, for which the existence of l-quasi-pyramidal optimal tour (for some fixed value of l) is proved.

1 Introduction

The Traveling Salesman Problem (TSP) is the famous combinatorial optimization problem having many valuable applications in operations research and attracting interest of scientists for decades (see e.g. [14,19,21]).

It is known that TSP is strongly NP-hard and hardly approximable in its general setting [22]. At the same time, the problem remains intractable in metric and Euclidean settings, but can be approximated well in these cases, admitting fixed-ratio algorithms for an arbitrary metric [10] and Polynomial Time Approximation Schemes (PTAS) for Euclidean spaces of any fixed dimension [1]. Many generalizations of TSP, e.g. Cycle Cover Problem [13,15,16], Peripatetic Salesman Problem [2,12], have the similar approximation behaviour.

Algorithmic issues of finding optimal restricted tours, for several kinds of restrictions, e.g. precedence constraints, are also actively investigated (see, e.g. [4,5,9]). Among others, restriction of TSP to considering so called *pyramidal tours* (see e.g. [8]) seems to be especially popular. A pyramidal tour respects the initial order defined on the nodeset of a given graph and, for some r, has the form

$$1 = v_{i_1}, v_{i_2}, \ldots, v_{i_r} = n, v_{i_{r+1}}, \ldots, v_{i_n},$$

where $v_{i_j} < v_{i_{j+1}}$ for any $j \in \{1, \ldots, r-1\}$ and $v_{i_j} > v_{i_{j+1}}$ for any $j \in \{r+1, \ldots, n-1\}$.

© Springer International Publishing AG 2017
X. Gao et al. (Eds.): COCOA 2017, Part I, LNCS 10627, pp. 265–277, 2017.
https://doi.org/10.1007/978-3-319-71150-8_23

It is widely known [18] that an optimal pyramidal tour can be found in time of $O(n^2)$ for any weighting function. Recently it was shown [6] that, for the Euclidean setting, an optimal pyramidal tour can be found in time $O(n \log^2 n)$. In papers [11,20], several generalizations of pyramidal tours, for which optimal tour can also be found efficiently were introduced. Despite their fame, pyramidal tours have one shortcoming. Known settings of TSP and its generalizations, for which the existence of optimal pyramidal tours is proven, remain very rare so far. Actually, they are mostly exhausted with settings satisfying the well known sufficient conditions by Demidenko and van der Veen (see e.g. [14]) and some other special cases [3,7,19].

The contribution of this paper is two-fold. First, we introduce (in Sect. 2) notion of generalized pyramidal tours, we call them *l-quasi-* and *l-pseudo-pyramidal*, extending the classic notion of pyramidal tours and results of [20] to the case of Generalized Traveling Salesman Problem (GTSP). We show that *l*-pseudo-pyramidal and *l*-quasi-pyramidal optimal tours can be found in time $O(2^l k^{l+4} n^3)$ and $O(4^l n^3)$, respectively. Then, in Sect. 3, we describe a non-trivial polynomially solvable subclass of GTSP, for which the existence of on optimal *l*-quasi-pyramidal tour (for some fixed *l*) is proved.

2 Generalized Pyramidal Tours

We proceed with the common setting of the Generalized Traveling Salesman Problem (GTSP). An instance of the GTSP is defined by a complete edge-weighted graph $G = (V, E, w)$ with weighting function $w: E \to \mathbb{R}_+$, and by a given partition $V_1 \cup \ldots \cup V_k = V$ of the nodeset $V = V(G)$ of the graph G. Feasible solutions are cyclic tours $\tau = (v_{i_1}, \ldots, v_{i_k})$ visiting each cluster V_i once. Hereinafter, we call such routes *Clustered Hamiltonian tours* or *CH-tours*. The problem is to find a CH-tour of the minimum weight[1].

In this section, we extend the well-known notion of a pyramidal tour to the case of partial orders defined implicitly by the orderings of clusters. Indeed, (linearly) ordered finite set $(V_1, \ldots, V_k)$ of clusters induces a partial order on the nodeset V of the graph G as follows: For any $u \in V_i$ and $v \in V_j$, $u \prec v$ if $i < j$.

Definition 1. *Let τ be a CH-tour*

$$v_1, v_{i_1}, \ldots, v_{i_r}, v_k, v_{j_{k-r-2}}, \ldots, v_{j_1}, \ 0 \leq r \leq k - 2$$

such that $v_t \in V_t$ for any t. We call τ an l-quasi-pyramidal tour, if $i_p - i_q \leq l$ and $j_{p'} - j_{q'} \leq l$ for any $1 \leq p < q \leq r$ and $1 \leq p' < q' \leq k - r - 2$.

The following theorems extend the results proposed in [20] for the classic TSP.

Theorem 1. *For any weighting function $w: E \to \mathbb{R}_+$, a minimum cost l-quasi-pyramidal CH-tour can be found in time of $O(4^l n^3)$.*

---

[1] In this paper, we restrict ourselves to the case of undirected graphs. Nevertheless, the analogous argument can be provided to the case of digraphs and asymmetric weighting functions w.

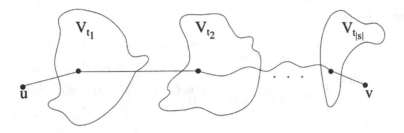

Fig. 1. Path from u to v through the clusters V_{t_j}, $t_j \in S$.

Proof. We start with some necessary notation. For any integers $i > j$, we use the common shortcuts $[j, i]$, $[j, i)$, and (j, i) for intersections with $\mathbb{N}$ of the sets $\{j, \ldots, i\}$, $\{j, \ldots, i-1\}$, $\{j+1, \ldots, i-1\}$, respectively. For any nodes $u \in V_i$ and $v \in V_j$, $i \neq j$, and an arbitrary subset $S \subset [i-l, i) \setminus \{1, j\}$ or $S \subset [j-l, j) \setminus \{1, i\}$, let $g(v, S, u)$ be the weight of a shortest $(|S| + 1)$-edge path from u to v visiting all the clusters $\{V_t : t \in S\}$ (see Fig. 1). Values of the function g can be easily calculated recursively, since $g(v, \varnothing, u) = w(\{v, u\})$ and

$$g(v, S, u) \begin{cases} = \min_{m \in S} \min_{v' \in V_m} \{g(v, S \setminus \{m\}, v') + w(\{v', u\})\}, & \text{if } S \subseteq [j-l, j) \setminus \{1, i\}, \\ = \min_{m \in S} \min_{v' \in V_m} \{w(\{v, v'\}) + g(v', S \setminus \{m\}, u)\}, & \text{if } S \subseteq [i-l, i) \setminus \{1, j\}. \end{cases}$$
(1)

Further, for any $1 \leq j < i \leq k$, let $f(u, v, T)$ be the weight of a shortest path P from $u \in V_i$ to $v \in V_j$ visiting all the clusters with numbers from $([1, i) \cup [1, j)) \setminus T$, where $T \subseteq ([i-l, i) \cup [j-l, j)) \setminus \{1, i, j\}$, and the path P has the form

$$u = v_{i_0}, v_{i_1}, \ldots, v_{i_r} = \bar{v} = v_{j_0}, v_{j_1}, \ldots, v_{j_s} = v,$$

for pairwise defferent indexes $i_0, \ldots, i_r, j_1, \ldots, j_s$, such that $\bar{v} \in V_1$, $i_t < i$ for $1 \leq t \leq r$, $j_{t'} < j$ for $0 \leq t' \leq s-1$, and

$$\begin{aligned} i_q - i_p &\leq l, \quad (0 < p < q \leq r), \\ j_{p'} - j_{q'} &\leq l, \quad (0 \leq p' < q' \leq s). \end{aligned}$$

As with the function g, values of the function f can be obtained recursively. We start with values $f(u, v, (1, t)) = w(\{u, v\})$ for any $u \in V_1$ and $v \in V_t$, $2 \leq t \leq l+2$. All other necessary values $f(u, v, T)$ for any $u \in V_i$, $v \in V_j$ and any $T \subset ([i-l, i) \cup [j-l, j)) \setminus \{1, i, j\}$ can be computed in ascending order by i and $j < i$ as follows. Let m be the maximum number of the cluster (excluding i and j) visited by the path P. If $m > j$, then $f(u, v, T)$ can be calculated by formula

$$f(u, v, T) = \min_{S \subseteq [m-l, m) \setminus (T \cup \{1, j\})} \min_{u' \in V_m} \{g(u, S, u') + f(u', v, T \cup S)\}, \quad (2)$$

and otherwise by

$$f(u,v,T) = \min \begin{cases} \min\limits_{S\subseteq[m-l,m)\setminus(T\cup\{1\})} \min\limits_{u'\in V_m} \{g(u,S,u') + f(u',v,T\cup S)\}, \\ \qquad\qquad\qquad\qquad \text{if } m \in \{i_1,\dots,i_{r-1}\} \\ \min\limits_{S\subseteq[m-l,m)\setminus(T\cup\{1\})} \min\limits_{u'\in V_m} \{f(u,u',T\cup S) + g(u',S,v)\}, \\ \qquad\qquad\qquad\qquad \text{if } m \in \{j_1,\dots,i_{s-1}\}. \end{cases} \qquad (3)$$

Finally, we obtain $f(u,v,T)$ for any $u \in V_k$ and $v \in V_{k-1}$ and for any $T \subseteq [k-l-1, k-1)$. Weight of an optimal l-quasi-pyramidal tour (see Fig. 2) is given by

$$\min_{T\subseteq[k-l-1,k-1)} \min_{u\in V_k} \min_{v\in V_{k-1}} \{f(u,v,T) + g(v,T,u)\}.$$

We compute a naïve upper bound for time complexity of the algorithm. At first, we calculate the necessary values $g(v,S,u)$ by formula (1) in time $O(2^l n^3)$. Then, the initial values $f(u,v,(1,t))$ can be computed in $O(n^2)$. Further, for any fixed u, v and T, the complexity of Eqs. (2) and (3) does not exceed $O(2^l n)$. Since, formulas (2) and (3) are invoked at most $O(2^l n^2)$ times, the overall time complexity bound is $O(4^l n^3)$, which completes our proof.

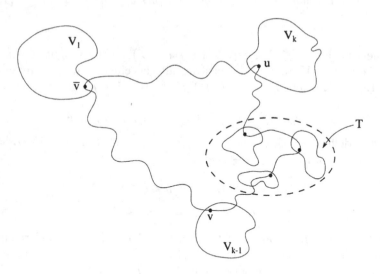

Fig. 2. Constructing a minimum weight l-quasi-pyramidal tour

Remark 1. Evidently, result of Theorem 1 can be considered in the context of parameterized complexity. Actually, Theorem 1 claims that, in the most general setting, GTSP is fixed-parameter tractable with respect to parametrization induced by quasi-pyramidal tours.

In Sect. 3, we describe a subclass of geometric GTSP, each of whose instance has l-quasi-pyramidal optimal tours for some fixed l. Nevertheless, this class seems to be very specific, and the scheme proposed can hardly be extended to more general settings. To overcome this gap, we propose a more common notion of pyramidal-like tours. We call them *pseudo-pyramidal*.

Definition 2. *Let τ be a CH-tour $v_1, v_{i_1}, \ldots, v_{i_r}, v_k, v_{j_{k-r-2}}, \ldots, v_{j_1}$ such that $v_t \in V_t$ for any t. We call τ an l-pseudo-pyramidal tour, if $i_p - i_{p+1} \leq l$ and $j_q - j_{q+1} \leq l$ for any $1 \leq p \leq r - 1$ and $1 \leq q \leq k - r - 3$.*

It easy to verify that any l-quasi-pyramidal tour is an l-pseudo-pyramidal as well.

Theorem 2. *For any weighting function $w \colon E \to \mathbb{R}_+$, a minimum cost l-pseudo-pyramidal CH-tour can be found in time of $O(2^l k^{l+4} n^3)$.*

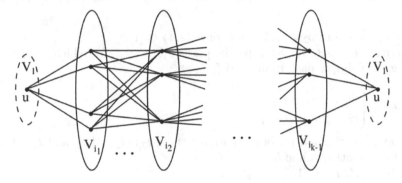

Fig. 3. Auxiliary graph $H_{\theta,u}$ induced by the tour $\theta = \{1, i_1, \ldots, i_{k-1}\}$ and the node $u \in V_1$

Proof. Our argument consists of two stages.

At the first stage, we enumerate all l-pseudo-pyramidal tours in an auxiliary complete graph $H = K_k$ with vertex set $\{1, \ldots, k\}$, which we call *graph of clusters*. Denote the set of these tours by Θ_l.

Then, at the second stage, for any tour $\theta = (1, i_1, \ldots, i_{k-1}) \in \Theta_l$ and any node $u \in V_1$, we find a shortest u-u-path $\rho(\theta, u)$ in the appropriate auxiliary $(k+1)$-partite graph $H_{\theta,u}$, which is defined as follows. Denote parts of $H_{\theta,u}$ by $\pi_0, \ldots, \pi_k$. Then, as it is shown at Fig. 3, $\pi_0 = \pi_k = \{u\}$ and, for any $j \in [1, k)$, π_j coincides with the cluster V_{i_j} of the graph G, i.e. $\pi_j = V_{i_j}$. For any $j \in [0, k)$, the subgraph $H_{\theta,u}\langle \pi_j \cup \pi_{j+1}\rangle$ induced by π_j and π_{j+1} is a complete bipartite graph. The edges of the graph $H_{\theta,u}$ inherit the edge weights of the given graph G.

Evidently, any u-u-path in the graph $H_{\theta,u}$ is equivalent to the appropriate CH-tour in the graph G. Therefore, a minimum cost l-pseudo-pyramidal CH-tour in the graph G is defined by a shortest path $\rho(\theta^*, u^*)$, i.e.

$$w(\rho(\theta^*, u^*)) = \min\{w(\rho(\theta, u)) \colon \theta \in \Theta_l, u \in V_1\}.$$

The time complexity of both stages does not exceed the product of the complexity $T(\Theta_l)$ of the l-pseudo-pyramidal tours enumeration procedure for the graph H (construction of the set Θ_l), the size of V_1, and the complexity of the shortest-path problem in graphs $H(\theta, u)$, i.e. $O(k \cdot n^2)$. Hence, the overall time complexity will be at most $T(\Theta_l) \cdot O(n^3)$, since, without loss of generality, we can assume that $|V_1| = \min\{|V_i| : i \in [1, k]\} \le n/k$.

To estimate $T(\Theta_l)$ we augment the dynamic programming procedure developed in [20].

We introduce two sets Θ_l^+ and Θ_l^- of partial simple (possibly closed) paths in the graph H. Each element of Θ_l^+ is a path $\theta^+ = (i_1, \ldots, i_c)$ such that $i_p - i_{p+1} \le l$ for any $p \in [1, c)$. Similarly, each element $\theta^- = (j_1, \ldots, j_d) \in \Theta_l^-$ satisfies the equation $j_{q+1} - j_q \le l$ for any $q \in [1, d)$.

The current state of the recursive procedure is encoded by a triple (i, S, E), whose entries is defined as follows. The number $i \in [1, k - 1]$ denotes depth of recursion. The set $S = \{p_1, \ldots, p_m\}$ consists of *signed* pairs $(i, j) \in [1, k]^2$ such that

(i) $p_1 = (1, s)^+$ and $p_2 = (t, 1)^-$ for some $\{s, t\} \subset [1, k]$
(ii) there exists a set $\Theta(i, S)$ of partial paths $\theta_1, \ldots, \theta_m$, for which
  – θ_a is a simple path from i_a to j_a

  – $\theta_a \in \begin{cases} \Theta_l^+, & \text{if } p_a = (i_a, j_a)^+ \\ \Theta_l^-, & \text{if } p_a = (i_a, j_a)^- \end{cases}$

  – let I_a be the nodeset of the path θ_a; then, $I_1 \cap I_2 = \{1\}$, and $I_a \cap I_b = \varnothing$ for any other a and b
  – $I_1 \cup \ldots \cup I_m = [1, i]$.

Finally, the set E is an arcset of the l-pseudo-pyramidal tour to be constructed (for the convenience, we store edges of this tour with their bypass directions). Denote

$$Q = \bigcup \{\{i_a, j_a\} : p_a \in S\}.$$

The recursion starts from the following set of initial states (see Fig. 4)

$$\{(k - 1, \{(1, s)^+, (t, 1)^-\}, \{(s, k), (k, t)\}) : \{s, t\} \subset [1, k]\}.$$

Any time, when $i > 1$, there are the following six options. Consider them separately.

Case 1. There exists $p = (i, i)^+ \in S$ (or $(i, i)^-$). In this simple case, we make a recursive call with the state $(i - 1, S \setminus \{p\}, E)$ immediately.

Case 2. There exists $p_a = (i, j)^+ \in S$. Then, in the path $\theta_a \in \Theta_l^+$, the node i has a successor $t \in [i - l, i - 1]$. We make a recursive call with the state

$$(i - 1, (S \cup \{(t, j)^+\}) \setminus \{p_a\}, E \cup \{(i, t)\})$$

for any $t \in [i - l, i - 1] \setminus (Q \setminus \{j\})$.

Case 3. There exists $p_a = (i,j)^- \in S$. Then, in the path $\theta_a \in \Theta_l^-$, there is a successor $t \in [1, i-1]$, and we call the recursion with the state

$$(i-1, (S \cup \{(t,j)^-\}) \setminus \{p_a\}, E \cup \{(i,t)\})$$

for each $t \in [1, i-1] \setminus (Q \setminus \{j\})$.

Case 4. There is $p = (j,i)^+ \in S$. This can be treated similarly to Case 3.

Case 5. There is $p = (j,i)^- \in S$. This is similar to Case 2.

Case 6. In this case, $i \notin Q$, and we should try iteratively all elements of the set S. Suppose, i belongs to path θ_a assigned to the pair $p_a = (i_a, j_a)^+ \in S$ such that $s \in [1, i-1]$ is its predecessor, and $t \in [i-l, i-1]$ is a successor. Then we should call the recursion with the state

$$(i-1, S \cup \{(i_a, s)^+, (t, j_a)^+\}, E \cup \{(s,i),(i,t)\})$$

for each $s \in [1, i-1] \setminus (Q \setminus \{i_a\})$ and $t \in [i-l, i-1] \setminus (Q \setminus \{j_a\})$. Similarly, for the pair $p_a = (i_a, j_a)^-$, we make a recursive call with states

$$(i-1, S \cup \{(i_a, s)^-, (t, j_a)^-\}, E \cup \{(s,i),(i,t)\})$$

for each $s \in [i-l, i-1] \setminus (Q \setminus \{i_a\})$ and $t \in [1, i-1] \setminus (Q \setminus \{j_a\})$.

If $i = 1$, then $S = \{(1,1)^+, (1,1)^-\}$, and the state $(1, S, E)$ is final. In this case, E contains arcs of an l-pseudo-pyramidal tour, which can be decoded in time $O(k)$.

Since, as it is shown in [20] the time complexity of the recursive procedure above is $O(2^l k^{l+3})$, the overall complexity bound of finding the minimum cost l-pseudo-pyramidal tour is

$$O(2^l k^{l+3}) \times O(k) \times O(n^3) = O(2^l k^{l+4} n^3).$$

Thus, the theorem is proved.

Remark 2. As with Theorem 1, Theorem 2 states that, for any weighting function, GTSP belongs to FPT with respect to parameters k and l. Also, since $O(2^l (\log n)^{l+4} n^3)$ asymptotically does not exceed $2^{O(l^3)} \cdot O(n^4)$, the problem has FPT algorithms with respect to parameter l only any time when $k = O(\log n)$.

3 Polynomial Time Solvable Subclass of GTSP on Grid Clusters

In this section, we describe a polynomially solvable subclass of the Generalized Traveling Salesman Problem on Grid Clusters, GTSP-GC for short. In this special case of the GTSP, an undirected edge-weighted graph $G = (V, E, w)$ is given

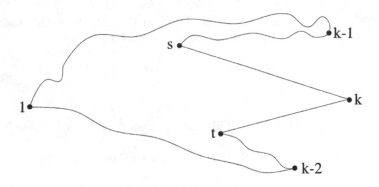

Fig. 4. Example of the initial recursion state

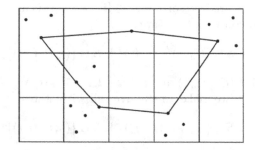

Fig. 5. An instance of the Euclidean GTSP-GC and its optimal solution

where the set of vertices V correspond to a set of points in the planar rectangular grid. Every nonempty 1×1 cell of the grid forms a cluster. The weighting function is induced by distances between the respective points with respect to some metric. To simplify it, we consider Euclidean distances, but similar results can be easily obtained for some other metrics, e.g. for l_1. In Fig. 5, we present an instance of the Euclidean GTSP-GC with 6 clusters.

For two special cases of the problem, when the number k of clusters is $O(\log n)$ or $n - O(\log n)$, polynomial time approximation schemes (PTAS) were proposed in [17]. Meanwhile, the question of a systematic description of polynomial time solvable subclasses of GTSP-GC, which is closely related to complexity analysis of the Hamiltonian cycle problem on grid graphs, is still far from its complete answer.

Let H and W be *height* and *width* (number of rows and columns) of the given grid, respectively. We consider a special case of the GTSP-GC, for which one of these parameters, say H does not exceed 2 (while the other one is unbounded). We call this case GTSP-GC(H2). We show that any instance of GTSP-GC(H2) has an l-quasi-pyramidal optimal CH-tour for some l independent on n. Therefore, this subclass of GTSP-GC is polynomially solvable due to Theorem 1.

Our argument is based on the *Tour straightening transformation* (Algorithm 1), which is closely related to the well-known class of *local search heuristics* and is introduced in the following.

Algorithm 1. Tour straightening transformation

Outer Parameter: t.
Input: an instance of GTSP-GC(H2) and a CH-tothe τ.
Output: a CH-tour τ' without t-zigzags.

1: set $\tau' := \tau$
2: **while** τ' has t-zigzag **do**
3:     assume that equation (5) is valid (without loss of generality, we assume that $c_p - c_q = t - 1$), the case of (6) can be treated similarly;
4:     let C be the set of columns with numbers $c_q, \dots, c_p$ (see Fig. 6);
5:     let $Y = (y_1, \dots, y_{2t+4})$ be ordinate sequence of the nodes visited by τ in C augmented by ordinates of left and right crossing points;
6:     find an optimal 2-medians the clustering for Y with medians m_1 and m_2;
7:     replace segments of tour τ' belonging to C by horizontal lines at height m_1 and m_2 connected to all points mentioned in Step 5 by line segments (Fig. 7)
8: **end while**
9: output the CH-tour τ'.

To describe the transformation, assign to columns of the grid defining the given instance of GTSP-GC(H2), integer numbers $1, 2, \dots, W$ (from the left to the right). Consider an arbitrary CH-tour τ. Assigning to each node v_i of τ the number c_i of the column it belongs to, obtain a sequence σ of column numbers presented in the order induced by the tour τ. Without loss of generality, assume that σ has the form

$$1 = c_1, c_2, \dots, c_r = W, c_{r+1} \dots, c_s \qquad (4)$$

for some appropriate numbers r and s.

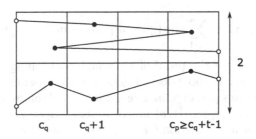

Fig. 6. Segment of τ with t-zigzag

Suppose, for some integer number t, whose value will be specified later, there exist indices

$$1 \leq p < q \leq r, \text{ such that } c_p - c_q \geq t - 1, \text{ or} \tag{5}$$
$$r + 1 \leq p' < q' \leq s, \text{ such that } c_{q'} - c_{p'} \geq t - 1. \tag{6}$$

In this case, we say that the tour τ has a t-zigzag (Fig. 6). Obviously, any l-quasi-pyramidal tour contains no t-zigzags, for $t \geq l$. Algorithm 1 replaces all segments of the tour τ having t-zigzags with subtours of a special kind.

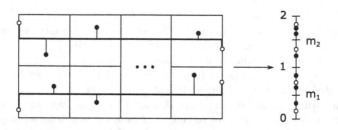

Fig. 7. Replacing t-zigzag with tour segments of the special kind

To specify the value of t, notice that the weight of eliminated segments of τ has an evident lower bound

$$t + 2(t - 1) + t - 2 = 4t - 4.$$

Meanwhile, the weight of their replacement in Step 7 at any iteration of Algorithm 1 is at most $2t + 2F(Y, S_2)$, where $F(Y, S_2)$ is an optimum value of the 2-medians clustering objective function for a sample Y taken from the line segment $S_2 = \{y : 0 \leq y \leq 2\}$.

To estimate an upper bound for $F(Y, S_2)$ we need the following technical lemma.

Lemma 1. *For any sample* $\xi = (p_1, \ldots, p_n)$, $p_i \in S_1 = \{p : 0 \leq p \leq 1\}$ *there exist numbers* m_1 *and* $m_2 \in S_1$ *such that*

$$F(\xi, S_1) = \sum_{i=1}^{n} \min\{|p_i - m_1|, |p_i - m_2|\} \leq n/6. \tag{7}$$

We give a short sketch of the proof of Lemma 1 postponing its full version to the forthcoming paper. Without loss of generality, we assume that any sample $\xi = (p_1, \ldots, p_n)$ contains points p_i in ascending order. Moreover, we assume that any cluster $C = \{i_1, \ldots, i_\mu\} \subset [1, n]$ inherits this property, i.e. $p_{i_1} \leq \ldots \leq p_{i_\mu}$ and $p_i \leq p_j$ holds for any partition $C_1 \cup C_2 = [1, n]$ and any $i \in C_1$ and $j \in C_2$. Then, for the median m of a μ-points cluster C we obtain

$$\sum_{t=1}^{\mu} |p_{i_t} - m| = \sum_{t=1}^{\lfloor \mu/2 \rfloor} (m - p_{i_t}) + \sum_{t=\lceil \mu/2 \rceil + 1}^{\mu} (p_{i_t} - m) = -\sum_{t=1}^{\lfloor \mu/2 \rfloor} p_{i_t} + \sum_{t=\lceil \mu/2 \rceil + 1}^{\mu} p_{i_t}.$$

Therefore, for a given sample ξ, the value $F(\xi, S_1)$ depends on the choice of a partition $C_1 \cup C_2 = [1, n]$ ultimately and obeys the equation

$$F(\xi, S_1) = \min \left\{ \sum_{i \in C_1} |p_i - m_1| + \sum_{i \in C_2} |p_i - m_2| : C_1 \cup C_2 = \mathbb{N}_n \right\}$$

$$= \min \left\{ - \sum_{i=1}^{\lfloor \mu_1/2 \rfloor} p_i + \sum_{i=\lceil \mu_1/2 \rceil + 1}^{\mu_1} p_i - \sum_{i=1}^{\lfloor \mu_2/2 \rfloor} p_{i+\mu_1} + \sum_{i=\lceil \mu_2/2 \rceil + 1}^{\mu_2} p_{i+\mu_1} : \mu_1 + \mu_2 = n \right\}.$$

Thus, $\sup_{\xi \in S_1^n} F(\xi, S_1)$ coincides with an optimum value $q^*(n, S_1)$ of linear program (8)

$$q^*(n, S_1) = \max q$$
$$s.t.$$

$$- \sum_{i=1}^{\lfloor \mu_1/2 \rfloor} p_i + \sum_{i=\lceil \mu_1/2 \rceil + 1}^{\mu_1} p_i$$

$$- \sum_{i=1}^{\lfloor \mu_2/2 \rfloor} p_{i+\mu_1} + \sum_{i=\lceil \mu_2/2 \rceil + 1}^{\mu_2} p_{i+\mu_1} \geq q, \quad (\mu_1 + \mu_2 = n),$$

$$0 \leq p_1 \leq \ldots \leq p_n \leq 1.$$
$$(8)$$

Applying to program (8) the recurrent variable elimination technique, it is easy to verify that $q^*(n, S_1) \leq n/6$, which completes the sketch of our proof.

Getting back to discussion of Algorithm 1, we obtain from Lemma 1 that

$$F(Y, S_2) \leq 2 \cdot q^*(2t + 4, S_1) \leq 2 \cdot (2t + 4)/6.$$

Therefore, at any iteration of Algorithm 1, the tour τ' becomes cheaper if

$$2t + 4t/3 + 8/3 \leq 4t - 4, \quad \text{i.e. } t \geq 10.$$

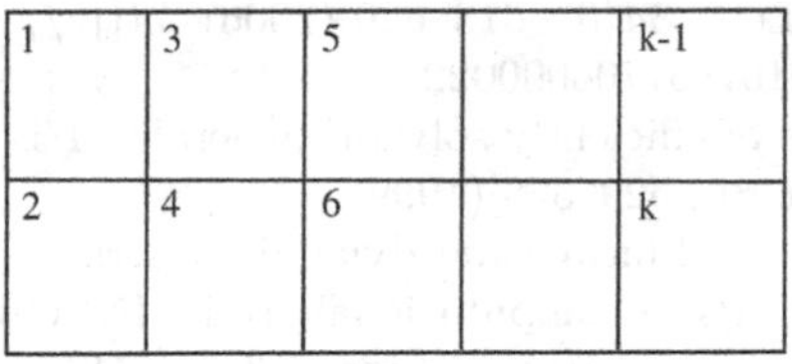

1	3	5		k-1
2	4	6		k

Fig. 8. Cluster ordering

Further, let the cells of the grid be ordered as in Fig. 8 (i.e., top-down and left-right). For $t = 10$, any CH-tour of the given GTSP-GC(H2) instance can be transformed to l-quasi-pyramidal CH-tour for $l = 20$ without increasing its weight. Hence, we have proved the following theorem.

Theorem 3. *Any instance of GTSP-GC(H2) has an optimal 20-quasi-pyramidal CH-tour.*

As a consequence of Theorems 1 and 3, we obtain that GTSP-GC(H2) can be solved to optimality in time $O(n^3)$.

4 Conclusion

In this paper, the new notions of l-quasi-pyramidal and l-pseudo-pyramidal tours extending the classic notion of pyramidal tours are introduced. We show that, similar to the case of pyramidal tours and TSP, an optimal l-quasi-pyramidal tour for the Generalized Traveling Salesman Problem can be found efficiently (for an arbitrary weighting function). Also, we describe a non-trivial polynomially solvable geometric special case of GTSP. Each instance of the problem in question has an l-quasi-pyramidal tour as an optimal solution. Actually, an instance of this problem is defined by the unit 2-row rectangular grid on the Euclidean plane. However, the trick with 2-medians can not be applied straightforward even to the case $h = 3$, we believe that we can soon prove the existence of l-pseudo-pyramidal optimal tours for the case of GTSP-GC(Hh) defined by a grid of an arbitrary fixed height h.

Acknowledgements. This research was supported by Russian Science Foundation, project no. 14-11-00109.

References

1. Arora, S.: Polynomial time approximation schemes for euclidean traveling salesman and other geometric problems. J. ACM **45**, 753–783 (1998)
2. Baburin, A., Della Croce, F., Gimadi, E.K., Glazkov, Y.V., Paschos, V.T.: Approximation algorithms for the 2-Peripatetic salesman problem with edge weights 1 and 2. Discrete Appl. Math. **157**(9), 1988–1992 (2009)
3. Baki, M.F.: A new asymmetric pyramidally solvable class of the traveling salesman problem. Oper. Res. Lett. **34**(6), 613–620 (2006). http://www.sciencedirect.com/science/article/pii/S0167637706000022
4. Balas, E.: New classes of efficiently solvable generalized Traveling Salesman Problems. Ann. Oper. Res. **86**, 529–558 (1999)
5. Balas, E., Simonetti, N.: Linear time dynamic-programming algorithms for new classes of restricted TSPs: a computational study. INFORMS J. Comput. **13**(1), 56–75 (2001). https://doi.org/10.1287/ijoc.13.1.56.9748
6. de Berg, M., Buchin, K., Jansen, B.M.P., Woeginger, G.: Fine-grained complexity analysis of two classic TSP Variants. In: Chatzigiannakis, I., Mitzenmacher, M., Rabani, Y., Sangiorgi, D. (eds.) 43rd International Colloquium on Automata, Languages, and Programming (ICALP 2016). Leibniz International Proceedings in Informatics (LIPIcs), vol. 55, pp. 5:1–5:14. Schloss Dagstuhl-Leibniz-Zentrum fuer Informatik, Dagstuhl, Germany (2016). http://drops.dagstuhl.de/opus/volltexte/2016/6277

7. Burkard, R.E., Glazkov, Y.V.: On the traveling salesman problem with a relaxed monge matrix. Inform. Process. Lett. **67**(5), 231–237 (1998). http://www.sciencedirect.com/science/article/pii/S0020019098001197

8. Burkard, R.E., Deineko, V.G., van Dal, R., van der Veen, J.A.A., Woeginger, G.J.: Well-solvable special cases of the traveling salesman problem: a survey. SIAM Rev. **40**(3), 496–546 (1998)

9. Chentsov, A.G., Khachai, M.Y., Khachai, D.M.: An exact algorithm with linear complexity for a problem of visiting megalopolises. Proc. Steklov Ins. Math. **295**(1), 38–46 (2016). https://doi.org/10.1134/S0081543816090054

10. Christofides, N.: Worst-case analysis of a new heuristic for the Traveling Salesman Problem. In: Symposium on New Directions and Recent Results in Algorithms and Complexity, p. 441 (1975)

11. Enomoto, H., Oda, Y., Ota, K.: Pyramidal tours with step-backs and the asymmetric Traveling Salesman Problem. Discrete Appl. Math. **87**(1–3), 57–65 (1998)

12. Gimadi, E.K., Glazkov, Y., Tsidulko, O.Y.: Probabilistic analysis of an algorithm for the m-planar 3-index assignment problem on single-cycle permutations. J. Appl. Ind. Math. **8**(2), 208–217 (2014)

13. Gimadi, E.K., Rykov, I.A.: On the asymptotic optimality of a solution of the euclidean problem of covering a graph by m nonadjacent cycles of maximum total weight. Dokl. Math. **93**(1), 117–120 (2016)

14. Gutin, G., Punnen, A.P.: The Traveling Salesman Problem and Its Variations. Springer, Boston (2007)

15. Khachai, M., Neznakhina, E.: Approximability of the problem about a minimum-weight cycle cover of a graph. Doklady Math. **91**(2), 240–245 (2015)

16. Khachay, M., Neznakhina, K.: Approximability of the minimum-weight k-Size cycle cover problem. J. Glob. Optim. **66**(1), 65–82 (2016). https://doi.org/10.1007/s10898-015-0391-3

17. Khachay, M., Neznakhina, K.: Towards a PTAS for the generalized TSP in grid clusters. AIP Conf. Proc. **1776**(1), 050003 (2016). https://doi.org/10.1063/1.4965324

18. Klyaus, P.: Generation of testproblems for the Traveling Salesman Problem. Preprint Inst. Mat. Akad. Nauk. BSSR (16) (1976). (in Russian)

19. Lawler, E.L., Lenstra, J.K., Rinnooy Kan, A.H.G., Shmoys, D.B.: The Traveling Salesman Problem: A Guided Tour of Combinatorial Optimization. Wiley Series in Discrete Mathematics & Optimization. Wiley, Chichester (1985)

20. Oda, Y., Ota, K.: Algorithmic aspects of pyramidal tours with restricted jump-backs. Interdisc. Inform. Sci. **7**(1), 123–133 (2001)

21. Pardalos, P., Du, D., Graham, R.: Handbook of Combinatorial Optimization. Springer, New York (2013). https://doi.org/10.1007/978-1-4419-7997-1

22. Sahni, S., Gonzales, T.: P-complete approximation problems. J. ACM **23**, 555–565 (1976)

The 2-Median Problem on Cactus Graphs with Positive and Negative Weights

Chunsong Bai[1] and Liying Kang[2(✉)]

[1] College of Mathematics, Fuyang Normal University,
Fuyang 236041, People's Republic of China
csbai@fync.edu.cn
[2] Department of Mathematics, Shanghai University,
Shanghai 200444, People's Republic of China
lykang@shu.edu.cn

Abstract. This paper studies the problem of locating two vertices in a cactus with positive and negative vertex weights. The problem has objective to minimize the sum of minimum weighted distances from every vertex of the cactus to the two medians. We develop an $O(n^2)$ algorithm for the 2-median problem, where n is the number of vertices of the cactus.

Keywords: Location problem · Median problem · Cactus graphs · Obnoxious facility

1 Introduction

For the classical p-median problem, we are given an undirected graph $G = (V, E)$ with vertex set V and edge set E. The aim is to locate p facilities on edges or vertices of G so as to minimize the overall sum of the weighted distances of the vertices to the respective closest facility. In reality, some vertices are desirable and others are undesirable, the problem is referred to as the semi-obnoxious location problem. Burkard et al. [4] considered 2-medians problems in trees with positive or negative (pos/neg-) weights, and formulated two objective functions: (1) the sum of the minimum weighted distances over all vertices (MWD); (2) the sum of the weighted minimum distances over all vertices (WMD). Benkoczi [1] and Benkoczi et al. [2] considered the pos/neg-weighted 2-median problem on trees.

In this paper, we consider the pos/neg-weighted 2-median problem on cacti for the MWD model. In Sect. 2, we give some notations and preliminaries, which will be used throughout this paper. Section 3 provides some results of 1-median problems on a kind of special graphs. In Sect. 4, we show that the 2-median problem can be solved in $O(n^2)$ time.

Research was partially supported by NSFC (grant numbers 11571222, 11471210, 61672006).

X. Gao et al. (Eds.): COCOA 2017, Part I, LNCS 10627, pp. 278–285, 2017.
https://doi.org/10.1007/978-3-319-71150-8_24

2   Notations and Preliminaries

Let $G = (V, E)$ be a connected graph. Each vertex $v \in V$ (or $v_i \in V$) has a real weight $w(v)$ (or w_i) and each edge $e = (v_i, v_j) \in E$ has a length $l(e) \geq 0$. If the weight of v is given by $w(v) + t$, where $t \geq 0$ is a parameter, then $w(v)$ and $w(v) + t$ are called the original and dynamic weight of v, respectively. A cycle is a sequence $(v_1, \ldots, v_k, v_{k+1} = v_1)$ of k, $k > 2$, distinct vertices supplemented by $v_{k+1} = v_1$ such that (v_j, v_{j+1}) is an edge, $j = 1, \ldots, k$. A graph G is a cactus graph is any two arbitrary cycles in G have at most one vertex in common.

Let $C = (V(C), E(C))$ be a cycle in the cactus G. Denote by $L(C) := \sum_{e \in E(C)} l(e)$ the length of C. Obviously, each vertex $v \in V(C)$ has exactly one point $op(v) \in C$ such that $d(v, op(v)) = \frac{L(C)}{2}$ holds, i.e., $op(v)$ is opposite to v in C. As in [5], the opposite points $op(v)$ for all vertices v which lie in a cycle can be found in $O(n)$ time.

For a given point x, and a pair $\{x, y\}$ of points, the sum of the minimum weighted distances are defined as follows:

$$f(x) := \sum_{i=1}^{n} w(v_i) d(v_i, x),$$

$$F(x, y) := \sum_{i=1}^{n} \min \left(w(v_i) d(v_i, x), w(v_i) d(v_i, y) \right).$$

The corresponding two optimization problems are:

(L_1) Find a location of the point x in G such that $f(x)$ is minimized.
(L_2) Find a location of points x, y in G such that $F(x, y)$ is minimized.

As shown in [5], the so-called *Vertex Optimality Property* holds for the problems L_1 and L_2 on a modified cactus graph G.

Given a tree T_0 rooted at v_1, in which v_i has a positive weight w_i while $i = 2, \ldots, n$, whereas v_1 has pos/neg-weight w_1. Let $e = (v_a, v_b)$ be an edge of T_0. Note that $T_0 \setminus e$ has exactly two connected components, whose vertex sets will be denoted by V_a and V_b respectively, where $v_a \in V_a$ and $v_b \in V_b$. Denote by $W(V_a)$ and $W(V_b)$ the total weights of vertices in V_a and V_b, respectively. Furthermore, we define

$$D(v_a, v_b) := W(V_a) - W(V_b).$$

Lemma 1. *Let* $e = (v_a, v_b)$ *be an edge of* T_0. *If* $v_1 \in V_a$ *and* $D(v_a, v_b) \leq \min\{0, 2w_1\}$, *then there exists a vertex* $v^* \in V_b$ *which is an optimal solution to the problem* L_1 *on* T_0.

Proof. By the definition of L_1, we have

$$
\begin{aligned}
f(v_b) - f(v_a) &= \sum_{u \in V_a \cup V_b} w(u)d(u, v_b) - \sum_{u \in V_a \cup V_b} w(u)d(u, v_a) \\
&= \sum_{u \in V_a} w(u)d(u, v_a) + W(V_a)l(e) + \sum_{u \in V_b} w(u)d(u, v_b) \\
&\quad - \Big(\sum_{u \in V_a} w(u)d(u, v_a) + \sum_{u \in V_b} w(u)d(u, v_b) + W(V_b)l(e) \Big) \\
&= D(v_a, v_b)l(e) \leq \min\{0, 2w_1\}l(e) \leq 0.
\end{aligned} \tag{1}
$$

Note that for each vertex $v \in V_a$, we have

$$
\begin{aligned}
f(v_a) - f(v) &= \sum_{u \in V_a} w(u)d(u, v_a) + \sum_{u \in V_b} w(u)d(u, v_a) \\
&\quad - \Big(\sum_{u \in V_a} w(u)d(u, v) + \sum_{u \in V_b} w(u)d(u, v_a) + \sum_{u \in V_b} w(u)d(v_a, v) \Big) \\
&= \sum_{u \in V_a} w(u)\big(d(u, v_a) - d(u, v)\big) - W(V_b)d(v_a, v) \\
&= \sum_{u \in V_a \setminus \{v_1\}} w(u)\big(d(u, v_a) - d(u, v)\big) + w_1\big(d(u, v_a) - d(u, v)\big) - W(V_b)d(v_a, v) \\
&\leq \sum_{u \in V_a \setminus \{v_1\}} w(u)d(v_a, v) + |w_1|d(v_a, v) - W(V_b)d(v_a, v) \\
&= \sum_{u \in V_a} w(u)d(v_a, v) - W(V_b)d(v_a, v) + (|w_1| - w_1)d(v_a, v) \\
&= D(v_a, v_b)d(v_a, v) + (|w_1| - w_1)d(v_a, v) \\
&\leq (\min\{0, 2w_1\} + |w_1| - w_1)d(v_a, v) = 0.
\end{aligned} \tag{2}
$$

Combining (1) with (2), we have

$$
f(v) \geq f(v_b)
$$

for each vertex $v \in V_a$. It follows that there exists a vertex $v^* \in V_b$ which is an optimal solution to the problem L_1 on T_0.

Theorem 1. *A vertex v^* is an optimal solution to the problem L_1 on T_0 if and only if, for each neighbor v_a of v^*: (1) $D(v^*, v_a) \geq \max\{0, -2w_1\}$ holds for the case $v_1 \in V_a$; and (2) $D(v^*, v_a) \geq 0$ holds for the case $v_1 \notin V_a$.*

3   Parametric Problems L_1 on Graphs

In this section, our aim is to develop algorithms for parametric problems L_1 on a cycle, a tree and a cactus, respectively.

3.1 A Parametric Problem L_1 on a Cycle

Given a cycle $C' = (V, E)$ with n vertices, in which v_i has a nonnegative weight w_i while $i = 2, \ldots, n$, whereas v_1 has a dynamic weight $w_1 + t$, $t \geq 0$. The task is to solve problems L_1 on C' for every parameter value t.

Recall that the objective function value of vertex v_i, for $i = 1, \ldots, n$ is given by

$$f_i(t) := \sum_{j=1}^{n} w_j d(v_i, v_j) + d(v_i, v_1) \cdot t.$$

Then the task of the parametric problem L_1 on C is to find the function

$$F(t) := \min_{i=1,\ldots,n} f_i(t)$$

and the corresponding arguments t.

Lemma 2. $F(t)$ *is a continuous, concave and piecewise linear function with at most n breakpoints.*

Based on Lemma 2, the parametric problem L_1 can be solved in linear time as in [6].

3.2 A Parametric Problem L_1 on a Tree

Given a tree $T' = (V, E)$ with n vertices, in which v_i has a nonnegative weight w_i while $i = 2, \ldots, n$, whereas v_1 has a dynamic weight $w_1 + t, t \geq 0$ and has the degree of one. The task is again to solve the problem L_1 on T' for every parameter value $t \geq 0$.

Lemma 3. *Let $v^* \in V$ be the optimal solution to the problem L_1 on T' for $t = 0$. Then, for every parameter value $t > 0$, there exists a vertex on the path from v^* to v_1 which is an optimal solution.*

Based on Lemma 3, in order to solve the parametric problem L_1, we only have to consider vertices on the path from v^* to v_1. It is easy to see that the parametric problem L_1 on T' can be solved in linear time.

3.3 A Parametric Problem L_1 on a Cactus

Finally, we show how to solve the considered problem on a cactus G with dynamic vertex v_1, by using of the idea for solving the parametric 1-median problem proposed in [6].

The method consists of two steps. In step 1, we solve the parametric problem L_1 on the tree structure of the cactus, and obtain optimal solutions for every parameter value $t \geq 0$. In step 2, we compute the corresponding optimal solution in G. Based on Theorem 1, if a vertex of the tree structure which was not a vertex in a cycle is optimal for all $t \in [t_1, t_2]$ (or $[t_1, \infty)$) the corresponding vertex in the cactus is also optimal for the same values of t. If v was obtained by shrinking some cycle C is optimal for all $t \in [t_1, t_2]$ (or $[t_1, \infty)$), then there exists a vertex in C which is optimal for a parameter value in $[t_1, t_2]$ (or $[t_1, \infty)$). In this case, we need to solve a parametric problem L_1 on C for all $t \in [t_1, t_2]$ (or $[t_1, \infty)$). It is shown in [6] that this problem can be solved in linear time for all cycles in the cactus. Therefore, it can be said that the considered parametric problem L_1 on a cactus can be solved in linear time.

4   Problems L_2 on Cactus Graphs

In this section we deal with the problem L_2 on a cactus $G = (V, E)$. Without loss of generality, we assume that each vertex of G is contained in at most one cycle. We first consider the case that the problem has an optimal solution consisting of two distinct vertices.

4.1   Local 1-Median Problems

Let $\{m_1, m_2\}$ be an optimal solution. Let m be the midpoint of the path from m_1 to m_2, and let $e = (v_i, v_j)$ be the edge that contains m. Assume that v_i lies on the path from v_j to m_1. Note that if the edge e is contained in a cycle C, then there exists another point m' in C such that $d(m_1, m') = d(m_2, m')$. Let $e' = (v_{i'}, v_{j'})$ be the edge that contains m'. Assume that $v_{i'}$ lies on the path from $v_{j'}$ to m_1. The edges e and e' are called *critical edges*.

Let $C = (V(C), E(C))$ be a cycle of G with l vertices. By deleting all edges $E(C)$ from G, we obtain l cacti $G_1 = (V_1, E_1), \ldots, G_l = (V_l, E_l)$, where we assume that $v_i \in V_i$ for $i = 1, \ldots, l$. Let V_i^+ and V_i^-, $i = 1, \ldots, l$, be the vertices in V_i with nonnegative weights and negative weights, respectively. For each v_i and G_i, $i = 1, \ldots, l$, we introduce two notations:

$$W_i^+ := \sum_{u \in V_i^+} w(u), \ W_i^- := \sum_{u \in V_i^-} w(u),$$

and two functions:

$$\bar{f}^+(v_i) := \sum_{u \in V_i^+} w(u)d(u, v_i), \ \bar{f}^-(v_i) := \sum_{u \in V_i^-} w(u)d(u, v_i).$$

It is easy to see that the values W_i^+ and W_i^- for all $i = 1, \ldots, l$ can be computed in linear time, and the values $\bar{f}^+(v_i)$ and $\bar{f}^-(v_i)$ for all $i = 1, \ldots, l$ can also be computed in linear time as in [3].

Furthermore, we introduce the following notations:

$$W^+(k_1, k_2) := \sum_{i=k_1}^{k_2} W_i^+, \ W^-(k_1, k_2) := \sum_{i=k_1}^{k_2} W_i^-,$$

where $1 \le k_1 \le k_2 \le l+1$. We can calculate all $W^+(k_1, k_2)$ and $W^-(k_1, k_2)$ for $1 \le k_1 \le k_2 \le l+1$ by passing through all G_i and cost $O(n^2)$ time.

For each pair of edges $e_1, e_k \in E(C), 2 \leq k \leq l$, construct two cacti G_k^1 and G_k^2, which differ from G only in the weights of the vertices, respectively. Denote by X_{2k} and $Y_{k+1,1}$ the sub-cacti of G which are obtained by deleting e_1 and e_k, where $v_2, v_k \in X_{2k}$ and $v_{k+1}, v_1 \in Y_{k+1,1}$. The weights of the vertices in G_k^1 are defined as

$$
w_i^1 = \begin{cases} w_i & \text{if } [(v_i \in X_{2k} \wedge w_i \geq 0) \text{ or } (v_i \in Y_{k+1,1} \wedge w_i \leq 0)], \\ 0 & \text{if } [(v_i \in X_{2k} \wedge w_i < 0) \text{ or } (v_i \in Y_{k+1,1} \wedge w_i > 0)]. \end{cases}
$$

Then the task of the *local 1-median problem* P_k^1 on G_k^1 is to find a vertex $v^* \in X_{2k}$ such that

$$
f(v^*) = \min_{v \in X_{2k}} f(v).
$$

Similarly, we can define the graphs G_k^2 and the corresponding problems P_k^2 for edges $e_1, e_k \in E(C), k = 2, \ldots, l$.

Furthermore, the parametric problems L_1 on G_i are solved for all $i = 1, 2, \ldots, l$ in a preprocessing procedure, where the weight of vertex $v_i \in C$ depends on the parameter $t > 0$, i.e., $w_i(t) := w_i + t$. It is easy to see that these problems can be solved in linear time.

4.2 Algorithm for Local 1-Median Problems

In this subsection we show how to solve the problems $P_k^1, 2 \leq k \leq l$ in linear time. In order to make this procedure more clear it is useful to define some function values

$$
\hat{f}_k^+(v_i) := \sum_{j=2}^{i-1} \sum_{u \in V_j^+} w(u)d(u, v_i), \quad \hat{f}_k^-(v_i) := \sum_{j=op(i)}^{l+1} \sum_{u \in V_j^-} w(u)d(u, v_i),
$$

and

$$
\check{f}_k^+(v_i) := \sum_{j=i+1}^{k} \sum_{u \in V_j^+} w(u)d(u, v_i), \quad \check{f}_k^-(v_i) := \sum_{j=k+1}^{op(i)-1} \sum_{u \in V_j^-} w(u)d(u, v_i),
$$

for all $i = 2, \ldots, k$. Using these definitions, the objective function value of the vertex $v_i \in V(C)$ is given by

$$
f_k(v_i) = \hat{f}_k^+(v_i) + \hat{f}_k^-(v_i) + \check{f}_k^+(v_i) + \check{f}_k^-(v_i) + \bar{f}^+(v_i) \tag{3}
$$

for all $i = 2, \ldots, k$.

Given a vertex $u \in V_i \setminus v_i$, the formula

$$
\begin{aligned}
f_k(u) = {} & \hat{f}_k^+(v_j) + \hat{f}_k^-(v_j) + \check{f}_k^+(v_j) + \check{f}_k^-(v_j) + \bar{f}^+(u) \\
& + \Big(W^+(2, i-1) + W^-(op(i+1), l+1) \\
& + W^+(i+1, k) + W^-(k+1, op(i)) \Big) \cdot d(v_i, u)
\end{aligned} \tag{4}
$$

holds for all $i = 2, \ldots, k$.

Note that the optimal solution to the problem P_2^1 on G_2^1 is already known from the preprocessing procedure and $\hat{f}_2^+(v_2) = \check{f}_2^+(v_2) = 0$ holds. Suppose that the problem P_k^1 on G_k^1 has already been solved and assume that the optimal solution v^* is in $G_{k'}$, $2 \leq k' \leq k$.

Consider the problem P_{k+1}^1 on G_{k+1}^1. Obviously, we have

$$\hat{f}_{k+1}^+(v_{k'}) = \hat{f}_k^+(v_{k'})$$

and

$$\hat{f}_{k+1}^-(v_{k'}) = \hat{f}_k^-(v_{k'}).$$

Moreover, we have

$$\check{f}_{k+1}^+(v_{k'}) = \check{f}_k^+(v_{k'}) + \bar{f}^+(v_{k+1}) + W_{k+1}^+ \cdot d(v_{k+1}, v_{k'})$$

and

$$\check{f}_{k+1}^-(v_{k'}) = \check{f}_k^-(v_{k'}) - \bar{f}^-(v_{k+1}) - W_{k+1}^- \cdot d(v_{k+1}, v_{k'}),$$

where $d(v_{k+1}, v_{k'})$ is the distance of the path from v_{k+1} to $v_{k'}$ in counterclockwise direction.

Note that, once all vertices in V_{k+1} with positive weights are added, the optimal solution v^{**} will in some sense move in the direction to the vertices in V_{k+1}. Furthermore, according to Lemma 1, v^{**} can never locate in $V_j \setminus v_j$ for $j = k'+1, \ldots, k$. Therefore, v^{**} can locate either in (i) $G_{k'}$; or (ii) vertices $v_{k'+1}, \ldots, v_k$; or (iii) G_{k+1}.

In the case (i), the optimal solution v^{**} can be found by solving the parametric problem L_1 on $G_{k'}$, where the dynamic weight of $v_{k'}$ is increased by

$$W_{k+1}^+ - W_{k+1}^-.$$

Then the optimal objective function value $f_{k+1}(v^{**})$ can be computed by using Eq. (3).

In the case (ii), we first compute the function values for $v_{k'+1}$:

$$\hat{f}_{k+1}^+(v_{k'+1}) = \hat{f}_{k+1}^+(v_{k'}) + \bar{f}^+(v_{k'}) + W^+(2, k') \cdot l(v_{k'}, v_{k'+1}), \tag{5}$$

$$\begin{aligned}\hat{f}_{k+1}^-(v_{k'+1}) = {}&\hat{f}_{k+1}^-(v_{k'}) - \bar{f}^-(op(v_{k'})) - W_{op(k')}^- \cdot \frac{L(C)}{2} \\ &+ W^-(op(k'+2), l+1) \cdot l(v_{k'}, v_{k'+1})\end{aligned} \tag{6}$$

and

$$\check{f}_{k+1}^+(v_{k'+1}) = \check{f}_{k+1}^+(v_{k'}) - \bar{f}^+(v_{k'+1}) - W^+(k'+1, k+1) \cdot l(v_{k'}, v_{k'+1}), \tag{7}$$

$$\begin{aligned}\check{f}_{k+1}^-(v_{k'+1}) = {}&\check{f}_{k+1}^-(v_{k'}) + \bar{f}^-(op(v_{k'})) + W_{op(k')}^- \cdot d(op(v_{k'}), v_{k'+1}) \\ &- W^-(k+2, op(k')) \cdot l(v_{k'}, v_{k'+1}).\end{aligned} \tag{8}$$

Combining Eqs. (5)–(8) with (4), the objective function value $f_{k+1}(v_{k'+1})$ can be obtained. This is repeated for the vertices $v_{k'+2}, \ldots, v_k$, and then choose a vertex from $v_{k'+1}, \ldots, v_k$ with minimal function value as v^{**}.

In the case (iii), it suffices to solve the parametric problem L_1 on G_{k+1}, where the dynamic weight of v_{k+1} is given by

$$w_{k+1}(t) = w_{k+1} + W^+(2, k) + W^-(k + 2, l + 1).$$

Then we need to add $\hat{f}^+_{k+1}(v_{k+1})$ and $\hat{f}^-_{k+1}(v_{k+1})$ to the optimal objective function value of the parametric problem to obtain the optimal value $f_{k+1}(v^{**})$. Thus, we have solved the 1-median problem on G^1_{k+1} and then we apply the same procedure for the graph $G^1_{k+2}, \ldots, G^1_l$.

Note that the updates given in Eqs. (5)–(8) can be done in constant time for each vertex. Moreover, all the remaining computations can be done in constant time. Thus, all corresponding median problems can be solved in linear time for the fixed critical edge e_1. Since there are $O(n)$ contained edges in a cycle, the computation of $\min_{v_s \in X} \sum_{v \in X^+ \cup Y^-} w(v)d(v, v_s) + \min_{v_t \in Y} \sum_{v \in Y^+ \cup X^-} w(v)d(v, v_t)$ can be completed for all pairs (X, Y) with two critical edges in $O(n^2)$ time.

In the case that a pair (X, Y) has just one bridge edge, we can solve the two corresponding pos/neg-weighted 1-median problems in linear time by using the algorithm proposed in [3]. Thus,

$$\min_{v_s \in X} \sum_{v \in X^+ \cup Y^-} w(v)d(v, v_s) + \min_{v_t \in Y} \sum_{v \in Y^+ \cup X^-} w(v)d(v, v_t)$$

can be solved for all pairs (X, Y) in $O(n^2)$ time. Summarizing we get the following theorem.

Theorem 2. *The problem L_2 on cactus graphs can be solved in $O(n^2)$ time.*

References

1. Benkoczi, R.: Cardinality constrained facility location problems in trees. Ph.D. Thesis, Simon Fraser University (2004)
2. Benkoczi, R., Breton, D., Bhattacharya, B.: Efficient computation of 2-medians in a tree network with positive/negative weights. Discrete Math. **306**(14), 1505–1516 (2006)
3. Burkard, R.E., Krarup, J.: A linear algorithm for the pos/neg-weighted 1-median problem on cactus. Computing **60**(3), 193–215 (1998)
4. Burkard, R.E., Çela, E., Dollani, H.: 2-median in trees with pos/neg weights. Discrete Appl. Math. **105**(14), 51–71 (2000)
5. Burkard, R.E., Hatzl, J.: Median problems with positive and negative weights on cycles and cacti. J. Comb. Optim. **20**(1), 27–46 (2010)
6. Hatzl, J.: Median problems on wheels and cactus graphs. Computing **80**(4), 377–393 (2007)

The Eigen-Distribution of Weighted Game Trees

Shohei Okisaka$^{(\boxtimes)}$, Weiguang Peng$^{(\boxtimes)}$, Wenjuan Li$^{(\boxtimes)}$,
and Kazuyuki Tanaka$^{(\boxtimes)}$

Mathematical Institute, Tohoku University, Sendai, Japan
shohei.okisaka@gmail.com, pwgedu@163.com, wenjuanli1701@gmail.com,
tanaka.math@tohoku.ac.jp

Abstract. This paper is devoted to the ongoing study on the equilibrium points of AND-OR trees. Liu and Tanaka (2007, 2007a) characterized the eigen-distributions that achieve the distributional complexity, and among others, they proved the uniqueness of eigen-distribution for a uniform binary tree. Later, Suzuki and Nakamura (2012) showed that the uniqueness fails if only directional algorithms are allowed. Peng *et al.* (2016) extended the studies on eigen-distributions to balanced multi-branching trees of height 2. But, it remains open whether the uniqueness still holds or not for general multi-branching trees. To this end, we introduce the weighted trees, namely, trees with weighted cost depending on the value of a leaf. Using such models, we prove that for balanced multi-branching trees, the uniqueness of eigen-distribution holds *w.r.t.* all deterministic algorithms, but fails *w.r.t.* only directional algorithms.

Keywords: Game trees with weight · Leaf cost function · Alpha-beta pruning algorithm · Computational complexity

1 Introduction

The AND-OR (OR-AND) game tree is a crucial and primary representative model for Boolean functions. Its root is labelled by AND (OR), internal nodes are alternatively labelled by either OR or AND, and leaves are associated with Boolean value 1 and 0. The value of each AND (OR) node is evaluated as the maximum (minimum) among all the values of its children. The value of a tree is the value of its root. The height of a node is the length of the path from the root to this node. Obviously, the height of the root is 0. The height of a tree is the largest height of the leaves.

An algorithm is carried out to query the leaves until it can determine the value of the tree. The standard cost to compute a tree is the total cost of all the leaves that are queried during the execution of an algorithm. The so-called alpha-beta pruning depth-first algorithm is characterized by following property: while it is querying the leaves of a subtree, it will stop querying it when and only when it determines the value of the subtree, and after that it will move on to querying other subtrees. See [4] for more details. Throughout this paper, by a deterministic algorithm, we always mean such an alpha-beta pruning depth-first algorithm.

© Springer International Publishing AG 2017
X. Gao et al. (Eds.): COCOA 2017, Part I, LNCS 10627, pp. 286–297, 2017.
https://doi.org/10.1007/978-3-319-71150-8_25

A randomized algorithm is a distribution over a family of deterministic algorithms. The randomized complexity is defined to be the minimum cost to compute the worst assignment for a tree over randomized algorithms. From a game-theoretic perspective, Yao [12] observed that the randomized complexity is equivalent to the distributional complexity, that is,

$$\underbrace{\min_{A_R} \max_{\omega} cost(A_R, \omega)}_{\text{Randomized complexity}} = \underbrace{\max_{d} \min_{A_D} cost(A_D, d)}_{\text{Distributional complexity}},$$

where A_R runs over randomized algorithms, ω over assignments for leaves (namely, the sequence of values for leaves), d over distributions on assignments and A_D over deterministic algorithms.

Subsequently, Liu and Tanaka [2] characterized the *eigen-distribution* that achieves the distributional complexity, namely a distribution δ which satisfies

$$\min_{A_D} cost(A_D, \delta) = \max_{d} \min_{A_D} cost(A_D, d).$$

Assuming that a probability distribution d on the assignments to the leaves is an independent distribution (ID), they claimed that for any uniform AND-OR tree, if such a distribution can achieve the equilibrium, it turns out to be an independent and identical distribution (IID). Suzuki and Niida [10] proved the claim for uniform binary trees. Recently, Peng et al. [6] proved that the claim holds for balanced multi-branching trees.

In [2,3], Liu and Tanaka defined i-set ($i = 0, 1$) for the reluctant assignments, and E^i-distribution as a distribution on i-set with the same cost for all deterministic algorithms. Then they proved that for a uniform binary tree, E^i-distribution is the unique eigen-distribution, which is the uniform distribution on the i-set (the Liu-Tanaka theorem).

In sequel, Suzuki and Nakamura [11] showed the Liu-Tanaka theorem holds for a class of algorithms that are closed under *transposition*, and they also showed that the uniqueness of eigen-distribution fails if only directional algorithms are considered. In fact, if we restrict the algorithms to be directional ones, there are uncountable many distributions that can achieve the distributional complexity.

Peng et al. [7] extended the studies on the uniqueness of eigen-distribution of [2,11] from uniform binary trees to balanced multi-branching trees. They showed the equivalence between the eigen-distribution and E^i-distribution for balanced multi-branching trees. For the uniqueness of eigen-distribution in this context, they only proved the height 2 case *w.r.t.* all deterministic algorithms, but the case of general height has not been solved. A survey concerning origin and recent studies on eigen-distribution can be found in [9].

Thus, the following problems have been open till now: for general balanced multi-branching trees, is the eigen-distribution unique *w.r.t.* all deterministic algorithms? Or not unique *w.r.t.* some class of algorithms?

Notice that the arguments for the uniform binary trees in [2,11] can not be applied directly to even n-branching trees for $n \geq 3$, where all internal nodes have n children. A general balanced multi-branching setting makes the proof rather intricate, since it requires induction not only on the height of a tree, but also the number of children of a node. In this paper, we solve the above problems by introducing the notion of balanced multi-branching weighted trees. By weighted trees, we mean the trees in which the cost is weighted depending on the value of the leaves. That is, for the leaves with value 1, they have cost weight $a > 0$, while with value 0 their cost weight is $b > 0$. Note that if we take $a = b = 1$, these trees are nothing but usual ones with the unit cost.

On the other hand, the weighted tree itself is also very interesting in the sense that, it characterizes the value dependent cost models. Such models have wide range of applications in both theoretical and applied computer science [1]. The weighted tree was used by Saks and Wigderson to compute the exact lower bound for randomized complexity [8], where they called them "trees with leaf-cost function pair". Cicalese and Milanic [1] treated the function evaluation problems when the cost of querying a variable depends on the value of the variable in the framework of priced information, and also suggested further investigation on value dependent models.

By $E^i(a, b)$-distribution, we denote the E^i-distribution for trees with cost weight pair (a, b), where $i = 0, 1$. Note that we can show that for AND-OR (resp., OR-AND) balanced multi-branching weighted trees, the eigen-distribution is equivalent to $E^1(a, b)$-distribution or $E^0(a, b)$-distribution by the same method in [7]. Therefore, in the following we just treat $E^i(a, b)$-distribution. We prove the uniqueness of $E^i(a, b)$-distribution holds *w.r.t.* deterministic algorithms, but fails if we restrict ourselves to directional algorithms for AND-OR (resp., OR-AND) balanced multi-branching weighted trees.

We again mention that the solutions to the above problems for general multi-branching trees follow from our results, by setting the weight pair $a = b = 1$. Recall that for AND-OR binary trees of height 2, E^0-distribution *w.r.t.* $\mathcal{A}_{dir}$ is unique [11]. However, our results imply that even for AND-OR balanced multi-branching trees of height 2, the uniqueness E^0-distribution no longer holds. More precisely, E^0-distribution is unique if and only if the root of an AND-OR multi-branching tree has no more than 2 children.

The remainder of this paper is organized as follows. Section 2 presents some basic terminology and notion. In Sect. 3, the first half is dedicated to the proof of the uniqueness of $E^i(a, b)$-distribution *w.r.t.* the class of all deterministic algorithms, while the second half shows the failure of the uniqueness of $E^i(a, b)$-distribution when we only consider directional algorithms.

2 Preliminary

To begin with, we define some terminology and notation.

Definition 1. *Given a tree $\mathcal{T}$, each node is labelled by a finite sequence from $\mathbb{N}$ as follows.*

- *The root is labelled by the empty sequence ε.*
- *For a nonterminal node labelled v, its n children are labelled $v0, v1, \ldots, v(n-1)$ from left to right.*

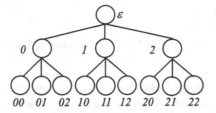

Fig. 1. The labelling for $\mathcal{T}_3^2$

We often identify the node with its label, so we can see $\mathcal{T}$ as a subset of finite sequences from $\mathbb{N}$ (Fig. 1).

An assignment for $\mathcal{T}$ is a function ω from the set of leaves to $\{0,1\}$. Note that by identifying each leaf with its value, an assignment ω can also be seen as a 0-1 sequence, whose length is the number of leaves.

An algorithm to evaluate $\mathcal{T}$ is called *directional* if it queries the leaves in a fixed order, independent from the query history [5]. By $\mathcal{A}_D$, we denote the set of all deterministic algorithms, and by $\mathcal{A}_{dir}$ the set of all directional algorithms. If an algorithm proceeds depending on its history, we say it is a non-directional algorithm.

Let $C(A, \omega)$ denote the cost of an algorithm A under an assignment ω. Given a set of assignments Ω, a distribution d on Ω (i.e., a function from Ω to $[0,1]$ such that $\sum_{\omega \in \Omega} d(\omega) = 1$), and $A \in \mathcal{A}_D$, then the expected cost by A w.r.t. d is defined by $C(A, d) = \sum_{\omega \in \Omega} d(\omega) \cdot C(A, \omega)$.

Definition 2 (*i*-set [2]). *Given $\mathcal{T}$, $i \in \{0,1\}$, i-set for $\mathcal{T}$ consists of assignments such that*

- *the root has value i, and*
- *if an AND-node has value 0 (or OR-node has value 1), just one of its children has value 0 (1), and all the other children have 1 (0).*

By $\Omega_i^{\wedge,h}$ and $\Omega_i^{\vee,h}$, we denote the *i*-set for AND-OR and OR-AND trees of height h, respectively.

Definition 3 (E^i-distribution [2]). *Suppose $\mathcal{A}$ is a subset of $\mathcal{A}_D$. A distribution d on i-set is called an E^i-distribution w.r.t. $\mathcal{A}$ if there exists $c \in \mathbb{R}$ such that for any $A \in \mathcal{A}$, $C(A, d) = c$.*

For general balanced multi-branching trees of height 2, we showed the uniqueness of E^i-*distribution w.r.t.* $\mathcal{A}_D$ in [7] as follows

Theorem 1 ([7]). *For any balanced multi-branching tree $\mathcal{T}$ of height 2, E^i-distribution w.r.t. $\mathcal{A}_D$ is unique.*

It is mentioned in [7] that for general height > 2, E^i-distribution w.r.t. $\mathcal{A}_D$ is also unique.

3   Main Results

In this section, we investigate the E^i-distribution for balanced multi-branching weighted trees.

Definition 4. *Let A be an algorithm, ω an assignment, $\sharp_1(A,\omega)$ (resp., $\sharp_0(A,\omega)$) denote the number of leaves probed by A and assigned 1 (resp., 0) on ω. For any positive real number a, b,*

$$C(A,\omega;a,b) := a \cdot \sharp_1(A,\omega) + b \cdot \sharp_0(A,\omega),$$

is called a generalized cost weighted with (a,b). It is obvious that $C(A,\omega) = C(A,\omega;1,1)$.

For a distribution d on Ω, the expected generalized cost

$$C(A,d;a,b) := \sum_{\omega \in \Omega} d(\omega) C(A,\omega;a,b).$$

We may say that $\mathcal{T}$ is a weighted tree if we consider the generalized cost.

Definition 5. *For any subset $\mathcal{A} \subseteq \mathcal{A}_D$, a distribution d on i-set is called an $E^i(a,b)$-distribution w.r.t. $\mathcal{A}$ if there exists $c \in \mathbb{R}$ such that for any $A \in \mathcal{A}$, $C(A,d;a,b) = c$.*

3.1   The Uniqueness of $E^i(a,b)$-Distribution *w.r.t $\mathcal{A}_D$*

Our goal of this subsection is to show that if d is an $E^i(a,b)$-distribution *w.r.t.* $\mathcal{A}_D$, then d is a uniform distribution on i-set ($i = 0,1$). For simplicity, we here only treat the n-branching trees, since our arguments and results can be extended straightforwardly to general balanced multi-branching weighted trees.

Definition 6. *Let A_i be a depth-first algorithm for the game tree $\mathcal{T}_n^{h_i}$ ($i = 1,2$). The depth-first algorithm $A_1 \times A_2$ for $\mathcal{T}_n^{h_1+h_2}$, as illustrated in Fig. 2, is defined as follows:*

(1) Evaluate the value of nodes at height h_2 according to the order of A_2.
(2) Probe the leaves of each subtree $\mathcal{T}_n^{h_1}$ according to the order of A_1.

Note that it suffices to use the above algorithm $A_1 \times A_2$ to demonstrate the uniqueness of $E^i(a,b)$-distribution *w.r.t.* $\mathcal{A}_D$, since if we assume $E^i(a,b)$, the choice of algorithm makes no difference to the cost. It is also worth remarking that such an algorithm in fact transfers the weight of leaves to the nodes at height h_2 so that all the leaves have weight pair (a,b) but the nodes at height h_2 are possibly equipped with a different weight pair (a',b').

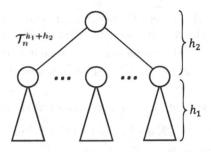

Fig. 2. A game tree of height $h_1 + h_2$

For an assignment $\omega \in \Omega^{h_1+h_2}$, we partition it into n^{h_2} assignments ω^j $(1 \leq j \leq n^{h_2})$ for $\mathcal{T}_n^{h_1}$, and denote $\omega = \omega^1\omega^2\ldots\omega^{n^{h_2}}$. If x_j is the value for $\mathcal{T}_n^{h_1}$ on ω^j $(1 \leq j \leq n^{h_2})$, $x_1x_2\ldots x_{n^{h_2}}$ defines the assignment for $\mathcal{T}_n^{h_2}$ induced by ω, denoted by $\omega \lceil_{h_2}$. For an $\omega_2 \in \Omega^{h_2}$, we set $\Omega_{\omega_2} := \left\{\omega \in \Omega^{h_1+h_2} : \omega \lceil_{h_2} = \omega_2\right\}$. To solve the final goal of this subsection, we only need to consider the case $h_1 = 1$ and $h_2 = h$.

We first show that if $\omega_1 = \omega_1^1\omega_1^2\ldots\omega_1^{n^h}$ and $\omega_2 = \omega_2^1\omega_2^2\ldots\omega_2^{n^h}$ are in Ω_i^{h+1} and the values of ω_1^j and ω_2^j are the same for all $j \leq n^h$, then $d(\omega_1) = d(\omega_2)$ for an $E^i(a,b)$-distribution d w.r.t. $\mathcal{A}_D$.

Lemma 1. *Let d be an $E^i(a,b)$-distribution w.r.t. $\mathcal{A}_D$ for $\mathcal{T}_n^{h+1}$. Suppose ω_1, $\omega_2 \in \Omega_i^{h+1}$ and $\omega_1 \lceil_h = \omega_2 \lceil_h$. Then $d(\omega_1) = d(\omega_2)$.*

Proof. W.l.o.g., we may assume that h is even and the label of root is $\wedge$. Suppose that $\omega_1, \omega_2 \in \Omega_i^{h+1}$ and $\omega_1 \lceil_h = \omega_2 \lceil_h$.

Let d be an $E^1(a,b)$-distribution on the 1-sets. For an assignment $\omega \in \Omega_1^{\wedge,h+1}$, we can find a directional algorithm A_ω which probes all the leaves with an assignment ω. We call such an algorithm *eigen*. Note that an eigen algorithm is not unique.

Let $\mathcal{T}_1, \ldots, \mathcal{T}_{n^h}$ be the subtrees with height 1 of $\mathcal{T}_n^{h+1}$, listed in the order where A_{ω_1} evaluates them on ω_1, and ω^j be the restriction of ω to $\mathcal{T}_j$. We say that ω^j is a 0-*part* of ω if it assigns 0 to the root of $\mathcal{T}_j$. We can see each ω^j as an element in $\Omega_0^{\wedge,1} \cup \Omega_1^{\wedge,1}$. So, if ω^j is a 0-part, it consists of one 0 and $(n-1)$ many 1's, and otherwise n many 1's. Moreover, the number of 0-parts does not depend on the choice of $\omega \in \Omega_1^{\wedge,h+1}$.

We show by induction on l that if $\omega_1^j = \omega_2^j$ for all $j \leq n^h - l$, and $\omega_1^j \sim \omega_2^j$ for all $j \geq n^h - l + 1$, then $d(\omega_1) = d(\omega_2)$. By $\omega \sim \omega'$, we mean that ω' is a permutation of ω.

For the case $l = 0$, the statement is trivial since $\omega_1 = \omega_2$. For the induction step $(l > 0)$, we assume that $\forall j \leq n^h - l$, $\omega_1^j = \omega_2^j$, $\omega_1^{n^h-l+1} \neq \omega_2^{n^h-l+1}$ and $\forall j \geq n^h - l + 1$ $(\omega_1^j \sim \omega_2^j)$. Next we describe the nondirectional algorithm A' as follows:

- A' works the same as A_{ω_1} except for assignments starting with $\omega_1^1 \omega_1^2 \ldots \omega_1^{n^h-l}$. Note the A' can check whether the first part of the assignment is $\omega_1^1 \omega_1^2 \ldots \omega_1^{n^h-l}$ or not while simulating A_{ω_1} on that part.
- Only if the first part is $\omega_1^1 \omega_1^2 \ldots \omega_1^{n^h-l}$, A' switches to $A_{\omega_2^{n^h-l+1}}$ and then back to A_{ω_1} to the end.
- Here we may assume that $A_{\omega_2^{n^h-l+1}}$ coincides with $A_{\omega_1^{n^h-l+1}}$ on any $\omega (\in \Omega_0^{\wedge,1}) \neq \omega_1^{n^h-l+1}, \omega_2^{n^h-l+1}$. Namely, the two algorithms exchange the probing order only where 0 appears in $\omega_1^{n^h-l+1}$ or $\omega_2^{n^h-l+1}$.

Since d is an $E^1(a,b)$-distribution, we have

$$C(A_{\omega_1}, d; a, b) - C(A', d; a, b)$$
$$= \sum_{\omega \in X} d(\omega) \Big(C(A_{\omega_1}, \omega; a, b) - C(A', \omega; a, b) \Big)$$
$$+ \sum_{\omega \in Y} d(\omega) \Big(C(A_{\omega_1}, \omega; a, b) - C(A', \omega; a, b) \Big)$$
$$= 0,$$

where

$$X := \left\{ \omega \in \Omega_1^{\wedge,h+1} : \forall j \leq n^h - l + 1 \ (\omega^j = \omega_1^j) \ \& \ \forall j > n^h - l + 1 \ (\omega^j \sim \omega_1^j) \right\},$$
$$Y := \left\{ \omega \in \Omega_1^{\wedge,h+1} : \forall j \leq n^h - l + 1 \ (\omega^j = \omega_2^j) \ \& \ \forall j > n^h - l + 1 \ (\omega^j \sim \omega_2^j) \right\}.$$

By the induction hypothesis, $d(\omega) = d(\omega_1)$ for $\omega \in X$ and $d(\omega) = d(\omega_2)$ for $\omega \in Y$. Moreover, there is a bijection from X to Y, then we can get

$$\sum_{\omega \in X} \Big(C(A_{\omega_1}, \omega; a, b) - C(A', \omega; a, b) \Big) = \sum_{\omega \in Y} \Big(C(A', \omega; a, b) - C(A_{\omega_1}, \omega; a, b) \Big).$$

Thus,

$$C(A_{\omega_1}, d; a, b) - C(A', d; a, b)$$
$$= \Big(d(\omega_1) - d(\omega_2) \Big) \sum_{\omega \in X} \Big(C(A_{\omega_1}, \omega; a, b) - C(A', \omega; a, b) \Big) = 0.$$

Because for any $\omega \in X$, $C(A_{\omega_1}, \omega; a, b) > C(A', \omega; a, b)$, we get $d(\omega_1) = d(\omega_2)$. Then the proposition follows from the case $l = n^h$.

To show the first proposition, we will give more observations.

Lemma 2. *Let A_i be a depth-first algorithm for $T_n^{h_i}$ ($i = 1, 2$). Let $\omega_2 \in \Omega_1^{\wedge,h_2}$. Then we have*

$$\sum_{\omega \in \Omega_{\omega_2}} C(A_1 \times A_2, \omega; a, b) = \sum_{\omega_1 \in \Omega_1^{*,h_1}} C(A_1, \omega_1; a, b) \frac{|\Omega_{\omega_2}|}{|\Omega_1^{*,h_1}|} \sharp_1(A_2, \omega_2)$$

$$+ \sum_{\omega_1 \in \Omega_0^{*,h_1}} C(A_1, \omega_1; a, b) \frac{|\Omega_{\omega_2}|}{|\Omega_0^{*,h_1}|} \sharp_0(A_2, \omega_2),$$

where $*$ is the label of nodes at height h_2.

Proof. Assume h_2 is even. In this case, $*$ is $\wedge$. We first observe that the number of subtrees of height h_1 checked by $A_1 \times A_2$ and their orders depend only on A_2 and $\omega_2 \in \Omega_1^{\wedge, h_2}$.

Suppose that ω_2 assigns 1 to the root of a subtree $T_i^{h_1}$. For an $\omega_1 \in \Omega_1^{\wedge, h_1}$, the number of $\omega \in \Omega_{\omega_2}$ which assigns ω_1 to the leaves of $T_i^{h_1}$ is $\frac{|\Omega_{\omega_2}|}{|\Omega_1^{\wedge, h_1}|}$. Since ω_1 runs over all the assignments in $\Omega_1^{\wedge, h_1}$, the total cost to evaluate each subtree of height h_1 whose root has value 1 is

$$\sum_{\omega_1 \in \Omega_1^{\wedge, h_1}} C(A_1, \omega_1; a, b) \frac{|\Omega_{\omega_2}|}{|\Omega_1^{\wedge, h_1}|}.$$

And similarly, the total cost to evaluate each subtree of height h_1 whose root has value 0 is

$$\sum_{\omega_1 \in \Omega_0^{\wedge, h_1}} C(A_1, \omega_1; a, b) \frac{|\Omega_{\omega_2}|}{|\Omega_0^{\wedge, h_1}|}.$$

Therefore we have the equation of the lemma.

The following proposition states the relation between the cost weight pair of leaves and the nodes at height h_2 as we mentioned in Definition 6 for the case of $h_2 = h$ and $h_1 = 1$. To show it, we also need the following notion. Let d be an $E^i(a, b)$-distribution on $\Omega_i^{\wedge, h+1}$. Now we define the distribution $\tilde{d}$ on $\Omega_i^{\wedge, h}$ by

$$\tilde{d}(\omega) := \sum_{\omega' \in \Omega_\omega} d(\omega')$$

where Ω_ω is the subset of $\Omega_i^{\wedge, h+1}$ that assigns ω to the sequence of nodes at height h.

By Lemma 1, we can also define $\tilde{d}(\omega) = |\Omega_\omega| d(\omega')$ for any $\omega' \in \Omega_\omega$.

Proposition 1. *If d is an $E^1(a, b)$-distribution on $\Omega_1^{\wedge, h+1}$, then $\tilde{d}$ is an $E^1(na, b + \frac{(n-1)a}{2})$-distribution on $\Omega_1^{\wedge, h}$.*

Proof. Let A be any (nondirectional) algorithm for T_n^h and A_0 the left-to-right algorithm for T_n^1. We define a (nondirectional) algorithm $\tilde{A} := A_0 \times A$ for T_n^{h+1}.

Then,

$$
C(\tilde{A}, d; a, b) = \sum_{\omega \in \Omega_1^{\wedge, h+1}} d(\omega) C(\tilde{A}, \omega; a, b) = \sum_{\omega \in \Omega_1^{\wedge, h}} \sum_{\omega' \in \Omega_\omega} d(\omega') C(\tilde{A}, \omega'; a, b)
$$

$$
= \sum_{\omega \in \Omega_1^{\wedge, h}} \frac{\tilde{d}(\omega)}{|\Omega_\omega|} \sum_{\omega' \in \Omega_\omega} C(\tilde{A}, \omega'; a, b)
$$

$$
\overset{(1)}{=} \sum_{\omega \in \Omega_1^{\wedge, h}} \frac{\tilde{d}(\omega)}{|\Omega_\omega|} \left(\sum_{\omega_1 \in \Omega_1^{\wedge, 1}} C(A_0, \omega_1; a, b) \frac{|\Omega_\omega|}{|\Omega_1^{\wedge, 1}|} \sharp_1(A, \omega) \right.
$$

$$
\left. + \sum_{\omega_1 \in \Omega_0^{\wedge, 1}} C(A_0, \omega_1; a, b) \frac{|\Omega_\omega|}{|\Omega_0^{\wedge, 1}|} \sharp_0(A, \omega) \right)
$$

$$
\overset{(2)}{=} \sum_{\omega \in \Omega_1^{\wedge, h}} \frac{\tilde{d}(\omega)}{|\Omega_\omega|} \left(na |\Omega_\omega| \sharp_1(A, \omega) + \left(b + \frac{(n-1)a}{2} \right) |\Omega_\omega| \sharp_0(A, \omega) \right)
$$

$$
= \sum_{\omega \in \Omega_1^{\wedge, h}} \tilde{d}(\omega) C(A, \omega; na, b + \frac{(n-1)a}{2}) = C(A, \tilde{d}; na, b + \frac{(n-1)a}{2})
$$

The equality (1) follows from Lemma 2, and (2) from

$$
C(A_0, \omega_1; a, b) = na \quad \text{for } \omega_1 \in \Omega_1^{\wedge, 1},
$$

$$
\sum_{\omega_1 \in \Omega_0^{\wedge, 1}} C(A_0, \omega_1; a, b) = nb + \frac{n(n-1)}{2} a, \quad \text{and}
$$

$$
|\Omega_0^{\wedge, 1}| = n.
$$

Since d is an $E^1(a, b)$-distribution, $\tilde{d}$ is an $E^1(na, b + \frac{(n-1)a}{2})$-distribution.

Note that the cases with an odd h and/or E^0-distribution can be treated in a similar way.

Theorem 2. *For any balanced multi-branching weighted tree T, the $E^i(a, b)$-distribution w.r.t. $\mathcal{A}_D$ is a uniform distribution on i-set.*

Proof. It is proved by induction on height h. Assume that a tree T is an AND-OR tree, and the OR-AND case can be treated similarly.

We first investigate the base case $h = 1$. Since the 1-set for T_n^1 is a singleton, the $E^1(a, b)$-distribution is trivially unique. The 0-set for T_n^1 consists of n assignments $\omega_1, \ldots, \omega_n$, where for each $i \leq n$, ω_i assigns 0 to the i-th leaf and 1 to the rest. Let d be an $E^0(a, b)$-distribution and $p_i := d(\omega_i)$. We will show for any i, $j \leq n$, $p_i = p_j$. Let $A(i, j)$ be the algorithm that probes the leaves from left to right skipping over the i-th and j-th leaves, and probes them at the end in this order: probes the i-th leaf next to last and finally the j-th one. For example, let $n = 6$, then $A(5, 3) = 124653$, as illustrated in Fig. 3.

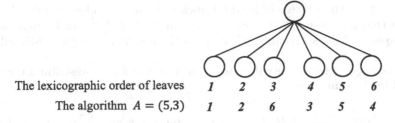

The lexicographic order of leaves   1   2   3   4   5   6
The algorithm $A = (5,3)$ 1   2   6   3   5   4

Fig. 3. The example of $A(i,j)$

Note that $C(A(i,j), \omega_i; a, b) = (n-2)a+b$, $C(A(i,j), \omega_j; a, b) = (n-1)a+b$ and for $k \neq i,j$, $C(A(i,j), \omega_k; a, b) = C(A(j,i), \omega_k; a, b)$. Since d is an $E^0(a,b)$-distribution, we have

$$C(A(i,j), d; a, b) - C(A(j,i), d; a, b)$$
$$= p_i \Big(C(A(i,j), \omega_i; a, b) - C(A(j,i), \omega_i; a, b) \Big)$$
$$+ p_j \Big(C(A(i,j), \omega_j; a, b) - C(A(j,i), \omega_j; a, b) \Big)$$
$$= a(p_j - p_i) = 0.$$

We get $p_i = p_j$ since $a \neq 0$.

For the induction step, we assume d is an $E^1(a,b)$-distribution on T of height $h+1$. By Proposition 1, $\widetilde{d}$ is an $E^1(na, b + \frac{(n-1)a}{2})$-distribution for T of height h. By the induction hypothesis, $\widetilde{d}$ is a uniform distribution on $\Omega_1^{\wedge,h}$. Since d can also be represented by $d(\omega) = \frac{\widetilde{d}(\omega')}{|\Omega_\omega|}$ for any $\omega' \in \Omega_\omega$, d is also uniform.

3.2 The Uniqueness of $E^i(a,b)$-Distribution Fails $w.r.t$ $\mathcal{A}_{dir}$

In this subsection, we will show that for balanced multi-branching trees, the uniqueness of $E^i(a,b)$-distribution no longer holds if we restrict the algorithms to $\mathcal{A}_{dir}$, and in particularly we show that E^0-distribution for AND-OR n-branching trees, namely T_n^h, is not unique for $h \geq 2$ and $n \geq 3$. It worth remarking that, however, for AND-OR trees T_2^2, E^0-distribution $w.r.t.$ $\mathcal{A}_{dir}$ turns out to be unique (Sect. 4 of [11]).

Theorem 3. *For any tree T_n^h ($h \geq 2$, $n \geq 3$), there are more than one $E^i(a,b)$-distributions $(i = 0, 1)$ $w.r.t.$ $\mathcal{A}_{dir}$.*

Proof. We show this by induction on height h of AND-OR trees, and the OR-AND case can be treated in the same way. Before giving the proof for case $h = 2$, we define some notations.

For $i = 1, \ldots, n$, let $\omega[i]$ denote the assignment for the subtree of height 1 such that value 1 only appears in the i-th position of $\omega[i]$, and $\omega[0]$ denote the assignment for the subtree of height 1 such that all the values are 0. Then we can see that any assignment $\omega = \omega_1 \ldots \omega_n$ in $\Omega_0^{\wedge,2} \cup \Omega_1^{\wedge,2}$ can be represented by a sequence from such $\omega[i]$'s, where ω_j is the assignment for the j-th subtree of T_n^2.

We only consider E^0-distribution, since the case for E^1-distribution can be treated in the same way. Let

$$\Omega_{k,j} := \left\{ \omega[i_1] \ldots \omega[i_n] \in \Omega_0^{\wedge,2} \; : \; i_j = 0, i_s \neq 0 \text{ if } s \neq j, \text{ and } \sum_{s=1}^{n} i_s = k \mod n \right\}$$

for $0 \leq k < n$.

Then $\Omega_0^{\wedge,2} = \bigsqcup_{j=1}^{n} \bigsqcup_{k=0}^{n-1} \Omega_{k,j}$ and we can easily observe that the number of ω in $\Omega_{k,j}$ such that ω_l is $\omega[i]$ does not depend on k for any i, j, l. Thus, we can show that for any k, k', j, and $A \in \mathcal{A}_{dir}$, $\sum_{\omega \in \Omega_{k,j}} C(A, \omega; a, b) = \sum_{\omega \in \Omega_{k',j}} C(A, \omega; a, b)$.

Let p_k $(k < n)$ be any reals such that $\sum_{k<n} p_k = \frac{1}{n^{n-1}}$. We define the distribution d on $\Omega_0^{\wedge,2}$ by $d(\omega) = p_k$ for $\omega \in \Omega_{k,j}$. Then, for any $A \in \mathcal{A}_{dir}$,

$$C(A, d; a, b) = \sum_{j=1}^{n} \sum_{k=0}^{n-1} \sum_{\omega \in \Omega_{k,j}} p_k C(A, \omega; a, b)$$

$$= \frac{1}{n^{n-1}} \sum_{j=1}^{n} \frac{1}{n} \sum_{k=0}^{n-1} \sum_{\omega \in \Omega_{k,j}} C(A, \omega; a, b)$$

$$= \frac{1}{n^n} \sum_{\omega \in \Omega_0^{\wedge,2}} C(A, \omega; a, b).$$

Since a uniform distribution on 0-set is an E^0-distribution, $C(A, d; a, b)$ does not depend on $A \in \mathcal{A}_{dir}$. Now p_k is any real such that $\sum_{k<n} p_k = \frac{1}{n^{n-1}}$, then there are uncountably many E^0-distribution w.r.t. $\mathcal{A}_{dir}$.

For the induction step, we consider OR-AND tree of height $h+1$. Let d_0 and d_1 be an $E^0(a, b)$-distribution and an $E^1(a, b)$-distribution for AND-OR tree T_n^h, respectively. Then we can construct an $E^i(a, b)$-distribution w.r.t. $\mathcal{A}_{dir}$ for an OR-AND tree T_n^{h+1} as follows.

- For $E^0(a, b)$-distribution, we define the distribution d on $\Omega_0^{\vee,h+1}$ by

$$d(\omega_0 \ldots \omega_{n-1}) = \prod_{i=0}^{n-1} d_0(\omega_i),$$

where each $\omega_i \in \Omega_0^{\wedge,h}$ is the assignment for the i-th subtree of height h. Then we can show d is an $E^0(a, b)$-distribution.

- For $E^1(a, b)$-distribution, d is defined by

$$d(\omega_0 \ldots \omega_{n-1}) = \frac{1}{n} d_1(\omega_i) \cdot \prod_{j \neq i} d_0(\omega_j),$$

where $\omega_i \in \Omega_1^{\wedge, h}$ and $\omega_j \in \Omega_0^{\wedge, h}$ for $j \neq i$.

By the induction hypothesis, the above d's are not uniform.

Acknowledgement. We would like to express our sincere appreciations for Prof. ChenGuang Liu (Northwestern Polytechnical University) for his original insight and helpful suggestions on this topic. We are also grateful to Prof. Yue Yang (National University of Singapore) for his useful discussions and valuable comments. This work was supported in part by the JSPS KAKENHI Grant Numbers 26540001 and 15H03634, and by National Natural Science Foundation of China Grant Number 11701438.

References

1. Cicalese, F., Milanic, M.: Computing with priced information: when the value makes the price. In: International Symposium on Algorithms and Computation, pp. 378–389 (2008)
2. Liu, C.G., Tanaka, K.: Eigen-distribution on random assignments for game trees. Inf. Process. Lett. **104**(2), 73–77 (2007a)
3. Liu, C.G., Tanaka, K.: The computational complexity of game trees by eigen-distribution. In: Dress, A., Xu, Y., Zhu, B. (eds.) COCOA 2007. LNCS, vol. 4616, pp. 323–334. Springer, Heidelberg (2007). https://doi.org/10.1007/978-3-540-73556-4_34
4. Knuth, D.E., Moore, R.W.: An analysis of alpha-beta pruning. Artif. Intell. **6**(4), 293–326 (1975)
5. Pearl, J.: Asymptotic properties of minimax trees and game-searching procedures. Artif. Intell. **14**(2), 113–138 (1980)
6. Peng, W., Peng, N., Ng, K., Tanaka, K., Yang, Y.: Optimal depth-first algorithms and equilibria of independent distributions on multi-branching trees. Inf. Process. Lett. **125**, 41–45 (2017)
7. Peng, W., Okisaka, S., Li, W., Tanaka, K.: The uniqueness of eigen-distribution under non-directional algorithms. IAENG Int. J. Comput. Sci. **43**(3), 318–325 (2016)
8. Saks, M., Wigderson, A.: Probabilistic Boolean decision trees and the complexity of evaluating game trees. In: Proceedings of 27th Annual IEEE Symposium on Foundations of Computer Science, pp. 29–38 (1986)
9. Suzuki, T.: Kazuyuki Tanaka's work on AND-OR trees and subsequent developments. Ann. Japan Assoc. Philos. Sci. **25**, 79–88 (2017)
10. Suzuki, T., Niida, Y.: Equilibrium points of an AND-OR tree: under constraints on probability. Ann. Pure Appl. Log. **166**(11), 1150–1164 (2015)
11. Suzuki, T., Nakamura, R.: The eigen-distribution of an AND-OR tree under directional algorithms. IAENG Int. J. Appl. Math. **42**(2), 122–128 (2012)
12. Yao, A.C.C.: Probabilistic computations: toward a unified measure of complexity. In: Proceedings of 18th Annual IEEE Symposium on Foundations of Computer Science, pp. 222–227 (1977)

A Spectral Partitioning Algorithm for Maximum Directed Cut Problem

Zhenning Zhang[1], Donglei Du[2], Chenchen Wu[3], Dachuan Xu[1],
and Dongmei Zhang[4(✉)]

[1] College of Applied Sciences, Beijing University of Technology, Beijing 100124,
People's Republic of China
[2] Faculty of Business Administration, University of New Brunswick,
Fredericton, NB E3B 5A3, Canada
[3] College of Science, Tianjin University of Technology, Tianjin 300384,
People's Republic of China
[4] School of Computer Science and Technology, Shandong Jianzhu University,
Jinan 250101, People's Republic of China
zhangdongmei@sdjzu.edu.cn

Abstract. In this paper, we study the maximum directed cut (MaxDC) problem. In the MaxDC, we are given a directed graph with nonnegative edge weights. Our goal is to obtain a bipartition of the vertices such that the total edge weight of the directed cut is maximized. By exploring the combinatorial characteristics of the optimal solution, we offer a 0.272-approximation algorithm based on the technique of spectral partitioning rounding.

Keywords: Maximum directed cut · Spectral graph theory · Spectral partitioning rounding · Approximation algorithm

1 Introduction

Semidefinite programming (SDP) is a powerful approach for combinatorial optimization, in particular graph partition problems. For the maximum cut (Max cut) problem, Goemans and Williamson [1] give a 0.87856-approximation algorithm by SPD, and this ratio cannot be improved under the unique game conjecture. Although SDP is solvable in polynomial time, it is highly time-demanding for large-scale problems. Therefore approximation algorithms without invoking SDP are sometimes more attractive and preferred, even with worse approximation ratios.

Approximation algorithms for the Max cut problem without using SDP include (1) Trevisan [2] who offers a 0.531-approximation algorithm; (2) Soto [3] who improves the bound to 0.614247; and (3) Kale and Seshadhri [4] who propose a combination approximation algorithm based on random walk with approximation ratio $\tilde{O}(n^b)$ ($b > 1.5$). For example, if the running times of the algorithm are $\tilde{O}(n^{1.6})$, $\tilde{O}(n^2)$ and $\tilde{O}(n^3)$, the corresponding approximations are 0.5051, 0.5155

© Springer International Publishing AG 2017
X. Gao et al. (Eds.): COCOA 2017, Part I, LNCS 10627, pp. 298–312, 2017.
https://doi.org/10.1007/978-3-319-71150-8_26

and 0.5727, respectively. If the value of b is big enough, the approximation ratio of the combination algorithm converges to that in [3]. Recently, Nikiforov [6] considers the bound of the spectral approximation algorithm.

In this work, we consider the maximum directed cut problem. Suppose that there is a directed graph $G = (V, A)$ with a weight $w_{ij} : A \to \mathbb{R}^+$ on each directed arc $(i, j) \in A$. The maximum directed cut problem is that of partitioning V into two sets S and $V - S$ such that the total weight of the edges with their tails in S and their heads in $V - S$ are maximized.

Several approximation algorithms exist for the maximum directed cut problem. Randomly assigning a node to a cut yields a simple 1/4-stochastic approximation algorithm. A 0.79607-approximation algorithm is available via SDP [1]. Using the prerounding rotations and skewed distributions of hyperplanes, Lewin et al. [5] improve the approximation ratio to 0.874. A special family of randomized algorithms for the maximum directed cut problem with an approximation ratio of at least 0.483 is also proposed in [7].

We investigate the maximum directed cut problems following the spectral partitioning method. The main idea is to design a rounding for the optimal solution of the relaxation of the MaxDC to derive a tripartition (V_+, V_0, V_-) of the set of vertices V at intermediate steps, where $V_0 \neq V$ is the set of "undecided" vertices. The recursive algorithm runs until every vertex is decided to yield a cut. For the tripartition of the directed graph at each step, we define the following notations.

- $A(V_+ : V_-)$: the set of the right cut edges from V_+ to V_-;
- $A(V_- : V_+)$: the set of the left cut edges from V_- to V_+;
- $A(V_+ : V_0) + A(V_0 : V_-)$: the set of the right cross edges from V_+ to V_0 and V_0 to V_-;
- $A(V_- : V_0) + A(V_0 : V_+)$: the set of the left cross edges from V_- to V_0 and V_0 to V_+;

Therefore, we introduce the cut edges (the union of the right cut and the left cut), the cross edges (the union of the right cross and the left cross edges) and the incident edges (at least one vertex involved in $V_+ \cup V_-$). We observe that the desired edges are at least half of the cut and cross edges at each iteration. We introduce the recoverable ratio to express the proportion of the desired edge weight to the incident edge weight.

Formulating the MaxDC in matrix form allows us to explore the properties of the optimal value of MaxDC in virtue of the corresponding eigenvector which can be derived in polynomial time. We then devise a spectral partitioning rounding for the MaxDC based on some special characteristics of the optimal solution. We show the recoverable ratio of the tripartition at every iteration, leading to the 0.272-approximation ratio of spectral partitioning algorithm for the MaxDC. The paper shows that the spectral partitioning technique is applicable to many graph partitioning problems.

The paper is organized as follows. In Sect. 2 we describe the MaxDC problem. In Sect. 3 we explore the properties of the optimal solution. Using the spectral partitioning rounding, we obtain the recoverable ratio at every iteration. In Sect. 4 we present the spectral partitioning algorithm and derive the 0.272-approximate ratio.

The omitted proofs in this paper are deferred to the journal version.

2   Maximum Directed Cut Problem

The maximum directed cut problem can be formulated as the following integer program:

$$\max_{x\in\{-1,1\}^{|V|+1}} \frac{1}{4} \sum_{(i,j)\in A} w_{ij}\left(1 + x_0 x_i - x_0 x_j - x_i x_j\right), \qquad (1)$$

where $x = (x_0, x_1, \ldots, x_n) \in \{-1,1\}^{|V|+1}$, x_0 is regarded as an index for the partition, $|V|$ is the number of the vertices.

Evidently, (1) is equivalent to the following quadratic integer program:

$$\max_{x\in\{-1,1\}^{V+1}} \frac{1}{8} \sum_{(i,j)\in A} w_{ij}\left(x_i^2 + x_j^2 + 2x_0 x_i - 2x_0 x_j - 2x_i x_j\right).$$

Or in matrix form:

$$\max_{x\in\{-1,1\}^{|V|+1}} \frac{1}{8} x^T M x,$$

where $M = \sum_{(i,j)\in A} M^{\{i,j\}}$, and each $M^{\{i,j\}} \in \mathbb{R}^{(n+1)\times(n+1)}$ is associated with the quadratic form:

$$x^T M^{\{i,j\}} x = w_{ij}\left(x_i^2 + x_j^2 + 2x_0 x_i - 2x_0 x_j - 2x_i x_j\right). \qquad (2)$$

According to (2), the real symmetric matrix $M^{\{i,j\}}$ is expressed as

$$M^{\{i,j\}} = \begin{array}{c} \\ 0 \\ \vdots \\ i \\ \vdots \\ j \\ \vdots \end{array} \begin{array}{c} 0 \quad\cdots\quad i \quad\cdots\quad j \quad\cdots \\ \left(\begin{array}{cccccc} 0 & \cdots & w_{ij} & \cdots & -w_{ij} & \cdots \\ \vdots & \vdots & \vdots & \vdots & \vdots & \vdots \\ w_{ij} & \cdots & \frac{1}{2}w_{ij} & \cdots & -w_{ij} & \cdots \\ \vdots & \vdots & \vdots & \vdots & \vdots & \vdots \\ -w_{ij} & \cdots & -w_{ij} & \cdots & \frac{1}{2}w_{ij} & \cdots \\ \vdots & \vdots & \vdots & \vdots & \vdots & \vdots \end{array}\right) \end{array}$$

with $i < j$, and

$$
M^{\{i,j\}} = \begin{array}{c} \\ 0 \\ \vdots \\ j \\ \vdots \\ i \\ \vdots \end{array}
\begin{array}{c} \begin{array}{cccccccc} 0 & \cdots & j & & \cdots & i & \cdots \end{array} \\
\left(\begin{array}{cccccccc}
0 & \cdots & -w_{ij} & \cdots & w_{ij} & \cdots \\
\vdots & \vdots & \vdots & \vdots & \vdots & \vdots \\
-w_{ij} & \cdots & \frac{1}{2}w_{ij} & \cdots & -w_{ij} & \cdots \\
\vdots & \vdots & \vdots & \vdots & \vdots & \vdots \\
w_{ij} & \cdots & -w_{ij} & \cdots & \frac{1}{2}w_{ij} & \cdots \\
\vdots & \vdots & \vdots & \vdots & \vdots & \vdots
\end{array} \right)
\end{array}
$$

with $j < i$.

$M = \sum_{(i,j)\in A} M^{\{i,j\}}$ can be considered as the generalized adjacent matrix of the directed graph G, which is given by

$$
M = \begin{pmatrix}
0 & d_1^{out} - d_1^{in} & \cdots & d_i^{out} - d_i^{in} & \cdots & d_n^{out} - d_n^{in} \\
d_1^{out} - d_1^{in} & \frac{1}{2}\left(d_1^{in} + d_1^{out}\right) & \cdots & -w_{1i} & \cdots & -w_{1n} \\
\vdots & \vdots & \ddots & \vdots & \vdots & \vdots \\
d_i^{out} - d_i^{in} & -w_{1i} & \cdots & \frac{1}{2}\left(d_i^{in} + d_i^{out}\right) & \cdots & -w_{ni} \\
\vdots & \vdots & \vdots & \vdots & \ddots & \vdots \\
d_n^{out} - d_n^{in} & -w_{1n} & \cdots & -w_{ni} & \cdots & \frac{1}{2}\left(d_n^{in} + d_n^{out}\right)
\end{pmatrix}, \quad (3)
$$

where d_i^{in} $(i = 1, 2, \cdots, n)$ denote the weight of the edges whose head vertex is i, and d_i^{out} $(i = 1, 2, \cdots, n)$ denote the weight of the edges whose tail vertex is i.

Analogously, the total weight of the directed graph is introduced by

$$
\begin{aligned}
W(A) &= \sum_{(i,j)\in A} w_{ij} \\
&= \frac{1}{2} \sum_{(i,j)\in A} w_{ij} \left(x_i^2 + x_j^2\right) \\
&= \frac{1}{2} x^T D x,
\end{aligned}
$$

where $D = \sum_{(i,j)\in A} D^{\{i,j\}}$ and each $D^{\{i,j\}} \in \mathbb{R}^{(n+1)\times(n+1)}$ is associated with the quadratic form:

$$
x^T D^{\{i,j\}} x = w_{ij} \left(x_i^2 + x_j^2\right). \tag{4}
$$

The diagonal matrix $D^{\{i,j\}}$ with respect to (4) is given by

$$
D^{\{i,j\}} = \begin{array}{c} \\ 0 \\ \vdots \\ i \\ \vdots \\ j \\ \vdots \end{array}
\begin{array}{c} 0 \;\cdots\; i \;\cdots\; j \;\cdots \\
\begin{pmatrix}
0 \cdots & 0 & \cdots & 0 & \cdots \\
\vdots \ddots & \vdots & \vdots & \vdots & \vdots \\
0 \cdots & w_{ij} & \cdots & 0 & \cdots \\
\vdots & \vdots & \vdots & \ddots & \vdots & \vdots \\
0 \cdots & 0 & \cdots & w_{ij} & \cdots \\
\vdots & \vdots & \vdots & \vdots & \vdots & \vdots
\end{pmatrix}
\end{array}
$$

with $i < j$, and

$$
D^{\{i,j\}} = \begin{array}{c} \\ 0 \\ \vdots \\ j \\ \vdots \\ i \\ \vdots \end{array}
\begin{array}{c} 0 \;\cdots\; j \;\cdots\; i \;\cdots \\
\begin{pmatrix}
0 \cdots & 0 & \cdots & 0 & \cdots \\
\vdots \ddots & \vdots & \vdots & \vdots & \vdots \\
0 \cdots & w_{ij} & \cdots & 0 & \cdots \\
\vdots & \vdots & \vdots & \ddots & \vdots & \vdots \\
0 \cdots & 0 & \cdots & w_{ij} & \cdots \\
\vdots & \vdots & \vdots & \vdots & \vdots & \vdots
\end{pmatrix}
\end{array}.
$$

with $j < i$.

As $D = \sum_{(i,j)\in A} D^{\{i,j\}}$, the real diagonal matrix $D \in \mathbb{R}^{(n+1)\times(n+1)}$ is

$$
D = diag\left\{0, d_1^{\text{in}} + d_1^{\text{out}}, \cdots, d_i^{\text{in}} + d_i^{\text{out}}, \cdots, d_n^{\text{in}} + d_n^{\text{out}}\right\}. \tag{5}
$$

2.1 Spectral Partitioning Rounding

In this section, we investigate the properties of the optimal solution. By using the spectral partitioning rounding technique, at every iteration we get the recoverable ratio of the tripartition (Lemma 3), which will be used later to obtain the approximation ratio for the spectral partitioning algorithm.

Lemma 1. *Let* $\text{OPT}_{\text{MaxDC}} \geq (1 - \varepsilon)W(A)$, *there exists a vector* $x = (x_0, x_1, \ldots, x_n)^T$ *whose first component* x_0 *is nonzero and the other components can not be zero simultaneously, such that*

$$
\frac{x^T M x}{x^T D x} \geq 4(1 - \varepsilon),
$$

that is,

$$
\sum_{(i,j)\in A} w_{ij}\left(x_i^2 + x_j^2 + 2x_0 x_i - 2x_0 x_j - 2x_i x_j\right) \geq 4(1 - \varepsilon) \sum_{(i,j)\in A} w_{ij}\left(x_i^2 + x_j^2\right).
$$

$$\tag{6}$$

Moreover, x_0 *can take any value in* $\mathbb{R}\backslash\{0\}$.

We now express an algorithm to show how to obtain a vector x satisfying (6).

Algorithm 1

Step 1. *For a parameter δ, it is possible to find a vector z in polynomial time such that*

$$z^T D^{-1/2} M D^{-1/2} z \geq \lambda_1 (1 - \delta) z^T z.$$

Step 2. *The associated vector $\tilde{x}$ is given by $\tilde{x} = D^{-1/2} z$*
Step 3. *Search every component of $\tilde{x}$ to find the smallest nonzero $|\tilde{x}_i|$; exchange the first component of $\tilde{x}$ by $\mathrm{sgn}(\tilde{x}_i) \min_{\tilde{x}} |\tilde{x}_i|$; and denote the new vector by x.*

We introduce the spectral partitioning rounding based on the associated vector x obtained by Algorithm 1.

Algorithm 2. *The spectral partitioning rounding:*

Input: *For a vector x given by Algorithm 1, without loss of generality, let $\|x\|_\infty = 1$.*
Output: *A vector $y \in \{\mathrm{sgn}(x_0), 0, -\mathrm{sgn}(x_0)\}^V \backslash \{0\}^V$.*
Step 1. *Generate a random variable t following the uniform distribution in $(0, 1)$.*
Step 2. *For each $i \in \{1, \ldots, |V|\}$, Set*

$$y_i = \begin{cases} \mathrm{sgn}(x_0), & x_i x_0 > \sqrt{t}, \\ -\mathrm{sgn}(x_0), & x_i x_0 < -\sqrt{t}, \\ 0, & |x_i x_0| \leq \sqrt{t}. \end{cases} \tag{7}$$

The set of vertices can be divided into three parts by Algorithm 2., that is,

$$\begin{aligned} V_+ &= \{i \mid y_i = \mathrm{sgn}(x_0)\}, \\ V_- &= \{i \mid y_i = -\mathrm{sgn}(x_0)\}, \\ V_0 &= \{i \mid y_i = 0\}. \end{aligned}$$

By the indicator vector y, the weight of the right cut edges, the right cross edges, the left cut edges, the left cross edges, the uncut edges and the incident edges are defined as follows,

$$\begin{aligned} \overrightarrow{\mathrm{Cut}}(y) &:= w[A(V_+ : V_-)], \\ \overrightarrow{\mathrm{Cross}}(y) &:= w[A(V_+ : V_0)] + w[A(V_0 : V_-)], \\ \overleftarrow{\mathrm{Cut}}(y) &:= w[A(V_- : V_+)], \\ \overleftarrow{\mathrm{Cross}}(y) &:= w[A(V_- : V_0)] + w[A(V_0 : V_+)], \\ \mathrm{Uncut}(y) &:= w^{\mathrm{out}}[A[V_+]] + w^{\mathrm{out}}[A[V_-]]. \end{aligned}$$

Let $\mathrm{Incident}(y)$ be the total weight of the directed edges whose endpoints (tail or head) lie in V_+ or V_- at least once, where $w^{\mathrm{out}}[A[V_+]]$ denotes the out-degree of the edges with both ends in V_+. Define further $\mathrm{Cut}(y)$ and $\mathrm{Cross}(y)$ as follows

$$\begin{aligned} \mathrm{Cut}(y) &:= \overrightarrow{\mathrm{Cut}}(y) + \overleftarrow{\mathrm{Cut}}(y), \\ \mathrm{Cross}(y) &:= \overrightarrow{\mathrm{Cross}}(y) + \overleftarrow{\mathrm{Cross}}(y). \end{aligned}$$

From the definition of the Incident(y), we have

$$\text{Incident}(y) = \text{Cut}(y) + \text{Cross}(y) + \text{Uncut}(y). \tag{8}$$

At every iteration, by the indicator vector y, the weight of desired edges are at least $\frac{1}{2}\left(\text{Cut}(y) + \frac{1}{2}\text{Cross}(y)\right)$. The recoverable ratio is defined as the weight of the desired edges with respect to the weight of the incident edges, that is

$$\frac{\text{Desired}(y)}{\text{Incident}(y)} \geq \frac{\frac{1}{2}\text{Cut}(y) + \frac{1}{4}\text{Cross}(y)}{\text{Incident}(y)}.$$

To find the optimal recoverable ratio of the spectral algorithm, we give the following lemma first.

Lemma 2. *Let*

$$C(i,j) = Pr\left\{ \begin{pmatrix} y_i \\ y_j \end{pmatrix} \in \left\{ \begin{pmatrix} \text{sgn}(x_0) \\ -\text{sgn}(x_0) \end{pmatrix}, \begin{pmatrix} -\text{sgn}(x_0) \\ \text{sgn}(x_0) \end{pmatrix} \right\} \right\}$$

and

$$X(i,j) = Pr\left\{ \begin{pmatrix} y_i \\ y_j \end{pmatrix} \in \left\{ \begin{pmatrix} \text{sgn}(x_0) \\ 0 \end{pmatrix}, \begin{pmatrix} 0 \\ -\text{sgn}(x_0) \end{pmatrix}, \begin{pmatrix} -\text{sgn}(x_0) \\ 0 \end{pmatrix}, \begin{pmatrix} 0 \\ \text{sgn}(x_0) \end{pmatrix} \right\} \right\}$$

be the probabilities that an edge is a cut edge or a cross edge induced by the indicator vector y as in (7). Then for all $0 \leq \beta \leq 1$, we have

$$C(i,j) + \beta X(i,j) \geq \beta(1 - \beta)x_0^2 \left(x_i^2 + x_j^2 + 2x_0x_i - 2x_0x_j - 2x_ix_j\right)$$
$$- \beta(1 - \beta)\left[2X(i,j) + 4C(i,j)\right]. \tag{9}$$

Proof. The probability of a cut edge is given by

$$C(i,j) = \mathbb{P}\left\{ \begin{pmatrix} y_i \\ y_j \end{pmatrix} \in \left\{ \begin{pmatrix} \text{sgn}(x_0) \\ -\text{sgn}(x_0) \end{pmatrix}, \begin{pmatrix} -\text{sgn}(x_0) \\ \text{sgn}(x_0) \end{pmatrix} \right\} \right\}$$

$$= \mathbb{P}\left\{ x_0x_i > \sqrt{t}, x_0x_j < -\sqrt{t} \text{ or } x_0x_i < -\sqrt{t}, x_0x_j > \sqrt{t} \right\} \tag{10}$$

$$= \mathbb{P}\left\{ \sqrt{t} < x_0x_i, \sqrt{t} < -x_0x_j \text{ or } \sqrt{t} < -x_0x_i, \sqrt{t} < x_0x_j \right\}$$

$$= \mathbb{P}\left\{ \sqrt{t} < \min\{x_0x_i, -x_0x_j\} \text{ or } \sqrt{t} < \min\{-x_0x_i, x_0x_j\} \right\}.$$

Similarly, the probability of a crossing edge can be obtained by

$$X(i,j) = \mathbb{P}\left\{ \begin{pmatrix} y_i \\ y_j \end{pmatrix} \in \left\{ \begin{pmatrix} \text{sgn}(x_0) \\ 0 \end{pmatrix}, \begin{pmatrix} 0 \\ -\text{sgn}(x_0) \end{pmatrix}, \begin{pmatrix} -\text{sgn}(x_0) \\ 0 \end{pmatrix}, \begin{pmatrix} 0 \\ \text{sgn}(x_0) \end{pmatrix} \right\} \right\}$$

$$= \mathbb{P}\left\{ x_0x_i > \sqrt{t}, |x_0x_j| \leq \sqrt{t} \text{ or } |x_0x_i| \leq \sqrt{t}, x_0x_j < -\sqrt{t} \right.$$

$$\left. \text{or } x_0x_i < -\sqrt{t}, |x_0x_j| \leq \sqrt{t} \text{ or } |x_0x_i| \leq \sqrt{t}, x_0x_j > \sqrt{t} \right\}$$

$$= \mathbb{P}\left\{ |x_0x_j| \leq \sqrt{t} < x_0x_i \text{ or } |x_0x_i| \leq \sqrt{t} < -x_0x_j \right.$$

$$\left. \text{or } |x_0x_j| \leq \sqrt{t} < -x_0x_i \text{ or } |x_0x_i| \leq \sqrt{t} < x_0x_j \right\}.$$

$$\tag{11}$$

In the following, we prove Eq. (9). Note that the vector x is derived by Algorithm 1. We will discuss the following three cases.

Case 1. If x_i and x_j $(i, j \in \{1, 2, \ldots, n\})$ are both equal to zero, (9) holds obviously.

Case 2. If exactly one of x_i and x_j is zero, then for symmetry, assume that x_i is zero.

Case 2.1. $x_j \neq 0$ and $x_j x_0 \geq 0$. We obtain $C(i, j) = 0$ and $X(i, j) = x_0^2 x_j^2$. Then, we have

$$
\begin{aligned}
C(i, j) + \beta X(i, j) &\geq \beta(1 - \beta) x_0^2 x_j^2 \\
&= \beta(1 - \beta) x_0^2 \left(x_j^2 - 2x_0 x_j \right) + 2\beta(1 - \beta) x_0^3 x_j \\
&\geq \beta(1 - \beta) x_0^2 \left(x_j^2 - 2x_0 x_j \right) - 2\beta(1 - \beta) x_0^2 x_j^2
\end{aligned}
$$

Case 2.2. $x_j \neq 0$ and $x_j x_0 \leq 0$. We easily obtain $C(i, j) = 0$ and $X(i, j) = x_0^2 x_j^2$. Since $-x_0 x_j \leq x_j^2$, we get

$$
\begin{aligned}
C(i, j) + \beta X(i, j) &\geq \beta(1 - \beta) x_0^2 x_j^2 \\
&= \beta(1 - \beta) x_0^2 (x_j^2 - 2x_0^2 x_j) + 2\beta(1 - \beta) x_0^3 x_j \\
&\geq \beta(1 - \beta) x_0^2 (x_j^2 - 2x_0^2 x_j) - 2\beta(1 - \beta) x_0^2 x_j^2
\end{aligned}
$$

Equation (9) holds in these cases.

Case 3. $x_i x_j \neq 0$. According to the signs and the sizes of x_i and x_j, we consider eight cases in the following.

Case 3.1. $x_0 x_i > 0$, $x_0 x_j > 0$, and $|x_0 x_j| \leq |x_0 x_i|(x_0 x_j \leq x_0 x_i)$. Then we have

$$
C(i, j) = 0,
$$

$$
X(i, j) = \mathbb{P}\left\{ |x_0 x_j| \leq \sqrt{t} < x_0 x_i \right\} = x_0^2 x_i^2 - x_0^2 x_j^2,
$$

and

$$
\begin{aligned}
C(i, j) + \beta X(i, j) &= \beta \left(x_0^2 x_i^2 - x_0^2 x_j^2 \right) \\
&= \beta \left(x_0 x_i + x_0 x_j \right) \left(x_0 x_i - x_0 x_j \right) \\
&\geq \beta(1 - \beta) \left(x_0 x_i - x_0 x_j \right)^2 \\
&= \beta(1 - \beta) \left(x_0^2 x_i^2 + x_0^2 x_j^2 - 2x_0^2 x_i x_j \right) \\
&= \beta(1 - \beta) x_0^2 \left(x_i^2 + x_j^2 + 2x_0 x_i - 2x_0 x_j - 2x_i x_j \right) \\
&\quad - \beta(1 - \beta) x_0^2 \left(2x_0 x_i - 2x_0 x_j \right).
\end{aligned}
$$

Since $x_0^2 \leq x_0 x_i + x_0 x_j$, we get

$$
\begin{aligned}
2x_0^2 \left(x_0 x_i - x_0 x_j \right) &\leq 2 \left(x_0 x_i + x_0 x_j \right) \left(x_0 x_i - x_0 x_j \right) \\
&= 2 \left(x_0^2 x_i^2 - x_0^2 x_j^2 \right) = 2X(i, j).
\end{aligned}
$$

Therefore,

$$-\beta(1-\beta)x_0^2\left(2x_0x_i - 2x_0x_j\right) \geq -\beta(1-\beta)2X(i,j)$$
$$= -\beta(1-\beta)2X(i,j) - \beta(1-\beta)4C(i,j).$$

Finally,

$$C(i,j) + \beta X(i,j) \geq \beta(1-\beta)x_0^2\left(x_i^2 + x_j^2 + 2x_0x_i - 2x_0x_j - 2x_ix_j\right)$$
$$- \beta(1-\beta)\left[2X(i,j) + 4C(i,j)\right].$$

Case 3.2. x_i, x_j and x_0 all have the same sign, and $|x_0x_i| \leq |x_0x_j|(x_0x_i \leq x_0x_j)$. Then, we have

$$C(i,j) = 0,$$
$$X(i,j) = \mathbb{P}\left\{|x_0x_i| \leq \sqrt{t} < x_0x_j\right\} = x_0^2x_j^2 - x_0^2x_i^2,$$

and

$$C(i,j) + \beta X(i,j) = \beta\left(x_0^2x_j^2 - x_0^2x_i^2\right)$$
$$= \beta\left(x_0x_j + x_0x_i\right)\left(x_0x_j - x_0x_i\right)$$
$$\geq \beta(1-\beta)\left(x_0x_j - x_0x_i\right)^2$$
$$= \beta(1-\beta)\left(x_0^2x_i^2 + x_0^2x_j^2 - 2x_0^2x_ix_j\right)$$
$$= \beta(1-\beta)x_0^2\left(x_i^2 + x_j^2 + 2x_0x_i - 2x_0x_j - 2x_ix_j\right)$$
$$- \beta(1-\beta)x_0^2\left(2x_0x_i - 2x_0x_j\right).$$

Since $x_0x_i - x_0x_j \leq 0$, we have $-\beta(1-\beta)x_0^2\left(2x_0x_i - 2x_0x_j\right) \geq 0$. Moreover, $-(x_0x_j + x_0x_i) \leq 0$. Therefore

$$-\beta(1-\beta)x_0^2\left(2x_0x_i - 2x_0x_j\right) \geq 0$$
$$\geq -\beta(1-\beta)x_0^2\left(2x_0x_i - 2x_0x_j\right)\left[-(x_0x_i + x_0x_j)\right]$$
$$\geq -\beta(1-\beta)2\left(x_0^2x_j^2 - x_0^2x_i^2\right)$$
$$= -\beta(1-\beta)2X(i,j)$$
$$= -\beta(1-\beta)\left[2X(i,j) + 4C(i,j)\right].$$

Finally,

$$C(i,j) + \beta X(i,j) \geq \beta(1-\beta)x_0^2\left(x_i^2 + x_j^2 + 2x_0x_i - 2x_0x_j - 2x_ix_j\right)$$
$$- \beta(1-\beta)\left[2X(i,j) + 4C(i,j)\right].$$

Case 3.3. $x_0x_i < 0$, $x_0x_j < 0$ and $|x_0x_j| \leq |x_0x_i|(x_0x_i \leq x_0x_j)$. Then, we have

$$C(i,j) = 0,$$
$$X(i,j) = \mathbb{P}\left\{|x_0x_j| \leq \sqrt{t} < -x_0x_i\right\} = x_0^2x_i^2 - x_0^2x_j^2,$$

and

$$C(i,j) + \beta X(i,j) = \beta \left(x_0^2 x_i^2 - x_0^2 x_j^2\right)$$
$$\geq \beta(1-\beta) \left|x_0 x_i + x_0 x_j\right| \left|x_0 x_i - x_0 x_j\right|$$
$$\geq \beta(1-\beta) \left(x_0 x_i - x_0 x_j\right)^2$$
$$= \beta(1-\beta) \left(x_0^2 x_i^2 + x_0^2 x_j^2 - 2x_0^2 x_i x_j\right)$$
$$= \beta(1-\beta) x_0^2 \left(x_i^2 + x_j^2 + 2x_0 x_i - 2x_0 x_j - 2x_i x_j\right)$$
$$- \beta(1-\beta) x_0^2 \left(2x_0 x_i - 2x_0 x_j\right).$$

Because of $\left|x_0 x_i - x_0 x_j\right| \leq \left|x_0 x_i + x_0 x_j\right|$, the third inequality follows from the second one. Since $(x_0 x_i - x_0 x_j) \leq 0$, we have

$$-\beta(1-\beta) x_0^2 \left(2x_0 x_i - 2x_0 x_j\right) \geq 0.$$

Moreover $(x_0 x_i + x_0 x_j) \leq 0$, we have

$$- \beta(1-\beta) x_0^2 \left(2x_0 x_i - 2x_0 x_j\right) \geq 0$$
$$\geq -\beta(1-\beta) x_0^2 \left(2x_0 x_i - 2x_0 x_j\right) \left(x_0 x_i + x_0 x_j\right)$$
$$\geq -\beta(1-\beta) 2 \left(x_0^2 x_i^2 - x_0^2 x_j^2\right)$$
$$= -\beta(1-\beta) \left[2X(i,j) + 4C(i,j)\right].$$

Therefore,

$$C(i,j) + \beta X(i,j) \geq \beta(1-\beta) x_0^2 \left(x_i^2 + x_j^2 + 2x_0 x_i - 2x_0 x_j - 2x_i x_j\right)$$
$$- \beta(1-\beta) \left[2X(i,j) + 4C(i,j)\right].$$

Case 3.4. $x_0 x_i < 0$, $x_0 x_j < 0$, and $\left|x_0 x_i\right| \leq \left|x_0 x_j\right| (x_0 x_j \leq x_0 x_i)$. Then, we have

$$C(i,j) = 0,$$
$$X(i,j) = \mathbb{P}\left\{\left|x_0 x_i\right| \leq \sqrt{t} < -x_0 x_j\right\} = x_0^2 x_j^2 - x_0^2 x_i^2,$$

and

$$C(i,j) + \beta X(i,j) = \beta \left(x_0^2 x_j^2 - x_0^2 x_i^2\right)$$
$$\geq \beta(1-\beta) \left|x_0 x_i + x_0 x_j\right| \left|x_0 x_i - x_0 x_j\right|$$
$$\geq \beta(1-\beta) \left(x_0 x_j - x_0 x_i\right)^2$$
$$= \beta(1-\beta) \left(x_0^2 x_i^2 + x_0^2 x_j^2 - 2x_0^2 x_i x_j\right)$$
$$= \beta(1-\beta) x_0^2 \left(x_i^2 + x_j^2 + 2x_0 x_i - 2x_0 x_j - 2x_i x_j\right)$$
$$- \beta(1-\beta) x_0^2 \left(2x_0 x_i - 2x_0 x_j\right)$$

Because of $\left|x_0 x_i - x_0 x_j\right| \leq \left|x_0 x_i + x_0 x_j\right|$, the third inequality follows from the second one. Meanwhile, from $x_0^2 \leq -\left(x_0 x_i + x_0 x_j\right)$ and

$$x_0^2 \left(2x_0 x_i - 2x_0 x_j\right) \leq -\left(x_0 x_i + x_0 x_j\right)\left(2x_0 x_i - 2x_0 x_j\right)$$
$$= 2 \left(x_0^2 x_j^2 - x_0^2 x_i^2\right)$$
$$= 2X(i,j),$$

we offer

$$C(i,j) + \beta X(i,j) \geq \beta(1-\beta)x_0^2 \left(x_i^2 + x_j^2 + 2x_0x_i - 2x_0x_j - 2x_ix_j\right)$$
$$- \beta(1-\beta)\left[2X(i,j) + 4C(i,j)\right].$$

Case 3.5. $x_0x_i > 0$, $x_0x_j < 0$, and $|x_0x_j| \leq |x_0x_i|(-x_0x_j \leq x_0x_i)$. Then, we have

$$C(i,j) = x_0^2x_j^2,$$
$$X(i,j) = \mathbb{P}\left\{|x_0x_j| \leq \sqrt{t} < x_0x_i\right\} = x_0^2x_i^2 - x_0^2x_j^2.$$

The inequality [8],

$$(1-\beta)a^2 + \beta b^2 \geq \beta(1-\beta)(a+b)^2$$

holds for $a, b \geq 0$, and $0 \leq \beta \leq 1$. Since $x_0x_i > 0$, $-x_0x_j > 0$, we obtain that

$$C(i,j) + \beta X(i,j) = x_0^2x_j^2 + \beta\left(x_0^2x_i^2 - x_0^2x_j^2\right)$$
$$= \beta x_0^2x_i^2 + (1-\beta)x_0^2x_j^2$$
$$\geq \beta(1-\beta)\left(x_0x_i - x_0x_j\right)^2$$
$$= \beta(1-\beta)\left(x_0^2x_i^2 + x_0^2x_j^2 - 2x_0^2x_ix_j\right)$$
$$= \beta(1-\beta)x_0^2\left(x_i^2 + x_j^2 + 2x_0x_i - 2x_0x_j - 2x_ix_j\right)$$
$$- \beta(1-\beta)x_0^2\left(2x_0x_i - 2x_0x_j\right).$$

Since $x_0x_i > 0$ and $x_0x_j < 0$, we have $x_0x_i - x_0x_j > 0$, $x_0x_i + x_0x_j > 0$, and

$$x_0^2\left(2x_0x_i - 2x_0x_j\right) \leq -x_0x_j\left(2x_0x_i - 2x_0x_j\right)$$
$$= 2\left(x_0^2x_j^2 + x_0x_i(-x_0x_j)\right)$$
$$\leq 2\left(x_0^2x_j^2 + x_0^2x_i^2\right))$$
$$= 2\left(x_0^2x_j^2 + x_0^2x_i^2 - x_0^2x_j^2 + x_0^2x_j^2\right))$$
$$= 2X(i,j) + 4C(i,j).$$

Therefore,

$$-\beta(1-\beta)x_0^2\left(2x_0x_i - 2x_0x_j\right) \geq -\beta(1-\beta)\left[2X(i,j) + 4C(i,j)\right].$$

Finally,

$$C(i,j) + \beta X(i,j) \geq \beta(1-\beta)x_0^2\left(x_i^2 + x_j^2 + 2x_0x_i - 2x_0x_j - 2x_ix_j\right)$$
$$- \beta(1-\beta)\left[2X(i,j) + 4C(i,j)\right].$$

Case 3.6. $x_0x_i > 0$, $x_0x_j < 0$, and $|x_0x_i| \leq |x_0x_j|(x_0x_i \leq -x_0x_j)$. Then, we have

$$C(i,j) = x_0^2x_i^2,$$
$$X(i,j) = \mathbb{P}\left\{|x_0x_i| \leq \sqrt{t} < -x_0x_j\right\} = x_0^2x_j^2 - x_0^2x_i^2,$$

and

$$
\begin{aligned}
C(i,j) + \beta X(i,j) &= x_0^2 x_i^2 + \beta \left(x_0^2 x_j^2 - x_0^2 x_i^2 \right) \\
&= \beta x_0^2 x_j^2 + (1 - \beta) x_0^2 x_i^2 \\
&\geq \beta(1 - \beta) \left(x_0 x_j - x_0 x_i \right)^2 \\
&= \beta(1 - \beta) \left(x_0^2 x_i^2 + x_0^2 x_j^2 - 2 x_0^2 x_i x_j \right) \\
&= \beta(1 - \beta) x_0^2 \left(x_i^2 + x_j^2 + 2 x_0 x_i - 2 x_0 x_j - 2 x_i x_j \right) \\
&\quad - \beta(1 - \beta) x_0^2 \left(2 x_0 x_i - 2 x_0 x_j \right).
\end{aligned}
$$

It is easy to get $x_0 x_i - x_0 x_j > 0$, $x_0 x_i + x_0 x_j > 0$, and

$$
\begin{aligned}
x_0^2 \left(2 x_0 x_i - 2 x_0 x_j \right) &\leq x_0 x_i \left(2 x_0 x_i - 2 x_0 x_j \right) \\
&= 2 \left(x_0^2 x_i^2 + x_0 x_i (-x_0 x_j) \right) \\
&\leq 2 \left(x_0^2 x_i^2 + x_0^2 x_j^2 \right) \\
&= 2 \left(x_0^2 x_i^2 + x_0^2 x_j^2 - x_0^2 x_i^2 + x_0^2 x_i^2 \right) \\
&= 2 X(i,j) + 4 C(i,j).
\end{aligned}
$$

Therefore,

$$
-\beta(1 - \beta) x_0^2 \left(2 x_0 x_i - 2 x_0 x_j \right) \geq -\beta(1 - \beta) \left[2 X(i,j) + 4 C(i,j) \right].
$$

Finally,

$$
\begin{aligned}
C(i,j) + \beta X(i,j) &\geq \beta(1 - \beta) x_0^2 \left(x_i^2 + x_j^2 + 2 x_0 x_i - 2 x_0 x_j - 2 x_i x_j \right) \\
&\quad - \beta(1 - \beta) \left[2 X(i,j) + 4 C(i,j) \right].
\end{aligned}
$$

Case 3.7. $x_0 x_i < 0$, $x_0 x_j > 0$, and $|x_0 x_j| \leq |x_0 x_i| (x_0 x_j \leq -x_0 x_i)$. Then, we have

$$
C(i,j) = x_0^2 x_j^2,
$$

$$
X(i,j) = \mathbb{P}\left\{ |x_0 x_j| \leq \sqrt{t} < -x_0 x_i \right\} = x_0^2 x_i^2 - x_0^2 x_j^2,
$$

and

$$
\begin{aligned}
C(i,j) + \beta X(i,j) &= x_0^2 x_j^2 + \beta \left(x_0^2 x_i^2 - x_0^2 x_j^2 \right) \\
&= \beta x_0^2 x_i^2 + (1 - \beta) x_0^2 x_j^2 \\
&\geq \beta(1 - \beta) \left(x_0 x_i - x_0 x_j \right)^2 \\
&= \beta(1 - \beta) \left(x_0^2 x_i^2 + x_0^2 x_j^2 - 2 x_0^2 x_i x_j \right) \\
&= \beta(1 - \beta) x_0^2 \left(x_i^2 + x_j^2 + 2 x_0 x_i - 2 x_0 x_j - 2 x_i x_j \right) \\
&\quad - \beta(1 - \beta) x_0^2 \left(2 x_0 x_i - 2 x_0 x_j \right).
\end{aligned}
$$

Since $x_0x_i + x_0x_j \leq 0$, we have

$$- \beta(1 - \beta)x_0^2 \left(2x_0x_i - 2x_0x_j\right) \geq 0$$
$$\geq -\beta(1 - \beta)x_0^2 \left(2x_0x_i - 2x_0x_j\right)\left(x_0x_i + x_0x_j\right)$$
$$\geq -\beta(1 - \beta)2\left(x_0^2x_i^2 - x_0^2x_j^2\right)$$
$$= -\beta(1 - \beta)2X(i,j)$$
$$\geq -\beta(1 - \beta)\left[2X(i,j) + 4C(i,j)\right].$$

Therefore,

$$C(i,j) + \beta X(i,j) \geq \beta(1 - \beta)x_0^2 \left(x_i^2 + x_j^2 + 2x_0x_i - 2x_0x_j - 2x_ix_j\right)$$
$$- \beta(1 - \beta)\left[2X(i,j) + 4C(i,j)\right].$$

Case 3.8. $x_ix_0 < 0$, $x_jx_0 > 0$, and $|x_0x_i| \leq |x_0x_j|(-x_0x_i \leq x_0x_j)$. Then, we have

$$C(i,j) = x_0^2x_i^2,$$
$$X(i,j) = \mathbb{P}\left\{|x_0x_i| \leq \sqrt{t} < -x_0x_j\right\} = x_0^2x_j^2 - x_0^2x_i^2,$$

and

$$C(i,j) + \beta X(i,j) = x_0^2x_i^2 + \beta\left(x_0^2x_j^2 - x_0^2x_i^2\right)$$
$$= \beta x_0^2x_j^2 + (1 - \beta)x_0^2x_i^2$$
$$\geq \beta(1 - \beta)\left(x_0x_j - x_0x_i\right)^2$$
$$= \beta(1 - \beta)\left(x_0^2x_i^2 + x_0^2x_j^2 - 2x_0^2x_ix_j\right)$$
$$= \beta(1 - \beta)x_0^2\left(x_i^2 + x_j^2 + 2x_0x_i - 2x_0x_j - 2x_ix_j\right)$$
$$- \beta(1 - \beta)x_0^2\left(2x_0x_i - 2x_0x_j\right).$$

Since $x_0x_i + x_0x_j \geq 0$, and $-\left(x_0x_i + x_0x_j\right) \leq 0$, we have

$$- \beta(1 - \beta)x_0^2\left(2x_0x_i - 2x_0x_j\right) \geq 0$$
$$\geq -\beta(1 - \beta)x_0^2\left(2x_0x_i - 2x_0x_j\right)\left[-\left(x_0x_i + x_0x_j\right)\right]$$
$$\geq -\beta(1 - \beta)2\left(x_0^2x_j^2 - x_0^2x_i^2\right)$$
$$= -\beta(1 - \beta)2X(i,j)$$
$$\geq -\beta(1 - \beta)\left[2X(i,j) + 4C(i,j)\right].$$

Therefore,

$$C(i,j) + \beta X(i,j) \geq \beta(1 - \beta)x_0^2\left(x_i^2 + x_j^2 + 2x_0x_i - 2x_0x_j - 2x_ix_j\right)$$
$$- \beta(1 - \beta)\left[2X(i,j) + 4C(i,j)\right]. \qquad \square$$

By virtue of Lemma 2, we can derive the lower bound of the recoverable ratio of the spectral partitioning algorithm at every iteration.

Lemma 3. *Given a vector x obtained by Algorithm 1, for some $0 \leq \varepsilon \leq 3/4$, the indicator vector $y \in \{\mathrm{sgn}(x_0), 0, -\mathrm{sgn}(x_0)\}^n \setminus \{0\}^n$ given by Algorithm 2. satisfies*

$$\frac{\mathrm{Desired}(y)}{\mathrm{Incident}(y)} \geq f(\varepsilon) = \begin{cases} \dfrac{1}{2\left(2\sqrt{2\varepsilon(1-2\varepsilon)}+1\right)}, & 0 < \varepsilon \leq 0.1773, \\[3mm] \dfrac{8(1-\varepsilon)^2+2(1-\varepsilon)\sqrt{16(1-\varepsilon)^2+1}}{(7-4\varepsilon)\sqrt{16(1-\varepsilon)^2+1}+16\varepsilon^2-44\varepsilon+29}, & 0.1773 \leq \varepsilon < 0.75. \end{cases}$$

$$(12)$$

Remark 1. From Lemmas 1 and 3, we obtain that for a given directed graph G with $\mathrm{OPT}_{\mathrm{MaxDC}} \geq (1 - \varepsilon)W(A)$, a tripartition (V_+, V_0, V_-) can be found by the spectral partitioning algorithm. The recoverable ratio of the tripartition is at least $f(\varepsilon)$. The function $f(\varepsilon)$ beats $1/4$ when $\varepsilon = 0.2113$.

3   The Spectral Partitioning Algorithm

The spectral partitioning algorithm for the maximum directed cut problems is depicted as Algorithm 3. below.

Algorithm 3. *We present the spectral partitioning algorithm of the MaxDC problem as follows.*

Input: *A directed graph (V, A), with nonnegative weight $w : E \to \mathbb{R}^+$.*
Output: *A bipartition (V_+, V_-) of V.*
Step 1. *Obtain a vector x by using Algorithm 1;*
Step 2. *Using   Algorithm   2,   we   get   an   indicator   vector*
  $y \in \{\mathrm{sgn}(x_0), 0, -\mathrm{sgn}(x_0)\}^V \setminus \{0\}^V$.
  1.   **If**

$$\frac{1}{2}\left(\mathrm{Cut}(y) + \frac{1}{2}\mathrm{Cross}(y)\right) < \frac{1}{4}\mathrm{Incident}(y)$$

  then *return a bipartition (V_+, V_-) of V such that the weight of desired edges is at least a quarter of $w(A)$ (a random cut suffices)* **else** *let (V_+, V_0, V_-) be the tripartition induced by y.*
  2.   **If** $V_0 = \emptyset$ **then** *return (V_+, V_-)* **else** *set $(W_+, W_-) \leftarrow \mathrm{ALG}(V_0, A[V_0])$.*
  3.   *Return the best of $(V_+ \cup W_+, V_- \cup W_-)$, $(V_+ \cup W_-, V_- \cup W_+)$, $(V_- \cup W_-, V_+ \cup W_+)$ and $(V_- \cup W_+, V_+ \cup W_-)$*
  4.   **end if**
  5.   **end if**

Theorem 1. *For a given directed graph with $\mathrm{OPT}_{\mathrm{MaxDC}} = (1 - \varepsilon)W(A) > 0$, a bipartition (V_+, V_-) can be finally returned by iterating Algorithm 1. The indicator vector $y \in \{\mathrm{sgn}(x_0), -\mathrm{sgn}(x_0)\}^V$ of the final bipartition satisfies*

$$\frac{\mathrm{Desired}(y)}{W(A)} \geq \int_0^1 \max\left(\frac{1}{4}, f\left(\frac{\varepsilon}{r}\right)\right) dr,$$

where $f(\cdot)$ is in the form of (12).

Theorem 2. *Algorithm 3. is a 0.272-approximation algorithm for the maximum directed cut problems.*

4   Discussions

We propose an approximation algorithm for the maximum directed cut problem based on spectral partitioning, which is expected to have wider application to other graph partitioning problems, such as Max-SAT.

Acknowledgments. The first author is supported by Beijing Excellent Talents Funding (No. 2014000020124G046). The second author's research is supported by Natural Sciences and Engineering Research Council of Canada (NSERC) grant 283106. The third author's research is supported by NSFC (No. 11501412). The fourth author's research is supported by NSFC (No. 11531014). The fifth author is supported by Shandong Jianzhu University grant Z0013.

References

1. Goemans, M.X., Williamson, D.P.: Improved approximation algorithms for maximum cut and satisfiability problems using semidefinite programming. J. ACM **42**, 1115–1145 (1995)
2. Trevisan, L.: Max cut and the smallest eigenvalue. SIAM. Comput. **41**, 1769–1786 (2012)
3. Soto, A.: Improved analysis of max-cut algorithm based on spectral partitioning. SIAM J. Diecrete Math. **29**, 259–268 (2015)
4. Kale, S., Seshadhri, C.: Combinatorial approximation algorithms for MaxCut using Random Walks. preprint, arXiv:1008.3938 (2010)
5. Lewin, M., Livnat, D., Zwick, U.: Improved rounding techniques for the MAX 2-SAT and MAX DI-CUT Problems. In: Cook, W.J., Schulz, A.S. (eds.) IPCO 2002. LNCS, vol. 2337, pp. 67–82. Springer, Heidelberg (2002). https://doi.org/10.1007/3-540-47867-1_6
6. Nikiforov, V.: Max k-cut and the smallest eigenvalue. Linear Algebra Appl. **504**, 462–467 (2016)
7. Feige, U., Jozeph, S.: Oblivious algorithms for the maximum directed cut problem. Algorithmica **71**, 409–428 (2015)
8. Beckenbach, E., Bellman, R.: An Introduction to Inequalities. Random House, New York (1961)

Better Approximation Ratios for the Single-Vehicle Scheduling Problems on Tree/Cycle Networks

Yuanxiao Wu[ID] and Xiwen Lu[✉][ID]

East China University of Science and Technology, Shanghai, China
yxwu0212@163.com, xwlu@ecust.edu.cn

Abstract. We investigate the single vehicle scheduling problems based on tree/cycle networks. Each customer, assumed as a vertex on the given network, has a release time and a service time requirements. The single vehicle starts from the depot and aims to serve all the customers. The objective of the problem is to find the relatively optimal routing schedule so as to minimize the makespan. We provide a $\frac{16}{9}$-approximation algorithm and a $\frac{48}{25}$-approximation algorithm for the tour-version and the path-version of single vehicle scheduling problem on a tree, respectively. For the tour-version of single vehicle scheduling problem on a cycle, we present a $\frac{5}{3}$-approximation algorithm.

Keywords: Vehicle · Routing · Scheduling · Network · Approximation algorithm

1 Introduction

The single vehicle scheduling problem (SVSP) consists of a set of customers situated at different vertices on a given network and a single vehicle initially located at a fixed depot. Each customer has a release time before which it cannot be served, and a service time which the vehicle has to spend in serving the customer. The vehicle, required to serve all the customers, takes a travel time when it travels from one customer to another. The completion time of a customer is defined as the time by which it has been served completely, while the completion time of the vehicle means the time by which it has served all the customers and returned to its initial location. A permutation of the customers, which implies the customer service order and can specify the routing of the vehicle, is considered as a schedule for the problem. The problem aims to find a schedule to minimize the makespan. We distinguish two versions. In the first one, which is called tour-version, the makespan is defined as the completion time of the vehicle. In the other one, which is known as path-version, the makespan means the completion time of the last served customer. For convenience, when the network is restricted to a line (resp. tree, cycle), we denote the single vehicle scheduling problem by L-SVSP (resp. T-SVSP, C-SVSP). When the service time of each

© Springer International Publishing AG 2017
X. Gao et al. (Eds.): COCOA 2017, Part I, LNCS 10627, pp. 313–323, 2017.
https://doi.org/10.1007/978-3-319-71150-8_27

customer is zero, SVSP is known as single vehicle routing problem (SVRP) in some paper. Thus we denote SVRP on a line, tree and cycle by L-SVRP, T-SVRP and C-SVRP, respectively.

There are plenty of results on VRP and VSP. [1] showed that both the tour-version and the path version of L-SVSP are ordinarily NP-hard. [2] presented polynomial time algorithms for both versions of L-SVRP. [3] provided a $\frac{3}{2}$-approximation algorithm for the tour-version of L-SVSP in the case that the vehicle initial locates at an extreme vertex, and [4] gave a $\frac{5}{3}$-approximation algorithm for the counterpart in which the initial location of the vehicle is arbitrary. [5,6] presented a $\frac{3}{2}$-approximation algorithm for the tour-version of L-SVSP and a $\frac{5}{3}$-approximation algorithm for the path-version of L-SVSP with no constraint on the initial location of the vehicle, and [6] also provided examples to show that the performance ratios are tight. When it comes to multi-vehicle scheduling problem (MVSP), [7] provided a 2-approximation algorithm for the path-version of the general L-MVSP in which the initial location of each vehicle is arbitrary. [8] showed that both the tour-version and path-version of L-MVRP can be solved in polynomial time.

It has been shown in [9] that both the tour-version and the path-version of T-SVRP (and hence T-SVSP) are ordinarily NP-hard. For the tour-version of T-SVSP, they showed that the problem can be exactly solved in $O(n \log n)$ time if adding a constraint that the vehicle has to process all tasks in a depth-first manner, and the $O(n \log n)$ time algorithm can be considered as a 2-approximation algorithm for the T-SVSP without the depth-first constraint. [10] ultimately proved that both versions of T-SVSP are strongly NP-hard. For the tour-version, they provided a $O(n^b)$ time DP algorithm where b is the number of leaves. [5] introduced an improved $\frac{11}{6}$-approximation algorithm for the tour-version of T-SVSP. In their algorithm, they partitioned the set of customers into two subsets. The vehicle first serves part of customers in one subset, and then serves all the remaining customers. They also presented a $\frac{9}{5}$-approximation algorithm with a similar method for the tour-version of C-SVSP. [11] provided a $\frac{9}{5}$-approximation algorithm and a $\frac{27}{14}$-approximation algorithm for the tour-version and the path-version of T-SVSP, respectively. For both tour-version and path-version of C-SVSP, they provided a $\frac{12}{7}$-approximation algorithm. [12] presented polynomial time approximation schemes for both versions of SVSP on a tree with a constant number of leaves. [13] presented a 3-approximation algorithm for MVSP on a tree and a $(5 - \frac{2}{m})$-approximation algorithm for that on a general network.

In this paper, we consider both the tour-version and the path-version of T-SVSP and the tour-version of C-SVSP. For each problem, we propose an approximation algorithm and prove its performance ratio. Our algorithms are improvements on those in [11].

The rest of this paper is structured as follows. In Sect. 2, we introduce the formulation of T-SVSP and some notations. In Sect. 3, we present approximation algorithms for two versions of T-SVSP and analyse the performance ratio. In Sect. 4, we describe an approximation algorithm for the tour-version of C-VSP and prove the performance ratio. Finally we give some concluding remarks in Sect. 5.

2   Problem Formulation and Preliminaries

The SVSP discussed in this paper is mathematically defined as follows. Let $G = (V \cup \{0\}, E)$ be an undirected network where $V = \{1, 2, ..., n\}$ is a vertex set and E is a set of edges. An edge $e \in E$ is an unordered pair (j, k) of two vertices in V, where j, k are called the endpoints of e. The travel time of the vehicle is $t_{j,k} \geq 0$, which is the time to traverse edge $e = (j, k)$ from j to k, and $t_{k,j} = t_{j,k}$. When $(j, k) \notin E$, $t_{k,j}$ denotes the travel time for the vehicle travelling along the shortest path from j to k. There is a unique customer i at each vertex $i \in V$. Unless ambiguity would result, we do not distinguish between vertex and customer. There is a vehicle initially located at the depot 0 to serve all the costumers. Each customer i is associated with a release time r_i and a service time p_i. It means that the vehicle cannot start serving customer i before r_i, and needs p_i time units to finish its service. The vehicle arriving at a vertex i before r_i either waits until r_i to serve the customer v_i, or moves to other vertices without serving v_i if it is more advantageous (in this case, the vehicle has to come back to i later to serve customer i). A routing schedule of the vehicle is specified by a sequence $\pi = (\pi(1), \pi(2), ..., \pi(n))$ of customers to be served, i.e. the vehicle travels along a shortest path from 0 to $\pi(1)$ in G, taking the travel time of the length of the path, waits until $r_{\pi(1)}$ if the arrival time is before $t = r_{\pi(1)}$ and serve the customer $\pi(1)$. After serving $\pi(1)$, it immediately moves to $\pi(2)$, waits until $r_{\pi(2)}$ if the arrival time is before $t = r_{\pi(2)}$ and serve the customer $\pi(2)$, and so on. In the following, for any feasible schedule π, we always assume $\pi(0) = \pi(n+1) = 0$. Let $C_{[i]}(\pi)$ denote the service completion time of customer $\pi(i)$ in π, and set $C_{[0]}(\pi) = 0$. Then, $C_{[i]}(\pi)$ equals to $\max\{r_{\pi(i)}, C_{[i-1]}(\pi) + t_{\pi(i-1),\pi(i)}\} + p_{\pi(i)}$ for all $i = 1, 2, ..., n$. The makespan of π is donoted by $C_{max}^{tour}(\pi)$ in the tour-version, and $C_{max}^{path}(\pi)$ in the path-version. Then, $C_{max}^{tour}(\pi) = C_{[n]}(\pi) + t_{0,\pi(n)}$, $C_{max}^{path}(\pi) = C_{[n]}(\pi)$.

We now introduce some notations to be used throughout the article. Let

$$t_{max} = max_{1 \leq i \leq n} t_{0,i}$$
$$L = \sum_{(i,j) \in E} t_{i,j}$$
$$r_{max} = max_{1 \leq i \leq n} r_i$$
$$P = \sum_{i=1}^{n} p_i$$

(1)

and for $0 \leq t \leq r_{max}$,

$$V(t) = \{i \in V \mid r_i \geq t\}$$
$$P(t) = \sum_{i \in V(t)} p_i$$
$$V'(t) = \{i \in V \mid r_i > t\}$$
$$P'(t) = \sum_{i \in V'(t)} p_i$$

(2)

Note that $P(t)$ and $P'(t)$ are piecewise constant functions of t, and different only at $r_j (j = 1, 2, ..., n)$.

For $0 \leq t \leq r_{max}$, let $v(t)$ and $v'(t)$ denote the farthest vertices from vertex 0 in $V(t)$ and $V'(t)$, respectively. We also define

θ: the shortest travelling tour over $V \cup \{0\}$

$\delta_0'(t)$: the shortest travelling path over $V'(t) \cup \{0\}$ starting from the depot 0.

When $G = (V \cup \{0\}, E)$ is a tree, for any $U \subset V \cup \{0\}$, we define the spanning subtree on U as the smallest subtree of G which contains all the vertices in U. Furthermore, for $0 \leq t \leq r_{max}$, we let $T(t)$, $T'(t)$, $\widehat{T}(t)$ and $\widehat{T}'(t)$ denote the spanning subtree on the vertex sets $V(t)$, $V'(t)$, $V(t) \cup \{0\}$ and $V'(t) \cup \{0\}$, respectively.

In the following, a symbol denoting a subgraph of G also be considered as its length, i.e. if G' is a subgraph of G, then $G' = \sum_{(j,k) \in G'} t_{j,k}$.

3   SVSP on Tree Network

In this section we consider T-SVSP. Without loss of generality, we consider the depot 0 as the root of the tree. A $\frac{16}{9}$-approximation algorithm for the tour-version is presented in Sect. 3.1, and a $\frac{48}{25}$-approximation algorithm for the path-version is provided in Sect. 3.2.

3.1   Tour-Version of T-SVSP

Karuno et al. [3] provided a $\frac{11}{6}$-approximation algorithm for the tour-version of T-SVSP. They first gave several different candidate schedules and then chose the best one as the approximation solution. Yu and Liu [6] proved that the tour version of T-SVSP has an r-approximation algorithm if the tour version of T-SVRP has. Then they presented a $\frac{9}{5}$-approximation algorithm for the tour-version of T-SVRP. We will adopt a similar approach and show the approximation ratio can be reduced to $\frac{16}{9}$.

Algorithm 1 for the tour-version of T-SVRP

Step 1. We define σ as a tour on the tree in which the vehicle starts out from the depot 0 and visits $v(0)$ first, and then visits other customers in a depth-first sequence, and finally returns to the depot. Construct a schedule π_1 such that the service order of the customers is the same as the visiting order of the customers in σ.

Step 2. Let x denote the point that is $L + \frac{t_{max}}{2}$ time units away from the depot 0 along σ. Let $t^* = max\{2L, r_{max}\}$. Construct a schedule π_2 such that the vehicle first waits at vertex 0 for $t^* - L - \frac{t_{max}}{2}$ time units, and then travels along σ without serving any customer until it arrives at $v(0)$, and then travels along σ from vertex $v(0)$ to point x to serve the customers in $V \backslash V'(t^* - L + \frac{t_{max}}{2})$, and then travels to vertex 0 to serve all the remaining customers in a depth-first sequence.

Step 3. Construct a schedule π_3 such that the vehicle first waits at vertex 0 for $t^* - L + \frac{t_{max}}{2}$ time units, and then travels reverse of σ to point x to serve the customers in $V \backslash V'(t^* - L + \frac{t_{max}}{2})$, and then travels to vertex 0 to serve all the remaining customers in a depth-first sequence.

Step 4. Choose the best one among π_1, π_2 and π_3 as the approximation solution π.

In the proof of the following theorem, we first propose several upper bounds on the makespan of π_1, π_2 and π_3, and then prove that Algorithm 1 is a $\frac{16}{9}$-approximation algorithm to the tour-version of T-SVSP. Let π^* denote the optimal schedule for the tour-version of T-SVRP.

Theorem 1. *Algorithm 1 is a $\frac{16}{9}$-approximation algorithm to the tour-version of T-SVRP.*

Proof. The proof will be presented in three steps.

Step 1: We prove an upper bound on $C_{max}^{tour}(\pi_1)$.

In π_1, the vehicle either waits at some customers for their release or doesn't wait at any vertex. If the vehicle waits at some customers, assume customer j is the last customer where the vehicle waits. Because the first served customer is $v(0)$, the travel time of the vehicle before it arriving at j is at least $t_{0,v(0)} + t_{v(0),j}$. The total travel time of the vehicle in π_1 is $2L$. Then, the travel time of the vehicle after it serves customer j is no more than $2L - t_{0,v(0)} - t_{v(0),j}$. Thus,

$$
\begin{aligned}
C_{max}^{tour}(\pi_1) &\leq r_j + 2L - t_{0,v(0)} - t_{v(0),j} \\
&= r_j + t_{0,j} + 2L - t_{0,v(0)} - t_{v(0),j} - t_{0,j} \\
&\leq 2C_{max}^{tour}(\pi^*) - 2t_{max}
\end{aligned}
\tag{3}
$$

where the last equality follows $r_j + t_{0,j} \leq C_{max}^{tour}(\pi^*)$, $2L \leq C_{max}^{tour}(\pi^*)$ and $t_{max} = t_{0,v(0)} \leq t_{v(0),j} + t_{0,j}$.

If the vehicle doesn't wait at any vertex, then

$$
\begin{aligned}
C_{max}^{tour}(\pi_1) &= 2L \\
&= C_{max}^{tour}(\pi^*) \\
&\leq 2C_{max}^{tour}(\pi^*) - 2t_{max}
\end{aligned}
\tag{4}
$$

Combing the above two cases, we conclude that

$$
C_{max}^{tour}(\pi_1) \leq 2C_{max}^{tour}(\pi^*) - 2t_{max}.
\tag{5}
$$

Step 2: We show an upper bound on $min\{C_{max}^{tour}(\pi_2), C_{max}^{tour}(\pi_3)\}$.

We first consider the total waiting time of the vehicle in π_2 and π_3. In both π_2 and π_3, the time of the vehicle arriving at point x is t^*. Since $t^* \geq r_{max}$, the vehicle does not wait at any customer in the later process. Thus, the total waiting time of the vehicle in π_2 is $t^* - L - \frac{t_{max}}{2}$, and that in π_3 is $t^* - L + \frac{t_{max}}{2}$. The total waiting time of the vehicle in π_2 and π_3 is $2t^* - 2L$.

Now we turns to provide an upper bound of the total travel time of the vehicle in π_2 and π_3. Let T_1 and T_2 be subtrees visited in π_2 and π_3, respectively. It is easy to see that the travel times of the vehicle in π_2 and π_3 are no more than $2L + 2((T_1 \backslash T_2) \cap \widehat{T}'(t^* - L + \frac{t_{max}}{2}))$ and $2L + 2((T_2 \backslash T_1) \cap \widehat{T}'(t^* - L + \frac{t_{max}}{2}))$, respectively. Since $2((T_1 \backslash T_2) \cap \widehat{T}'(t^* - L + \frac{t_{max}}{2})) + 2((T_2 \backslash T_1) \cap \widehat{T}'(t^* - L + \frac{t_{max}}{2})) \leq 2\widehat{T}'(t^* - L + \frac{t_{max}}{2})$, the total travel time of the vehicle in π_2 and π_3 is at most $4L + 2\widehat{T}'(t^* - L + \frac{t_{max}}{2})$, which is no more than $4L + 2\widehat{T}'(L + \frac{t_{max}}{2})$ for $t^* - L \geq L$.

Combing the above two discussion,

$$C_{max}^{tour}(\pi_2) + C_{max}^{tour}(\pi_3) \leq 2t^* - 2L + 4L + 2\widehat{T}'(L + \tfrac{t_{max}}{2})$$
$$= 2t^* + 2L + 2\widehat{T}'(L + \tfrac{t_{max}}{2}) \tag{6}$$

A straightforward conclusion of the inequality above is showed as follows.

$$\min\{C_{max}^{tour}(\pi_2), C_{max}^{tour}(\pi_3)\}$$
$$\leq \tfrac{1}{2}(C_{max}^{tour}(\pi_2) + C_{max}^{tour}(\pi_3))$$
$$\leq t^* + L + \widehat{T}'(L + \tfrac{t_{max}}{2}) \tag{7}$$
$$\leq t^* + L + (C_{max}^{tour}(\pi^*) - L - \tfrac{t_{max}}{2} + t_{0,v(L + \tfrac{t_{max}}{2})})/2$$
$$\leq \tfrac{7}{4}C_{max}^{tour}(\pi^*) + \tfrac{t_{max}}{4}$$

where the second inequality follows $C_{max}^{tour}(\pi^*) \geq t + 2\widehat{T}(t) - t_{0,v(t)}$ and $\widehat{T}'(t) \leq \widehat{T}(t)$, the last inequality follows $t_{0,v(L + \tfrac{t_{max}}{2})} \leq t_{max}$ and $C_{max}^{tour}(\pi^*) \geq t^* \geq 2L$.

Step 3: We prove the correctness of Theorem 1.

Combining the conclusion of the above two steps,

$$\min\{C_{max}^{tour}(\pi_1), C_{max}^{tour}(\pi_2), C_{max}^{tour}(\pi_3)\}$$
$$\leq \min\{2C_{max}^{tour}(\pi^*) - 2t_{max}, \tfrac{7}{4}C_{max}^{tour}(\pi^*) + \tfrac{t_{max}}{4}\} \tag{8}$$
$$\leq \tfrac{16}{9}C_{max}^{tour}(\pi^*)$$

This completes the proof of Theorem 1.   □

[11] showed that, in linear time, each instance $\mathcal{I}$ of the tour-version of T-SVSP can be transformed into an instance $\mathcal{I}'$ of the tour-version of T-SVRP such that any r-approximation schedule π' of $\mathcal{I}'$ also can be transformed into an r-approximation schedule of $\mathcal{I}$. Therefore, we design the following $\tfrac{16}{9}$-approximation algorithm for the tour-version of T-SVSP.

Algorithm 2 for the tour-version of T-SVSP

Step 1. Given an instance $\mathcal{I} = (T = (V \cup \{0\}, E), r, t, p)$ of T-SVSP, construct an auxiliary instance $\mathcal{I}'(T' = (V' \cup \{0\}, E'), r', t')$ of T-SVRP as follows. $V' = V \cup \bigcup_{i=1}^{n} n+i$, $E' = E \cup \bigcup_{i=1}^{n} (i, n+i)$, $t'_{j,k} = t_{j,k}$ for each edge $(j,k) \in E$, $t'_{i,n+i} = \tfrac{p_i}{2}$, $r'_i = r_i$ and $r'_{n+i} = r_i + \tfrac{p_i}{2}$ for each $i \in V$.

Step 2. Call Algorithm 1 to solve auxiliary instance $\mathcal{I}'$ of T-SVRP and obtain a schedule π'.

Step 3. Construct a schedule π such that customer i is served before customer j if customer $n+i$ is served before customer $n+j$ in π' for each pair $i, j \in V$.

Theorem 2. *Algorithm 2 is a $\tfrac{16}{9}$-approximation algorithm to the tour-version of T-SVSP.*

Proof. Theorem 2 is a direct inference of Theorem 1.   □

3.2 Path-Version of T-SVSP

When it comes to the path-version of T-SVSP, notice that the gap between the makespan of a schedule for the tour-version of T-SVSP and that of the same schedule for the path-version of T-VSP is no more than t_{max}. Therefore, the approximation algorithm for the tour-version of T-VSP can also be used to solve the path-version of T-SVSP.

Algorithm 3 for the path-version of T-SVSP

Step 1. Call Algorithm 2 to obtain a schedule π_1 for the corresponding tour-version of T-SVSP.

Step 2. Let $P_{0,v(0)}$ indicate the unique path between the vertices 0 and $v(0)$ in the tree G. It is easy to see that $t_{0,v(0)} = t_{max}$. We assume that there are $m+2$ vertices $0, 1, 2, ..., m + 1 = v(0)$ on path $P_{0,v(0)}$. Then, deleting the edges in $P_{0,v(0)}$, we obtain $m+2$ subtrees, which can be described as $T^i (0 \leq i \leq m+1)$ such that T^i is connected with vertex i. Let V^i denote the vertex set of T^i. Solving the auxiliary L-SVRP on the path $P_{0,v(0)}$, where the release time of vertex i is redefined as $max\{r_j \mid j \in V^i\}$, by the dynamic programming algorithm of [2], we obtain a service order of $V^i s$. Then, serving the customers in each V^i in an arbitrary depth-first order, we obtain a schedule π_2.

Step 3. Choose the best one between π_1 and π_2 as the approximate solution π.

In the proof of the following theorem, we first propose several upper bounds on the makespan of π_1 and π_2, and then prove that Algorithm 3 is a $\frac{48}{25}$-approximation algorithm to the path-version of T-SVSP. Let π' and π^* denote the optimal schedule for the tour-version of T-SVSP and the path-version of T-SVSP, respectively.

Theorem 3. *Algorithm 3 is a $\frac{48}{25}$-approximation algorithm to the path-version of T-SVSP.*

Proof. The proof will be presented in three steps.

Step 1. We prove an upper bound on $C_{max}^{path}(\pi_1)$.

Let j denote the last customer served by the vehicle in π^*. It is easy to see that $C_{max}^{tour}(\pi^*) - C_{max}^{path}(\pi^*) \leq t_{0,j}$, and $t_{0,j} \leq t_{max}$. Then, we obtain $C_{max}^{tour}(\pi^*) \leq C_{max}^{path}(\pi^*) + t_{max}$. Thus,

$$
\begin{aligned}
C_{max}^{path}(\pi_1) &\leq C_{max}^{tour}(\pi_1) \leq \tfrac{16}{9} C_{max}^{tour}(\pi') \\
&\leq \tfrac{16}{9} C_{max}^{tour}(\pi^*) \\
&\leq \tfrac{16}{9} C_{max}^{path}(\pi^*) + \tfrac{16}{9} t_{max}
\end{aligned}
\tag{9}
$$

Step 2. We show an upper bound on $C_{max}^{path}(\pi_2)$.

It is easy to see that the optimal makespan of the auxiliary L-SVRP is a lower bound of $C_{max}^{path}(\pi^*)$. Compared with the optimal makespan of L-SVRP, the makespan of π_2 increases at most $2(L - t_{max}) + P$ time units for travelling the subtrees $T^0, T^1, ..., T^{m+1}$ and serving all the customers. Then, we obtain

$$C_{max}^{path}(\pi_2) \leq C_{max}^{path}(\pi^*) + 2(L - t_{max}) + P$$
$$\leq 2C_{max}^{path}(\pi^*) - t_{max} \tag{10}$$

Step 3. We prove the correctness of Theorem 3.

$$\min\{C_{max}^{path}(\pi_1), C_{max}^{path}(\pi_2)\}$$
$$\leq \min\{\tfrac{16}{9}C_{max}^{path}(\pi^*) + \tfrac{16}{9}t_{max}, 2C_{max}^{path}(\pi^*) - t_{max}\}$$
$$\leq \tfrac{48}{25}C_{max}^{path}(\pi^*) \tag{11}$$

This completes the proof of Theorem 3.   □

4   SVSP on Cycle Network

We now turn to the tour version of C-SVSP. Let $G = (V \cup \{0\})$ be a cycle, where the vertices in $V \cup \{0\}$ are numbered increasingly in the counterclockwise order. In the following of this section, we consider vertex $n + 1$ as vertex 0.

Recall the definitions of $L, \theta, \delta_0'(t)$ in Sect. 2. We assume that there is no edge $(i, i+1)$ such that $t_{i,i+1} \geq \tfrac{1}{2}L$. Otherwise there exists an optimal schedule which never goes through the edge $(i, i+1)$, thus the tour-version of C-SVSP can be considered as a tour-version of L-SVSP. Since [6] showed a $\tfrac{3}{2}$-approximation algorithm for the tour-version of L-SVSP, we focus on the situation satisfying $t_{i,i+1} < \tfrac{1}{2}L$ for all $i = 0, 1, ..., n$, which implies $\theta = L$. Suppose that $V'(t) = \{i_1, i_2, ..., i_k\}$ with $i_1 < i_2 < ... < i_k$, and $i_0 = i_{k+1} = 0$. Then $\delta_0'(t)$ is the shortest one among the following paths:

(i) for $0 \leq j \leq k - 1$, the paths first from i_0 to i_j in the counterclockwise direction, and then from i_j to i_{j+1} in the clockwise direction;
(ii) for $2 \leq j \leq k + 1$, the paths first from i_0 to i_j in the clockwise direction, and then from i_j to i_{j-1} in the counterclockwise direction.

Then, we provide a $\tfrac{5}{3}$-approximation algorithm for the tour-version of C-SVSP as follows.

Algorithm 4 for the tour-version of C-SVSP

Step 1. Solve the corresponding C-SVRP by the dynamic programming algorithm of [10] to generate a schedule π_1.

Step 2. Find $t^*(0 \leq t^* \leq r_{max})$ such that $P'(t^*) + \delta_0'(t^*) \leq 2t^* \leq P(t^*) + \delta_0(t^*)$. Partition the customers into $V \backslash V'(t^*)$ and $V'(t^*)$. Let $v_{\delta_0'(t^*)}$ denote the other end point differing from vertex 0 in path $\delta_0'(t^*)$. Construct π_2 in which the vehicle first travels to $v_{\delta_0'(t^*)}$ and waits until t^*(if necessary), then goes through the cycle to serve all the customers in $V \backslash V'(t^*)$ before it arrives $v_{\delta_0'(t^*)}$ again, and travels along $\delta_0'(t^*)$ to serve the customers in $V'(t^*)$.

Step 3. Choose the best one among π_1 and π_2 as the approximate solution π.

In the proof of the following theorem, we first propose several upper bounds on the makespan of π_1 and π_2, and then prove that Algorithm 4 is a $\frac{5}{3}$-approximation algorithm to the tour-version of T-SVSP. Let π^* denote the optimal schedule for the tour-version of C-SVSP.

Theorem 4. *Algorithm 4 is an $\frac{5}{3}$-approximation algorithm to the tour-version of C-SVSP.*

Proof. The proof will be presented in three steps.

Step 1. We show that $C_{max}^{tour}(\pi_1) \leq C_{max}^{tour}(\pi^*) + P$.

The optimal makespan of the C-SVRP is a lower bound of $C_{max}^{tour}(\pi^*)$. As a solution to C-SVSP, π_1 increases at most P time units in makespan than as an optimal solution to C-SVRP. Then, $C_{max}^{tour}(\pi_1) \leq C_{max}^{tour}(\pi^*) + P$.

Step 2. We prove that $C_{max}^{tour}(\pi_2) \leq max\{\frac{7}{3}C_{max}^{tour}(\pi^*) - P, \frac{5}{2}C_{max}^{tour}(\pi^*) - \frac{3}{2}P, \frac{5}{3}C_{max}^{tour}(\pi^*)\}$.

The vehicle does not wait at any customer in $V \backslash V'(t^*)$, because it starts out from vertex $v_{\delta'_0(t^*)}$ at or later than time t^*. If the vehicle does not wait at any customer in $V'(t^*)$, we have

$$\begin{aligned} C_{max}^{tour}(\pi_2) &\leq max\{t^*, t_{0, v_{\delta'_0(t^*)}}\} + \theta + \delta'_0(t^*) + P \\ &\leq max\{\tfrac{1}{3}C_{max}^{tour}(\pi^*), \tfrac{1}{2}\theta\} + 2\theta + P \qquad (12) \\ &\leq max\{\tfrac{7}{3}C_{max}^{tour}(\pi^*) - P, \tfrac{5}{2}C_{max}^{tour}(\pi^*) - \tfrac{3}{2}P\} \end{aligned}$$

where the second inequality follows from $C_{max}^{tour}(\pi^*) \geq t^* + \delta_0(t^*) + P(t^*) \geq 3t^*$, $t_{0, v'(t^*)} \leq \frac{1}{2}\theta$ and $\delta'_0(t^*) \leq \theta$, and the last inequality follows from $C_{max}^{tour}(\pi^*) \geq \theta + P$.

If the vehicle waits at some customer in $V'(t^*)$, let k be the last customer where the vehicle waits. Then we have

$$\begin{aligned} C_{max}^{tour}(\pi_2) &\leq r_k + P'(t^*) + \delta'_0(t^*) \\ &\leq \tfrac{5}{3}C_{max}^{tour}(\pi^*) \end{aligned} \qquad (13)$$

where the last inequality follows from $C_{max}^{tour}(\pi^*) \geq r_i + t_{0,i}$, *for any* $i \in V$ and $C_{max}^{tour}(\pi^*) \geq t^* + \delta_0(t^*) + P(t^*) \geq \frac{3}{2}(\delta'_0(t^*) + P'(t^*))$.

Step 3. We prove the minimum one between the makespans of π_1 and π_2 is at most $\frac{5}{3}C_{max}^{tour}(\pi^*)$.

$$\begin{aligned} &min\{C_{max}^{tour}(\pi_1), C_{max}^{tour}(\pi_2)\} \\ &\leq min\{C_{max}^{tour}(\pi^*) + P, max\{\tfrac{7}{3}C_{max}^{tour}(\pi^*) - P, \tfrac{5}{2}C_{max}^{tour}(\pi^*) - \tfrac{3}{2}P, \tfrac{5}{3}C_{max}^{tour}(\pi^*)\}\} \\ &= \tfrac{5}{3}C_{max}^{tour}(\pi^*) \end{aligned}$$

$$(14)$$

This completes the proof. $\square$

5   Conclusions

In this paper, we consider the single-vehicle scheduling problems on a tree and a cycle, and all of these problems are known to be NP-hard. For the tour-version and the path-version of T-SVSP, we present a $\frac{16}{9}$-approximation algorithm and a $\frac{48}{25}$-approximation algorithm, respectively. We also consider the tour-version of single-vehicle scheduling problem on a cycle, and give a $\frac{5}{3}$-approximation algorithm. Our algorithms improve the previous best results in the literature. The main idea used in the improved algorithms is to utilize the structure characteristics of tree and cycle to antedate the department time of the vehicle.

There are some issues that are still unsolved. We would like to know whether the approximation bounds obtained in this paper are tight. As a natural extension of this paper, researchers may study SVSP on a general network. If the general network satisfies the triangle inequality, it is straightforward to design a $\frac{5}{2}$-approximation algorithm. A better approximation bound is desirable.

Acknowledgement. The authors would like to thank the associated editor and the anonymous referees for their constructive comments and kind suggestions. This research was supported by the National Natural Science Foundation of China under Grant No. 11371137.

References

1. Tsitsiklis, J.N.: Special cases of traveling salesman and repairman problems with time windows. Networks **22**, 263–282 (1992). https://doi.org/10.1002/net.3230220305
2. Psaraftis, H.N., Solomon, M.M., Magnanti, T.L., Kim, T.-U.: Routing and scheduling on a shoreline with release times. Manage. Sci. **36**, 212–223 (1990). https://doi.org/10.1287/mnsc.36.2.212
3. Karuno, Y., Nagamochi, H., Ibaraki, T.: Better approximation ratios for the single-vehicle scheduling problems on line-shaped networks. Networks **39**(4), 203–209 (2002). https://doi.org/10.1002/net.10028
4. Gaur, D.R., Gupta, A., Krishnamurti, R.: A $\frac{5}{3}$-approximation algorithm for scheduling vehicles on a path with release and handling times. Inform. Process. Lett. **86**, 87–91 (2003). https://doi.org/10.1016/S0020-0190(02)00474-X
5. Bhattacharya, B., Carmi, P., Hu, Y., Shi, Q.: Single vehicle scheduling problems on Path/Tree/Cycle Networks with release and handling times. In: Hong, S.-H., Nagamochi, H., Fukunaga, T. (eds.) ISAAC 2008. LNCS, vol. 5369, pp. 800–811. Springer, Heidelberg (2008). https://doi.org/10.1007/978-3-540-92182-0_70
6. Yu, W., Liu, Z.: Single vehicle scheduling problems with release and service times on a line. Networks **57**, 128–134 (2011). https://doi.org/10.1002/net.20393
7. Karuno, Y., Nagamochi, H.: 2-approximation algorithms for the multi-vehicle scheduling prbolem on a path with release and handling times. Discrete Appl. Math. **129**, 433–447 (2003). https://doi.org/10.1016/S0166-218X(02)00596-6
8. Yu, W., Liu, Z.: Vehicle routing problems on a line-shaped network with release time constraints. Oper. Res. Lett. **37**, 85–88 (2009). https://doi.org/10.1016/j.orl.2008.10.006
9. Karuno, Y., Nagamochi, H., Ibaraki, T.: Vehicle scheduling on a tree with release and handling times. In: Ng, K.W., Raghavan, P., Balasubramanian, N.V., Chin, F.Y.L. (eds.) ISAAC 1993. LNCS, vol. 762, pp. 486–495. Springer, Heidelberg (1993). https://doi.org/10.1007/3-540-57568-5_280

10. Nagamochi, H., Mochizuki, K., Ibaraki, T.: Complexity of the single vehicle scheduling problem on graphs. Inform. Syst. Oper. Res. **35**, 256–276 (1997). https://doi.org/10.1080/03155986.1997.11732334

11. Bao, X., Liu, Z.: Approximation algorithms for single vehicle scheduling problems with release and service times on a tree or cycle. Theoret. Comput. Sci. **434**, 1–10 (2012). https://doi.org/10.1016/j.tcs.2012.01.046

12. Augustine, J.E., Seiden, S.S.: Linear time approximation schemes for vehicle scheduling problems. Theoret. Comput. Sci. **324**, 147–160 (2004). https://doi.org/10.1016/j.tcs.2004.05.013

13. Bhattacharya, B., Hu, Y.: Approximation algorithms for the multi-vehicle scheduling problem. In: Cheong, O., Chwa, K.-Y., Park, K. (eds.) ISAAC 2010. LNCS, vol. 6507, pp. 192–205. Springer, Heidelberg (2010). https://doi.org/10.1007/978-3-642-17514-5_17

An Efficient Primal-Dual Algorithm for Fair Combinatorial Optimization Problems

Viet Hung Nguyen[1] and Paul Weng[2,3,4(✉)]

[1] Sorbonne Universités, UPMC Univ Paris 06, UMR 7606, LIP6, Paris, France
Hung.Nguyen@lip6.fr
[2] SYSU-CMU Joint Institute of Engineering, Guangzhou, China
[3] School of Electronics and Information Technology, SYSU, Guangzhou, China
paweng@mail.sysu.edu.cn
[4] SYSU-CMU Joint Research Institute, Shunde, China

Abstract. We consider a general class of combinatorial optimization problems including among others allocation, multiple knapsack, matching or travelling salesman problems. The standard version of those problems is the maximum weight optimization problem where a sum of values is optimized. However, the sum is not a good aggregation function when the fairness of the distribution of those values (corresponding for example to different agents' utilities or criteria) is important. In this paper, using the Generalized Gini Index (GGI), a well-known inequality measure, instead of the sum to model fairness, we formulate a new general problem, that we call fair combinatorial optimization. Although GGI is a non-linear aggregating function, a $0, 1$-linear program (IP) can be formulated for finding a GGI-optimal solution by exploiting a linearization of GGI proposed by Ogryczak and Sliwinski [21]. However, the time spent by commercial solvers (e.g., CPLEX, Gurobi...) for solving (IP) increases very quickly with instances' size and can reach hours even for relatively small-sized ones. As a faster alternative, we propose a heuristic for solving (IP) based on a primal-dual approach using Lagrangian decomposition. We demonstrate the efficiency of our method by evaluating it against the exact solution of (IP) by CPLEX on several fair optimization problems related to matching. The numerical results show that our method outputs in a very short time efficient solutions giving lower bounds that CPLEX may take several orders of magnitude longer to obtain. Moreover, for instances for which we know the optimal value, these solutions are quasi-optimal with optimality gap less than 0.3%.

Keywords: Fair optimization · Generalized Gini Index · Ordered weighted averaging · Matching · Subgradient method

1 Introduction

The solution of a weighted combinatorial optimization problem can be seen as the selection of n values in a combinatorial set $\mathcal{X} \subset \mathbb{R}^n$. The maximum weight version of such a problem consists in maximizing the sum of these n

© Springer International Publishing AG 2017
X. Gao et al. (Eds.): COCOA 2017, Part I, LNCS 10627, pp. 324–339, 2017.
https://doi.org/10.1007/978-3-319-71150-8_28

values (e.g., $\sum_{i=1}^{n} u_i$). For instance, in a matching problem on a graph, the sum of weights that is optimized corresponds to the sum of weights of the edges selected in a matching. In practice, the vector of weights $(u_1, u_2, \ldots, u_n)$ could receive different interpretations depending on the actual problem. In a multi-agent setting, each value u_i represents the utility of an agent i, as in a bi-partite matching problem where n objects have to be assigned to n agents. In a multi-criteria context, those n values can be viewed as different dimensions to optimize. For example, in the travelling salesman problem (TSP) with n cities, a feasible solution (i.e., Hamiltonian cycle) is valued by an n-dimensional vector where each component represents the sum of the distances to reach and leave a city.

In both interpretations, it is desirable that the vector of values $(u_1, u_2, \ldots, u_n)$ be both Pareto-optimal (i.e., not improvable on all components at the same time) and balanced (or fair). We call optimization with such concerns *fair optimization* by adopting the terminology from multi-agent systems. In this paper, we focus on the fair optimization version of a class of combinatorial problems (including allocation, general matching, TSP...). Note that optimizing the sum of the values (i.e., maximum weight problem) yields a Pareto-optimal solution, but does not provide any guarantee on how balanced the vector solution would be.

Various approaches have been proposed in the literature to provide such a guarantee with different models for fairness or "balancedness" (see Sect. 2 for an overview). In this paper, our approach is based on an inequality measure called *Generalized Gini Index* (GGI) [29], which is well-known and well-studied in economics and can be used to control for both Pareto-efficiency and fairness. Indeed, fairness has naturally been investigated in economics [17]. In this literature, two important requirements have been identified as essential for fairness: equal treatment of equals and efficiency. The first notion implies that two agents with the same characteristics (notably the same preferences) have to be treated the same way, while the second entails that a fair solution should be Pareto-optimal. GGI satisfies both requirements, as it is symmetric in its arguments and increasing with Pareto dominance. The notion of fairness that GGI encodes is based on the Pigou-Dalton transfer principle, which states that a small transfer of resource from a richer agent to a poorer one yields a fairer distribution.

To the best of our knowledge, fair optimization in such general combinatorial problems has not been considered so far, although the GGI criterion has been investigated before in some specific problems (allocation [12], capital budgeting [11], Markov decision process [19,20]...). The difficulty of this combinatorial optimization problem lies in the fact that the objective function is non-linear. The contribution of this paper is fourfold: (1) we introduce a new general combinatorial problem (e.g., fair matching in general graph or fair TSP have not been studied so far); (2) we provide an optimality condition and an approximation ratio; (3) we propose a fast general heuristic method based on a primal-dual approach and on Lagrangian decomposition; (4) we evaluate this method on several problems related to matching to understand its efficiency. Although our general combinatorial formulation covers problems whose maximum weight version is

NP-hard, we leave for a follow-up work the integration of our fast heuristic with approximation algorithms to solve those NP-hard problems.

The paper is organized as follows. Section 2 gives an overview of related work. Section 3 provides a formal definition of our problem, which can be solved by a 0, 1-linear program. As a faster alternative, we present a heuristic primal-dual solving method based on Lagrangian decomposition in Sect. 4 and evaluate it experimentally in Sect. 5. Finally, we conclude in Sect. 6.

2   Related Work

Fair optimization is an active and quite recent research area [14, 18] in multi-objective optimization. Fairness can be modeled in different ways. One simple approach is based on maxmin, so called Egalitarian approach, where one aims at maximizing the worse-off component (i.e., objective, agent...). Due to the drowning effects of the min operator, vectors with the same minimum cannot be discriminated. A better approach [24] is based on the lexicographic maxmin, which consists in considering the minimum first when comparing two vectors, then in case of a tie, focusing on the second smallest values and so on. However, due to the noncompensatory nature of the min operator, vector $(1, 1, \ldots, 1)$ would be preferred to $(0, 100, \ldots, 100)$, which may be debatable. To take into account this observation, one can resort to use a strictly increasing and strictly Schur-concave (see Sect. 3.2 for definition) aggregation function f (see [18] for examples) that evaluates each vector such that higher values are preferred.

In this paper, we focus on the Generalized Gini Index (GGI) proposed in the economics literature [29], because it satisfies natural properties for encoding fairness. GGI is a particular case of a more general family of operators known as Ordered Weighted Averaging (OWA) [31]. Much work in fair optimization has applied the OWA operator and GGI in multiobjective (continuous and combinatorial) optimization problems. To cite a few, it was used in network dimensioning problems [22], capital budgeting [11], allocation problems [12], flow optimization in wireless mesh networks [10] and multiobjective sequential decision-making under uncertainty [19, 20]. One common solving technique is based on a linearization trick of the nonlinear objective function based on GGI [21]. Recently, [9] considered a similar setting to ours, but tries to solve its continuous relaxation.

In multicriteria decision-making, fair optimization is related to compromise optimization, which generally consists in minimizing a distance to an ideal point [27]. More generally, the ideal point can be replaced by any reference point that a decision maker chooses, as in the reference point method [30]. In this context, a judiciously chosen reference point can help generate a solution with a balanced profile on all criteria. One main approach is based on minimizing the augmented weighted Tchebycheff distance. This method has been applied in many multicriteria problems, for instance, in process planning [25], in sequential decision-making under uncertainty [23], in discrete bicriteria optimization problems [6], in multiobjective multidimensional knapsack problems [15].

Note that our combinatorial optimization problem should not be confused with the multicriteria version of those problems where each scalar weight becomes vectorial and the value of a solution is obtained by aggregating the selected weight vectors with a componentwise sum. For instance, Anand [1] investigated a multicriteria version of the matching problem and proved that the egalitarian approach for vector-valued matching leads to NP-hard problems. In our problem, the weights are scalar and the value of a solution is *not* obtained by summing its scalar weights, but by aggregating them with GGI.

3   Model

In this section, we formally describe the general class of combinatorial problems considered in this paper and provide some concrete illustrative examples in this class. Then we recall the generalized Gini index as a measure of fairness and define the fair combinatorial optimization problems tackled in this paper. We start with some notations. For any integer n, $[n]$ denotes the set $\{1, 2, \ldots n\}$. For any vector $\boldsymbol{x}$, its component is denoted x_i or x_{ij} depending on its dimension.

3.1   General Model

We consider a combinatorial optimization problem (e.g., allocation, multiple knapsack, matching, travelling salesman problem...), whose feasible solutions $\mathcal{X} \subseteq \{0, 1\}^{n \times m}$ can be expressed as follows:

$$\boldsymbol{Az} \leq \boldsymbol{b}$$
$$\boldsymbol{z} \in \{0, 1\}^{n \times m}$$

where $\boldsymbol{A} \in \mathbb{Z}^{p \times (nm)}$, $\boldsymbol{b} \in \mathbb{Z}^p$, n, m and p are three positive integers, and $\boldsymbol{z}$ is viewed as a one-dimensional vector $(z_{11}, \ldots, z_{1m}, z_{21}, \ldots, z_{2m}, \ldots, z_{n1}, \ldots, z_{nm})^{\mathsf{T}}$.

Let $u_{ij} \in \mathbb{N}$ be the utility of setting z_{ij} to 1. The *maximum weight problem* defined on combinatorial set $\mathcal{X}$ can be written as a 0, 1-linear program (0, 1-LP):

$$\text{max.} \sum_{i \in [n]} \sum_{j \in [m]} u_{ij} z_{ij}$$
$$\text{s.t.} \, \boldsymbol{z} \in \mathcal{X}$$

Because this general problem includes the travelling salesman problem (TSP), it is NP-hard in general. As mentioned before, this objective function provides no control on the fairness of the obtained solution. Although possibly insufficient, one simple approach to fairness consisting in focusing on the worse-off component is the *maxmin problem* defined on set $\mathcal{X}$, which can also be written as a 0, 1-LP:

$$\text{max.} \, v$$
$$\text{s.t.} \, v \leq \sum_{j \in [m]} u_{ij} z_{ij} \quad \forall i \in [n]$$
$$\boldsymbol{z} \in \mathcal{X}$$

Even for some polynomial problems like allocation, this version is NP-hard in general [4]. To avoid any confusion, in this paper, allocation refers to matching on a bi-partite graph and matching generally implies a complete graph.

For illustration, we now present several instantiations of our general model on allocation and matching problems, some of which will be used for the experimental evaluation of our proposed methods in Sect. 5.

Example 1 (Allocation). *Let $G = (V_1 \cup V_2, E, u)$ be a valued bipartite graph where V_1 and V_2 are respectively an n-vertex set and an m-vertex set with $V_1 \cap V_2 = \emptyset$, $E \subseteq \{\{x, y\} \mid (x, y) \in V_1 \times V_2\}$ is a set of non-directed edges and $u : E \to \mathbb{R}$ defines the nonnegative utility (i.e., value to be maximized) of an edge. As there is no risk of confusion, we identify V_1 to the set $[n]$ and V_2 to the set $[m]$. An allocation of G is a subset of E such that each vertex i in V_1 is connected to α_i to β_i vertices in V_2 and each vertex in V_2 is connected to α'_j to β'_j vertices in V_1 where $(\boldsymbol{\alpha}, \boldsymbol{\beta}) \in \mathbb{N}^{n \times n}$ and $(\boldsymbol{\alpha'}, \boldsymbol{\beta'}) \in \mathbb{N}^{m \times m}$.*

The assignment problem where n tasks need to be assigned to n agents is a special case where $n = m$ and $\alpha_i = \beta_i = \alpha'_j = \beta'_j = 1$ for $i \in [n]$ and $j \in [n]$. The conference paper assignment problem where m papers needs to be reviewed by n reviewers such that each paper is reviewed by 3 reviewers and each reviewer receives at most 6 papers can be represented with $\alpha_i = 0$, $\beta_i = 6$, $\alpha'_j = 3$ and $\beta'_j = 3$ for $i \in [n]$ and $j \in [m]$. The Santa Claus problem [3] where m toys needs to be assigned to n children with $n \leq m$ is also a particular case with $\alpha_i = 0$, $\beta_i = m$, $\alpha'_j = \beta'_j = 1$ for $i \in [n]$ and $j \in [m]$.

The maximum weight problem can be solved with the following $0, 1$-LP:

$$\max. \sum_{i \in [n]} \sum_{j \in [m]} u_{ij} z_{ij}$$

$$\text{s.t.} \, \alpha_i \leq \sum_{j \in [m]} z_{ij} \leq \beta_i \, \forall i \in [n] \tag{3a}$$

$$\alpha'_j \leq \sum_{i \in [n]} z_{ij} \leq \beta'_j \, \forall j \in [m] \tag{3b}$$

$$z \in \{0, 1\}^{n \times m}$$

Interestingly, its solution can be efficiently obtained by solving its continuous relaxation because the matrix defining its constraints (3a)–(3b) is totally unimodular [26]. However, the maxmin version is NP-complete [4].

Example 2 (Matching). *Let $G = (V, E, u)$ be a valued graph where V is a $2n$-vertex set (with $n \in \mathbb{N}\backslash\{0\}$), $E \subseteq \{\{x, y\} \mid (x, y) \in V^2, x \neq y\}$ is a set of non-directed edges and $u : E \to \mathbb{R}$ defines the nonnegative utility of an edge. A matching M of G is a subset of E such that no pair of edges of M are adjacent, i.e., they do not share a common vertex: $\forall (e, e') \in E^2, e \neq e' \Rightarrow e \cap e' = \emptyset$. A perfect matching M is a matching where every vertex of G is incident to an edge of M. Thus, a perfect matching contains n edges. Without loss of generality, we identify V to the set $[2n]$ and denote $\forall e = \{i, j\} \in E, u_{ij} = u(e)$ when convenient.*

The standard maximum weight perfect matching problem aims at finding a perfect matching for which the sum of the utilities of its edges is maximum. Let $\delta(i) = \{\{i,j\} \in E \mid j \in V\backslash\{i\}\}$ be the set of edges that are incident on vertex i. It is known [13] that this problem can be formalized as a $0,1$-LP (where z_{ij}'s for $i > j$ are unnecessary and can be set to 0):

$$(\mathcal{P}_M) \begin{cases} \text{max.} & \displaystyle\sum_{i \in [2n]} \sum_{j \in [2n], j > i} u_{ij} z_{ij} & & \text{(4a)} \\[2ex] \text{s.t.} & \displaystyle\sum_{\{i,j\} \in \delta(k), i < j} z_{ij} = 1 & \forall k \in [2n] & \text{(4b)} \\[2ex] & z_{ij} \in \{0, 1\} & \forall i \in [2n], j = i+1, \ldots, 2n & \text{(4c)} \end{cases}$$

where (4b) states that in a matching only one edge is incident on any vertex.

This problem can be solved as an LP by considering the continuous relaxation of $\mathcal{P}_M$ and adding the well-known blossom constraints (5b) in order to remove the fractional solutions introduced by the relaxation:

$$(\mathcal{RP}_M) \begin{cases} \text{max.} & \displaystyle\sum_{i \in [2n]} \sum_{j \in [2n], j > i} u_{ij} z_{ij} & & \\[2ex] \text{s.t.} & \displaystyle\sum_{\{i,j\} \in \delta(k), i < j} z_{ij} = 1 \quad \forall k \in [2n] & & \text{(5a)} \\[2ex] & z(\delta(S)) \geq 1 & \forall S \subset V, |S| \text{ odd}, |S| \geq 3 & \text{(5b)} \\[2ex] & 0 \leq z_{ij} \leq 1 & \forall i \in [2n], j = i+1, \ldots, 2n & \text{(5c)} \end{cases}$$

where $z(\delta(S)) = \sum_{\{i,j\} \in \delta(S), i < j} z_{ij}$ and $\delta(S) = \{\{i,j\} \in E \mid i \in S \text{ and } j \in V\backslash S\}$. Constraints (5a)–(5c) define the so-called perfect matching polytope. *In practice, this problem can be efficiently solved with the Blossom algorithm proposed by Edmonds [8]. To the best of our knowledge, the maxmin version of the matching problem (on complete graph) has not been investigated so far.*

In this paper we focus on a variant of those combinatorial problems: search for a solution z whose distribution of values $\left(\sum_{j \in [m]} u_{ij} z_{ij}\right)_{i \in [n]}$ is fair to its components (e.g., different agents' utilities or criteria). To model fairness we use a special case of the ordered weighted averaging operator that we recall next.

3.2   Ordered Weighted Average and Generalized Gini Index

The *Ordered Weighted Average* (OWA) [31] of $v \in \mathbb{R}^n$ is defined by:

$$OWA_{\boldsymbol{w}}(v) = \sum_{k \in [n]} w_k v_k^{\uparrow}$$

where $\boldsymbol{w} = (w_1, \ldots, w_n) \in [0,1]^n$ is the OWA weight vector and $v^{\uparrow} = (v_1^{\uparrow}, \ldots, v_n^{\uparrow})$ is the vector obtained from v by rearranging its components in an increasing order. OWA defines a very general family of operators, e.g., the sum (for $w_k = 1$, $\forall k \in [n]$), the average, the minimum (for $w_1 = 1$ and $w_k = 0$, $\forall k > 1$), the maximum (for $w_n = 1$ and $w_k = 0$, $\forall k < n$), the leximin when differences between OWA weights tends to infinity or the augmented weighted Tchebycheff distance [20].

Let the *Lorenz components* [2] of v be denoted by $(L_1(v), \ldots, L_n(v))$ and be defined by $\forall k \in [n]$, $L_k(v) = \sum_{i \in [k]} v_i^\uparrow$. Interestingly, OWA can be rewritten as:

$$OWA_w(v) = \sum_{k \in [n]} w_k' L_k(v) \tag{6}$$

where $\forall k \in [n]$, $w_k' = w_k - w_{k+1}$ and $w_{n+1} = 0$. With this rewriting, one can see that OWA is simply a weighted sum in the space of Lorenz components.

The notion of fairness that we use in this paper is based on the *Pigou-Dalton principle* [16]. It states that, all other things being equal, we prefer more "balanced" vectors, which implies that any transfer (called *Pigou-Dalton transfer*) from a richer component to a poorer one without reversing their relative positions yields a preferred vector. Formally, for any $v \in \mathbb{R}^n$ where $v_i < v_j$ and for any $\epsilon \in (0, v_j - v_i)$ we prefer $v + \epsilon 1_i - \epsilon 1_j$ to v where 1_i (resp. 1_j) is the canonical vector, null everywhere except in component i (resp. j) where it is equal to 1.

When the OWA weights are strictly decreasing and positive [29], OWA is called the *Generalized Gini Index* (GGI) [29] and denoted G_w. It encodes both:

efficiency: G_w is increasing with respect to Pareto-dominance (i.e., if $v \in \mathbb{R}^n$ Pareto-dominates[1] $v' \in \mathbb{R}^n$, then $G_w(v) > G_w(v')$); and
fairness: G_w is strictly Schur-concave, i.e., it is strictly increasing with Pigou-Dalton transfers ($\forall v \in \mathbb{R}^n$, $v_i < v_j$, $\forall \epsilon \in (0, v_j - v_i)$, $G_w(v + \epsilon 1_i - \epsilon 1_j) > G_w(v)$).

The classic Gini index, which is a special case of GGI with $w_i = (2(n - i) + 1)/n^2$ for all $i \in [n]$, enjoys a nice graphical interpretation (see Fig. 1). For a given distribution $v \in \mathbb{R}_+^n$, let $\bar{v}$ denote the average of the components of v, i.e., $\bar{v} = \frac{1}{n} \sum_{i=1}^n v_i$. Distribution v can be represented by the curve going through the points $(0, 0)$ and $(\frac{k}{n}, L_k(v))$ for $k \in [n]$. The most equitable distribution with the same total sum as that of v (i.e., $n\bar{v}$) can be represented by the straight line going through the points $(0, 0)$ and $(\frac{k}{n}, k\bar{v})$ for $k \in [n]$. The value $1 - G_w(v)/\bar{v}$ is equal to twice the area between the two curves.

Interestingly, the Lorenz components of a vector can be computed by LP [21]. Indeed, the k-th Lorenz component $L_k(v)$ of a vector v can be found as the solution of a knapsack problem, which is obtained by solving the following LP:

$$(\mathcal{LP}_k) \begin{cases} \text{min.} & \sum_{i \in [n]} a_{ik} x_i \\ \text{s.t.} & \sum_{i \in [n]} a_{ik} = k \\ & 0 \leq a_{ik} \leq 1 & \forall i \in [n] \end{cases}$$

---

[1] Vector v Pareto-dominates vector v' if $\forall i \in [n]$, $v_i \geq v_i'$ and $\exists j \in [n]$, $v_j > v_j'$.

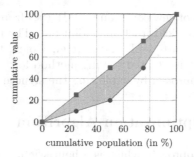

Fig. 1. Lorenz curves

Equivalently, this can be solved by its dual:

$$(\mathcal{DL}_k) \begin{cases} \text{max.} & kr_k - \sum_{i \in [n]} d_{ik} \\ \text{s.t.} & r_k - d_{ik} \leq v_i & \forall i \in [n] \\ & d_{ik} \geq 0 & \forall i \in [n] \end{cases}$$

The dual formulation is particularly useful. Contrary to the primal, it can be integrated in an LP where the v_i's are also variables [21]. We will use this technique to formulate a $0,1$-LP to solve our general combinatorial optimization problem.

3.3 Fair Combinatorial Optimization

The problem tackled in this paper is defined by using GGI as objective function:

$$\text{max. } G_{\boldsymbol{w}}\Big(\big(\sum_{j \in [m]} u_{ij} z_{ij}\big)_{i \in [n]}\Big) \quad \text{s.t. } \begin{cases} \boldsymbol{Az} \leq \boldsymbol{b} \\ \boldsymbol{z} \in \{0,1\}^{n \times m} \end{cases}$$

Following Ogryczak and Sliwinski [21], we can combine the rewriting of OWA based on Lorenz components (6) and LPs $(\mathcal{DL}_k)$ for $k \in [2n]$ to transform the previous non-linear optimization program into a $0,1$-LP:

$$\text{max.} \sum_{k \in [n]} w'_k \big(kr_k - \sum_{i \in [n]} d_{ik}\big) \tag{9a}$$

$$\text{s.t. } \boldsymbol{Az} \leq \boldsymbol{b} \tag{9b}$$

$$\boldsymbol{z} \in \{0,1\}^{n \times m} \tag{9c}$$

$$r_k - d_{ik} \leq \sum_{j \in [m]} u_{ij} z_{ij} \qquad \forall i \in [n], \forall k \in [n] \tag{9d}$$

$$d_{ik} \geq 0 \qquad \forall i \in [n], \forall k \in [n] \tag{9e}$$

Due to the introduction of new constraints (9d)–(9e) from LPs $(\mathcal{DL}_k)$, the relaxation of this $0, 1$-LP may yield fractional solutions. The naive approach to solve it would be to give it to a $0, 1$-LP solver (e.g., Cplex, Gurobi...). Our goal in this paper is to propose an adapted solving method for it, which would be much faster than the naive approach by exploiting the structure of this problem.

4   Alternating Optimization Algorithm

Before presenting our approach, which is a heuristic method based on a primal-dual technique using a Lagrangian decomposition, we first make an interesting and useful observation. The dual of the continuous relaxation of the previous $0, 1$-LP (9) is given by:

$$\min. \boldsymbol{b}^{\mathsf{T}}\boldsymbol{v} + \sum_{i \in [n]} \sum_{j \in [m]} t_{ij} \tag{10a}$$

$$\text{s.t. } (\boldsymbol{v}^{\mathsf{T}}\boldsymbol{A})_{ij} + t_{ij} - \sum_{k \in [n]} u_{ij} y_{ik} \geq 0 \qquad \forall i \in [n], \forall j \in [m] \tag{10b}$$

$$\sum_{i=1}^{n} y_{ik} = k w'_k \qquad \forall k \in [n] \tag{10c}$$

$$0 \leq y_{ik} \leq w'_k \qquad \forall i \in [n], \forall k \in [n] \tag{10d}$$

$$v_j \geq 0 \qquad \forall j \in [p] \tag{10e}$$

$$t_{ij} \geq 0 \qquad \forall i \in [n], \forall j \in [m] \tag{10f}$$

Interestingly, with fixed y_{ik}'s, the dual of the previous program can be written in the following form, which is simply the continuous relaxation of the original program with modified weights:

$$\max. \sum_{i \in [n]} \Big(\sum_{k \in [n]} y_{ik} \Big) \sum_{j \in [m]} u_{ij} z_{ij} \tag{11a}$$

$$\text{s.t. } \boldsymbol{A}\boldsymbol{z} \leq \boldsymbol{b} \tag{11b}$$

$$\boldsymbol{z} \in [0, 1]^{n \times m} \tag{11c}$$

Therefore, solving this program with discrete $\boldsymbol{z}$ yields a feasible solution of the original problem. We denote $(P_{\boldsymbol{y}})$ the $0, 1$-LP (11) defined with $\boldsymbol{y} = (y_{ik})_{i \in [n], k \in [n]}$.

4.1   Optimality Condition and Approximation Ratio

Next we express an optimality condition so that an integer solution $\boldsymbol{z}^*$ computed from a dual feasible solution $\boldsymbol{y}^*$ of (10) is optimal for program (9). First, note that any extreme solution $(\boldsymbol{v}, \boldsymbol{t}, \boldsymbol{y})$ of program (10) is such that either $y_{ik} = 0$ or $y_{ik} = w'_k$ for all $i \in [n]$ and $k \in [n]$.

Theorem 1. *Let (v, t, y^*) be an extreme solution of (10) and let z^* be the optimal solution of program (P_{y^*}). Let $T_i^* = \sum_{j \in [m]} u_{ij} z_{ij}^*$ for all $i \in [n]$ and assume without loss of generality that $T_1^* \geq T_2^* \geq \ldots \geq T_n^*$.*
If for all $k \in [n]$, $y_{ik}^* = w_k'$ for all $i \geq n + 1 - k$ and $y_{ik}^* = 0$ for all $i \in [n - k]$ then z^* is an optimal solution of program (9).

Proof. Let (v^*, t^*) be the dual optimal solution associated with z^* when solving (P_{y^*}). Composing them with y^*, we obtain a feasible solution (v^*, t^*, y^*) of (10). By duality theory of linear programming, the objective value of this solution is equal to $\sum_{i \in [n]} (\sum_{j \in [i]} w_{n+1-j}') T_i^*$. Let us now build a feasible solution (r^*, d^*, z^*) of (9) based on z^* as follows. For all $k \in [n]$,

- $r_k^* = T_{n+1-k}^*$ and
- $d_{ik}^* = \begin{cases} r_k^* - T_i^* & \text{if } i \geq n + 1 - k \\ 0 & \text{otherwise} \end{cases}$ for all $i \in [n]$.

We now show that (r^*, d^*) satisfy constraints (9d). For any $i \in [n]$ and $k \in [n]$, if $i \leq n + 1 - k$ then as $r_k^* = T_{n+1-k}^* \leq T_i^*$ and $d_{ik}^* = 0$, we have

$$r_k^* - d_{ik}^* \leq T_i^* = \sum_{j \in [m]} u_{ij} z_{ij}^*.$$

If $i \geq n + 1 - k$ then as $d_{ik}^* = r_k^* - T_i^*$, $r_k^* - d_{ik}^* = T_i^* = \sum_{j \in [m]} u_{ij} z_{ij}^*$. Hence (r^*, d^*, z^*) is a feasible solution of (9). For any $k \in [n]$, $k r_k^* - \sum_{i \in [n]} d_{ik}^* = k r_k^* - \sum_{i=n+1-k}^{n} d_{ik}^* = k r_k^* - (k r_k^* - \sum_{i=n+1-k}^{n} T_i^*) = \sum_{i=n+1-k}^{n} T_i^*)$. Then it is easy to see that the objective value of this solution, which is $\sum_{k \in [n]} w_k' (k r_k^* - \sum_{i \in [n]} d_{ik}^*)$ is equal to $\sum_{k \in [n]} w_k' \sum_{i=n-k+1}^{n} T_i^*$. This sum is just a rewriting of $\sum_{i \in [n]} (\sum_{j \in [i]} w_{n+1-j}') T_i^*$. Thus, by duality of linear programming, the solution (r^*, d^*, z^*) is optimal for program (9). □

Theorem 1 provides an optimality condition for any feasible solution z^*, but does not indicate how to find "good" solutions. Yet, one may be interested in the quality of some special solutions, e.g., the optimal solution of the maximum weight version. The following theorem establishes an approximation ratio for the latter, which also applies to our method as discussed later.

Theorem 2. *Let $\bar{z}$ be an optimal solution of the maximum weight version. Let $\bar{T}_i = \sum_{j \in [m]} u_{ij} \bar{z}_{ij}$ for all $i \in [n]$ and assume without loss of generality that $\bar{T}_1 \geq \bar{T}_2 \geq \ldots \geq \bar{T}_n$. Let $w_{\max}' = \max_{k \in [n]} w_k'$. Then the GGI value of $\bar{z}$ is at worst $\max(\frac{2 w_n'}{(n+1) w_{\max}'}, \frac{n \bar{T}_n}{(\sum_{i \in [n]} \bar{T}_i)})$ of the optimal objective value of program (9).*

Proof. Let vector $\bar{y} \in \mathbb{R}^{n \times n}$ be defined as $\bar{y}_{ik} = \frac{k}{n} w_k'$ for $i, k \in [n]$, which is feasible for program (10). The objective function of $(P_{\bar{y}})$ satisfies:

$$\sum_{i \in [n]} (\sum_{k \in [n]} \bar{y}_{ik}) \sum_{j \in [m]} u_{ij} z_{ij} = \sum_{i \in [n]} \sum_{j \in [m]} (\sum_{k \in [n]} \frac{k}{n} w_k') u_{ij} z_{ij}$$

$$\leq \sum_{i \in [n]} \sum_{j \in [m]} (\sum_{k \in [n]} \frac{k}{n} w_{\max}') u_{ij} z_{ij} \qquad (12)$$

Program (11) with objective (12) corresponds to the maximum weight version scaled by a constant. It is equal to $\sum_{k\in[n]}(kw'_{max}/n)\times \sum_{i\in[n]}(\bar{T}_i)$ for solution $\bar{z}$, which is an upperbound of the objective value associated with $\bar{y}$ of (10) and hence an upperbound for the optimal value of (9).

Proceeding as for Theorem 1, we define a feasible solution of (9) based on $\bar{z}$:

- $\bar{r}_k = \bar{T}_{n+1-k}$ for all $k \in [n]$, and
- $\bar{d}_{ik} = \begin{cases} \bar{r}_k - \bar{T}_i & \text{if } i \geq n+1-k \\ 0 & \text{otherwise} \end{cases}$ for all $i \in [n]$, for all $k \in [n]$.

The objective value of this solution $\sum_{i\in[n]}(\sum_{j\in[i]} w'_{n+1-j})\bar{T}_i$ (see proof of Theorem 1) is to be compared with upperbound $\sum_{i\in[n]}(\sum_{k\in[n]} kw'_{max}/n)\bar{T}_i$.

By comparing term by term w.r.t. $\bar{T}_i$ for $i \in [n]$, we can see that the worst case happens to the term associated with $\bar{T}_1$ with the ratio $w'_n/(\sum_{k\in[n]} kw'_{max}/n)$. Therefore, we obtain the ratio $\frac{2w'_n}{(n+1)w'_{max}}$. This ratio is consistent since when $n = 1$, the optimal solution of the maximum weight version coincides with the optimum solution of (9).

By comparing term by term with respect to w'_k for $k \in [n]$, we can see that the worst case happens to the term associated with w'_1 with the ratio $n\bar{T}_n/(\sum_{i=1}^{n} \bar{T}_i)$, which can be interpreted as the ratio of the smallest utility over the average utility in the optimal solution of the maximum weight version. This ratio is consistent since in the case of equal utilities in the optimal solution of the maximum weight version, the latter coincides with the optimum solution of (9). □

4.2 Iterative Algorithm

The previous discussion motivates us to design an alternating optimization algorithm that starts with a feasible y for (10), computes the associated z and uses the latter to iteratively improve y. Formally, it can be sketched as follows:

```
1: t ← 0
2: compute y^(0)
3: repeat
4:     t ← t + 1
5:     solve 0, 1-LP (P_{y^{t-1}}) to obtain feasible solution z^(t)
6:     update y^(t) based on y^(t-1) and z^(t)
7: until max iteration has been reached or change on y_{ik}^(t) is small
8: return z^(t) with highest GGI
```

Interestingly, lines 2 and 6 can be performed in different ways. For line 2, an initial $y^{(0)}$ can be obtained by solving the dual LP (10). Another approach is to solve the maximum weight version of our combinatorial problem and get the dual solution variables for $y^{(0)}$. Note that Theorem 2 then provides a guarantee on the final solution, as it is at least as good as that of the maximum weight problem. For line 6, one approach is to solve (9) with z fixed to $z^{(t)}$ in order to get dual solution variables $y^{(t)}$. A better approach as observed in the experiments and explained next is based on Lagrangian relaxation.

The Lagrangian relaxation of (9) with respect to constraint (9d) can be written as follows with Lagrangian multipliers $\lambda = (\lambda_{ik})_{i \in [n], k \in [n]}$:

$$\mathcal{L}(\lambda) = \quad \max. \sum_{k \in [n]} (w'_k k - \sum_{i \in [n]} \lambda_{ik}) r_k - \sum_{k \in [n]} \sum_{i \in [n]} (w'_k - \lambda_{ik}) d_{ik} \qquad (13a)$$

$$+ \sum_{i \in [n]} (\sum_{k \in [n]} \lambda_{ik}) \sum_{j \in [m]} u_{ij} z_{ij} \qquad (13b)$$

$$\text{s.t. } Az \le b \qquad (13c)$$

$$z \in \{0,1\}^{n \times m} \qquad (13d)$$

$$d_{ik} \ge 0 \qquad \forall i \in [n], \forall k \in [n] \qquad (13e)$$

The Lagrangian dual of (13) is then given by:

$$\min. \mathcal{L}(\lambda) \quad \text{s.t. } \lambda_{ik} \ge 0 \quad \forall i \in [n], \forall k \in [n] \qquad (14)$$

For an optimal solution z^*, r^*, d^* of the 0,1-LP (9), we have for any $\lambda \in \mathbb{R}_+^{n \times n}$:

$$\sum_{k \in [n]} w'_k (k r_k^* - \sum_{i \in [n]} d_{ik}^*) \le \sum_{k \in [n]} (w'_k k - \sum_{i \in [n]} \lambda_{ik}) r_k - \sum_{k \in [n]} \sum_{i \in [n]} (w'_k - \lambda_{ik}) d_{ik}$$

$$+ \sum_{i \in [n]} (\sum_{k \in [n]} \lambda_{ik}) \sum_{j \in [m]} u_{ij} z_{ij} \le \mathcal{L}(\lambda)$$

The first inequality holds because of the nonnegativity of λ and the feasibility of z^*, r^*, d^*. The second is true because of the maximization in (13). Therefore the best upperbound is provided by the solution of the Lagrangian dual (14), though this problem is not easy to solve due to the integrality condition over z.

An inspection of program (13) leads to two observations: (i) it can be decomposed into two maximization problems, one over z and the other over r and d; (ii) for program (13) to yield a useful upperbound, λ should satisfy two constraints (otherwise $\mathcal{L}(\lambda) = \infty$):

$$\sum_{i \in [n]} \lambda_{ik} = k w'_k \quad \forall k \in [n] \quad \text{and} \quad \lambda_{ik} \le w'_k \quad \forall i \in [n], \forall k \in [n]$$

Interestingly, in the above decomposition, the maximization problem over z corresponds to (P_λ) and therefore λ can be identified to the dual variable y.

Based on those observations, line 6 can be performed as follows. Given λ (or y), the upperbound $\mathcal{L}(\lambda)$ can be improved by updating λ so as to decrease (13a), which can be simply done by a projected sub-gradient step:

$$\lambda'_{ik} \leftarrow \lambda_{ik} - \gamma(r_k - d_{ik} - \sum_{j \in [m]} u_{ij} z_{ij}) \qquad \forall i \in [n], k \in [n] \qquad (15)$$

$$\lambda \leftarrow \arg \min_{\lambda \in \mathbb{L}} ||\lambda' - \lambda|| \qquad (16)$$

where γ is the sub-gradient step and (16) is the Euclidean projection of $\boldsymbol{\lambda}'$ on $\mathbb{L} = \{\boldsymbol{\lambda} \in \mathbb{R}_+^{n \times n} \mid \forall k \in [n], \sum_{i \in [n]} \lambda_{ik} = k w_k', \forall i \in [n], \lambda_{ik} \leq w_k'\}$.

Projection (16) can be performed efficiently by exploiting the structure of $\mathbb{L}$:

$$\arg\min_{\boldsymbol{\lambda} \in \mathbb{L}} ||\boldsymbol{\lambda}' - \boldsymbol{\lambda}|| = \arg\min_{\boldsymbol{\lambda} \in \mathbb{L}} ||\boldsymbol{\lambda}' - \boldsymbol{\lambda}||^2 = \arg\min_{\boldsymbol{\lambda} \in \mathbb{L}} \sum_{i \in [n]} \sum_{k \in [n]} (\lambda_{ik}' - \lambda_{ik})^2$$

$$= \Big(\arg\min_{\boldsymbol{\lambda}_k \in \mathbb{L}_k} \sum_{i \in [n]} (\lambda_{ik}' - \lambda_{ik})^2 \Big)_{k \in [n]} = \Big(\arg\min_{\boldsymbol{\lambda}_k \in \mathbb{L}_k} \sum_{i \in [n]} (\frac{\lambda_{ik}'}{w_k'} - \frac{\lambda_{ik}}{w_k'})^2 \Big)_{k \in [n]} \quad (17)$$

where $\mathbb{L}_k = \{\boldsymbol{\lambda}_k \in \mathbb{R}_+^n \mid \sum_{i \in [n]} \lambda_{ik}/w_k' = k, \forall i \in [n], \lambda_{ik}/w_k' \leq 1\}$. Equation (17) states that projection (16) can be efficiently performed by n projections on capped simplices [28]. The complexity of this step would be in $O(n^3)$, which is much faster than solving the quadratic problem (16). Besides, the n projections can be easily computed in a parallel way.

We can provide a simple interpretation to the variable $\boldsymbol{\lambda}$ (or $\boldsymbol{y}$). Considering programs (10) and (11), we can observe that $\boldsymbol{y}$ corresponds to an allocation of weights w_k''s over the different component i's. Indeed, an optimal solution of (10) would yield an extreme point of $\mathbb{L}$ (for a given $k \in [n]$, exactly k terms among $(y_{1k}, \ldots, y_{nk})$ are equal to w_k' and the other ones are null). The projected sub-gradient method allows to search for an optimal solution of our fair combinatorial problem by moving inside the convex hull of those extreme points.

5   Experimental Results

We evaluated our method on two different problems: assignment and matching. The LPs and 0, 1-LPs were solved using CPLEX 12.7 on a PC (Intel Core i7-6700 3.40 GHz) with 4 cores and 8 threads and 32 GB of RAM. Default parameters of CPLEX were used with 8 threads. The sub-gradient step γ_t is computed following the scheme: $\gamma_t := \frac{(val(\boldsymbol{z}_t) - bestvalue)\rho_t}{sqn}$ where $val(\boldsymbol{z}_t)$ is the objective value of the program (11) with solution $\boldsymbol{z}_t$, $bestvalue$ is the best known objective value of the program (9) so far and sqn is the square of the Euclidean norm of the subgradient vector. The parameter ρ_t is divided by two every 3 consecutive iterations in which the upperbound $\mathcal{L}(\boldsymbol{\lambda})$ has not been improved. The GGI weights were defined as follows: $w_k = 1/k^2$ for $k \in [n]$ so that they decrease fast in order to enforce more balanced solutions.

Assignment. To demonstrate the efficiency of our heuristic method, we generate hard random instances for the assignment problem. A random instance of this problem corresponds to a random generation of the u_{ij}'s, which are generated as follows. For all $i \in [n]$, u_{i1} follows a uniform distribution over [100] and for all $j \in [n]$, $u_{ij} = u_{i1} + \epsilon$ where ϵ is a random variable following a uniform distribution over integers between $-d$ and d (with d a positive integer parameter). With such a generation scheme, agents' preferences over objects are positively correlated and the solution of the fair optimization problem is harder due to the difficulty of finding a feasible solution that satisfies everyone.

Table 1. Numerical results for (left) assignment and (right) general matching problems

Instance	CPLEX		AlterOpt		Instance	CPLEX		AlterOpt	
	CPU1	CPU2	CPU	Gap		CPU1	CPU2	CPU	Gap
v50-20	1.02	1.02	0.23	0%	v50-30	0.86	0.86	0.79	0%
v50-30	3.14	3.14	0.26	0%	v50-40	2.43	2.43	1.42	0%
v50-40	64.95	14.26	0.45	0.28%	v50-50	5.14	5.14	2.67	0%
v50-50	1054.14	100.23	0.65	0.26%	v50-60	148.5	25.45	13.43	0.01%
v30-20	0.89	0.89	0.2	0%	v50-70	2406.02	1282.8	17.71	0.005%
v30-30	8.83	8.83	0.3	0.015%	v30-30	1.15	1.15	0.78	0%
v30-40	590.66	45.93	0.48	0.13%	v30-40	7.13	7.13	1.44	0%
v10-20	1.55	1.55	0.18	0%	v30-50	81.75	75.5	2.45	0.01%
v10-30	342.78	342.78	0.94	0%	v30-60	1003.69	615.16	12.8	0.036%
					v10-30	5.33	5.33	0.76	0%
					v10-40	1325.7	806.8	1.4	0.06%
					v10-50	29617.78	3370.7	2.48	0.053%

Matching. We use the lemon library [7] for solving the maximum weight matching problem. For the generation of the matching problem (in a complete graph with $2n$ nodes), we follow a similar idea to the assignment problem. Recall we only need u_{ij} (and z_{ij}) for $i < j$. For all $i \in [n]$, for all $j \in [n]$ with $i < j$, $u_{ij} = -1000$. For all $i \in [n]$, $u_{i,n+1}$ follows a uniform distribution over $[100]$ and for all $j \geq \max(i+1, n+2)$, $u_{ij} = u_{i,n+1} + \epsilon$ where ϵ is defined as above.

Explanations. The name of the instances is of the form "vd-x" where d denotes the deviation parameter mentioned above and x the number of the vertices of the graphs (i.e., $n = x/2$). Column "CPLEX" regroups CPLEX's results. Subcolumn "CPU1" reports the time (in seconds) that CPLEX spent to solve program (9) to optimal. Subcolumn "CPU2" reports the times needed by the primal heuristic of CPLEX to obtain a feasible integer solution that is better than or equal to the solution given by our algorithm. Column "AlterOpt" reports our algorithm's results. Subcolumn "CPU" is the time spent by our algorithm. Subcolumn "Gap" reports the gap in percentage between *Sol* and *Opt*, which is equal to $(Opt - Sol) \times 100/Opt\%$ where *Opt* is the optimal value and *Sol* is the value of the solution given by our algorithm. The times and the gaps reported are averaged over 10 executions corresponding to 10 random instances.

Table 1 shows that the CPU time spent by CPLEX (subcolumn CPU1) for solving program (9) increases exponentially with n and can quickly reach up to around 10 h. Moreover, the smaller the deviation x, the more difficult the problem. For example, for $x = 50$, we cannot solve instances with more than 50 and more than 70 vertices for respectively the fair assignment and general matching problems within 10 h of CPU time. For $x = 10$, this limit is respectively 30 and 50 vertices. In contrast, the CPU time spent by our algorithm (subcolumn CPU) seems to increase linearly with n and remains within tens or so seconds. The quality of the solutions output by our algorithm is very good as the gap is at maximum around 0.3% for fair assignment. This is even better for fair general matching, in all cases the gap is smaller than 0.1%. Moreover, the CPU time that CPLEX needs to find a feasible integer solution of similar quality by primal heuristic is much longer than the CPU time of our algorithm (up to hundreds times longer). It is interesting to notice that the fair assignment seems to be more difficult in our experiments than the fair general matching. This contrasts with the classical maximum weight version where the assignment problem is generally easier than the general maximum matching.

6   Conclusion

We formulated the fair optimization with the Generalized Gini Index for a large class of combinatorial problem for which we proposed a primal-dual algorithm based on a Lagrangian decomposition. We demonstrated its efficiency on several problems. We also provided some theoretical bounds on its performance. As future work, we plan to improve those bounds and investigate other updates for the Lagrangian multipliers. Another interesting direction is to consider other linearization techniques such as the one proposed by Chassein and Goerigk [5]. Finally, we will also apply our method to problems whose maximum weight version is NP-hard.

References

1. Anand, S.: The multi-criteria bipartite matching problem (2006)
2. Arnold, B.: Majorization and the Lorenz Order. Springer, New York (1987). https://doi.org/10.1007/978-1-4615-7379-1
3. Bansal, N., Sviridenko, M.: The Santa Claus problem. In: STOC, pp. 31–40 (2006)
4. Bezakova, I., Dani, V.: Allocating indivisible goods. ACM SIGecom Exch. **5**(3), 11–18 (2005)
5. Chassein, A., Goerigk, M.: Alternative formulations for the ordered weighted averaging objective. Inf. Process. Lett. **115**, 604–608 (2015)
6. Dachert, K., Gorski, J., Klamroth, K.: An augmented weighted Tchebycheff method with adaptively chosen parameters for discrete bicriteria optimization problems. Comput. Oper. Res. **39**(12), 2929–2943 (2012)
7. Dezs, B., Juttner, A., Kovacs, P.: LEMON - an open source C++ graph template library. Electron. Notes Theor. Comput. Sci. **264**(5), 23–45 (2011)
8. Edmonds, J.: Maximum matching and a polyhedron with 0, 1-vertices. J. Res. Natl. Bur. Stand. **69B**, 125–130 (1965)
9. Gilbert, H., Spanjaard O.: A game-theoretic view of randomized fair multi-agent optimization. In: IJCAI Algorithmic Game Theory Workshop (2017)
10. Hurkala, J., Sliwinski, T.: Fair flow optimization with advanced aggregation operators in wireless mesh networks. In: Federated Conference on Computer Science and Information Systems, pp. 415–421 (2012)
11. Kostreva, M., Ogryczak, W., Wierzbicki, A.: Equitable aggregations and multiple criteria analysis. Eur. J. Oper. Res. **158**, 362–367 (2004)
12. Lesca, J., Perny, P.: LP solvable models for multiagent fair allocation problems. In: ECAI (2011)
13. Lovész, L., Plummer, M.: Matching Theory. North Holland, Amsterdam (1986)

14. Luss, H.: Equitable Resource Allocation. Wiley, Hoboken (2012)
15. Lust, T., Teghem, J.: The multiobjective multidimensional knapsack problem: a survey and a new approach. Int. Trans. Oper. Res. **19**(4), 495–520 (2012)
16. Moulin, H.: Axioms of Cooperative Decision Making. Cambridge University Press, Cambridge (1988)
17. Moulin, H.: Fair Division and Collective Welfare. MIT Press, Cambridge (2004)
18. Ogryczak, W., Luss, H., Pióro, M., Nace, D., Tomaszewski, A.: Fair optimization and networks: a survey. J. Appl. Math. **2014**, 25 (2014)
19. Ogryczak, W., Perny, P., Weng, P.: On minimizing ordered weighted regrets in multiobjective Markov decision processes. In: Brafman, R.I., Roberts, F.S., Tsoukiàs, A. (eds.) ADT 2011. LNCS (LNAI), vol. 6992, pp. 190–204. Springer, Heidelberg (2011). https://doi.org/10.1007/978-3-642-24873-3_15
20. Ogryczak, W., Perny, P., Weng, P.: A compromise programming approach to multiobjective Markov decision processes. IJITDM **12**, 1021–1053 (2013)
21. Ogryczak, W., Sliwinski, T.: On solving linear programs with the ordered weighted averaging objective. Eur. J. Oper. Res. **148**, 80–91 (2003)
22. Ogryczak, W., Sliwinski, T., Wierzbicki, A.: Fair resource allocation schemes and network dimensioning problems. J. Telecom. Inf. Technol. **2003**(3), 34–42 (2003)
23. Perny, P., Weng, P.: On finding compromise solutions in multiobjective Markov decision processes. In: ECAI (short paper) (2010)
24. Rawls, J.: The Theory of Justice. Havard University Press, Cambridge (1971)
25. Rodera, H., Bagajewicz, M.J., Trafalis, T.B.: Mixed-integer multiobjective process planning under uncertainty. Ind. Eng. Chem. Res. **41**(16), 4075–4084 (2002)
26. Schrijver, A.: Theory of Linear and Integer Programming. Wiley, Chichester (1998)
27. Steuer, R.: Multiple Criteria Optimization. Wiley, New York (1986)
28. Wang, W., Lu, C.: Projection onto the capped simplex (2015). arXiv:1503.01002
29. Weymark, J.: Generalized Gini inequality indices. Math. Soc. Sci. **1**, 409–430 (1981)
30. Wierzbicki, A.: A mathematical basis for satisficing decision making. Math. Model. **3**, 391–405 (1982)
31. Yager, R.: On ordered weighted averaging aggregation operators in multi-criteria decision making. IEEE Trans. Syst. Man Cyber. **18**, 183–190 (1988)

Efficient Algorithms for Ridesharing
of Personal Vehicles

Qian-Ping Gu[1(✉)], Jiajian Leo Liang[1], and Guochuan Zhang[2]

[1] School of Computing Science, Simon Fraser University, Burnaby, Canada
{qgu,jjl24}@sfu.ca
[2] College of Computer Science and Technology,
Zhejiang University, Hangzhou, China
zgc@zju.edu.cn

Abstract. Given a set of trips in a road network, where each trip has
a vehicle, an individual and other requirements, the ridesharing problem
is to deliver all individuals to their destinations by a subset of vehicles
satisfying the requirements. Minimizing the total travel distance of the
vehicles and minimizing the number of vehicles are major optimization
goals. These minimization problems are complex and NP-hard because
each trip may have many requirements. We study simplified minimization
problems in which each trip's requirements are specified by the source,
destination, vehicle capacity, detour distance and preferred path parameters. We show that both minimization problems can be solved in polynomial time if all of the following conditions are satisfied: (1) all trips
have the same destination; (2) no detour is allowed and (3) each trip
has one unique preferred path. It is known that both minimization problems are NP-hard if any one of the three conditions is not satisfied. Our
results and the NP-hard results suggest a clear boundary between the
polynomial time solvable cases and NP-hard cases for the minimization
problems.

Keywords: Ridesharing problem · Optimization problems · Polynomial
time algorithms

1 Introduction

We consider the following ridesharing problem: given a set of trips, where each
trip has a vehicle, an individual (the owner of the vehicle), a source and destination in a road network, and other requirements, select a subset of vehicles
to deliver all individuals to their destinations satisfying the requirements of the
trips. When a vehicle is selected in the delivery, the owner of the vehicle is called
a *driver* and an individual other than a driver is called a *passenger*. Minimizing the total travel distance of drivers and minimizing the number of drivers
are major optimization problems in the ridesharing. The ridesharing has many
advantages including saving the total cost of all trips and reducing the traffic
congestion, fuel consumption and air pollution [6,13,14]. According to [3], the

© Springer International Publishing AG 2017
X. Gao et al. (Eds.): COCOA 2017, Part I, LNCS 10627, pp. 340–354, 2017.
https://doi.org/10.1007/978-3-319-71150-8_29

estimated cost of congestion in the United States is around \$121 billion per year. The congestion roughly translates to 5.5 billion hours of time wasted in traffic and 2.9 billion gallons of fuel burned. The side effects of the congestion are the extra vehicular emissions.

Despite these advantages, the ridesharing was usually organized in an ad hoc way in small scales and the ridesharing coordination was not fully regulated and organized in industry. The shared use of personal vehicles has decreased in the past decades [7] and the average occupancy rate of personal vehicles in the United States is 1.6 persons per vehicle based on reports in 2011 [8,16]. Major technical challenges for organizing the ridesharing in a large scale include efficient algorithms for selecting drivers and scheduling the delivery, and convenient communication schemes among the individuals of the trips. Privacy, safety, pricing and social discomfort may be the other hurdles for the ridesharing in large scale. Recently, systems for ridesharing in large scale emerged due to improvements in communication technologies, for example, GPS-enabled mobile phones with wireless networks make the communication among the ridesharing participants much easier. Based on these infrastructures, there are systems known as mobility on demand (MoD) systems by companies such as Uber and Lyft for ridesharing in large scale [3,12,15]. Developing efficient algorithms for selecting drivers remains a challenging issue for the ridesharing in large scale. The surveys in [2,7] review the methods for general ridesharing and approaches for encouraging the participation of the ridesharing.

The minimization problems in the ridesharing are complex and NP-hard because each trip may have many parameters. A general approach for solving the problems formulates the problems as an Integer Programming (IP) or Mixed Integer Programming (MIP) problem and solves the IP or MIP problem by an exact method or heuristics [3,4,11]. Many heuristics for solving the minimization problems in a practical environment are also known [1,12,15]. Computational studies on comparing heuristics with some general approaches such as branch and bound algorithms and IP (MIP) are given in [12]. The computational studies are based on a data set from Shanghai taxis. Computational studies on a data set from New York taxis are given in [3,15]. IP or MIP based exact algorithms are time consuming and not practical for ridesharing problems of large scale, while heuristics do not have a guarantee on the quality of solutions. Many previous works also focus on developing efficient exact algorithms for simplified variants of the minimization problems such as a single passenger at a time, at most two passengers of a driver's trip, single pick-up of a driver's trip, ignoring social parameters like pricing and so on [3–5]. A recent work in [9] gives an algorithmic analysis on exploring to what extend the simplification would make the minimization problems polynomial time solvable.

In general, a trip in the ridesharing problem has a vehicle, an individual (the owner of the vehicle), and requirements specified by many parameters such as source/destination in a road network, departure/arrival time, preferred path of the vehicle owner, distance/time detour limit the vehicle owner can tolerate for serving passengers, vehicle capacity, price, and so on. To realize a trip is to arrange a vehicle to deliver the individual in the trip to his/her destination satisfying the trip requirements. Due to a large number of parameters, the minimization problems are complex and NP-hard. The problem can be simplified by dropping some parameters. However, most previous works do not have a clear model for analyzing the relations between the complexity of the problem and the parameters. Recently, a new model is introduced in [9] for analyzing the relations between the complexity of the problem and the parameters of source, destination, vehicle capacity, detour distance limit, and preferred paths. This model provides a formal platform for algorithmic analysis of the simplified problem specified by the above parameters. It is shown in [9] that if any one of the following three conditions is not satisfied then both minimization problems remain NP-hard: (1) all trips have the same destination; (2) no detour is allowed and (3) each trip has a unique preferred path. When all of three conditions are satisfied, the work in [9] shows that a special case of minimizing the number of drivers can be solved in polynomial time: in this special case, the preferred paths of all trips lie on a same path of the road network. It is open whether each of the minimization problems is polynomial time solvable or not when all three conditions above are satisfied. In this paper, we follow the model in [9] and give positive answers for the open problems. More specifically, we give efficient algorithms for both minimization problems. Let M be the size of a ridesharing instance which contains a road network and l trips, every trip has a source, a destination, a vehicle capacity, a distance detour limit and a preferred path (a sequence of links in the road network). Our results are as follows:

- There is a dynamic programming algorithm that, given a ridesharing instance of l trips and size M satisfying Conditions (1), (2) and (3), finds a solution for the instance with the minimum total travel distance in $O(M + l^3)$ time. The algorithm also finds a solution for the instance with the minimum number of drivers in $O(M + l^3)$ time.
- There is a greedy algorithm that, given a ridesharing instance of l trips and size M satisfying Conditions (1), (2) and (3), finds a solution for the instance with the minimum number of drivers in $O(M + l^2)$ time.

There are some novel ideas in the dynamic programming algorithm: We introduce a serve relation between trips and process each trip to select drivers in an order based on the serve relation. A trip is called a source trip if it can not be served by any other trip. We work out partial solutions by selecting drivers from a subset R_1 of trips (starting from R_1 with a single source trip) and then expand the solutions to partial solutions by drivers from a subset R_2 with $R_1 \subset R_2$. Our processing order guarantees that an optimal solution can be found by at most l solution expansions. To bound the number of solutions in each expansion, we introduce a dominating relation between the solutions: a solution S dominates another solution S' if S serves at least as many trips as S' does and the total travel distance of S is at most that of S'. Then we only need to consider the solutions that are not dominated in each expansion. This makes each expansion in polynomial time possible. In the greedy algorithm, we find one partial solution by selecting drivers from a subset R_1 of trips (starting from R_1 with a single source trip). Then we try to expand the partial solution to another partial solution to serve more trips by including a new driver with the largest capacity selected from a subset R_2 of trips with $R_1 \subset R_2$. This reduces the running time by a factor of l compared to the dynamic programming approach.

Our algorithms and the NP-hardness results in [9] give a clear boundary between the polynomial time solvable cases and NP-hard cases of the minimization problems. Our algorithms may be applied to applications such as the mixed evacuation scheduling [10]. Our algorithms may also be applied to improve heuristics for more complex ridesharing problems by providing driver-passenger matchings for a subset of participants.

The rest of this paper is organized as follows. Section 2 gives the preliminaries of the paper. In Sect. 3, we show a dynamic programming algorithm for minimizing the total travel distance. Section 4 gives a greedy algorithm for minimizing the number of drivers. The final section concludes the paper.

2   Preliminaries

A graph G (undirected) consists of a set $V(G)$ of vertices and a set $E(G)$ of edges, where each edge $\{u, v\}$ of $E(G)$ is a set of two vertices in $V(G)$. A digraph H consists of a set $V(H)$ of vertices and a set $E(H)$ of arcs, where each arc (u, v) of $E(H)$ is an ordered pair of vertices in $V(H)$. A graph G (digraph H) is weighted if every edge of G (arc of H) is assigned a real number as the edge length. When the edge (arc) length is not specified, the length is one. A *path* between vertex v_0 and vertex v_k in graph G is a sequence $e_1, .., e_k$ of edges, where $e_i = \{v_{i-1}, v_i\} \in E(G)$ for $1 \leq i \leq k$ and no vertex appears more than once in the sequence. A path from vertex v_0 to vertex v_k in a digraph H is defined similarly with each $e_i = (v_{i-1}, v_i)$ an arc in H. The *length* of a path, denoted by $\mathrm{dist}(P)$, is the sum of the lengths of the edges (arcs) in P. We express a road network by a weighted graph G with non-negative edge length: $V(G)$ is the set of locations in the network, an edge $\{u, v\}$ is a road between u and v, and the length of $\{u, v\}$ is the cost to use the road (e.g., the length of the road).

In the ridesharing problem, we assume that the individual of every trip can be a driver or passenger. In general, in addition to a vehicle and individual, each trip has a source, a destination, an earliest departure time, a latest arrival time, a preferred path (e.g., a shortest path) to reach the destination, a limit on the detour distance/time from the preferred path to serve other individuals and a price limit the individual can pay if served. If all of these parameters are considered, the ridesharing problem is complex. The ridesharing problem can be simplified by considering a subset of the parameters. In an algorithmic analysis of the ridesharing problem [9], a simplified variant of the ridesharing problem is defined, where each trip has only the source, destination, vehicle capacity, distance detour limit and preferred path parameters. For the simplified problem, each trip is expressed by an integer label i and specified by $(s_i, t_i, n_i, d_i, \mathcal{P}_i)$, where

- s_i is the source (start location) of i (a vertex in G),
- t_i is the destination of i (a vertex in G),
- n_i is the number of seats (capacity) of i available for passengers,
- d_i is the detour distance limit i can tolerate for offering services, and

– $\mathcal{P}_i$ is a set of preferred paths of i from s_i to t_i in G.

We use the ridesharing problem for the simplified ridesharing problem in the rest of the paper unless otherwise stated.

When the vehicle of a trip i delivers the individual of a trip j, we say trip i *serves* trip j and call i a *driver* and j a *passenger*. A trip i can serve i itself and can serve a trip $j \neq i$ if i and j can arrive at their destinations such that j a passenger of i and the detour of i is at most d_i. A trip i can serve a set $\sigma(i)$ of trips if trip i can serve all trips of $\sigma(i)$ and the total detour of i is at most d_i. At any specific time point, a trip i can serve at most $n_i + 1$ trips. If trip i serves some trips after serving some other trips (known as *re-take passengers* in previous studies), trip i may serve more than $n_i + 1$ trips. Let (G, R) be an instance of the ridesharing problem, where G is a weighted graph and $R = \{1, .., l\}$ is a set of trips. (S, σ), where $S \subseteq R$ is a set of drivers and σ is a mapping $S \rightarrow 2^R$, is a partial solution of (G, R) if

– for each $i \in S$, i can serve $\sigma(i)$,
– for each pair $i, j \in S$ with $i \neq j$, $\sigma(i) \cap \sigma(j) = \emptyset$, and
– $\sigma(S) = \cup_{i \in S} \sigma(i) \subseteq R$.

When $\sigma(S) = R$, (S, σ) is called a solution of (G, R). For a (partial) solution (S, σ) we sometimes simply call S a (partial) solution when σ is clear from the context or not related to the discussion.

Given a (partial) solution (S, σ) of an instance, for every driver $i \in S$, let dist(i) be the travel distance of i for serving $\sigma(i)$ and dist$(S) = \sum_{i \in S}$ dist(i) be the total travel distance of S. We consider the problem of minimizing dist(S) (the total travel distance of the drivers) and the problem of minimizing $|S|$ (the number of drivers) in a solution (S, σ) for the ridesharing problem. To analyze the time complexity of the minimization problems, the following conditions are introduced in [9]:

(1) Unique destination: all trips have the same destination, that is, $t_i = D$ for every $i \in R$.
(2) Zero detour: each trip can only serve others on his/her preferred path, that is, $d_i = 0$ for every $i \in R$.
(3) Fixed path: $\mathcal{P}_i$ has a unique preferred path P_i.

It is shown in [9] that both minimization problems are NP-hard if any of Conditions (1), (2) and (3) is not satisfied. Notice that it is implicit in [9] that the *serve relation* is transitive, that is, if trip i can serve trip j and j can serve trip k then i can serve k. The NP-hard results hold when the serve relation is transitive. In this paper, we assume that the serve relation is transitive.

To deal with the trips with zero vehicle capacity ($n_i = 0$), we introduce a *pseudo serve relation* in an instance (G, R): trip i can pseudo serve trip j if

– $n_i > 0$ and i can serve j or
– $n_i = 0$ and a preferred path of j is a subpath of a preferred path of i.

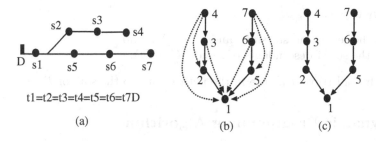

Fig. 1. (a) A set $R = \{i|1 \le i \le 7\}$ of seven trips with the same destination D, (b) the digraph with the short cuts (expressed by dashed arcs) and (c) the inverse tree in the digraph H_R.

The pseudo serve relation becomes the serve relation when $n_i > 0$ for every $i \in R$. The pseudo serve relation can be expressed by a digraph such that the digraph has R as the vertex set and there is an arc (i, j) in the digraph if trip i can pseudo serve trip j. An arc (i, j) in the digraph is called a *short cut* if after removing (i, j) from the digraph, there is a path from i to j in the digraph. We remove all short cuts from the digraph to get a digraph H_R to express the pseudo serve relation in R: trip i can pseudo serve trip j if there is a path from i to j in H_R. A trip i in H_R is called a *source* (resp. *sink*) if there is no arc (j, i) (resp. (i, j)) in H_R. A connected component of H_R is called a tree if the graph obtained from replacing every arc (i, j) in the component by an edge $\{i, j\}$ is a tree. When Conditions (2) and (3) are satisfied, if i can pseudo serve j then the preferred path of j is a subpath of the preferred path of i, implying the pseudo serve relation is transitive. When all of Conditions (1), (2) and (3) are satisfied, every connected component of H_R is a tree and we call the component an *inverse tree* as it has one unique sink and at least one source. Figure 1 gives a set R of trips, the digraph with the short cuts and the inverse tree in the digraph H_R for the pseudo serve relation in R.

Given a ridesharing instance (G, R), for any two connected components T_1 and T_2 in the pseudo serve relation digraph H_R, any trip in T_1 can not pseudo serve any trip in T_2 and vice versa. Let (S_1, σ_1) and (S_2, σ_2) be optimal solutions for trips in T_1 and trips in T_2, respectively, for each of the minimization problems. Let $S = S_1 \cup S_2$ and $\sigma(i) = \sigma_1(i)$ for $i \in S_1$ and $\sigma(i) = \sigma_2(i)$ for $u \in S_2$. Then (S, σ) is an optimal solution for trips in T_1 and trips in T_2 for each of the minimization problems. So we assume that H_R has one connected component T $(V(T) = R)$ because if H_R has more than one connected component, we can solve the minimization problems for each component independently.

The following notations will be used. Let T be a component of H_R for a ridesharing instance (G, R). For trips i and j in T,

- i is a *parent* of j if arc (i, j) is in T;
- i is an *ancestor* of j if there is a path from i to j in T;
- i is a *child* of j if arc (j, i) is in T (each trip has at most one child) and
- i is a *descendant* of j if there is a path from j to i in T.

For every trip i in T,

- A_i is the set of ancestors of i and i ($i \in A_i$), and
- D_i is the set of descendants of i and i ($i \in D_i$).

Notice that all trips of D_i are in the path from i to the sink of T.

3   Dynamic Programming Algorithm

Given an instance (G, R) of the ridesharing problem, where $R = \{1, .., l\}$, and the digraph H_R for the pseudo serve relation in R, let T be the connected component (an inverse tree) of H_R. We rearrange the labels of the trips in T by Procedure Preprocessing in Fig. 2. In the rest of this section, each trip in T is expressed by the label assigned in the procedure. Notice that trip l is a source in T, trip 1 is the sink in T, and if trip i can pseudo serve trip j then $i > j$ for every $i \neq j$. The following notations will be used in this section.

- For $R = \{1, .., l\}$ and $1 \le i \le j \le l$, $R(i, j) = \{i, i + 1, .., j\}$.
- For a set S of drivers, $S(i, j) = S \cap R(i, j)$.
- For a trip $i \in R$, v_i is the ancestor of i with the largest label.

Notice that $A_i = R(i, v_i)$ and v_i is a source. $R(i, v_i)$ is called a *branch* of T. A trip i is called a *merge point* if i has at least two parents in T. A branch is *simple* if it does not have any merge point. A branch $R(i, v_i)$ is *maximal* if the child of i is a merge point or $i = 1$ is the sink. For a (partial) solution (S, σ) of R and $1 \le i \le j \le l$, S is called a solution of $R(i, j)$ if $R(i, j) \subseteq \sigma(S)$.

We give a dynamic programming algorithm for minimizing the total travel distance. We process trips to find solutions for simple branches of T, expand these solutions to solutions of branches consisting of simple branches, and expand

Procedure Preprocess
Input: An inverse tree T of H_R of l trips.
Output: A distinct integer label i, $1 \le i \le l$, for each trip in T.
begin
    $i := l$; let ST be a stack; push the sink of T into ST;
    mark every arc in T un-visited;
    while ST $\neq \emptyset$ **do**
        let u be the trip at the top of ST;
        if there is an arc (v, u) in T un-visited **then**
            push v into ST; mark (v, u) visited;
        else
            remove u from ST; assign u integer label i; $i := i - 1$;
        endif
    endwhile
end.

Fig. 2. Procedure for assigning integer labels to trips in T.

solutions of branches to solutions of larger branches. We process trips in the decreasing order of their labels. This processing order guarantees that at most l expansions is enough. To make the number of solutions in each expansion small, we introduce a dominating relation between the solutions and only keep the non-dominated solutions for each expansion. For a solution (S, σ) of $R(i, j)$, let $\text{dist}(S) = \sum_{i \in S} \text{dist}(P_i)$ be the *cost* of S, where $\text{dist}(P_i)$ is the length of the preferred path P_i of trip i. For two solutions (S, σ) and (S', σ') of $R(i, j)$, S *dominates* S' if $|\sigma(S)| \geq |\sigma(S')|$ and $\text{dist}(S) \leq \text{dist}(S')$. Two solutions are *non-dominating* if none of them dominates the other. A set $\mathcal{X}$ of solutions is *non-dominating* if every pair of solutions in $\mathcal{X}$ is non-dominating.

3.1   Algorithm

There are three major functions in our algorithm:

- Process each trip i from l to 1. If i is a source, a set $\mathcal{X}(i, i)$ of one solution $S = \{i\}$ is computed. When we process i which is not a source, a set $\mathcal{X}(i+1, v_{i+1})$ of solutions for $R(i+1, v_{i+1})$ has been computed. For each $S \in \mathcal{X}(i+1, v_{i+1})$, we compute a solution (S', σ') with $S' = S \cup \{i\}$ to find a set $\mathcal{X}(i, v_i)$ of solutions for $R(i, v_i)$. Our algorithm makes $\sigma'(i)$ to serve as many trips in D_i that are not in $\sigma(S)$ and are closest to i as possible. More formally, we define $N(i, c, S)$ to be the set of c trips in D_i such that $N(i, c, S) \subseteq (D_i \setminus \sigma(S))$ and for any trip u in $N(i, c, S)$ and any trip v in $D_i \setminus (\sigma(S) \cup N(i, c, S))$, $\text{dist}(i \rightarrow u) < \text{dist}(i \rightarrow v)$, where $i \rightarrow u$ ($i \rightarrow v$) is the path from i to u (from i to v) in T. Let $c = \min\{n_i + 1, |D_i \setminus \sigma(S)|\}$. In our algorithm, $\sigma'(i) = N(i, c, S)$.
- Merge solutions. By processing trips, we can find a set $\mathcal{X}(i, v_i)$ of solutions for each maximal simple branch $R(i, v_i)$. For a merge point $i - 1$, let $i_1, .., i_r$ be the parents of $i - 1$ such that $i_a < i_b$ if $a < b$ (see Fig. 3). Notice that $v_{i_r} = v_{i-1}$ and $i_1 = i$. After trip i is processed, all of $\mathcal{X}(i_a, v_{i_a})$, $1 \leq a \leq r$, have been computed. We merge $\mathcal{X}(i_a, v_{i_a})$, $1 \leq a \leq r$, into a set $\mathcal{X}(i, v_{i-1})$ of solutions for $R(i, v_{i-1})$. The merge is realized by including $S'' = S \cup S'$ in $\mathcal{X}(i_{a-1}, v_{i-1})$ for every $S \in \mathcal{X}(i_{a-1}, v_{i_{a-1}})$ and every $S' \in \mathcal{X}(i_a, v_{i-1})$ for $1 < a \leq r$ (see Fig. 3).
  By processing trips and merging solutions, we find a set $\mathcal{X}(i, v_i)$ of solutions for each maximal branch $R(i, v_i)$ and finally a set $\mathcal{X}(1, l)$ of solutions for R.
- Remove dominated solutions. When we compute a set of solutions, we remove each solution in the set that is dominated by another one in the same set.

The pseudo code of our algorithm is given in Fig. 4. Below is an example explaining major steps of the algorithm. Let $R = \{i | 1 \leq i \leq 7\}$ be the set of trips in Fig. 1. Assume that the capacity n_i and travel distance $\text{dist}(P_i)$ are as follows: $(n_1 = 1, \text{dist}(P_1) = 2), (n_2 = 0, \text{dist}(P_2) = 4), (n_3 = 1, \text{dist}(P_3) = 4.5), (n_4 = 1, \text{dist}(P_4) = 5), (n_5 = 0, \text{dist}(P_5) = 2.5), (n_6 = 1, \text{dist}(P_6) = 4), (n_7 = 1, \text{dist}(P_7) = 5)$. The algorithm processes branch $R(5, 7)$ first and branch $R(2, 4)$ next. For $i = 7$ (source), $\mathcal{X}(7, 7)$ has one solution $(S = \{7\}, \sigma(7) = $

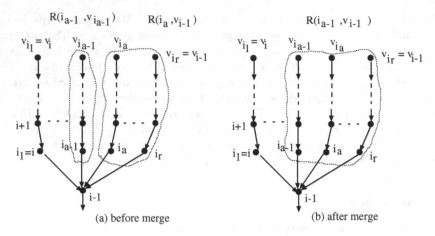

Fig. 3. Merge $\mathcal{X}(i_{a-1}, v_{i_{a-1}})$ (solutions of $R(i_{a-1}, v_{i_{a-1}})$) and $\mathcal{X}(i_a, v_{i-1})$ (solutions of $R(i_a, v_{i-1})$) into $\mathcal{X}(i_{a-1}v_{i-1})$ (solutions of $R(i_{a-1}, v_{i-1})$) for some $1 < a \leq r$. (a) $R(i_{a-1}, v_{i_{a-1}})$ and $R(i_a, v_{i-1})$, and (b) $R(i_{a-1}, v_{i-1})$.

$\{7, 6\})\}$ denoted by $(\{7\}; \{7, 6\})$. For $i = 6$ (non-source), $\mathcal{X}(6, v_6) = \mathcal{X}(6, 7)$, $\mathcal{X}(7, v_7) = \mathcal{X}(7, 7)$ and solution $(\{7\}; \{7, 6\})$ is included in $\mathcal{X}(6, 7)$. Then solution $(\{7, 6\}; \{7, 1\}, \{6, 5\})$ is included in $\mathcal{X}(6, 7)$. Since the two solutions of $\mathcal{X}(6, 7)$ are non-dominating, they are kept in $\mathcal{X}(6, 7)$. For $i = 5$ (non-source), $\mathcal{X}(5, v_5) = \mathcal{X}(5, 7)$. Solutions of $\mathcal{X}(6, 7)$ are included in $\mathcal{X}(5, 7)$. Then solutions $(\{7, 5\}; \{7, 6\}, \{5\})$ and $(\{7, 6, 5\}; \{7, 1\}, \{6\}, \{5\})$ are included in $\mathcal{X}(5, 7)$. Next, solutions $(\{7\}; \{7, 6\})$ (not a solution of $R(5, 7)$) and $(\{7, 6, 5\}; \{7, 1\}, \{6\}, \{5\})$ (dominated by the solution $(\{7, 6\}; \{7, 1\}, \{6, 5\})$) are removed from $\mathcal{X}(5, 7)$. So $\mathcal{X}(5, 7)$ has two solutions $(\{7, 5\}; \{7, 6\}, \{5\})$ and $(\{7, 6\}; \{7, 1\}, \{6, 5\})$. By processing branch $R(3, 4)$, the algorithm computes $\mathcal{X}(2, 4)$ which has two solutions $(\{4, 2\}; \{4, 3\}, \{2\})$ and $(\{4, 3\}; \{4, 1\}, \{3, 2\})$. The solutions of $\mathcal{X}(2, 4)$ and solutions of $\mathcal{X}(5, 7)$ are merged to get solutions of $\mathcal{X}(2, 7)$. In the merge, four driver sets $\{4, 2\} \cup \{7, 5\}, \{4, 2\} \cup \{7, 6\}, \{4, 3\} \cup \{7, 5\}$ and $\{4, 3\} \cup \{7, 6\}$ are computed. Solutions with driver sets $\{4, 2, 7, 6\}$ and $\{4, 3, 7, 6\}$ are dominated and removed, and $\mathcal{X}(2, 7)$ has two solutions $(\{4, 2, 7, 5\}; \{4, 3\}, \{2\}, \{7, 6\}, \{5\})$ and $(\{4, 3, 7, 5\}; \{4, 1\}, \{3, 2\}, \{7, 6\}, \{5\})$. Finally, for $i = 7$, $\mathcal{X}(1, 7)$ has one solution $(\{4, 3, 7, 5\}; \{4, 1\}, \{3, 2\}, \{7, 6\}, \{5\})$ which is an optimal solution for R.

3.2 Analysis of Algorithm

Lemma 1. *Let $R(i, v_i)$ be any maximal simple branch in T. For any solution S^* of R, there is a solution S in $\mathcal{X}(i, v_i)$ such that S dominates $S^*(i, v_i)$.*

Proof. We prove the lemma by induction for $i \leq a \leq v_i$. Since v_i is a source in T, v_i can be served only by itself. So any solution S^* of R contains v_i, implying $S^*(v_i, v_i) = \{v_i\}$. From this and $S = \{v_i\} \in \mathcal{X}(v_i, v_i)$, the lemma holds for $a = v_i$. Assume that the lemma is true for $i < a \leq v_i$ and we prove it for

Algorithm 1 Find Minimum Cost Solution
Input: An inverse tree T of H_R of l trips.
Output: A solution (S, σ) for R with dist(S) minimized.
begin
    for $i := l$ to 1 **do** /* process every trip of T, l is a source */
        if i is a source of T **then** /* process a source trip*/
            $S := \{i\}$; $c := \min\{n_i + 1, |D_i \setminus \sigma(S)|\}$; $\sigma(i) := N(i, c, S)$; $\mathcal{X}(i, i) := \{(S, \sigma)\}$;
        else /* process a non-source trip */
            $\mathcal{X}(i, v_i) := \mathcal{X}(i + 1, v_{i+1})$; /* compute $\mathcal{X}(i, v_i)$ */
            for every $(S, \sigma) \in \mathcal{X}(i + 1, v_{i+1})$ **do**
                $S' := S \cup \{i\}$; $\sigma'(S') :=$ Serve(i, S, σ); $\mathcal{X}(i, v_i) := \mathcal{X}(i, v_i) \cup \{(S', \sigma')\}$;
            endfor /* end of computing $\mathcal{X}(i, v_i)$ */
            for every solution S in $\mathcal{X}(i, v_i)$ **do** /* make $\mathcal{X}(i, v_i)$ non-dominating */
                if S is not a solution of $R(i, v_i)$ **then** remove S from $\mathcal{X}(i, v_i)$;
                if S is dominated by some S' in $\mathcal{X}(i, v_i)$ **then** remove S from $\mathcal{X}(i, v_i)$;
            endfor
            if $i - 1$ is a merge point **then** /* merge solutions */
                let $i_1, .., i_r$ be the parents of $i - 1$ with $i_a < i_b$ for $a < b$;
                for $a := r$ to 2 **do** $\mathcal{X}(i_{a-1}, v_{i-1}) :=$ Merge$(\mathcal{X}(i_{a-1}, v_{i_{a-1}}), \mathcal{X}(i_a, v_{i-1}))$;
            endif
        endif /* end of processing a non-source trip */
    endfor /* end of processing every trip of T */
    Let (S, σ) be a solution in $\mathcal{X}(1, l)$ with the minimum dist(S);
end.
Procedure Serve(i, S, σ)
begin
    $\sigma'(j) := \sigma(j)$ for every $j \in S$; $c := \min\{n_i + 1, |D_i \setminus \sigma(S)|\}$; $\sigma'(i) := N(i, c, S)$;
    if $i \in \sigma(S)$ **then**
        let $k \in S$ s.t. $i \in \sigma(k)$; $\sigma'(k) := \sigma'(k) \setminus \{i\}$; $\sigma'(k) := \sigma'(k) \cup N(k, 1, S')$;
    endif
end.
Procedure Merge$(\mathcal{X}(i_{a-1}, v_{i_{a-1}}), \mathcal{X}(i_a, v_{i-1}))$
begin
    include $S'' = S \cup S'$ in $\mathcal{X}(i_{a-1}, v_{i-1})$ for $S \in \mathcal{X}(i_{a-1}, v_{i_{a-1}})$ and $S' \in \mathcal{X}(i_a, v_{i-1})$;
    dist$(S'') :=$ dist$(S) +$ dist(S');
    set $|\sigma''(S'')|$ to $\min\{|R(i_{a-1}, v_{i-1})| + |D_{i-1}|, |\sigma(S)| + |\sigma(S')|\}$;
    remove S'' from $\mathcal{X}(i_{a-1}, v_{i-1})$ if S'' is dominated;
    for every $S'' \in \mathcal{X}(i_{a-1}, v_{i-1})$ **do**
        $\sigma''(j) := \sigma'(j)$ for $j \in S'$; $\sigma''(j) := \sigma(j) \setminus \sigma'(S')$ for $j \in S$;
        for every $j \in S$ **do** $c_j := |\sigma(j) \cap \sigma'(S')|$ and $\sigma''(j) := \sigma''(j) \cup N(j, c_j, S'')$;
    endfor
end.

Fig. 4. Algorithm for finding a solution of R with the minimum dist(S).

$a - 1$. By the induction hypothesis, there is a solution S in $\mathcal{X}(a, v_i)$ such that S dominates $S^*(a, v_i)$. Let $S' = S \cup \{a-1\}$ be the solution obtained in Algorithm 1. If $a - 1 \in S^*(a - 1, v_i)$ then from the fact that S dominates $S^*(a, v_i)$,

$$|\sigma(S')| = \min\{|D_{v_i}|, |\sigma(S)| + n_{a-1} + 1\}$$
$$\geq \min\{|D_{v_i}|, |\sigma^*(S^*(a, v_i))| + n_{a-1} + 1\} \geq |\sigma^*(S^*(a - 1, v_i))|$$

and

$$\text{dist}(S') = \text{dist}(S) + \text{dist}(a - 1)$$
$$\leq \text{dist}(S^*(a, v_i)) + \text{dist}(a - 1) = \text{dist}(S^*(a - 1, v_i)).$$

Therefore, S' dominates $S^*(a - 1, v_i)$.

Assume that $a-1 \notin S^*(a-1, v_i)$. Because S^* is a solution of R, $a-1$ is served by some trip in S^*. Further, $a - 1$ can be served only by trips in $R(a - 1, v_{a-1})$ and $v_{a-1} = v_i$. Therefore, $S^*(a, v_i)$ is a solution of $R(a-1, v_i)$. Since S dominates $S^*(a, v_i)$, S is a solution of $R(a - 1, v_i)$. If S is in $\mathcal{X}(a - 1, v_i)$ then the lemma is true. Otherwise, S is removed from $\mathcal{X}(a - 1, v_i)$ because S is dominated by a solution $S' \in \mathcal{X}(a - 1, v_i)$. This implies that S' dominates $S^*(a - 1, v_i)$ and the lemma is proved. □

Lemma 2. *Let $i - 1$ be a merge point in T such that there is no merge point in $A_{i-1} \setminus \{i - 1\}$. For any solution S^* of R, there is a solution S in $\mathcal{X}(i - 1, v_{i-1})$ computed by Algorithm 1 such that S dominates $S^*(i - 1, v_{i-1})$.*

Proof. Let $i_1, .., i_r$ be the parents of $i - 1$ such that $i_a < i_b$ if $a < b$. Then v_{i_a} is the unique source ancestor of i_a for each $1 \leq a \leq r$, $v_{i_r} = v_{i-1}$ and $i_1 = i$. We prove the following statement by induction: for every a with $1 \leq a \leq r$, there is a solution $S \in \mathcal{X}(i_a, v_{i-1})$ such that S dominates $S^*(i_a, v_{i-1})$. For $a = r$, from Lemma 1, the statement holds. Assume that the statement is true for $1 < a \leq r$ and we prove it for $a - 1$. From Lemma 1, there is a solution S in $\mathcal{X}(i_{a-1}, v_{i_{a-1}})$ such that S dominates $S^*(i_{a-1}, v_{i_{a-1}})$. From the induction hypothesis, there is a solution $S' \in \mathcal{X}(i_a, v_{i-1})$ such that S' dominates $S^*(i_a, v_{i-1})$. Let $S'' = S \cup S'$ as computed in Algorithm 1. Let $c = |R(i_{a-1}, v_{i-1})| + |D_{i-1}|$. Then

$$|\sigma''(S'')| = \min\{c, |\sigma(S)| + |\sigma(S')|\}$$
$$\geq \min\{c, |\sigma^*(S(i_{a-1}, v_{i_{a-1}}))| + |\sigma^*(S^*(i_a, v_{i-1}))|\} = |\sigma^*(S^*(i_{a-1}, v_{i-1}))|$$

and

$$\text{dist}(S'') = \text{dist}(S) + \text{dist}(S')$$
$$\leq \text{dist}(S^*(i_{a-1}, v_{i_{a-1}})) + \text{dist}(S^*(i_a, v_{i-1})) = \text{dist}(S^*(i_{a-1}, v_{i-1})).$$

That is, S'' dominates $S^*(i_{a-1}, v_{i-1})$. Therefore, there is a solution S in $\mathcal{X}(i, v_{i-1})$ such that S dominates $S^*(i, v_{i-1})$. By a similar argument for proving Lemma 1, there is a solution S in $\mathcal{X}(i - 1, v_{i-1})$ such that S dominates $S^*(i - 1, v_{i-1})$. □

Lemma 3. *For any solution* (S^*, σ^*) *of* R*, there is a solution* $(S, \sigma) \in \mathcal{X}(1, l)$ *computed by Algorithm 1 such that* S *dominates* S^*.

Proof. If there is no merge point in T, then by Lemma 1, the lemma holds. If there is one merge point $i - 1$ in T, then by Lemma 2, there is a solution $S \in \mathcal{X}(i - 1, v_{i-1})$ such that S dominates $S^*(i - 1, v_{i-1})$. Since $i - 1$ is the only merge point of T, $v_{i-1} = l$, D_{i-1} is the path consisting of all trips from $i - 1$ to 1. By a similar argument for proving Lemma 1, there is a solution S in $\mathcal{X}(1, l)$ such that S dominates S^*.

Assume that $u_1, .., u_s$, $1 < s$, are the merge points in T such that $u_a < u_b$ if $a < b$. For each u_a, $1 \leq a \leq s$, if the child of u_a is a merge point then let $w_a = u_a$, otherwise let w_a be the trip in D_{u_a} such that $R(w_a, v_{w_a})$ is a maximal branch and there is no merge point other than u_a in the path from u_a to w_a in T. We prove the following statement by induction: for $1 \leq a \leq s$, there is a solution S in $\mathcal{X}(w_a, v_{w_a})$ such that S dominates $S^*(w_a, v_{w_a})$. For $a = s$, $A_{u_a} \setminus \{u_a\}$ does not contain any merge point. By Lemma 2, there is a solution $S \in \mathcal{X}(u_a, v_{u_a})$ such that S dominates $S^*(u_a, v_{u_a})$. Then by a similar argument for proving Lemma 1, there is a solution S in $\mathcal{X}(w_a, v_{w_a})$ such that S dominates $S^*(w_a, v_{w_a})$, implying the induction base. Assume that the statement holds for $1 < a \leq s$ and we prove it for $a - 1$. If $A_{u_{a-1}} \setminus \{u_{a-1}\}$ does not have any merge point then by Lemma 2 and a similar argument for proving Lemma 1, the statement holds for $a - 1$. Assume that $A_{u_{a-1}} \setminus \{u_{a-1}\}$ has some merge point. Notice that for every merge point $u_j \in A_{u_{a-1}} \setminus \{u_{a-1}\}$, $a \leq j$. Let $i_1, .., i_r$ be the parents of u_{a-1}. By the induction hypothesis and a similar argument for proving Lemma 1, for every $1 \leq b \leq r$, there is a solution S in $\mathcal{X}(i_b, v_{i_b})$ such that S dominates $S^*(i_b, v_{i_b})$. By a similar argument for proving Lemma 2, there is a solution S in $\mathcal{X}(u_{a-1}, v_{u_{a-1}})$ such that S dominates $S^*(u_{a-1}, v_{u_{a-1}})$. Then by a similar argument for proving Lemma 1, there is a solution S in $\mathcal{X}(w_{a-1}, v_{w_{a-1}})$ such that S dominates $S^*(w_{a-1}, v_{w_{a-1}})$. Therefore, the statement holds, implying the lemma. $\square$

Theorem 1. *There is an algorithm that, given a ridesharing instance* (G, R) *of size* M *and* l *trips satisfying all of Conditions (1), (2) and (3), computes a solution* (S, σ) *for* R *with* $\text{dist}(S)$ *minimized in* $O(M + l^3)$ *time.*

Proof. Let S^* be a solution for R with the minimum $\text{dist}(S^*)$. By Lemma 3, Algorithm 1 finds a solution S with $\text{dist}(S) \leq \text{dist}(S^*)$ in $\mathcal{X}(1, l)$. Algorithm 1 computes $\mathcal{X}(i, v_i)$ for every trip i. Since $\mathcal{X}(i + 1, v_{i+1})$ is non-dominating, the solutions of $\mathcal{X}(i + 1, v_{i+1})$ can be listed as $S_1, S_2, ..$ such that $|\sigma(S_a)| < |\sigma(S_b)|$ and $\text{dist}(S_a) > \text{dist}(S_b)$ for $a < b$. So there are $O(l)$ solutions in $\mathcal{X}(i + 1, v_{i+1})$ because $|\sigma(S_a)| \leq l$ for every S_a in $\mathcal{X}(i+1, v_{i+1})$. For every $S_a \in \mathcal{X}(i + 1, v_{i+1})$, the solution $S'_a \in \mathcal{X}(i, v_i)$ can be computed in $O(l)$ time. So, it takes $O(l^2)$ time to compute all solutions in $\mathcal{X}(i, v_i)$. It takes $O(l)$ time to make $\mathcal{X}(i, v_i)$ non-dominating. Therefore, it takes $O(l^3)$ time to process all of the l trips.

In Algorithm 1, Procedure Merge is called $O(l)$ times. We play a small trick to reduce the running time of the procedure: we compute $|\sigma''(S'')|$ before $\sigma''(j)$ is actually decided for each $j \in S''$. This allows us to get a non-dominating set of solutions. Then we decide $\sigma''(j)$ only for non-dominating solutions ($O(l)$

many) instead of all $S'' = S \cup S'$ ($O(l^2)$ many). In each call, it takes $O(l)$ time to compute $|R(i_{a-1}, v_{i-1})| + |D_{i-1}|$ and there are $O(l^2)$ solutions $S \cup S'$ for $S \in \mathcal{X}(i_{a-1}, v_{i_{a-1}})$ and $S' \in \mathcal{X}(i_a, v_{i-1})$ because each of $\mathcal{X}(i_{a-1}, v_{i_{a-1}})$ and $\mathcal{X}(i_a, v_{i-1})$ is non-dominating and thus has $O(l)$ solutions. It takes $O(1)$ time to compute $\text{dist}(S'')$ and $|\sigma''(S'')|$. Therefore, it takes $O(l^2)$ time to compute $O(l^2)$ solutions for $\mathcal{X}(i_{a-1}, v_{i-1})$ and $O(l^2)$ time to make $\mathcal{X}(i_{a-1}, v_{i-1})$ non-dominating. Then there are $O(l)$ non-dominating solutions in $\mathcal{X}(i_{a-1}, v_{i-1})$. For each $S'' \in \mathcal{X}(i_{a-1}, v_{i-1})$, it takes $O(l)$ time to compute $\sigma''(j)$ for every $j \in S''$. Therefore, each execution of Procedure Merge takes $O(l^2)$ time and total time for merge operations is $O(l^3)$ time. So the total time of Algorithm 1 is $O(l^3)$.

It takes $O(M)$ time to compute H_R. The preprocessing for rearranging labels of trips takes $O(l)$ time. Therefore, the theorem holds.   □

If we set $\text{dist}(P_i) = 1$ for every $i \in R$ then by Theorem 1, the following result holds.

Theorem 2. *There is an algorithm that, given a ridesharing instance (G, R) of size M and l trips satisfying all of Conditions (1), (2) and (3), computes a solution (S, σ) for R with $|S|$ minimized in $O(M + l^3)$ time.*

4   Greedy Algorithm for Minimizing Number of Drivers

We give an $O(M + l^2)$ time algorithm for minimizing the number of drivers. This algorithm is more efficient than the $O(M + l^3)$ time algorithm in Theorem 2. Given a component T of H_R and a partial solution (S, σ) for R, let A_T be the set of sources in T. Our algorithm processes every trip in T starting from a trip in A_T. When a trip x in A_T is processed, x is included in a partial solution (S, σ) and x serves as many trips in D_x that are not served by S and are closest to x as possible. Recall that $N(x, c, S)$ is the set of c such trips as defined in Sect. 3.1. When x is processed, $\sigma(x) = N(x, c, S)$, where $c = \min\{n_x + 1, |D_x \setminus \sigma(S)|\}$. A trip x is *marked* if x is assigned as a driver or a passenger by the algorithm. Each trip $x \notin A_T$ is processed only if all ancestors of x have been marked by the algorithm and $|\sigma(v)| = n_v + 1$ for every $v \in S \cap A_x$. When trip x is processed, a trip u with the largest capacity n_u is selected from $A_x \setminus S$ as a driver and included in S. Our algorithm makes u serve as many trips in D_u that are not served by S and are closest to u as possible, that is, $\sigma(u) = N(u, c, S)$, where $c = \min\{n_u + 1, |D_u \setminus \sigma(S)|\}$. At any execution point of the algorithm, whenever a trip is assigned to be a driver, it remains as a driver throughout the algorithm. On the other hand, an assigned passenger can be changed to a driver when a new trip is processed. The pseudo code of the algorithm is given in Fig. 5. For the algorithm, we have the following result (proof omitted due to space limit).

Lemma 4. *Given a ridesharing instance (G, R) satisfying all of Conditions (1), (2) and (3), Algorithm 2 finds a solution (S, σ) for R with the minimum $|S|$.*

Algorithm 2 Find Minimum Number of Drivers.
Input: An inverse tree T of H_R of l trips.
Output: A solution (S, σ) for T with $|S|$ minimized.
begin
    $S := \emptyset$; let A_T be the set of sources in T;
    for every trip $v \in A_T$ **do**
        $x := v$; /* initialization */
        while $(x \in A_T)$ **or** (all trips in $A_x \setminus \{x\}$ are marked) **do**
            let u be a trip in $A_x \setminus S$ with the largest n_u;
            if $u \neq x$ **then**
                let $k \in S$ s.t. $u \in \sigma(k)$; $\sigma(k) := (\sigma(k) \setminus \{u\}) \cup \{x\}$; mark x;
            endif
            $c := \min\{n_u + 1, |D_u \setminus \sigma(S)|\}$; $S := S \cup \{u\}$;
            $\sigma(u) := N(u, c, S)$; mark all trips in $\sigma(u)$;
            if (all trips in D_u are marked) **then**
                break the **while** loop;
            else
                let x be the unmarked trip in D_u with the minimum dist$(u \to x)$ in T;
            endif
        endwhile
    endfor
end.

Fig. 5. Algorithm for minimizing the number of drivers.

Theorem 3. *There is an algorithm that, a ridesharing instance of size M and l trips satisfying Conditions (1), (2) and (3), computes a solution for the instance with the minimum number of drivers in $O(M + l^2)$ time.*

Proof. By Lemma 4, Algorithm 2 finds an optimal solution for an input instance. It takes $O(l)$ time to find an unmarked x, $O(l)$ time to check if all vertices in $A_x \setminus \{x\}$ are marked, $O(l)$ time to find u with the largest n_u and $O(l)$ time to compute $\sigma(u)$ in each iteration of the while loop. Therefore, the running time of Algorithm 2 is $O(l^2)$. It takes $O(M)$ time to compute H_R. Thus, the theorem holds. $\square$

5   Concluding Remarks

We proposed polynomial time algorithms for minimizing the total travel distance of drivers and minimizing the number of drivers for the simplified ridesharing problem satisfying Conditions (1), (2) and (3). It is known that if any one of the three conditions is not satisfied then both minimization problems are NP-hard. These results suggest a clear boundary between the polynomial time solvable cases and NP-hard cases of simplified ridesharing problems. It is worth developing approximation algorithms for the NP-hard cases and exploring the algorithmic complexity of other simplified variants of ridesharing problem.

It is interesting to study the applications of our algorithms to more complex ridesharing problems and other problems such as the evacuation problems.

Acknowledgement. The authors thank anonymous reviewers for their constructive comments. The work was partially supported by Canada NSERC Engage/Discovery Grants and China NSFC Grant 11531014.

References

1. Agatz, N., Erera, A., Savelsbergh, M., Wang, X.: Dynamic ride-sharing: a simulation study in metro atlanta. Transp. Res. Part B **45**(9), 1450–1464 (2011)
2. Agatz, N., Erera, A., Savelsbergh, M., Wang, X.: Optimization for dynamic ridesharing: a review. Eur. J. Oper. Res. **223**, 295–303 (2012)
3. Alonso-Mora, J., Samaranayake, S., Wallar, A., Frazzoli, E., Rus, D.: On-demand high-capacity ride-sharing via dynamic trip-vehicle assignment. Proc. Natl. Acad. Sci. (PNAS) **114**(3), 462–467 (2017)
4. Baldacci, R., Maniezzo, V., Mingozzi, A.: An exact method for the car pooling problem based on Lagrangean column generation. Oper. Res. **52**(3), 422–439 (2004)
5. Calvo, R.W., de Luigi, F., Haastrup, P., Maniezzo, V.: A distributed geographic information system for daily car pooling problem. Comput. Oper. Res. **31**(13), 2263–2278 (2004)
6. Chan, N.D., Shaheen, S.A.: Ridesharing in North America: past, present, and future. Transp. Rev. **32**(1), 93–112 (2012)
7. Furuhata, M., Dessouky, M., Ordóñez, F., Brunet, M., Wang, X., Koenig, S.: Ridesharing: the state-of-the-art and future directions. Transp. Res. Part B Methodol. **57**, 28–46 (2013)
8. Ghoseiri, K., Haghani, A., Hamedi, M.: Real-time rideshare matching problem. Final Report of UMD-2009-05, U.S. Department of Transportation (2011)
9. Gu, Q.-P., Liang, J.L., Zhang, G.: Algorithmic analysis for ridesharing of personal vehicles. In: Chan, T.-H.H., Li, M., Wang, L. (eds.) COCOA 2016. LNCS, vol. 10043, pp. 438–452. Springer, Cham (2016). https://doi.org/10.1007/978-3-319-48749-6_32
10. Hanawa, Y., Higashikawa, Y., Kamiyama, N., Katoh, N., Takizawa, A.: The mixed evacuation problem. In: Chan, T.-H.H., Li, M., Wang, L. (eds.) COCOA 2016. LNCS, vol. 10043, pp. 18–32. Springer, Cham (2016). https://doi.org/10.1007/978-3-319-48749-6_2
11. Herbawi, W., Weber, M.: The ridematching problem with time windows in dynamic ridesharing: a model and a genetic algorithm. In: Proceedings of ACM Genetic and Evolutionary Computation Conference (GECCO), pp. 1–8 (2012)
12. Huang, Y., Bastani, F., Jin, R., Wang, X.S.: Large scale real-time ridesharing with service guarantee on road networks. Proc. VLDB Endow. **7**(14), 2017–2028 (2014)
13. Kelley, K.: Casual carpooling enhanced. J. Pub. Transp. **10**(4), 119–130 (2007)
14. Morency, C.: The ambivalence of ridesharing. Transportation **34**(2), 239–253 (2007)
15. Santi, P., Resta, G., Szell, M., Sobolevsky, S., Strogatz, S.H., Ratti, C.: Quantifying the benefits of vehicle pooling with shareability networks. Proc. Natl. Acad. Sci. (PNAS) **111**(37), 13290–13294 (2014)
16. Santos, A., McGuckin, N., Nakamoto, H.Y., Gray, D., Liss, S.: Summary of travel trends: 2009 national household travel survey. Technical report, US Department of Transportation Federal Highway Administration (2011)

Cost-Sharing Mechanisms for Selfish Bin Packing

Chenhao Zhang and Guochuan Zhang[✉]

College of Computer Science, Zhejiang University, Hangzhou, China
{zchrea,zgc}@zju.edu.cn

Abstract. The selfish bin packing problem (SBP) considers the classical bin packing problem in a game-theoretic setting where each item is controlled by a selfish agent. It is well-known that the classical bin packing problem admits an asymptotic fully polynomial-time approximation scheme (AFPTAS). However, all previously-studied cost-sharing mechanisms for the selfish bin packing problem (SBP) have PoA greater than 1.6. Obviously, there is quite a big gap between the results of the two highly-related problems. In this paper, we revisit the SBP and find more efficient mechanisms for SBP to narrow the gap. We first present a simple mechanism with $PoA = 1.5$, which significantly improves previous bounds. We observe that for a large class of mechanisms for the SBP, 1.5 is actually a lower bound of PoA. Finally, we propose new rules for the SBP which lead a better mechanism with $PoA \leq 1.467$.

1 Introduction

Suppose we are given a set of items $N = \{1, 2, \ldots, n\}$ and sufficiently many bins of unit capacity 1. The item i has size $s_i \in (0, 1]$, for $i = 1, 2, \ldots, n$. The load $s(B)$ of a bin B is defined to be the total size of items packed in it. In the classical bin packing problem which was first introduced in the 1970s [7,11], we need to pack the items into as few bins as possible while ensuring the load of each bin does not exceed its capacity. The classical bin packing problem is shown to be NP-hard, yet there exists an asymptotic fully polynomial-time approximation scheme [6], which has an asymptotic approximation ratio arbitrarily close to 1.

The selfish bin packing problem (SBP) considers bin packing in a game-theoretic setting. In the SBP, each bin has unit cost 1 which is shared by all items in it. Each item is controlled by a selfish agent who wants to minimize the shared cost. An item chooses freely a bin it resides in as long as the capacity constraints are not violated. In order to minimize the social cost (i.e. the number of bins used), the design of cost-sharing mechanisms is critical.

To measure the efficiency of a mechanism, the price of anarchy (PoA) [8] is commonly used. The PoA is defined to be the supremum of the ratio between the social cost of the worst equilibrium and the optimal solution without selfish behavior over all instances, which resembles the approximation ratio in the

Research partially supported by NSFC (11531014, 11671355).

X. Gao et al. (Eds.): COCOA 2017, Part I, LNCS 10627, pp. 355–368, 2017.
https://doi.org/10.1007/978-3-319-71150-8_30

analysis of the approximation algorithms. Let BP be the set of all instances of the SBP. Let $OPT(G)$ be the number of bins used in the optimal solution of the instance G, $n(\pi)$ be the number of bins used in the packing π and $NE_{\mathbf{c}}(G)$ be the set of all Nash equilibria (NE) of G under a given mechanism $\mathbf{c}$. The PoA of the mechanism $\mathbf{c}$ is defined formally [4] as

$$PoA(\mathbf{c}) = \limsup_{OPT(G) \to \infty} \sup_{G \in BP} \max_{\pi \in NE_{\mathbf{c}}(G)} \frac{n(\pi)}{OPT(G)}.$$

Previous Results. The SBP was first studied by Bilò [2] with applications in the bandwidth cost sharing problem in non-cooperative networks. He analyzed the proportional cost-sharing rule, under which an item with size s in a bin B has a cost $s/s(B)$. He proved that any feasible packing always converges to an NE in finite steps and the PoA is in between 1.6 and 1.67. Epstein and Kleiman [4] later improved the by showing that PoA is in between 1.6416 and 1.6428. Yu and Zhang [12] also independently obtained the same lower bound and an upper bound of 1.6575. Epstein et al. [5] proved that for the parametric case where all items have size at most $\frac{1}{t}$ (t is a positive integer), PoA is upper bounded by

$$\frac{2t^3 + t^2 + 2}{(2t + 1)(t^2 - t + 1)}.$$

Particularly, if all items have size no greater than $\frac{1}{2}$, i.e., when $t = 2$, the upper bound of PoA is $22/15 \approx 1.467$.

Ma et al. [10] studied the equally sharing mechanism for SBP, under which the cost of a bin is equally shared among all items in it regardless of their size. They proved that the mechanism has $PoA = 1.7$.

There are other research topics [1,3,9] focusing on different aspects or variants of the problem. However, none of the previous research provides a mechanism with $PoA \leq 1.64$ for the original SBP, despite that $1 + \varepsilon$ is well known to be asymptotically reachable in non-selfish environments. Therefore, designing mechanisms for SBP with a smaller PoA is of great significance.

Our Contribution. Note that the proportional cost-sharing rule works better if item sizes are small. It advises us to pay attention to large items (those of size larger than $1/2$). More specifically, we aim at mechanisms under which small items are willing to share a bin with a large item. Along this line, in this paper, we first propose a simple mechanism with $PoA = 1.5$, which significantly improves the previous results. We strengthen the result by proving that 1.5 is a lower bound PoA for a large class of mechanisms. We then figure out several key properties which a mechanism with a smaller PoA needs to satisfy. Based on this, we derive a better mechanism achieving a $PoA \leq 1.467$, which clearly moves the bottleneck of the general case to the special case with only small items.

2   A Simple Mechanism

Let the item set be $N = \{1, 2, \ldots, n\}$, where the item i has size $s_i \in (0, 1]$. In a slight abuse of notation, we also use s_i to denote the item i. A packing π is defined to be a partition of the items $N = B_{(1)} \sqcup B_{(2)} \cdots \sqcup B_{(m)}$ such that $s(B_{(i)}) \leq 1$, $i = 1, \ldots, m$ where $s(B_{(i)}) = \sum_{s \in B_{(i)}} s_i$. Each $B_{(i)}$ denotes a bin and also the set of items in that bin. A bin B is *full* if $s(B) = 1$. We use B_j and B'_j to denote, respectively, the bin in two packings π and π' in which j resides. Let $m = n(\pi)$ be the social cost of the packing π.

Given a mechanism, let c_i denote the cost of item i under the mechanism in the corresponding packing. In this paper, all mechanisms we discussed are cost-sharing mechanisms which satisfy $\sum_{i \in B} c_i = 1$ for all packings.

Let π' be a packing and c'_i be the cost of item i in π'. π' is said to be a *selfish improvement* of another packing π with respect to s_i if

$$B'_j = B_j, \ \forall j \neq i \ \text{and} \ c'_i < c_i,$$

in other words, s_i can pay strictly less by unilaterally moving from B_i to B'_i. A packing π^* is an NE under the mechanism if it has no selfish improvement with respect to any item.

We call an item s_i a *large item* if $s_i > \frac{1}{2}$, otherwise we call it a *small item*. We call a bin B a *large bin* if B contains a large item, otherwise we call it a *small bin*. These definitions apply to any feasible packing.

We propose the Large-Pay-Pubic (LPP) mechanism that encourages small items to be packed with the large ones by requiring the large items to pay for the unoccupied space of the bins they reside in. The cost-sharing scheme of the LPP mechanism is stated as follows.

Large-Pay-Pubic Mechanism

Given a packing, if s_i is packed in a large bin B_i,

$$c_i = \begin{cases} 1 - s(B_i) + s_i & s_i > \frac{1}{2} \\ s_i & s_i \leq \frac{1}{2}. \end{cases}$$

If s_i is packed in a small bin B_i,

$$c_i = \frac{s_i}{s(B_i)}.$$

The intuition of the mechanism is simple: For every large bin, the large item pays for the public space (the unoccupied space) and its private space. Each small item pays for its private space. For every small bin, the cost is shared using the proportional rule. From another point of view, every large item is required to pay for a whole bin first, then it "subleases" the remaining space to small items.

Remark. Under the LLP mechanism, a small item prefers to stay in a large bin than any non-full small bin. For two small bins, a small item prefers to stay in the one with greater load. A large item always prefers to stay in a bin with greater load.

We now show that, an NE always exists under the LPP mechanism.

Theorem 1. *Under the LPP mechanism, any packing π will converge to an NE packing π^* in a finite number of selfish improving steps.*

Proof. We define a vector-valued potential function $P(\cdot)$ on the set of all packings. Given any packing $\pi = \{B_{(1)}, B_{(2)}, \ldots, B_{(m)}\}$,

$$P(\pi) = (s(B'_{(1)}), \ldots, s(B'_{(l)}), s(B'_{(l+1)}), \ldots, s(B'_{(m)})),$$

where $B'_{(1)}, \ldots, B'_{(l)}$ are large bins in non-increasing order of their load and $B'_{(l+1)}, \ldots, B'_{(m)}$ are small bins in non-increasing order of their load.

We define an order $\succ$ on the set of vectors. For two vectors $u = (u_1, \ldots, u_s)$ and $v = (v_1, \ldots, v_t)$, $u \succ v$ if $s < t$ or $s = t$ but u is lexicographically greater than v.

If π is not an NE, let π' be a selfish improvement of π with respect to s_i. If s_i is a large item, then $P(\pi') \succ P(\pi)$ since it must be the case that $s(B'_i) > s(B_i)$. If s_i is a small item, then either B'_i is large and B_i is small or $s(B'_i) > s(B_i)$. This also means that $P(\pi') \succ P(\pi)$. Since the number of packings is finite, π will always converge to an NE in a finite number of steps.

It is not hard to show that the LPP mechanism has $PoA \leq 1.5$. Let G be any instance of the SBP and $OPT(G)$ be number of bins used by an optimal solution.

Lemma 1. *In any NE packing, all small bins except the one with smallest load contain at least two items.*

Proof. If the lemma is not true, there is a bin B other than the one $\hat{B}$ with smallest load, containing a single item. Both of B and $\hat{B}$ have load at most $\frac{1}{2}$. Hence, the items in $\hat{B}$ have incentive to move to B. It is a contradiction.

Lemma 2. *The LPP mechanism has $PoA \leq 1.5$.*

Proof. Consider any NE packing π^*. Let n_L denote the number of large bins and n_S denote the number of small bins used in π^*.

Case 1. $n_L \geq 2n_S$. Since $OPT(G) \geq n_L$, we have

$$n(\pi^*) = n_L + n_S \leq \frac{3}{2}n_L \leq \frac{3}{2}OPT(G).$$

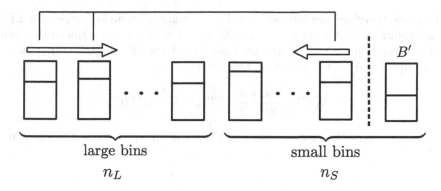

Fig. 1. Reordered bins

Case 2. $n_L \leq 2n_S - 1$. We reorder all bins in π^* as in Fig. 1: We place all large bins before the small bins and arrange the small bins in non-increasing order of their load. We call the small bin containing a single item B' and put it aside if such a bin exists.

We group two bins from the left and one bin from the right in the list until there are at most 2 bins left. Let k be the number of groups. We claim that each such group has a total load at least 2. Since $n_L \leq 2n_S - 1$, there are at most 2 large bins in any group and hence the rightmost bin in each group must be a small bin.

If the rightmost bin in a group is full, it is obvious that the group has a load greater than 2 as the left two bins in the group must have a total load greater than 1.

Now suppose the rightmost bin in a group is not full. By Lemma 1, it contains at least 2 items. Since π^* is an NE, each item in the rightmost bin does not fit in any of the left bins, otherwise they would have incentive to move under the LPP mechanism. Hence the total load of the group must be greater than 2.

Therefore, the total size of items in the k groups is larger than $2k$. We now take the ungrouped bins into account.

1. Suppose there is no bin except B' left ungrouped. The total size of the items is strictly greater than $2k$ taking the item in B' into account. Thus, $OPT(G) \geq 2k + 1$ and $n(\pi^*) = n_L + n_S = 3k + 1$. Therefore

$$n(\pi^*) \leq \frac{3k+1}{2k+1}OPT(G) \leq \frac{3}{2}OPT(G)$$

2. Suppose there is one bin left besides B'. Since π^* is an NE under the LPP mechanism, the item in B' can not fit in the bin left. Hence the total size of the items is strictly greater than $2k + 1$. Thus, $OPT(G) \geq 2k + 2$ and $n(\pi^*) = n_L + n_S = 3k + 2$. Therefore

$$n(\pi^*) \leq \frac{3k+2}{2k+2}OPT(G) \leq \frac{3}{2}OPT(G)$$

3. Suppose there are two bins left besides B'. Since π^* is an NE under the LPP mechanism, the item in B' can not fit in either of the bins. Hence the total size of the items is strictly greater than $2k+1$ too. Thus, $OPT(G) \geq 2k+2$ and $n(\pi^*) = n_L + n_S = 3k+3$. Therefore

$$n(\pi^*) \leq \frac{3k+3}{2k+2}OPT(G) \leq \frac{3}{2}OPT(G)$$

It is easy to show the same result using the similar arguments if B' does not exist.

3   A Lower Bound of PoA

By giving a family of instances, we now show that the $PoA = 1.5$ is tight for the LPP mechanism. This family of instances also applies to a large class of mechanisms, providing a lower bound of PoA for the class.

Consider the following instances: for any positive integer k, let $\varepsilon = \frac{1}{2k}$. The item set contains $2k$ large items with size $\frac{1}{2} + \varepsilon$ and $2k^2$ small items with size ε.

There exists an NE packing π^* that uses $3k$ bins as shown in Fig. 2. In π^*, each large item occupies a bin alone and every set of $2k$ small items stays together in a full bin. The cost of each small item is ε, which is exactly the same as its size. Thus, none of small items will be strictly better off by moving to the bins with large items under the LLP mechanism.

However, the optimal packing shown in Fig. 3 only uses $2k+1$ bins. In this packing, each large item is packed with $k-1$ small items. The remaining $2k$ small items have total size $2k \cdot \frac{1}{2k} = 1$ and they fit in a single bin.
Since

$$\lim_{k \to \infty} \frac{n(\pi^*)}{OPT(G)} = 1.5,$$

we conclude the following theorem.

Lemma 3. *The LLP mechanism has $PoA = 1.5$.*

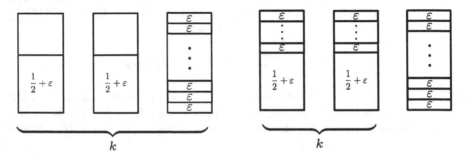

Fig. 2. $n(\pi^*) = 3k$ Fig. 3. $OPT(G) = 2k+1$

It is easy to see that, the family of instances above also holds for any mechanism under which an item has cost no smaller than its size.

Theorem 2. *Any cost-sharing mechanism under which $c_i \geq s_i, \forall i$ in all packings has PoA ≥ 1.5.*

Proof. We only need to show that π^* in the previous instance is an NE for all such mechanisms as well. Obviously, any large item can move nowhere to reduce its cost. If any small item in a full bin has cost at least ε, then the cost must be exactly ε, otherwise the total cost of items in that bin would exceed 1, which contradicts the property of cost-sharing mechanisms. Hence, all small items already have the smallest possible cost under such mechanisms, and they have no incentive to move.

In order to design better mechanisms, we need to find out what kind of properties such a mechanism should have. The following property is justified by our previous discussions.

Observation 1. A small item should be offered a discount, i.e., it should have cost strictly less than its size, when packed with a large item.

Now suppose we let a small item pay an amount less than its size when packed with a large item. However, it does not work if we simply let the small item pay nothing when packed with a large item. As we shall see, a mechanism still has $PoA \geq 1.5$ as long as a small item gets a constant discount regardless of the load of the bin it resides in.

Consider the following family of instances: let k be any positive integer and $\varepsilon = \frac{1}{2k}$. The item set contains $2k$ large items with size $\frac{1}{2} + \varepsilon$, $2k$ small items with size $\frac{1}{2} - \varepsilon$ and $2k$ small items with size ε.

The packing π^* shown in Fig. 4 uses $3k$ bins. It's not hard to see that π^* is an NE for any mechanism under which a small item gets a constant discount when staying with a large one. The items with size ε have no incentive to get together while the items with size $\frac{1}{2} - \varepsilon$ do not fit in the large bins.

We have the optimal packing shown in Fig. 5 which uses $2k + 1$ bins. In this packing, the $2k$ small items with size ε are packed in a single bin as their total size is 1. Therefore, we know that 1.5 is still a lower bound of the PoA.

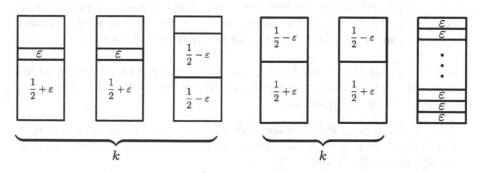

Fig. 4. $n(\pi^*) = 3k$ **Fig. 5.** $OPT(G) = 2k + 1$

Observation 2. A small item should get greater discount when it is packed in a large bin with greater load.

The two properties above are all in the interest of small items. However, if we can not guarantee that a large item gets better off when the load of the bin it resides gets greater, it may "escape" to a bin with smaller load to reduce its cost. Some small items, of course, will then follow it to the new bin. The large item may then escape again. The instability of such mechanisms may lead to the non-existence of NE. Hence, in addition to the two properties above, we also force mechanisms to have the property that a large item pays less when it resides in a bin with greater load.

Observation 3. The cost of a large item should be lower when it is packed in a bin with greater load.

4   A Better Mechanism

Based on our discussions in the previous section, we now propose the Discount-Sharing (DS) mechanism for the SBP.

Discount-Sharing Mechanism

Let $\beta \in (0, \frac{2}{3}]$ and the discount factor $f(x) = 1 - \beta x$. Given a packing, if s_i is packed in a large bin B_i,

$$c_i = \begin{cases} 1 - f(s(B_i)) \cdot (s(B_i) - s_i) & s_i > \frac{1}{2} \\ s_i \cdot f(s(B_i)) & s_i \leq \frac{1}{2}. \end{cases}$$

If s_i is packed in a small bin B_i,

$$c_i = \frac{s_i}{s(B_i)}.$$

Under the DS mechanism, a small item in a large bin pays the amount of cost that equals its size multiplied by the discount factor. The discount factor decreases as the load of the bin increases. The large item in the bin pays the remaining cost. The cost in a small bin is still shared by all its small items using the proportional rule.

We claim that the DS mechanism has the following properties that we discussed in the last section (assuming B_i and B_i' are the two bins accommodating the item s_i in different packings):

1. For a small item s_i, if B_i is a large bin and B_i' is a small bin then $c_i < c_i'$.
2. For a small item s_i, if B_i and B_i' are of the same type (both are large or small), then $s(B_i) > s(B_i') \Leftrightarrow c_i < c_i'$.
3. For a large item s_i, $s(B_i) > s(B_i') \Leftrightarrow c_i < c_i'$.

The first two properties hold trivially according to the definition of the DS mechanism. It remains to show that the property 3 holds.

Lemma 4. *For any large item s_i under the DS mechanism, if $s(B_i) > s(B_i')$, then $c_i < c_i'$.*

Proof. Consider the cost c_i of the large item s_i. Let $b = s(B_i)$.

$$c_i = 1 - (1 - \beta b) \cdot (b - s_i) = \beta b^2 - (1 + \beta s_i)b + s_i + 1.$$

Differentiate c_i with respect to b

$$\frac{dc_i}{db} = 2\beta b - (1 + \beta s_i).$$

Since $s_i > \frac{1}{2}$, we have

$$\frac{dc_i}{db} = 2\beta b - (1 + \beta s_i) < 2\beta b - (1 + \beta \cdot \frac{1}{2}).$$

From $b \leq 1$, we know that

$$\frac{dc_i}{db} < 2\beta - (1 + \beta \cdot \frac{1}{2}) = \frac{3}{2}\beta - 1.$$

As $\beta \in (0, \frac{2}{3}]$, we have

$$\frac{dc_i}{db} \leq 0,$$

which implies that the cost decreases as the bin load increases. The lemma holds.

By using the same potential function as Sect. 2, it's easy to show that any packing under the DS mechanism will converge to an NE in a finite number of selfish-improving steps. Note that the DS mechanism is reduced to the known proportional cost-sharing rule if only small items are present. Recall that the proportional rule has a $PoA \leq 1.467$ for this special case [5]. In this section, we show that the DS mechanism has a $PoA \leq 1.467$ as well for the general case, which pushes the bottleneck to the special case.

We use similar techniques as that in [5]. The main tool is the *weight function method*. Given any NE packing, we assign a weight to each item according to its size and the bin it resides in. We then prove that all but a constant number of bins in the NE packing each have a total weight at least 1. Since an optimal solution must also pack all the items, we get an upper bound of the PoA by upper-bounding the average weight of each bin in the optimal solution.

Consider any NE packing π^* under the DS mechanism. We classify the bins of π^* as follows: for a large bin, we call it an *L-bin* if it is occupied by a large item alone, otherwise we call it an *M-bin*. For a small bin, we call it an *X-bin* if it has load no less than $\frac{5}{6}$, otherwise we call it a *Y-bin*.

Lemma 5. *In an NE packing, all but at most 1 large bins with load not exceeding $\frac{3}{4}$ are L-bins.*

Proof. Suppose there are two large bins with load not exceeding $\frac{3}{4}$ and neither of them is an L-bin. Each of them contains a small item besides the large item. Each small item has size at most $\frac{3}{4} - \frac{1}{2} = \frac{1}{4}$. Hence, the small item in the bin with smaller load will move to the one with greater load, contradicting the property of an NE.

Case 1. All but at most two small bins have load at least $\frac{3}{4}$. We leave the two special bins out if necessary. We define the following weighting function on the items in the rest bins.

$$w(s_i) = \begin{cases} 1 & B_i \text{ is an L-bin} & \text{(I)} \\ 1 - \frac{8}{9}(s(B_i) - s_i) & s_i > \frac{1}{2}, B_i \text{ is an M-bin} & \text{(II)} \\ \frac{8}{9}s_i & s_i \leq \frac{1}{2}, B_i \text{ is an M-bin} & \text{(III)} \\ \frac{4}{3}s_i & s_i \leq \frac{1}{2}, B_i \text{ is a small bin} & \text{(IV)} \end{cases}$$

First we claim that all bins (except for a constant number of bins) in NE has a weight at least 1.

- For any L-bin, the claim holds trivially.
- For any M-bin, its weight is competed to 1.
- For any small bin, the claim holds as it has load at least $\frac{3}{4}$.

Next we consider bins in OPT and show that any bin in OPT has a weight at most $13/9 < 1.467$.

For a small bin in OPT, as all small items from (III) or (IV) have weight no greater than $\frac{4}{3} \approx 1.334$, the total weight is bounded by $\frac{4}{3}$. For a large bin in OPT, we consider the following cases.

Case 1a. The large item comes from an L-bin in the NE.

The large item can not share the bin with any item from (IV), otherwise such a small item would fit in the L-bin in NE. Hence, only items from (III) can share the same bin with it. The total weight of the bin is at most

$$1 + \frac{8}{9} \cdot \frac{1}{2} = \frac{13}{9} \approx 1.445.$$

Case 1b. The large item comes from an M-bin in the NE.

We need to determine the largest possible weight such a bin may have. Since items from (IV) have larger weights than items from (III), the remaining space of the bin should all be filled with items from (IV).

Let s_i be the size of the large item and B_i be the M-bin it comes from. Given s_i and B_i, the weight of the bin is

$$w \leq 1 - \frac{8}{9}(s(B_i) - s_i) + \frac{4}{3}(1 - s_i) = \frac{7}{3} - \frac{8}{9}s(B_i) - \frac{4}{9}s_i.$$

By Lemma 5, all but at most one M-bin have load greater than $\frac{3}{4}$. Hence we have $s(B_i) > \frac{3}{4}$. Since $s_i > \frac{1}{2}$, the largest possible weight of such bins is at most

$$\frac{7}{3} - \frac{8}{9} \cdot \frac{3}{4} - \frac{4}{9} \cdot \frac{1}{2} = \frac{13}{9} \approx 1.445.$$

Case 2. All but at most two small bins have load at least $\frac{17}{24}$ and there exists at least three small bins with load less than $\frac{3}{4}$.

Lemma 6. *In any NE packing, there are at most two small bins with load less than $\frac{2}{3}$.*

Proof. Suppose there are three small bins with load less than $\frac{2}{3}$. Two of them must have load greater than $\frac{1}{2}$, otherwise items from the bin with smallest load will move. Since these two small bins have load greater than $\frac{1}{2}$ (but less than $\frac{2}{3}$), they both must contain at least two items. One of the two items has size less than $\frac{1}{3}$. Therefore, an item in the bin with smaller load will move to the bin with greater load, contradicting the fact that the packing is an NE.

Consider the small bins with load less than $\frac{3}{4}$. By Lemma 6, all but at most two of them have loads at least $\frac{2}{3}$. We leave the two special bins out if necessary. Since small bins contain only small items, the smallest item in the bin has size at most $\frac{3}{4} \cdot \frac{1}{2} = \frac{3}{8}$. Hence, all L-bins must have load at least $1 - \frac{3}{8} = \frac{5}{8}$, otherwise such a small item would move there.

$$
w(s_i) = \begin{cases}
1 & B_i \text{ is an L-bin} & \text{(I)} \\
1 - \frac{96}{85}(s(B_i) - s_i) & s_i > \frac{1}{2}, B_i \text{ is an M-bin} & \text{(II)} \\
\frac{96}{85} s_i & s_i \leq \frac{1}{2}, B_i \text{ is an M-bin} & \text{(III)} \\
\frac{24}{17} s_i & s_i \leq \frac{1}{2}, B_i \text{ is a small bin} & \text{(IV)}
\end{cases}
$$

By the same argument as Case 1 we can prove that all bins in NE have a weight at least 1.

For a small bin in OPT, since all small items from (III) or (IV) have a weight no greater than $\frac{24}{17} \approx 1.412$. The weight of the small bin is also upper bounded by $\frac{24}{17}$.

For a bin in OPT that contains a large item, we consider the following cases.

Case 2a. The large item comes from an L-bin in the NE. The large item can not share the bin with any item from (IV), otherwise such a small item would fit in the L-bin in NE. Hence, only items from (III) can share the same bin with it.

Since all L-bins have load at least $\frac{5}{8}$ by the discussion above, the total weight of the bin is at most

$$
1 + \frac{96}{85} \cdot \left(1 - \frac{5}{8}\right) = 1 + \frac{96}{85} \cdot \frac{3}{8} = \frac{121}{85} \approx 1.424.
$$

Case 2b. The large item comes from an M-bin in the NE.

Given the large item s_i and the M-bin B_i where it comes from, by the same argument as Case 1b, the largest possible weight of such bins should be

$$
w = 1 - \frac{96}{85}(s(B_i) - s_i) + \frac{24}{17}(1 - s_i) = \frac{41}{17} - \frac{96}{85}s(B_i) - \frac{24}{85}s_i.
$$

Again, by Lemma 5, all but at most one M-bin have load greater than $\frac{3}{4}$. Hence we have $s(B_i) > \frac{3}{4}$. Since $s_i > \frac{1}{2}$, the largest possible weight of such bins is at most

$$
w = \frac{41}{17} - \frac{96}{85} \cdot \frac{3}{4} - \frac{24}{85} \cdot \frac{1}{2} = \frac{121}{85} \approx 1.424.
$$

Case 3. There exists at least three small bins with load less than $\frac{17}{24}$.

Lemma 7 ([5]). *In an NE packing, all bins that are filled by less than $\frac{5}{6}$, except for maybe a constant number of bins, contain exactly two items with size in $(\frac{7}{24}, \frac{1}{2}]$.*

Since the cost of each small bin is still shared using the proportional rule under the DS mechanism, the above lemma still holds for Y-bins in any NE under the DS mechanism.

We define the following weighting function on the items.

$$
w(s_i) = \begin{cases}
1 & s_i \text{ is in an L-bin} & \text{(I)} \\
1 - \dfrac{6}{5}(s(B_i) - s_i) & s_i > \frac{1}{2}, \ B_i \text{ is an M-bin} & \text{(II)} \\
\dfrac{6}{5}s_i & s_i \leq \frac{1}{2}, \ B_i \text{ is an M-bin} & \text{(III)} \\
\dfrac{6}{5}s_i & s_i \leq \frac{1}{2}, \ B_i \text{ is an X-bin} & \text{(IV)} \\
\dfrac{6}{5}s_i + \dfrac{1 - \frac{6}{5}s(B_i)}{2} & s_i \leq \frac{1}{2}, \ B_i \text{ is a Y-bin} & \text{(V)}
\end{cases}
$$

First we claim that all bins (except for maybe a constant number of bins) in NE has weight ≥ 1.

 - For any L-bin, the claim holds trivially.
 - For any M-bin, its weight is completed to 1
 - For X-bins and Y-bins, the claim holds by Lemma 7 [5].

We now show that in OPT, the average weight of bins does not exceed 1.467 as $OPT(G) \to \infty$.

We first consider all small bins in OPT. Items of a small bin in OPT come from (III), (IV) or (V). We assign items from (III) the same weight as items from (IV). By the same argument as [5], the average weight of a small bin is at most $\frac{22}{15} \approx 1.467$ as $OPT(G) \to \infty$.

For a bin in OPT that contains a large item, we consider the following cases.

Case 3a. The large item comes from an L-bin in the NE. The large item can not share the bin with any item from (IV) and (V), since otherwise such a small item would fit in the L-bin in NE. Hence, only items from (III) can share the same bin with it.

Now consider a bin with load less than $\frac{17}{24}$. By Lemma 7, we know that it contains exactly 2 items. Since it has load less than $\frac{17}{24}$, the size of the smaller item of the two must be less than $\frac{17}{48}$. Hence, the unoccupied space of any L-bin must be less than $\frac{17}{48}$. Therefore, the total weight of the bin is at most

$$
w = 1 + \frac{17}{48} \cdot \frac{6}{5} = 1.425.
$$

Case 3b. The large item comes from an M-bin in the NE.

Let s_i denote the size of the large item and B_i denote the bin it comes from. If the remaining space of the bin are all occupied by small items from (III) or (IV), by the same argument as Case 1b or Case 2b, the largest possible weight of the bin should be

$$w = 1 - \frac{6}{5}(s(B_i) - s_i) + \frac{6}{5}(1 - s_i) = \frac{11}{5} - \frac{6}{5}s(B_i) \leq 1.3.$$

The last equation is due to Lemma 5, saying that the M-bin has a load greater than $\frac{3}{4}$.

We now turn to the situation where items from (V) are involved. All Y-bins (except for a constant number of bins) contain exactly two items with sizes in $(\frac{7}{24}, \frac{1}{2}]$. Only one of such items can fit in the remaining space of the bin mentioned above. Using Lemma 6, we know that the additive term in the weight of that item is at most $\frac{1-\frac{2}{3}}{2} = \frac{1}{6}$. Hence, the weight of items in the remaining space is at most

$$w_r = \frac{6}{5}(1 - s_i) + \frac{1}{6},$$

and the total weight of the bin is

$$w = 1 - \frac{6}{5}(s(B_i) - s_i) + w_r = 1 - \frac{6}{5}s(B_i) + \frac{41}{30} \leq \frac{22}{15}.$$

It has been shown that in any case, the average weight of the bins in OPT is at most $22/15$. We thus arrive at the following conclusion.

Theorem 3. *The Discount-Sharing mechanism has* $PoA \leq 22/15 \approx 1.467$.

5   Concluding Remarks

In this paper, we discussed the cost-sharing mechanism design for the selfish bin packing problem. We designed the LPP mechanism with $PoA = 1.5$ which significantly improves the previous results. We also proved that 1.5 is a lower bound of PoA for a large class of mechanisms. We studied the properties to further improve the bound. Finally, we proposed a better mechanism with $PoA \leq 1.467$ which pushes the bottleneck of the SBP to the case with only small items.

An obvious question is left to deal with small items more efficient than the proportional rule. A more fundamental question asks the possibility to design a family of mechanisms that have PoA arbitrarily close to 1, which resembles the AFPTAS in environments without selfish behaviors.

In addition, if we are not strictly restricted to the cost-sharing mechanisms, say, items in a bin may be charged a little more than the cost of the bin, does there exist any more efficient mechanism?

References

1. Adar, R., Epstein, L.: Selfish bin packing with cardinality constraints. Theor. Comput. Sci. **495**, 66–80 (2013)
2. Bilò, V.: On the packing of selfish items. In: Proceedings of 20th International Parallel and Distributed Processing Symposium, 9 pp. IEEE (2006)
3. Cao, Z., Yang, X.: Selfish bin covering. Theor. Comput. Sci. **412**(50), 7049–7058 (2011)
4. Epstein, L., Kleiman, E.: Selfish bin packing. In: Halperin, D., Mehlhorn, K. (eds.) ESA 2008. LNCS, vol. 5193, pp. 368–380. Springer, Heidelberg (2008). https://doi.org/10.1007/978-3-540-87744-8_31
5. Epstein, L., Kleiman, E., Mestre, J.: Parametric packing of selfish items and the subset sum algorithm. In: Leonardi, S. (ed.) WINE 2009. LNCS, vol. 5929, pp. 67–78. Springer, Heidelberg (2009). https://doi.org/10.1007/978-3-642-10841-9_8
6. de La Vega, W.F., Lueker, G.S.: Bin packing can be solved within $1+ \varepsilon$ in linear time. Combinatorica **1**(4), 349–355 (1981)
7. Johnson, D.S., Demers, A., Ullman, J.D., Garey, M.R., Graham, R.L.: Worst-case performance bounds for simple one-dimensional packing algorithms. SIAM J. Comput. **3**(4), 299–325 (1974)
8. Koutsoupias, E., Papadimitriou, C.: Worst-case equilibria. In: Meinel, C., Tison, S. (eds.) STACS 1999. LNCS, vol. 1563, pp. 404–413. Springer, Heidelberg (1999). https://doi.org/10.1007/3-540-49116-3_38
9. Li, W., Fang, Q., Liu, W.: An incentive mechanism for selfish bin covering. In: Chan, T.-H.H., Li, M., Wang, L. (eds.) COCOA 2016. LNCS, vol. 10043, pp. 641–654. Springer, Cham (2016). https://doi.org/10.1007/978-3-319-48749-6_46
10. Ma, R., Dósa, G., Han, X., Ting, H.-F., Ye, D., Zhang, Y.: A note on a selfish bin packing problem. J. Glob. Optim. **56**(4), 1457–1462 (2013)
11. Ullman, J.D.: The performance of a memory allocation algorithm. Technical report 100, Princeton University, Princeton (1971)
12. Yu, G., Zhang, G.: Bin packing of selfish items. In: Papadimitriou, C., Zhang, S. (eds.) WINE 2008. LNCS, vol. 5385, pp. 446–453. Springer, Heidelberg (2008). https://doi.org/10.1007/978-3-540-92185-1_50

Application

Modelling and Solving Anti-aircraft Mission Planning for Defensive Missile Battalions

Trang T. Nguyen[✉], Trung Q. Bui[✉], Bang Q. Nguyen[✉], and Su T. Le[✉]

Command, Control, Communications, Computers, and Intelligent Department,
Viettel Research and Development Institute, Hanoi, Vietnam
{trangnt11,trungbq5,bangnq,sult2}@viettel.com.vn

Abstract. The theater defense distribution is an important problem in the military that determines strategies against a sequence of offensive attacks in order to protect his targets. This study focuses on developing mathematical models for two defense problems that generate anti-aircraft mission plans for a group of missile battalions. While the Anti-aircraft Mission Planning problem maximizes the defender's effectiveness using all his available battalions, the Inverse Anti-aircraft Mission Planning problem computes necessary weapon resources (battalions and their missiles) to obtain a given defensive effectiveness value. The proposed formulations are Mixed Integer Programs that describe not only the positions of missile battalions, but also engage battalions to fleets of attacking aircrafts. We additionally prove that these problems are NP-hard. A comprehensive set of experiments is then evaluated to show that these proposed programs can be applied to solve fast real-life instances to optimality.

Keywords: Military · Theater defense distribution · Anti-aircraft mission planning · Mixed integer program

1 Introduction

The Theater Distribution Model (TDM) is specifically designed to support the combatant commander to ensure his effective plans within area of operations [12]. At the operational level of war, the senior commander is responsible for development of the distribution and ultimately for transportation system. A theater distribution system is comprised of facilities, installations, methods, and procedures designed to receive, store, maintain, distribute, and control the flow of materials between exogenous inflows to that system and distribution to end-user activities and units within the theater. Most of researches in the literature focus on the military theater distribution problem associated with determining positions of defender's missile battalions within a potentially geographical region [11].

1.1 Related Works

The TDM has been motivated and validated for both defensive side and offensive side in general. Some of these studies have been developed into mathematical

© Springer International Publishing AG 2017
X. Gao et al. (Eds.): COCOA 2017, Part I, LNCS 10627, pp. 371–385, 2017.
https://doi.org/10.1007/978-3-319-71150-8_31

models which can be classified as the Theater Distribution and Vehicle Routing Problem (TDVRP), the Theater Attacking Model (TAM), and the Defender-Attacker Model (DAM).

One of the most popular transportation model in the TDM is the TDVRP, a generalization of vehicle routing problem, that is concerned with constructing optimal routes and schedules to satisfy transportation requests at multiple locations. An exact formulation for the TDVRP is introduced in [9] that has a very large number of variables and constraints. In order to solve the TDVSP effectively in practice, the authors utilize advanced Tabu search techniques, including reactive Tabu search and group theory, to develop a heuristic procedure for solving specific situation of location routing pickup and delivery problem. The TDVRP is also considered in [3], in which the authors describe a flexible group theoretic Tabu search framework which evaluates the routing and scheduling of theater transportation assets at the individual asset operational level to provide efficient time-definite delivery of cargo to customers.

Among attacking models, the authors in [6] describe TAM as a large-scale linear program used to aid commanders in making tough budget procurement decisions for the United States Air Force. The objective function of this model maximizes the total target value destroyed by the aircraft and munitions in its scenarios. The model uses decision variables to represent each sortie which is defined by a particular aircraft and ordnance combination to maximize target value destroyed for a given target, weather, time period, and distance to the target. In TAM model, the aircraft sorties are not grouped into strike packages and as a result, the advantages of mutual support and mass are not presented in this model. The Air Force Studies and Analysis Agency [13] describes TAC Thunder which is a combat model simulating air war, ground war, and resupply. The TAC Thunder's network linear program and meta-heuristics are designed to optimize sorties' allocations in term of mission effectiveness against a target list. The TAC Thunder model is then applied as a package to the Future Theater Level Model (FTLM) in [1]. The air mission planning algorithm for FTLM results in a linear program that allocates the optimal number and type of aircraft and munitions against each target. Although the air mission planning algorithm provides fast, approximated solution; the FTLM problem omits many details in actual aircrafts.

Another important class of the TDM that has been studied in the literature, is the DAM. An instance of DAM in a fast theater model is introduced in [10], which is built upon the existing air model. In this research, the authors study a joint theater level attrition model combining ground combat with optimized air strikes. The air strike attacker's main objective is maximizing target value destroyed by killing as many targets with high values as possible, while the ground combat wants to minimize its own losses. The resulting model is a Mixed Integer Program (MIP) finding an optimal, actively defensive actions by the ground force that can significantly reduce the air attacker's effectiveness. The defender-attacker (pursuer-evader) problem is continued studying in [8], in which a new methodology for strategy optimization under uncertainty has been proposed. The authors describe the implementation of a genetic programming algorithm to determine an optimized evasion strategy for the extended two-dimensional pursuer-evader problem under conditions of uncertainty about the type of pursuer. The DAM model is also applied to defender's risk assessment and mitigation. For instance, [2] formulates a defender-attacker-mitigator problem as a min-max-min model, in which defender minimizes an objective function of a maximization problem, the optimal solution invests to reduce the expected damage, given the future mitigation capability. Beside the DAM, a defensive model is also studied such as the Joint Theater Ballistic Missile Defense model in [4], in which the authors express the enemy course of action as a mathematical optimization to maximize expected damage, then use Bender's decomposition technique to optimize the defensive interceptor pre-positioning to minimize the maximum achievable expected damage.

1.2   Objective, Contribution, and Outline

In most cases, the defender faces with so many risks from terrorist attacks of all kinds. The work we research here is directly motivated by such one risk assessment: surface-to-air missile battalions defense against many fleets of penetrating aircraft. Beside the role of specializing anti-aircraft, missile battalions in an air defense system also protect a target such as capital, political area, economy or military center. The Department of Air Defense and Air Force has conducted some risk assessment exercises and estimated risks of many attack possibilities. One area that is still undeveloped is an algorithm for determining an intelligent package of missile battalion's actions to aircraft attack. A missile battalion package consisting of only one missile type might be ineffective but when several types of missiles are combined properly, they confidently destroy the attacking aircraft and protect their target. More precisely, we develop mathematical formulations for anti-aircraft mission planning effectively for a group of missile battalions. A missile battalion is considered as a fundamental tactical building block which recruits or conscripts in one geographical area assigned by a feudal lord. Given some threat, the defender must decide where to locate defensive missile battalions among potential locations and how they should engage fleets of attacking aircraft. We express the defender's courses of action as following mathematical optimization problems. Given an attacking plan, Anti-aircraft Mission Planning problem (AMP) finds optimal battalions' locations, and a launching assignment (i.e., number of missiles should be launched from a battalion to a fleet), such that the defensive effectiveness is maximized. While Inverse Anti-aircraft Mission Planning problem (IAMP) computes necessary weapon resources, battalions' locations, and a launching assignment to obtain a given effectiveness threshold.

The most valuable contribution of this paper comes from the statements and practical mathematical formulations for two crucial problems in TDM class. While the AMP arises naturally in the context of limited weapon resources, the IAMP is necessarily tackled when the protected target should not be fallen in any cases. Although these problems are proven NP-hard, the computational results show the efficient of our formulation since it can be applied directly to solve fast real-life instances to optimality. The message here is that, with the validation of Vietnamese veterans, we have gained confidence that the formulations have significant implementation in any defender's combat field.

The rest of paper is organized as follows. In Sect. 2, we state the problems, formulate them as mixed integer programs and as well as prove their hardness. The experimental results are reported and analyzed in Sect. 3. Finally, we conclude the paper and draw some future directions in Sect. 4.

2 Problem Formulations

In this section, we state and formulate the anti-aircraft mission planning problems. The complexity of these problems is also discussed.

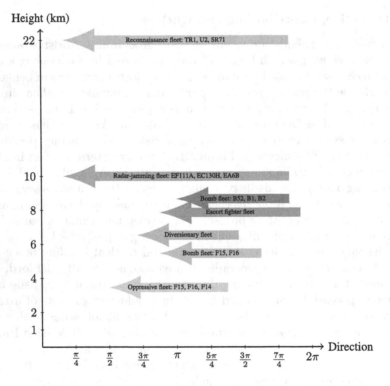

Fig. 1. An example of an attacking plan

2.1 Problem Statement

Suppose that the attacker's plan can be observed by an intelligence system of the defender and is described as follows. In the offensive side, the attacker strikes the target by a group of fleets of attacking aircraft. For a sake simplicity, from

Fig. 2. An example of an attacking fleet

now on, the term "fleet" is used stead of "fleet of attacking aircraft". Each fleet is organized by a group of aircraft which have same missions such as carrying bombs or making radar noise; enter the theater at same height, direction; and fly with same velocity. Each fleet is associated with a weight of importance depending mainly on its mission. Figure 1 illustrates an attacking plan, in which the horizontal axis represents the directions of the fleets, considered as angles between attacking directions and a predefined axis; while the vertical axis represents the height of the fleets. There are seven fleets at different heights, directions and velocities, drawn by seven arrows. The color of the arrows reflects the fleets' weights. For instance, the darkest arrow describes the fleet with the highest weight, carrying bombs such as B52, B1, B2. Further, flying positions of aircraft in a fleet must be captured in detail, for example, a flying position of a four-aircraft fleet is illustrated in Fig. 2.

In the defensive side, the defender's responsibility is to engage fleets to protect its point target. In order to formulate the AMP and the IAMP, we take into account following factors. The first one is *critical radius* corresponding to each fleet that defines a critical circle centered at the target. The defender has to make a defensive plan such that no attacking aircraft in that fleet is able to get inside that circle. This critical radius is computed depending on the height, velocity of that fleet and type of bombs carried by that fleet. For instance, in Fig. 3, the point target is described by the green rectangle with critical radius OX corresponding to a fleet coming from AO direction. The second factor is *action range* of a battalion corresponding to a fleet that can be understood as a fleet's flying path where the fleet will be intercepted by that battalion. In Fig. 3, the action range of battalion B_1 is CD segment, where D is the intersection of attacking direction AO and the critical circle, and C is the intersection of the direction AO and the circle centered at B_1 of radius B_1C which is defined as the long range of missiles belonging to that battalion. Similarly, the action range of battalion B_2 is BD segment. The third factor *maximum launches* representing firepower capabilities of a battalion can be seen as maximum number of missiles that can be launched from the battalion to the fleet in its action range. This number is calculated basing on the type of missile, number of missiles in that battalion, as well as the shortest time between two successive launches. The fourth factor is *expected number of killed aircraft* of a fleet caused by a battalion in a number of launches. This value can be estimated by number of missiles, probability of kill, defensive mode related to each fleet, and set of coefficients corresponding to each missile battalion such as technical coefficient, control coefficient, and complex coefficient of combat. Lastly, a minimum distance between each pair of

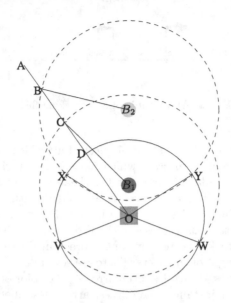

Fig. 3. An example of influence of battalion's locations on their firepower capabilities (Color figure online)

battalions is required to avoid radar jamming between missiles in the battalions. Note that if two battalions locate at a same position, this constraint can be ignored.

As a defender, we would like to measure the result of our defense. An effective criterion is then introduced as *defensive effectiveness*, calculated by fraction between value of killed aircraft and value of all aircraft in attacking fleets. The effectiveness is strongly influenced by battalions' locations, that motivates us to study an adaptable, efficient, and cost-effective process to analyse missile battalion defense against aircraft threats. Suppose that in both the AMP and the IAMP, the defender observes an attacking plan that includes all information about the offensive side stated as above. In the AMP, the defender seeks to allocate a limited number of battalions and missiles to distribute a defensive mission plan that maximizes his defensive effectiveness. We consider now somewhat inverse version of the AMP, the IAMP. Assume, in particular, that the defender has a set of battalions, in stead of using all battalions to maximize his defensive effectiveness, one generates a defensive mission plan such that the corresponding effectiveness is greater than or equal to a given value. Intuitively, one would like to locate defensive battalions as well as compute number of missiles with the lowest cost such that its effectiveness is at least γ (such as $\gamma = 0.5, 0.6, ...$). The detailed mathematical formulations for these problems are proposed as follows.

2.2 Mathematical Formulations

For a simplicity of presentation, we first introduce some notations.

Notations

- B: set of missile battalions;
- F: set of fleets of attacking aircraft;
- L: set of potential locations for battalions;
- For a battalion $b \in B$, denote $m(b)$ and $c(b)$ by number of missiles distributed to b and cost of a missile of b, respectively;
- For a fleet $f \in F$, denote $n(f)$ and $w(f)$ by number of aircraft in fleet f and weight of that fleet, respectively;
- For each pair (b, l) of a battalion $b \in B$ and a location $l \in L$, $c(b, l)$ refers to cost of allocating battalion b at location l;
- For each pair of locations (l_i, l_j) $(l_i, l_j \in L)$, denote $d(l_i, l_j)$ by geometric distance between l_i and l_j; for a pair of battalions (b_i, b_j) $(b_i, b_j \in B)$, denote $\overline{d}(b_i, b_j)$ by minimum distance between b_i and b_j if they are located at two different positions;
- For each triple (b, l, f) $(b \in B, l \in L, f \in F)$, denote $t(b, l, f)$ by maximum number of missiles that battalion b located at l is able to launch to fleet f in its action range;
- Last, for a triple (b, f, t) $(b \in B, f \in F$ and $t \in \mathbb{Z}^+)$, denote $e(b, f, t)$ by expected number of killed aircraft in fleet f caused by t missiles launched from battalion b to fleet f.

AMP Mathematical Formulation

This formulation defines three categories of variables. First, binary variables $x_{b,l}$ where $b \in B, l \in L$, indicate whether battalion b locates at location l or not. Second, binary variables $y_{b,l,f,t}$ where $b \in B, l \in L, f \in F$ and $t \in \mathbb{Z}^+$, state if t missiles are launched from battalion b located at location l to fleet f. Lastly, continuous variables u_f where $f \in F$, represent expected number of killed aircraft in fleet f.

$$Max \sum_{f \in F} \frac{w(f)u_f}{w(f)n(f)} \tag{1a}$$

$$s.t. \sum_{l \in L} x_{b,l} = 1, \quad \forall b \in B \tag{1b}$$

$$\sum_{l \in L} \sum_{f \in F} \sum_{t=1}^{t(b,l,f)} t y_{b,l,f,t} \leq m(b), \quad \forall b \in B \tag{1c}$$

$$(1 - x_{b_i,l_i})\infty + (1 - x_{b_j,l_j})\infty + x_{b_i,l_i}d(l_i, l_j) \geq \overline{d}(b_i, b_j),$$
$$\forall b_i, b_j \in B, b_i \neq b_j, \forall l_i, l_j \in L, l_i \neq l_j \tag{1d}$$

$$u_f - \sum_{l \in L} \sum_{b \in B} \sum_{t=1}^{t(b,l,f)} e(b, f, t)y_{b,l,f,t} = 0, \quad \forall f \in F \tag{1e}$$

$$u_f \leq n(f), \quad \forall f \in F \tag{1f}$$

$$\sum_{f \in F} \sum_{t=1}^{t(b,l,f)} y_{b,l,f,t} \leq x_{b,l}, \quad \forall b \in B, l \in L \tag{1g}$$

$$x_{b,l}; y_{b,l,f,t} \in \{0,1\}, \quad \forall b \in B, l \in L, f \in F, t \in \{1,\ldots,t(b,l,f)\} \tag{1h}$$

$$u_f \geq 0, \quad \forall f \in F. \tag{1i}$$

In this formulation, the defender's objective is maximizing its effectiveness (1a) which is the sum of fractions of expected number of killed aircraft u_f and number of aircraft $n(f)$ for all fleets while considering additionally the importance weights of these fleets. Constraints (1b) simply limit each missile battalion to one location. Constraints (1c) stipulate that the number of launches from each battalion should not be greater than its given number of missiles. Constraints (1d) show that if two missile battalions b_i and b_j are at locations l_i and l_j where $l_i \neq l_j$, respectively, the distance between these battalions must be greater than a required minimum distance $\bar{d}(b_i, b_j)$. Constraints (1e) express u_f as the expected number of killed aircraft in fleet f, while constraints (1f) define upper bounds for u_f, $f \in F$, which are number of aircraft in these fleets. Constraints (1g) tell us that there are some missiles launched from a location to a fleet if and only if there exists at least one missile battalion located at that location. Lastly, constraints (1h) and (1i) indicate that $x_{b,l}, y_{b,l,f,t}$ where $b \in B, l \in L, f \in F, t \in \{1,\ldots,t(b,l,f)\}$ are binary variables, while u_f where $f \in F$ are non-negative continuous variables.

IAMP Mathematical Formulation

In addition to variables used in the AMP mathematical formulation, the IAMP mathematical formulation introduces more integer variables $z_b \in \mathbb{Z}^+$, indicating number of missiles launched from battalion $b \in B$.

$$Min \sum_{b \in B} \sum_{l \in L} c(b,l) x_{b,l} + \sum_{b \in B} c(b) z_b \tag{2a}$$

$$s.t. \sum_{l \in L} x_{b,l} \leq 1, \quad \forall b \in B \tag{2b}$$

$$Constraints(1d), (1e), (1f), (1g), (1h), (1i) \tag{2c}$$

$$\sum_{f \in F} \frac{w(f) u_f}{w(f) n(f)} \geq \gamma \tag{2d}$$

$$z_b - \sum_{l \in L} \sum_{f \in F} \sum_{t=1}^{t(b,l,f)} t y_{b,l,f,t} = 0, \quad \forall b \in B \tag{2e}$$

$$z_b \leq m(b) \sum_{l \in L} x_{b,l}, \quad \forall b \in B \tag{2f}$$

$$z_b \in \mathbb{Z}^+, \quad \forall b \in B \tag{2g}$$

where $\gamma \in [0, 1]$ is a given effectiveness value.

The objective of this formulation is to minimize the total cost (2a) that is calculated by sum of establishing battalion cost and launched missile cost on these battalions. Constraints (2b) are different from (1b) since not every missile battalion is required to locate at some location. Constraint (2d) requires the obtained effectiveness be at least a given value, γ. The number of missiles on each battalion is set at constraints (2e) while its upper bound has been given in constraints (2f). The other constraints are similar to ones in the AMP formulation, except that number of launched missiles are integer variables as in (2g).

Note here that, the AMP always results in an optimal solution with the effectiveness belonging to $[0, 1]$, while the IAMP sometimes returns no solution in case the weapon resources are not sufficiently available to reach the expected effectiveness.

2.3 NP-Hardness

The NP-completeness is indicated as the complexity of the AMP and the IAMP in this section.

Theorem 1. *The IAMP is NP-complete.*

Proof. To prove this, we reduce the 3-Partition Problem (3PP) to the IAMP since the 3PP is known as NP-complete [5].

Recall that in the 3PP, we are given a multi-set S of $3b$ positive integers $i_1, i_2, \ldots, i_{3b}$, where the value of every element in the set belongs to interval $(\frac{C}{4}, \frac{C}{2})$, for a positive integer C; and we are asked to decide if there are b disjoint subsets of S such that sum of all elements in each subset equals to C.

For any instance of the 3PP, an instance of the IAMP is simultaneously created as follows: The set of battalion B is initialized by b missile battalions; each battalion is allocated 3 missiles; the set of potential locations L just contains only one location; the set of fleets includes $3b$ fleets, $f_1, f_2, \ldots, f_{3b}$ (each positive integer in the 3PP instance corresponds to a fleet in the IAMP instance); each fleet is formed by just one aircraft; each battalion can launch at most one missile to each fleet; the expected number of killed aircraft is $\frac{i_k}{C}$ ($\frac{i_k}{C} \in (\frac{1}{4}, \frac{1}{2})$) when one missile is launched from any battalion to a fleet f_k; for all $k \in [1, 2, \ldots, 3b]$, and the lower bound on the effectiveness is set to 1.

As a result, the 3PP instance has a partition into b disjoint subsets, sum of elements in each subset is equal to C, if and only if all battalions located at the unique location, each battalion launches exactly 3 missiles to 3 fleets and the defender's effectiveness reaches to its ideal value of 1 or all fleets are destroyed completely. Thus 3PP reduces to the question whether the IAMP has a solution or not. Thus, a pseudo polynomial reduction has been given, showing that the IAMP is NP-complete. □

A similar pseudo polynomial reduction is also applied to prove the NP-completeness of AMP by reducing the 3PP to the decision problem of the AMP.

3 Experiments

This section dedicates to report and analyze experimental results on variety of instances for both AMP and IAMP. For each problem, we generated randomly

20 instances with the help of experienced soldiers to make them real and reliable. The coefficients used to generate instances were carefully extracted from the historical data and technical guides of missiles S-75 and S-125. The mathematical formulations for the AMP and the IAMP were implemented in C++ programming language, using IBM Ilog Cplex Concert Technology, version 12.5. The standard cuts of Cplex were automatically added. Since the number of constraints (1c) is large and they belong to Miller-Tucker-Zemlin class [7], these constraints are treated as lazy constraints to reduce the computation time as follows. At the beginning of resolution, these constraints are all relaxed. Each time, an integer feasible solution is counted at a node in the search tree, this solution is checked for the satisfiability of all these constraints, all violated constraints are injected more in the model. This cut-addition strategy always ensures the returning solution is correct since all constraints (1c) are strictly respected. All computations were performed in multi-thread mode on a computer with an Intel Core i7-479 CPU, 3.6 GHz, and 4 GB RAM running Ubuntu version 16.4, 64 bits. A time limit of 2 h of CPU time was set for each instance resolution.

All computational results for the AMP and the IAMP are respectively indicated in Tables 1 and 2. The columns in the tables have the following meanings.

- *No.*: Instance number
- *# Battalions*: Number of missile battalions
- *# Missiles*: Number of missiles of each battalion
- *# Locations*: Number of potential locations
- *# Fleets*: Number of attacking aircraft fleets
- *Gap*: Gap between the optimal solution of the integral relaxation of an integer program and integer feasible solution of that program found so far
- *Effectiveness*: Defensive effectiveness value
- *# Nodes*: Number of nodes in the search tree
- *Time (s)*: Overall CPU time in seconds

In reality, the practical AMP usually allocates a missile regiment of 4 to 6 missile battalions to protect an important target which is attacked by 5 to 8 attacking aircraft fleets. Instances of those sizes can be solved so fast by our formulation as report in the first five rows of Table 1, specially instances (1, 4, 5) of these sizes were solved in less than one second. This result motivates us to increase the size of test instances up to 10 battalions, 100 locations and 10 fleets to evaluate the tackling ability of the formulation. In the time limit of 2 h, the 10 easy instances among 15 instances were completely solved to optimality, most of them take less than 1 min; while the 5 hard instances were solved nearly to optimality since the gaps are so small (0.01%–0.07%). The resolution of the hard instances generated huge search trees containing a millions of nodes, for example there are 6328326 nodes in the search tree in instance 9. Although 15 last test instances have same sizes in term of the number of battalions, the number of fleets and the number of locations, the computational difficulty is different. This diversification comes from the geographic of the potential locations. In comparison to defensive effectiveness of easy instances, the hard ones have greater effectiveness.

Table 1. Experimental results for AMP

Instances					Experimental results			
No.	# Battalions	# Missiles	# Locations	# Fleets	Gap (%)	Effectiveness	# Nodes	Time (s)
1	7	12	38	8	0	0.72	0	0.6
2	6	12	80	5	0	0.92	304447	508.9
3	6	12	58	8	0	0.62	2969	1.5
4	7	12	50	5	0	1.00	0	0.5
5	5	12	83	8	0	0.49	0	0.7
6	10	12	100	10	0	1.00	0	7.7
7	10	12	100	10	0	0.67	197	5.3
8	10	12	100	10	0	0.70	327289	212.9
9	10	12	100	10	0.02	0.91	6328326	72001.4
10	10	12	100	10	0	0.63	21336	49.8
11	10	12	100	10	0	0.78	35296	172.6
12	10	12	100	10	0	0.70	327289	212.9
13	10	12	100	10	0	0.70	2940	7.1
14	10	12	100	10	0	0.90	30780	11.6
15	10	12	100	10	0.05	0.94	835158	7211.4
16	10	12	100	10	0	0.69	3555	30.1
17	10	12	100	10	0.07	0.91	2683626	7217.5
18	10	12	100	10	0	0.74	49427	8.9
19	10	12	100	10	0	0.73	49952	34.8
20	10	12	100	10	0.01	0.71	2622249	7201.8

Sets of available missile battalions in IAMP instances were generated around 20 battalions that are larger than the ones used in Table 1. Among 20 instances in Table 2, there are 12 easy instances solved to optimality in less than 30 s, while 7 hard instances were not completely solved in the time limit, and one instance (No. 18) returned to "Out of memory" error during the resolution. Similar to hard AMP instances, the hard IAMP instances generate very large search trees that may cause no feasible solution. It is essentially seen that the easy IAMP instances correspond to reasonable effectiveness values, while the hard IAMP instances usually are associated with high effectiveness values. These experimental results were validated and recommended by the experienced soldiers that this formulation should be packaged for the purpose of training and integrated into a C4I system.

We next evaluate the performance of the defensive effectiveness value in the IAMP programs on his weapon resources in the instance No. 15 with 20 available battalions, recorded in Table 3. The effectiveness values were generated uniformly at random between 0 and 1. It is obviously that the total cost increases when the defensive effectiveness increases, but it is not linear dependent. Further, we observe that when the effectiveness is less than 0.6, it takes less than 8 min to obtain its optimal values, and the number of battalions needed is also less than 7,

Table 2. Experimental results for IAMP

Instances					Experimental results		
No.	# Battalions	# Missiles	# Locations	# Fleets	Effectiveness	Gap (%)	# Nodes	Time (s)
1	21	12	56	11	0.72	0	7337	18.6
2	21	12	56	11	0.82	0	30706	26.7
3	21	12	41	11	0.81	0	0	5.6
4	22	12	42	12	0.97	0.83	356981	7209.9
5	21	12	46	11	0.74	0	41059	15.0
6	20	12	45	10	0.52	0	3231	11.7
7	21	12	51	11	0.57	0	2245	20.5
8	24	12	54	14	0.98	0.05	9374445	7213.7
9	22	12	52	12	0.84	0.20	1554149	7227.9
10	23	12	48	13	0.46	0	0	9.8
11	23	12	58	13	0.91	0.12	1897796	7210.6
12	22	12	47	12	0.71	0	14763	22.1
13	21	12	41	11	0.56	0	5293	29.8
14	23	12	48	13	0.94	0.12	1609379	7225.7
15	20	12	55	10	0.84	0.09	2305688	7217.6
16	22	12	47	12	0.44	0	614	8.6
17	22	12	47	12	0.65	0	26379	15.2
18	22	12	57	12	0.60	Out of memory		
19	20	12	55	10	0.60	0.03	3006774	7203.5
20	22	12	47	12	0.45	0	474	11.2

Table 3. Influences of weapon resources on effectiveness

Effectiveness	Objective value	# Battalions	# Total missiles	Gap	# Nodes	Time (s)
0.1	220	1	12	0	68	157.0
0.2	440	2	24	0	0	102.2
0.3	703	3	35	0	685	337.7
0.4	1039	5	50	6.42	59620	7214.6
0.5	1337	6	66	0	16811	472.7
0.6	1664	7	94	0	1775	131.9
0.7	2087	9	101	1.77	53792	7210027.0
0.8	2567	11	123	3.57	10405	7280.6
0.9	3156	13	151	2.11	7900	7331.5
1.0	4660	19	215	4.89	7449	7287.2

except the instance with the effectiveness of 0.4. And, if the required effectiveness is greater than 0.7 while the available weapon resources are extremely excessive, the optimal solution can not be obtained in the time limit. Note here that in the last row of Table 3, the defender must use almost all his weapon resources to completely protect his target. The nonlinear dependence of the effectiveness on the weapon resources in IAMP can be essentially explained by the diversity of launching modes.

4 Conclusion

In this paper, we formulated the AMP and the IAMP problems, the defensive missile battalions mission planning against aircraft attack model, that support defensive decision makers not only decide where to locate their missile battalions, but also point out that how many missiles should be launched from each battalion to each attacking aircraft fleet. The mathematical formulations are MIPs, proved as NP-hard. These mathematical programs were implemented and experimented on test instances generated basing on the help of experienced veterans, in which parameters on probability of kill, maximum number of launches, as well as minimum distance between two battalions, were pre-processed for a particular set of defensive battalions and aircraft attack. The numerical results provide the incidence that the proposed formulations should be widely applied in real-life combat field. As future works, we intend to formulate and tackle other variance of the defensive distribution models.

Acknowledgment. We would like to thank Mr. Dung Nguyen, the advisor of Vietnamese Department of Defense Air and Air Force, and his team for their support in the problem definition and the result validation.

A Appendix

A.1   Compute $e(b, f, t)$

Suppose that a battalion $b \in B$ plan to launch t missiles to fleet $f \in F$ that has $n(f)$ aircrafts. We are given coefficient corresponding to each missile battalion $b \in B$, $c^b = c_t^b c_c^b c_d^b$, where c_t^b is technical coefficient, c_c^b is control coefficient and c_d^b is combat complex coefficient. The probability of kill of each missile launched from battalion b to fleet f is known as p ($p \in [0, 1]$). Based on defensive mode, we consider following situations:

1. Disperse mode: Suppose that each time a battalion decides to launch 2 missiles to an aircraft of a fleet. Since the probability of kill is $p(b, f) = p$ for all $b \in B, f \in F$, expected number of killed aircraft is $e(b, f, t) = t(1 - (1 - p)^2)$.
2. Focus mode: Suppose that battalion b launches t times focusing on fleet f, where $t = n(f)t_1 + t_2$, then probability of kill on each aircraft in t_1 launches is $1 - (1 - p)^{t_1}$. Battalion b has $t_2(t_2 < n(f))$ launches left, inferring probability of kill on one aircraft in each launch is $p(1 - p)^{n(f)-1}$. Then, expected number of killed aircraft can be estimated by $eb, f, t = n(f)(1 - (1 - p)^{t_1}) + t_2 p(1 - p)^{n(f)-1}$.

3. Random mode: Let X_i where $i = 1, 2, ..., n(f)$, be random variables defined
by $X_i = \begin{cases} 1 & \text{if aircraft } i \in f \text{ is killed} \\ 0 & \text{otherwise} \end{cases}$

While probability of kill on aircraft i in fleet f is $1 - (1 - \frac{p}{n(f)})^t$, expected number of killed aircraft can be approximated as $e(b, f, t) = E(\sum_{i=1}^{n(f)} X_i) = n(f)E(X_i) = n(f)(1 - (1 - \frac{p}{n(f)})^t)$.

A.2   Compute $t(b, l, f)$

Value $t(b, l, f)$ is maximum number of launches that a battalion b located at location l can launch to fleet f. This number depends on following quantities. For a fleet f, we let $v(f)$, $h(f)$ and $l(f)$ be its velocity, height and length, respectively. In an attack, fleet brings different type of bomb that can be verified as $tb(f) = 1$ if fleet f brings nuclear bomb and $tb(f) = 0$ if fleet f brings regular bomb. For a battalion b, we denote d_{max} and r_b by long range of missile on battalion b and distance between that battalion and the target, respectively. We suppose that the shortest time between two consecutive launches, t_{as}, as well as obscured coefficient, δ, are known. Furthermore, angle of battalion location, α_b, and angle of in-coming fleet, α_f, are parameters. Function $t(b, l, f)$ can be computed as follows:

1. Compute critical radius $r_s = 5000tb(f) + v(f)\sqrt{\frac{2h(f)}{g}} - \Delta$ where $5000\,\text{m}$ is active radius of nuclear bomb, $g \approx 9.8\,\text{m/s}^2$ is gravity acceleration, $\Delta = 0.25h(f)$ if $v(f) \leq 300\,\text{m/s}$, $\Delta = 0.4h(f)$ if $v(f) > 300\,\text{m/s}$.
2. Compute shape time of fleet t_{fs}: $t_{fs} = \frac{l(f)}{v(f)}$
3. Compute launching time of battalion t_{bs}:
   (a) Angle between battalion's location and fleet φ: $\varphi = |\alpha_f - \alpha_b|$.
   (b) If $(r_b + r_s > d_{max}$ and $r_b + d_{max} > r_s$ and $r_s + d_{max} > r_b$) then
      i. If $\varphi > \varphi^*$ then $t(b, l, f) = 0$ where $\varphi* = \arccos(\frac{r_b^2 + r_s^2 - d_{max}^2}{2r_b r_s})$.
      ii. If $\varphi \leq \varphi^*$ then $t_{bs} = \frac{x - r_s}{v(f)}$ where x is root of equation $x^2 + r_b^2 - d_{max}^2 = 2xr_b \cos\varphi$.
   (c) If $(d_{max} \geq r_b + r_s$) then $t_{bs} = \frac{y - r_s}{v_f}$ where y is root of equation $y^2 + r_b^2 - d_{max}^2 = 2yr_b \cos\varphi$.
   (d) If $(r_s \geq d_{max} + r_b)$ then $t(b, l, f) = 0$.
   (e) If $(r_b \geq d_{max} + r_s)$ then
      i. If $\varphi > \varphi^*$ then $t(b, l, f) = 0$ where $\varphi* = \arcsin(\frac{d_{max}}{r_b})$.
      ii. If $\varphi \leq \varphi^*$ then $t_{bs} = \frac{2\sqrt{d_{max}^2 - r_b^2 \sin^2\varphi}}{v_f}$.
4. Compute $t(b, l, f) = 1 + \frac{\delta t_{bs} + t_{fs}}{t_{as}}$.

References

1. Brian, J.: An air mission planning algorithm for a theater level combat model. Master thesis, Air force Institute of Technology (1994)
2. Brown, G., Carlyle, M., Wood, K.: Applying defender-attacker optimization to terror risk assessment and mitigation. Calhoun, The NPS Institutional Archive (2008)
3. Crino, J., Moore, J.: Solving the theater distribution vehicle rounting and scheduling problem using group theoretic tabu search. Math. Comput. Model. **39**(6), 599–616 (2004)
4. Diehl, D.D.: How to optimized joint theater ballistic missile defense. Master Thesis, Naval Postgraduate School Monterey, CA (2004)
5. Garey, M.R., Johnson, D.S.: Computers and Intractability; A Guide to the Theory of NP-Completeness. W. H. Freeman & Co., New York (1990)
6. Jackson, J.: A taxonomy of advanced linear programming techniques and the theater attack model. Master thesis, Air Force Institute of Technology (1989)
7. Miller, C.E., Tucker, A.W., Zemlin, R.A.: Integer programming formulation of traveling salesman problems. J. ACM **7**(4), 326–329 (1960)
8. Moore, F.W.: A methodology for missile countermeasures optimization under uncertainty. Evol. Comput. **10**(2), 129–149 (2002)
9. Robert, E.: An adaptive tabu search heuristic for the location rounting pickup and delivery problem with time windows, a theater distribution application. Doctoral Thesis, Air force Institute of Technology (2006)
10. Seichter, S.: The fast theater model optimization of air-to ground engagements as a defender-attacker model. Master thesis, Naval Postgraduate School (2005)
11. Shalikashvili, G.: Joint tactics, techniques, and procedures for movement control. Chairman of the Joint Chiefs of StaffJoint Publication, Washington 4–01.2 (1993)
12. Shalikashvili, G.: Joint tactics, techniques, and procedures for movement control. Chairman of the Joint Chiefs of StaffJoint Publication, Washington 4–01.3 (6 1996)
13. Air Force Studies and Analysis Agency: TAC thunder analysis manual. CACI Products Company, Arlington (1992)

Perspectives of Big Data Analysis in Urban Railway Planning: Shenzhen Metro Case Study

Keke Peng[1], Caiwei Yuan[2], and Wen Xu[3(✉)]

[1] Shenzhen City Traffic Planning Design Research Center, Guangdong, China
jack_gaga@126.com
[2] Department of Computer Science and Technology,
Harbin Institute of Technology (Shenzhen), Shenzhen, Guangdong, China
caiwei_yuan@163.com
[3] Department of Mathematics and Computer Science,
Texas Woman's University, Denton, Texas 76204, USA
wxu1@twu.edu

Abstract. Urban railway system is of great importance to public transportation and economic development. However, due to the fast development of urban cities and time-consuming construction, it is quite challenging to plan a successful metro railway system beforehand. In this paper, we propose perspectives of evaluating traffic efficiency of metro railway systems from various factors such as the total railway traffic flow, the structure of the traffic system and the spatial distribution of work-and-home. We evaluate the implementation effect of Shenzhen railway system (particular the second phase construction) based on historical and real-time data reported by 28,000 passengers, which will provide insightful suggestions for Shenzhen metro construction in the future.

Keywords: Traffic efficiency · Railway evaluation · City dynamics

1 Introduction

Since 2000s, urban railway investment increased significantly in China. According to statistics, by 2013, there are 16 cities in China whose railway transit length has reached 2213 km, 35 cities whose rail transit length under construction reached 2760 km. Beijing, Shanghai, Guangzhou, Shenzhen and other major large cities are speeding up the following network planning and construction.

These investments were in general planned due to the acceleration of urbanization and the increasing demand for residents traveling, but very few have been successful in improving the transport and the urban environment. Previous research has shown that while most of the new generation urban rail systems could be enhanced if the co-ordination between transport planning and urban planning be stronger. However, coordination is very difficult to achieve within the constantly changing social dynamics and fragmented planning system.

© Springer International Publishing AG 2017
X. Gao et al. (Eds.): COCOA 2017, Part I, LNCS 10627, pp. 386–400, 2017.
https://doi.org/10.1007/978-3-319-71150-8_32

Recently, big data reflecting city dynamics have become widely available [2,5,10], e.g., traffic flow, human mobility, and population survey, enabling us to solve this challenging problem from a data perspective. According to existing studies [13], these data have a strong correlation with railway successfulness.

This paper explores ways of making new urban rail systems more successful using big data analysis. It develops a methodology for analyzing the success of systems, identifying the factors behind the success, and enhancing the success. Taking Shenzhen City as an example, the paper utilizes both survey and real-time data of Phase II of rail transit reported by 28,000 passengers for analysis, an evaluation framework is developed based on the analysis. From the perspectives of the utilization efficiency of the railway system, the organizational efficiency of the urban transportation and residential structure [1], the framework helps planners to evaluate and increase the success of their systems [4].

The framework has two main functions: it evaluates the success of railway systems, and makes recommendations on how their success can be enhanced. While the framework addresses many factors that may affect success, there is a special focus on exploring methods for providing and sustaining co-ordination between transport and urban planning.

The main contributions of this article are:

- We propose specific criteria to measure the performance of the system. The more criteria the system fulfill, the more successful it is regarded as in this work. The criteria are presented in Sect. 3 based on the indicators listed above. Other potential indicators of success are also discussed in order to try to avoid any bias that may result from the choice of criteria.
- During the analysis of the indicators that the criteria are based on, the performance of Shenzhen urban rail systems has been evaluated comprehensively based on historical and real-time data reported by 28,000 passengers.
- Throughout the analysis, possible reasons for success or failure are discussed, and links are suggested between the success of the systems and the various factors reflecting the planning background and the planning process.

2   Urban Railway Datasets

As the first special economic zone in China, Shenzhen's urban transportation infrastructure has maintained a rapid development trend. The city vigorously promoted the construction of urban rail transit and opened 5 railway lines with the length about 178 km in June 2011 for the World University Games.

Shenzhen railway transit construction has gone through a long period. The "overall urban plan" constructed in 1995 proposed a long-term railway network and Phase I route scheme. The Railway Phase I was opened to public in 2004, including Metro Line 1 (Luohu Port - Window of the World), Line 4 (Fukuda Port Station - Children's Palace Station), with overall length about 21.4 km, 20 stations, and daily traffic about 42 million people. In June 2011, Railway Phase II was opened to public, including the extension part of Metro Line 1 and Line 4,

Table 1. Summary of Shenzhen metro rail lines.

Line numer	Line name	Length (km)	Station number	Starting point–end point	Description
1	Luobao line	40.9	30	Lo Wu Station to Airport East Station	A trunk line which goes through the central city from east to west, connecting the city center and the western development axis.
2	Shekou line	35.8	29	Xinxiu station to Chiwan station	A line which connects the city center area, Nanshan sub-central area and Shekou area.
3	Longgang line	41.6	30	Shuang Long station to Yitian station	The passenger trunk which connects the city center and the eastern development axis.
4	Longhua line	20.3	15	Futian Port Station to Qinghu Station	The north-south trunk line which connects the Futian Port, the downtown area, and the Longhua Development Area.
5	Ring center line	40	27	Huang Beiling to Qianhaiwan Station	A line which connects the West, Central and East development axises of Shenzhen, including Shenzhen logistics center, high-tech area, University Town. etc.
Total	178	131			

and part of Metro Line 2, Line 3 and Line 5, with overall length about 156.5 km. After Phase II, a 178 km rail transit network with 131 stations has been formed. The details of five metro lines in Shenzhen are summarized in Table 1.

In order to accurately evaluate the implementation effect, this study selected 52 sites of the network (see Fig. 1) and took the sample size by 2% of the traffic during working days. The survey investigation was conducted on 21, 23, 25 and 26 May 2012, in which 3 days for working days, 1 day for weekend day. 27,300 questionnaires were distributed and 23,300 valid questionnaires were collected.

The data used in this study are basically from the following two categories: (1) Survey data, which include the full travel chain information of 2.3 million railway passengers, and traffic data of 20 roads parallel to the railway. (2) Dynamic card data, which are provided by Shenzhen Railway Company.

2.1 Data of Passenger Travel Behavior

(1) Travel Chain Data. A trip of the railway passenger can be described in the following travel chain mode: Departure - Railway site - Transfer site (1-n) - Railway site - Arrival. The data describes inside trips, lines transfers, outbound connections, vehicle changes and personal characteristics.

(2) Parallel Traffic Data. Road sections paralleled with the railway line may decentralize passenger flow, therefore the corresponding road traffic data are collected and analyzed.

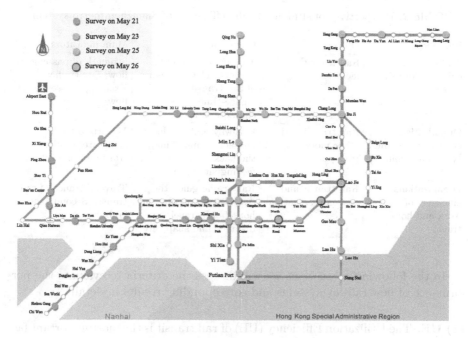

Fig. 1. Shenzhen metro transit map studied.

2.2 Dynamic Railway Card Data

Shenzhen Railway Company provides electronic card data of the railway passengers, which includes unique passenger identifier, credit card transaction types, inbound sites, inbound times, outbound sites, and outbound time.

3 Methodology

In this section, we present a set of indicators and criteria, which help to evaluate the efficiency of new railway system. A general framework is also proposed to evaluate the success of railway systems.

In this paper, we mainly adopt the perspective of traffic efficiency and define it from three aspects. First, from the perspective of the rail traffic itself, we evaluate the passenger flow growth, the ability that the new rail transit attracts residents to use it [14,19], for which the main indicators used is the rail transit strength. Secondly, from the perspective that the rail transit contributes to the entire public transport system, we analyze the relationship between the rail transit and the conventional traffic, the relationship between the rail transit and the urban traffic structure. Finally, from the perspective of rail transit to guide the urban travelling structure and urban spatial structure, we analyze the organizational efficiency of new urban traffic distribution. Three aspects to evaluate traffice efficiency are summerized in Table 2.

Table 2. Perspectives of studying traffic efficiency of Shenzhen metro system.

Efficiency	Object	Connotation	Main indicators
The utilization efficiency of rail transit	Rail transit facilities	The utilization and intensity characteristics of rail transit (space-time characteristics)	Strength of passenger flow; Site traffic; Exchange passenger flow
Organizational efficiency of traffic system matching	Rail transit and other modes of transport	Pathway and bus lines and access matching, the backbone and feeding situation	Motorized traffic structure; Public transport connection
Organizational efficiency of work-and-home travel	Rail transit and space organization	Rail traffic guide the distribution of jobs and household, travel distance and shape etc	Travel distance of rail transit; Second Line Hub

In the following subsections we will discuss the criteria to evaluate the performance of new railway systems and corresponding results in detail.

(1) UE. The Utilization Efficiency (UE) of rail transit is the most important factor of performance of railway systems and is mainly considered in three aspects: Strength of passenger flow (SPF); Site traffic (ST) and Exchange passenger ratio (EPR).

The SPF of a city can be calculated as follow:

$$SPF = \frac{Passenger\ Traffic(million\ passengers/day)}{Length(km)}$$

where the Passenger Traffic refers to the total passenger flow volume of the whole city's rail system during one day and Length refers to the length of the rail system.

The ST of a city can be calculated as follow:

$$ST = \frac{Passenger\ Traffic(million\ passengers/day)}{Station(s)}$$

where Station refers to the total number of stations in railway system.

The EPR of a city can be calculated as follow:

$$EPR = \frac{1}{n}\sum_{i=1}^{n} TR_i$$

where TR_i is the Transfer Ratio of the i-th transfer station, and n is the total number of transfer stations.

Thus, the utilization efficiency (UE) of the city can be calculated as:

$$UE = SPF + ST + EPR$$

the bigger UE means the better utilization efficiency of railway system

(2) OETSM. The Organization Efficiency of Traffic System Matching (OETSM) is also an important factor of railway systems performance and is mainly considered in two aspects: Motorized traffic structure (MTS); Public transport connection (PTC). While the Non-motorized connection factor is also discussed.

The traffic structure determines the operational efficiency of urban traffic. Higer proportion of public transportation would leads to the shirnk of individual traffic, smoother road system [7], and reduce traffic congestion [17]. The MTS of a city is defined as follow:

$$MTS = CB + RT \times 2 - IT$$

where the CB refers to the ratio of conventional bus, RT refers to the ratio of railway transportation and IT refers to the ratio of individual transportation $(CB + RT + IT = 1)$. Bigger MTS means better road condition.

The PTC of a station is defined as follow:

$$PTC = Accommodate - (Opened + Passby \times 0.2)$$

where Accommadate refers to the amount of lines that station can accommodate, Opened refers to the number of lines station has been opened, and Passby means the number of lines just passby the station. The PTC of a station greater than 0 means the space resources are underutilized, and when the value is less then 0 means the road is overload and may prone to traffic jam phenomenon. The PTC of a city is calculated by the standard deviation of every railway station's PTC.

(3) DOW. The Distribution of Work-and-home (DOW) affects residents' daily travel distance directly, and due to the Second Line Hub, the traffic pressure of original SAR and Baoan District ect. is kind of heavy. So it's necessary to evaluate the railway system from these aspects to figure out whether the Second Phase meet the daily needs of residents or not. The imbalance of daily traffic is also discussed in this part.

3.1 Utilization Efficiency of Rail Transit Analysis

(1) Strength of Passenger Flow (SPF). The Strength of Passenger Flow (SPF) is an important factor to reflect the situation of railway utilization. The SPF of Shenzhen is growing rapidly after the operation of Phase II. Compared with the cities that have been operating for many years [2,6], the scale of the initial passenger flow capacity is basically the same or even slightly better, but the traffic utilization efficiency still can be improved.

One year after Phase II opened, the average daily traffic flow of Shenzhen track increased from 610,000 to 2.13 million. The increase percentage reached 249%. Compared to Beijing, Shanghai, Guangzhou and other cities that have run railways for many years with initial passenger flow around 2 million person per day, Shenzhen has reached 213 million passengers per day only one year after its track opening (see Table 3). It is obvious that Shenzhen passenger growth rate is higher than the above-mentioned cities in China.

Table 3. Summary of the development of major cities in 2011.

Index	City					
	Shanghai	Beijing	Guangzhou	Hong Kong	Tokyo	Shenzhen
Number of lines (bars)	11	15	8	9	13	5
Length (km)	424	370	236	175	305	178
Station (s)	266	191	144	82	250	118
Passenger traffic (million passengers/day)	640	693	520	440	1091	213
SPF (million passengers/km)	1.5	1.9	2.2	2.5	3.6	1.2
ST (million passengers/station)	2.4	3.6	3.6	5.4	4.4	1.8

(2) Site Traffic (ST). Comparing with some international track developed cities, there is a quite big space for Shenzhen railway system to grow in terms of the track utilization efficiency. In 2010, the passenger track strength of Shenzhen is 18,000 passengers/km, while after opening the whole network the passenger track strength is 12,000 passengers/km. Compared with Hong Kong and Tokyo, which are dominated by public transport, Shenzhen traffic flow is only 48% of Hong Kong and 33% of Tokyo, and at the same time, with site traffic (ST) only 33% of Hong Kong and 41% of Tokyo (see Table 3).

The reason of reduction is, on the one hand, the running time of Shenzhen Phase II railway is not long and the passenger flow has potential to grow, on the other hand, there are large differences among the five lines of the new railway in terms of passenger flow efficiency. From Table 4, it is clearly to see that the Luo Bao line has the highest passenger strength which has reached 19,000 passengers/km. That's because this line is coupled with the city axis, the area with the largest population and job gathering. However other lines has a big gap in the coverage of the population [8,14], the use of land, and access conditions compared with BaoLuo line. Especially for the SheKou line, which connects Nanshan and the central city, the passenger strength of it is only 0.6 million people/km.

(3) Exchange Passenger Ratio (EPR). The lack of transfer facilities is an important factor affecting the operation of rail transit. After the Phase II opened, the rapid growth of passenger flow caused great pressure on the transfer facilities. The transfer coefficient of the railway network increased from 1.05 to 1.47, the passenger flow increased from 30,000 per day to 1.33 million. Especially at some transfer hubs, the transfer passenger flow is more than inbound (outbound) passenger flow. For instance, the average transfer passenger traffic of the Convention and Exhibition Center is 120,000 per day, which is four times as the inbound (outbound) passenger flow of the station. At other transfer hubs such

Table 4. Statistics of length and passenger strength of five metro lines.

Name of lines	Length (km)	Strength of passenger flow (Million passengers/km)
Luo Bao line	41[a]	1.9
Shekou line	36	0.6
Longgang line	41	1.1
Longhua line	20	1.3
Ring center line	40	0.8
total	178	1.2

[a]The data is based on the AFC Auto-ticket-selling data provided by Shenzhen Tong Company on May 20, 2012.

as Old street, Phuket, Shenzhen North Station, the daily transfer passenger flow is more than 60,000. Therefore in order to make railway system run successfully [18], it is necessary to improve the station transfer facilities, such as reducing the length of transferring channels, and increasing number of elevators. The exchange passenger ratio of several transfer station are summarized in Table 5, and the total EPR of Shenzhen can be caculated: $EPR = \frac{1}{13} \sum_{i=1}^{13} TR_i = 0.69$. And at the same time, the utilization efficiency(UE) of Shenzhen can be calculated from Tables 3 and 5: $UE = SPF + ST + EPR = 1.2 + 1.8 + 0.69 = 3.69$.

3.2 Organization Efficiency of Traffic System Matching

(1) Motorized Traffic Structure (MTS). After the opening of the track [16], the proportion of rail transit in motorized transportation rose from 3% to 9%. The percentage of public transportation (conventional bus plus rail traffic) in the motorized transportation increased to 43%. The percentage of private car use in the motorized transportation reduced to 40% [15]. Furthermore, the proportion of public transportation exceeded personal travel for the first since 2005. The average speed of road traffic during peak hours (morning and night) in central city is 1.0 km/h and 0.8 km/h faster respectively. Hence the new railway system plays a critical role in improving the average speed of road traffic. The MTS before Phase II in 2010 is: $MTS_{2010} = PT + RT \times 2 - IT = -0.21$, and the MTS after Phase II in 2012 is: $MTS_{2010} = PT + RT \times 2 - IT = -0.04$. The opening of Phase II optimized the traffic structure of Shenzhen in a certain extend (Table 6).

(2) Public Transport Connection (PTC). After the opening of Phase II, there are totally 23 bus stations and 30 bike stations built around the track sites, which basically cover all bus stops.

The Table 7 shows the statistics of connecting stations of the Shenzhen railway system. According to Table 7, the PTC of the stations varies from −6.6 to

Table 5. The exchange passenger flow of transfer station.

Serial number	Transfer station	The number of people entering this site	Actual passenger volume	Transfer ratio	Transfer number (Allocation method)	Transfer number (Rail Office)
1	Lao Jie station	50651	139008	64%	88357	120000
2	Exhibition center station	25198	112555	78%	87357	120000
3	Baoan center station	7813	68321	89%	60508	58000
4	Bu Ji station	17147	76540	78%	59393	60000
5	Shenzhen north station	18943	78286	76%	59343	60000
6	Shimin center station	7794	46121	83%	38327	30000
7	Children's palace station	12836	50780	75%	37944	30000
8	Huang Beling station	17144	49612	65%	32468	40000
9	Window of the world	28796	61118	53%	32322	45000
10	Grand theater station	47793	77862	39%	30069	30000
11	Fu Tian station	9345	29006	68%	19661	20000
12	Shopping park station	23948	42789	44%	18841	10000
13	Qianhaiwan station	209	1389	85%	1180	10000

4.2, which means that the structure of public transport connection is kind of imbalance, and still can be optimized. There has opened 50 connecting lines, while the station can accommodate up to 64 lines, which means the site resources have not been fully exploited [9]. On the other hand, there are a large number of buses passing by the stations (113 lines), causing the congestion of the road traffic to a certain extent. In order to reduce waste of resources and relieve bus traffic, it is highly recommended to fully utilize connecting stations and bike usage [9].

In addition to the transfers inside railway stations, the transfer between railway stations and bus stations also increased rapidly. The transfer passenger flow between railway stations and bus stations reached 320,000 people per day in 2012. The growth is nearly 10 times compared to time when system not opened.

In Shenzhen railway system, the transfer distance between railway station and bus station is generally within 250 m. For example, the distance of Window of the World, Airport station in Line 1 and connecting bus terminal (including

Table 6. The percentages of motorized transportation of Shenzhen over years.

Year	Public transit(%)			Car (%)	Other (%)[b]	Motor vehicle (Ten thousand)
	Conventional bus	Railway	Total			
2000	37%	-	37%	18%	45%	32
2005	42%	2%	44%	35%	21%	81
2010	35%	3%	38%	44%	18%	171
2012	34%	9%	43%	40%	17%	223

[b]The other(%) from the table includes motorcycles, electric cars, taxis and work unit buses etc. The data in the table comes from the Shenzhen residents travel survey in 2001, 2005, 2010, and survey after the Railway Phase II opened in 2011. The motor vehicle data in 2012 is as of November 2012.

Table 7. The number of lines of the popular stations of Shenzhen metro system Phase I and Phase II.

Station Name	The amount of lines that station can accommodate	The number of lines station has been opened	The number of lines passby the station
Window of the world	8	8	9
Shopping park	6	10	13
Science Museum	3	3	9
Huaqiang North	2	3	2
Xin'an	5	0	4
Airport East	6	7	7
Nanlian	3	3	15
Hongshan	3	2	4
Qinghu	3	4	6
Shangtang	4	1	3
Minzhi	2	0	1
Dayun	6	7	15
Baigelong	4	0	4
Liuyue	3	2	13
Gushu	6	0	8
total	64	50	113

the vertical distance and horizontal distance) is within 200 m. But there are also several bus connecting stations are far from the rail stations with about 400 m due to the reasons for the use of land. Long transfer distance between railway stations and bus connecting stations greatly increased the transfer time and decrease the transfer efficiency directly. Therefore the connection facility and the transfer distance should be taken into account, in order to make the whole railway system perform more efficiently.

(3) Non-motorized Connection. The rapid growth of the transition between the railway and walk & bike has a great impact on the efficiency of the traffic organization as well [12]. 78.1% of passenger flow chose walking as a transiting and connecting way. Observing peak hours of passenger flow, 20% of the site passengers all day long may be concentrated in an hour. That is, the number of pedestrians on the road around the railway sites in peak hour will increase dramatically. It can be foreseen that the larger number of the road bikes, electric cars, and pedestrians in peak hours will undoubtedly bring a severe test to the surrounding road traffic organization [11].

3.3 Organization Efficiency from the Distribution of Work-and-Home

(1) Travel Distence of Rail Transit. After the Phase II opened, long distance travel become dominant in all travels. Transfer travels account for 60% of the total track travel. The average track travel distance is increased from 5.8 km to 12.9 km. The rail traffic shares daily access of all passenger flow of the second line by 20%–33%. Figure 3 shows the distribution of residents travel distance, the orbital travel is mainly distributed in the center area within 20–30 km range.

At present, the average travel time between two railway stations in the whole city is about 30 min. 92% of the railway travel takes within one hour (see Fig. 3).

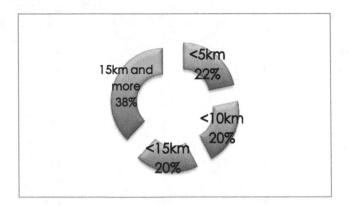

Fig. 2. Residents travel distance distribution

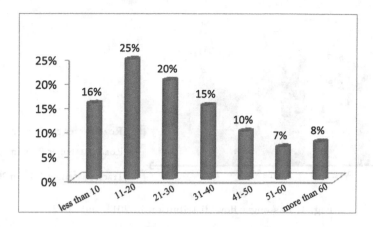

Fig. 3. Track trave time (Min)

The railway can meet the service requirement of one-hour trip. And the railway backbone is becoming more and more prominent.

Learning from the experience of foreign cities such as Tokyo, the development of urban space and the backbone of city transportation should be matched with each other. Railway traffic is the main transportation for the commuters of Tokyo. Among the citizens working in the central city (Tokyo district), 85% of the people live within 30 km to the city center. The commute travel is basically dependent on the railway line solution; and the average length of railway lines is 24.3 km.

When the city Shenzhen expanding to the Pingshan, Guangming and Dakonggang and other peripheral areas in the future, residents' travel distance will be further grow, the existing speed and structure of railway system will be difficult to meet the requirement of one hour trip from city center to peripheral area. So faster railways or new lines should be considered to meet the fast travel needs of residents.

(2) Second Line Hub. With the accelerated integration in Shenzhen, the inside and outside of original SAR are getting closer, average daily travels across the second line hub increased to 2.9 million people, almost 10 times from 2000.

As shown in Fig..4, the opening of the second phase greatly relieved the daily traffic pressure of the Second Line Hub. According to the survey, LuoBao Line and the Ring Center Line share 33% daily passenger flow of the western Second Line Hub. Longhua Line shares 20% daily passenger flow of the central Second Line Hub. Longgang Line and Ring Center Line share 26% daily passenger flow of the eastern Second Line Hub, reducing the road traffic of the western and central Second Line Hub by 16% and 8%. However the road traffic increases by 23% in eastern area due to strong demand growth.

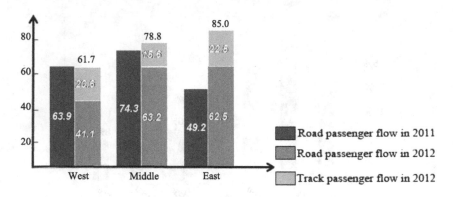

Fig. 4. Passenger flow distribution in 2011 and 2012.

(3) Imbalance. Due to the tide and peak accumulation of passenger flow, the utilization of axial railway is low. On the one hand, the main trip purpose of passengers is simply commuting to work, leading to the huge passenger flow in peak hours. Take Longhua Line for example, the passenger flow in the four hours during morning and evening peak takes about two-thirds of all day traffic. On the other hand, a large number of tidal commuters come and go between the city center and new town area. As a result, during the morning peak hour, the imbalance coefficient of Longhua Line is 7:1, and the imbalance coefficient of Longgang Line is 5:1.

The imbalance distribution of living and working areas along the stations of railways of the original SAR is the main reason for the passenger tide and peak accumulation. According to statistics, within 500 m of the original SAR site, the office and commercial area takes 35%. 500 m away from the original SAR, residence buildings take 70%, while office and commercial areas take only 9% [19].

From Table 8 we can see that there is a big difference in site traffic during peak hours [20]. Peak hour coefficient is 11%–13%. Different types of sites vary widely on coefficient, with the maximum 23% and the minimum 5%.

Table 8. Peak hour traffic of metro lines.

Line name	Peak hour average coefficient	Peak hour coefficient range
Luobao line	13%	6%–23%
Shekou line	13%	5%–21%
Longgang line	11%	6%–22%
Longhua line	13%	8%–16%
Huanzhong line	12%	9%–22%

4 Conclusion

The rail transit network needs to be improved continuously to cover the city's major transportation hubs. Based on condition that the basic network of city railway system is established, set target of establish the overall rail transit trunk plus the connecting public transport system, exploit the advantages of public transport to the full, and enhance the service level of public transport. To achieve such a goal, it needs to make efforts in the backbone, access, and transfer, making each link interlocking and collectively effective.

The next step is integrate the public transportation fare system and promote traffic facilities according to residents needs. The recent proposal is to further expand the peak commuting time, reduce peak intension, reduce traffic pressure, and improve the utilization rate of axial orbit.

References

1. Zhang, Y., Guo, L.: Study on coordinated relationship between urban railtransit and land-use. LISS, pp. 1–5 (2016). https://doi.org/10.1109/LISS.2016.7854339
2. Yang, X., Chen, A., Ning, B., Tang, T.: Measuring route diversity for urban rail transit networks: a case study of the beijing metro network. IEEE Trans. Intell. Transp. Syst. **18**, 259–268 (2017)
3. He, Z., Huang, J., Du, Y., Wang, B., Yu, H.: The prediction of passenger flow distribution for urban rail transit based on butil-factor model. ICITE, pp. 128–132 (2016). https://doi.org/10.1109/ICITE.2016.7581320
4. Zhang, H., Song, M., Zhang, M.: An energy efficient optimized control algrorithm for urban rail transit system. CCC (2016). https://doi.org/10.1109/ChiCC.2016.7554972
5. Guan, H., Yin, Y., Yan, H., Han, Y., Qin, H.: Urban railway accessibility. Tsinghua Science and Technology, vol. 12, pp. 192–197 (2007)
6. Caracciolo, F., Fumi, A., Cinieri, E.: Managing the italian high-speed railway network: provisions for reducing interference between electric traction systems. IEEE Electrification Mag. **4**, 42–47 (2016)
7. Boreiko, O., Teslyuk, V.: Structural model of passenger counting and public transport tracking system of smart city. In: 2016 XII International Conference on Perspective Technologies and Methods in MEMS Design (2016)
8. Huang, R., Liu, Z., Wang, D., Ma, L.: Organization mode of suburban railways in urban rail transit system. In: 5th Advanced Forum on Transportation of China (2009). https://doi.org/10.1049/cp.2009.1601
9. Tian, Z., Weston, P., Hillmansen, S., Roberts, C., Zhao, N.: System energy optimisation of metro-transit system using Monte Carlo Algorithm. ICIRT (2016). https://doi.org/10.1109/ICIRT.2016.7588768
10. Liu, L.: How does rail transit promote the sustainable development of Beijing metropolitan area? IEIS. IEEE Conference Publications (2016). https://doi.org/10.1109/IEIS.2016.7551863
11. Li, H.: Dynamic location optimization methodology for urban transfer centers. In: 2015 Ninth International Conference on Frontier of Computer Science and Technology (2015). https://doi.org/10.1109/FCST.2015.19

12. Meng, M., Li, S., Lam, S.H., Wong, Y.D.: Public transit coordination under different strategies between operators. MT-ITS (2015). https://doi.org/10.1109/MTITS.2015.7223276
13. Meng, B., Zheng, L., Yu, H., Me, G.: Spatial characteristics of the residents' commuting behavior in Beijing. In: 2011 19th International Conference on Geoinformatics (2011). https://doi.org/10.1109/GeoInformatics.2011.5981020
14. Zhao, K., Tarkoma, S., Liu, S., Vo, H.: Urban human mobility data mining: an overview. In: 2016 IEEE International Conference on Big Data (Big Data) (2016). https://doi.org/10.1109/BigData.2016.7840811
15. Li, C., Chiang, A., Dobler, G., Wang, Y., Xie, K., Ozbay, K., Ghandehari, M., Zhou, J., Wang, D.: Robust vehicle tracking for urban traffic videos at intersections. AVSS (2016). https://doi.org/10.1109/AVSS.2016.7738075
16. Glickenstein, H.: New developments in land transportation [Transportation Systems]. IEEE Veh. Technol. Mag. 5, 17–20 (2010)
17. Yang, X., Li, X., Ning, B., Tang, T.: A survey on energy-efficient train operation for urban rail transit. IEEE Trans. Intell. Transp. Syst. 17(1), 2–13 (2016)
18. Cadarso, L., Maróti, G., Marín, Á.: Smooth and controlled recovery planning of disruptions in rapid transit networks. IEEE Trans. Intell. Transp. Syst. 16, 2192–2202 (2015)
19. Souza, E.S., Barbosa, J.D.C., Millian, F.M., Torres, M., Ambrosio, P.S.: Tracking system for urban buses with people flow management. IEEE Lat. Am. Trans. 16, 944–949 (2011)
20. Hong, L., Li, Y., Xu, Z., Jiang, Y., Li, F., Lin, L., Ling, J., Chen, X.: A service benefit analysis of the urban rail transit. ICSSSM (2015). https://doi.org/10.1109/ICSSSM.2015.7170163

Cloning Automata: Simulation and Analysis of Computer Bacteria

Chu Chen[1], Zhenhua Duan[1(✉)], Cong Tian[1(✉)], and Hongwei Du[2]

[1] ICTT and ISN Laboratory, Xidian University, Xi'an 710071,
People's Republic of China
zhhduan@mail.xidian.edu.cn, ctian@mail.xidian.edu.cn
[2] Department of Computer Science and Technology, Harbin Institute of Technology,
Shenzhen 518055, People's Republic of China

Abstract. In order to simulate the self-replication of computer bacteria, a new model named cloning automata is put forward. It can simulate the self-replication in two different ways: fusion and fission. Properties such as the self-replicating velocity and the threshold time for denial of service are analyzed. Also, methods for interrupting the self-replicating behavior are presented. As an example, a concrete computer bacterium i.e. fork bomb is simulated and analyzed with cloning automata.

Keywords: Self-replication · Computer bacteria · Model · Cloning automata · Fusion · Fission

1 Introduction

Malicious software (Malware) such as computer bacteria (also known as germs), worms, viruses, spyware, Trojan horses, ransomware and so forth are main threats to computer and cyber systems in the world today [4,5]. Different malware adopt different malicious behavior for different purposes. As the origin of malware, computer bacteria replicate themselves continually just like the reproduction of biological rabbits. The self-replication consumes system resources such as the time of a central processing unit (CPU), the space of a random access memory (RAM) and the space of a disk [12]. The continual self-replication makes computer systems become ever slower and denial of service (DoS) occurs when systems cannot respond to users before a deadline. Worms replicate themselves through networks and keep one copy active in the memory of each host machine [14]. Thus worms consume both network and system resources to some extent. Unlike bacteria or worms, viruses often insert possibly evolved copies of themselves at the beginning, in the middle or at the end of host files and do some damage [7,17]. Spyware try their best to be invisible to users and steal important files or data [19] while Trojan horses usually work differently from what they look like [16]. Different from computer bacteria, worms and viruses, neither spyware nor Trojan horses replicate themselves. Others such as ransomware [8], downloader [15] and rootkit [9] behave differently and do not replicate themselves.

© Springer International Publishing AG 2017
X. Gao et al. (Eds.): COCOA 2017, Part I, LNCS 10627, pp. 401–416, 2017.
https://doi.org/10.1007/978-3-319-71150-8_33

In order to understand structures and working principles of malware, it is necessary to model them mathematically. However, malware are so capricious as mentioned above that it is too hard to build one universal model to simulate all of them. One feasible way is to build a model for one category of malware with similar behavior. As for self-replication, John Von Neumann [11] and Stanislaw Ulam [13] built the cellular automata model for biological cells in the 1940s. In the 1980s, Stephen Wolfram studied models produced by 256 rules of elementary cellular automata [18]. Different from cellular automata, a new model with name "cloning automata" is put forward to simulate the continual self-replicating behavior of computer bacteria.

The remainder is arranged as follows. The new model and two different ways to realize the self-replication are introduced in the following section. After that, properties of cloning automata including the self-replicating velocity and the threshold time for DoS are analyzed in Sect. 3. Methods for interrupting the process of self-replication are presented according to this model in Sect. 4. As a concrete example of computer bacteria, fork bomb is simulated with cloning automata via fusion and the analysis of its important properties is made in Sect. 5. Finally, Sect. 6 gives a brief conclusion of this work.

2   Cloning Automata

Cloning automata (CA) are systems used to model the self-replicating behavior of computer bacteria. Every element of this system is a single cloning automaton which comprises two parts within its structure. Each part consists of an automaton with a stack which is mainly used to store the other part. This structure offers the opportunity to realize the self-replication either in the way of fusion or in the way of fission. The formal definition of this new model is given as follows. Cloning automata are defined as a 5-tuple:
$CA = (\bigcup_{i=2^0}^{2^n} Q_{2i-1} \cup Q_{2i}, \bigcup_{i=2^0}^{2^n} \Sigma_{2i-1} \cup \Sigma_{2i}, \bigcup_{i=2^0}^{2^n} \Gamma_{2i-1} \cup \Gamma_{2i}, \bigcup_{i=2^0}^{2^n} \delta_{2i-1,2i}, S)$,
where (1) $Q_{2i-1} \cup Q_{2i}$ is a non-empty set of states of the ith cloning automaton which is denoted by CA_i; (2) $\Sigma_{2i-1} \cup \Sigma_{2i}$ is an input alphabet of CA_i; (3) $\Gamma_{2i-1} \cup \Gamma_{2i}$ is a stack alphabet of CA_i: $\forall j \in \{2i-1, 2i\}$, $\Gamma_j = \Sigma_j \cup \{\#, \$\} \cup \{A_k | k = j-1$, if $(j \mod 2) = 0; k = j+1$, otherwise$\}$; (4) $\delta_{2i-1,2i}$ is a transition function: $Q_{2i-1} \times (\Sigma_{2i-1} \cup \{\epsilon\}) \times \Gamma_{2i-1} \times Q_{2i} \times (\Sigma_{2i} \cup \{\epsilon\}) \times \Gamma_{2i} \rightarrow 2^{Q_{2i-1} \times \Gamma_{2i-1}^* \times Q_{2i} \times \Gamma_{2i}^*}$; (5) S is the set of initial states. The internal automaton A_k with its stack can be defined by $A_{2i-1} = (Q_{2i-1}, \Sigma_{2i-1}, \Gamma_{2i-1}, \delta_{2i-1}, p_0^{2i-1})$ and $A_{2i} = (Q_{2i}, \Sigma_{2i}, \Gamma_{2i}, \delta_{2i}, q_0^{2i})$. For brevity, superscripts and subscripts are omitted where there is no ambiguity. In general, δ can be defined as $\delta(p_m, a, b; q_n, c, d) = \{(p_s, \alpha; q_t, \beta)\}$, which means that when A_{2i-1} reads the input a at the state p_m with b on the top of A_{2i-1}'s stack and A_{2i} reads the input c at the state q_n with d on the top of A_{2i}'s stack, A_{2i-1} goes to the state p_s with its stack changed to α and A_{2i} goes to the state q_t with its stack changed to β. Under the assumption that α' is the content of A_{2i-1}'s stack and that β' is the content of A_{2i}'s stack before this transition, α and β may be changed from α' and β' by executing several actions such as pop and push, respectively. Sometimes, more attention is paid

to the result of reading a string than the detailed process of each symbol of this string. At this time, the detail is neglected and δ is written in brief like this: $\delta^*(p_m, \omega_1, b; q_n, \omega_2, d) = \{(p_s, \alpha; q_t, \beta)\}$, which can be obtained by several standard δ's. Semicolons are used here to separate A_{2i-1}'s part from A_{2i}'s part. Contrary to this brief form, sometimes the detail is more concerned and δ can be written in detail. For example, $\delta(p_m, \epsilon, top(\alpha); q_n, \epsilon, top(\beta)) = \{(p_s, \gamma; q_t, nil)\}$ after $\delta^*(p_x, \omega_1, b; q_y, \omega_2, d) = \{(p_m, \alpha; q_n, \beta)\}$ expresses that the contents of A_{2i-1}'s stack is changed from α to γ and that the content of A_{2i}'s stack from β to empty. The symbol "nil" here indicates that the corresponding stack is empty. For convenience, details of stack operations are neglected except for some operations related to important symbols. The detailed definition of δ is based on the way in which one CA replicates itself. With respect to CA, fusion and fission are two ways to realize the self-replication, thus two δ's are defined in different ways in the following two subsections.

2.1 Fusion

As for the structure of a CA, each part can use its stack to store a copy of the other part and these two parts which are stored in stacks can be combined to produce a new CA with the original CA's structure left unchanged. This way of self-replication is called fusion. With the CA model considered only, the corresponding δ which reflects this process is defined as follows. $\forall p_j \in Q_{2i-1}, q_j \in Q_{2i}, \exists p_{(j+1) \bmod 3} \in Q_{2i-1}, q_{(j+1) \bmod 3} \in Q_{2i}, j = 0, 1, 2$: (1) $\delta(p_0, \epsilon, nil; q_0, \epsilon, nil) = \{(p_1, A_{2i}\#; q_1, A_{2i-1}\#)\}$; (2) $\delta(p_1, \epsilon, A_{2i}; q_1, \epsilon, A_{2i-1}) = \{(p_2, \#A_{2i-1}A_{2i}\#; q_2, nil)\}$; (3) $\delta(p_2, \epsilon, \#; q_2, \epsilon, nil) = \{(p_0, nil; q_0, nil)\}$. The first part of the definition of δ specifies conditions under which copies of A_{2i-1} and A_{2i} can be stored in stacks. The second part of the definition of δ combines two parts which are stored in stacks by moving the part in A_{2i}'s stack into A_{2i-1}'s stack. The third one pops up the combined parts as a new CA. The same result can be arrived at by moving the part in A_{2i-1}'s stack into A_{2i}'s stack and then popping up it. The corresponding definition of δ is: $\delta(p_1, \epsilon, A_{2i}; q_1, \epsilon, A_{2i-1}) = \{(p_2, nil; q_2, \#A_{2i}A_{2i-1}\#)\}$ and $\delta(p_2, \epsilon, nil; q_2, \epsilon, \#) = \{(p_0, nil; q_0, nil)\}$. In order to avoid chaos, the combination operation will be done in A_{2i-1}'s stack in the following. Additionally, combinational stack operations instead of basic ones are used for brevity. For instance, (push $A_{2i}\#$, stack$_{A_{2i-1}}$) is used instead of (push $\#$, stack$_{A_{2i-1}}$) and then (push A_{2i}, stack$_{A_{2i-1}}$), similarly (pop stack$_{A_{2i-1}}$, $\#A_{2i-1}A_{2i}\#$) instead of (pop stack$_{A_{2i-1}}$, $\#$), (pop stack$_{A_{2i-1}}$, A_{2i-1}), (pop stack$_{A_{2i-1}}$, A_{2i}) and then (pop stack$_{A_{2i-1}}$, $\#$). Parentheses are used here to delimit every operation.

The self-replication via fusion is shown by Algorithm 1. The process is written in an infinite loop and it can be executed if there are enough resources available for a new CA and the process table is not saturated. The condition that the process table is not saturated always holds if the number of processes is not limited.

Figure 1 briefly shows the first cycle of this algorithm. When both stacks of A_1 and A_2 ($i = 1$) are empty, under the assumption that A_1's state is p_0 and

Algorithm 1: FUSION

Input: one cloning automaton
Output: numerous cloning automata

1 **while** *1* **do**
2   **if** available($resources$)$\geq$size(CA) **and** unsaturated($process\_table$) **then**
3   (push $A_{2i}\#$, stack$_{A_{2i-1}}$) and (push $A_{2i-1}\#$, stack$_{A_{2i}}$) in parallel if (empty(stack(A_{2i-1})) and empty(stack(A_{2i})));
4   (pop stack$_{A_{2i}}$, $A_{2i-1}\#$) if (top(stack$_{A_{2i}}$)==A_{2i-1});
5   (push $\#A_{2i-1}$, stack$_{A_{2i-1}}$);
6   (pop stack$_{A_{2i-1}}$, $\#A_{2i-1}A_{2i}\#$) if (top(stack$_{A_{2i-1}}$)==$\#$);
7   **end**
8 **end**

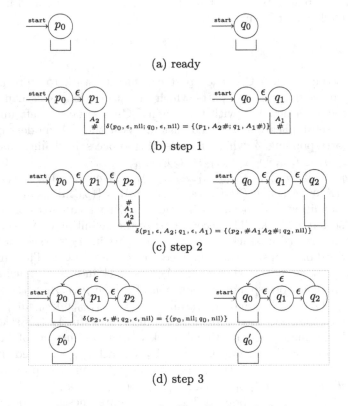

(a) ready

(b) step 1

(c) step 2

(d) step 3

Fig. 1. A process of self-replication via fusion

that A_2's state is q_0, A_1 pushes $\#$ and a copy of A_2 into A_1's own stack and goes to the state p_1, A_2 pushes $\#$ and a copy of A_1 into A_2's own stack and goes to the state q_1. Here, $\#$ is at the bottom of each stack. When A_2 is on the top of stack of A_1 and A_1 is on the top of stack of A_2, A_2 pops up A_1 and $\#$ from its stack and pushes them into A_1's stack in order, with states of A_1 and

A_2 changed to p_2 and q_2 respectively. At this point, $\#$ under A_2 and $\#$ above A_1 can be regarded as the right and the left boundaries of "A_1A_2", respectively. When $\#$ is on the top of A_1's stack, A_1 pops up "$\#A_1A_2\#$" from its stack and a new CA comes into being, with states of A_1 and A_2 changed to p_0 and q_0. At this time, these two $\#$'s "tie" "A_1A_2" up as a whole and distinguish it from any other CA.

2.2 Fission

Similar to the self-replication via fusion, each part of a CA uses its stack to store a copy of the other part. Different from the fusion, each part with its stack storing a copy of the other part breaks away from the original CA and constitutes a new CA with the other part popped up from its stack. Thus two new cloning automata are created with the original CA's structure broken up. This way of self-replication is referred to as fission. With the CA model considered only, the corresponding δ which reflects this process is defined as follows. (1) For $p_0 \in Q_{2i-1}, q_0 \in Q_{2i}, \exists p_1 \in Q_{2i-1}, q_1 \in Q_{2i} : \delta(p_0, \epsilon, \text{nil}; q_0, \epsilon, \text{nil}) = \{(p_1, A_{2i}; q_1, A_{2i-1})\}$; (2) for $p_1 \in Q_{2i-1}, q_1 \in Q_{2i}, \exists q'_0 \in Q'_{2i}, p'_0 \in Q'_{2i-1} : \delta(p_1, \epsilon, A_{2i}; q_1, \epsilon, A_{2i-1}) = \{(p_0, \text{nil}; q'_0, \text{nil}), (p'_0, \text{nil}; q_0, \text{nil})\}$. The first part of the definition of δ is similar in conditions to that of Sect. 2.1. The second part of the definition of δ splits the original CA into two new cloning automata. In the second part of the definition, q'_0 and Q'_{2i} are the initial state and the set of states of the other part popped up from A_{2i-1}'s stack, respectively. Similarly, p'_0 and Q'_{2i-1} are the initial state and the set of states of the other part popped up from A_{2i}'s stack, respectively. The symbol "$'$" here is used as a mark to distinguish the states and sets from those which exist before this break-up. For brevity, combinational stack operations as mentioned in Sect. 2.1 are used in the following.

The self-replication via fission is shown by Algorithm 2. The process is also written in an infinite loop and is similar in conditions to that of Algorithm 1.

Algorithm 2: FISSION

Input: one cloning automaton
Output: numerous cloning automata

```
1 while 1 do
2   │  if available(resources)≥size(CA) and unsaturated(process_table) then
3   │  │  (push A_2i, stack_{A_2i-1}) and (push A_2i-1, stack_{A_2i}) in parallel if
      │  │  (empty(stack(A_2i-1)) and empty(stack(A_2i)));
4   │  │  A_2i-1 and A_2i break away from the original CA if
      │  │  ((top(stack_{A_2i-1})==A_2i) and (top(stack_{A_2i})==A_2i-1));
5   │  │  (pop stack_{A_2i-1}, A_2i) and (pop stack_{A_2i}, A_2i-1) in parallel;
6   │  end
7 end
```

The first cycle of this algorithm is shown briefly in Fig. 2. When both stacks of A_1 and A_2 ($i = 1$) are empty, under the assumption that A_1's state is p_0 and that A_2's state is q_0, A_1 pushes a copy of A_2 into A_1's own stack and goes to the state p_1; A_2 pushes a copy of A_1 into A_2's own stack and goes to the state q_1. When A_2 is on the top of stack of A_1 and A_1 is on the top of stack of A_2, A_1 and A_2 break away from the original CA and the original CA does not exist anymore. A_1 pops up A_2 from its stack and goes to the state p_0. A_1 and the new A_2 which is just popped up from A_1's stack constitute a new CA. A_2 pops up A_1 from its stack and goes to the state q_0. A_2 and the new A_1 which is just popped up from A_2's stack form the other new CA. Thus, two new cloning automata are created.

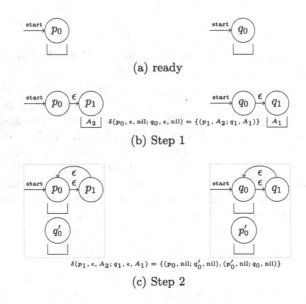

(a) ready

(b) Step 1

$\delta(p_0, \epsilon, \text{nil}; q_0, \epsilon, \text{nil}) = \{(p_1, A_2; q_1, A_1)\}$

(c) Step 2

$\delta(p_1, \epsilon, A_2; q_1, \epsilon, A_1) = \{(p_0, \text{nil}; q'_0, \text{nil}), (p'_0, \text{nil}; q_0, \text{nil})\}$

Fig. 2. A process of self-replication via fission

3   Analysis of Cloning Automata

An analysis of malware can help us to understand them better and further to fight against them more effectively. As abstract representations of concrete malware, models of malware are convenient to analyze with precision for general purposes. CA is one useful model of self-replicating malware and the analysis of CA is the basis of analyses of concrete computer bacteria. As for computer bacteria, properties such as the self-replicating cycle, frequency, velocity, volume and the time for DoS are concerned. These properties will be analyzed theoretically according to the CA model. Different ways to realize self-replication lead to slightly different analytical results. So, analyses will be made respectively.

For convenience, symbols to be used are explained in Table 1.

Table 1. Explanation of symbols

Symbol	Explanation
T	One self-replicating cycle;
$\Delta t_{\mathrm{push}(target)}$	The time needed to push $target$ into a stack;
$\Delta t_{\mathrm{push}(target_1\|\|target_2)}$	The maximum time needed to push $target_1$ and $target_2$ into stacks in parallel;
$\Delta t_{\mathrm{pop}(target)}$	The time needed to pop up $target$ from a stack;
Δt_{mal}	The time needed for other malicious behavior;
f	The self-replicating frequency;
v_{one}	The self-replicating velocity of one father CA;
$vol(t)$	The volume or the total size at the time t;
$v_i(t)$	The instantaneous self-replicating velocity of the CA system;
$a(t)$	The self-replicating acceleration of the CA system;
$v_a(t)$	The average self-replicating velocity of the CA system;
s	The size of a CA;
r	The threshold ratio of DoS;
max	The maximum resources available;

3.1   Analysis of Fusion

As defined in Sect. 2.1 and shown in Fig. 1, one cycle of fusion consists of three steps: the first step is to push $\#$ and a copy of A_2 into A_1's stack and to push $\#$ and a copy of A_1 into A_2's stack in parallel; the second step is to pop up A_1 and $\#$ from A_2's stack and then to push them into A_1's stack; the third step is to pop up $\#$, A_1, A_2 and $\#$ from A_1's stack. The time needed for the first step is $\Delta t_{\text{push}(A_2\#\|A_1\#)}$. Serial can be regarded as a special kind of parallel and $\Delta t_{\text{push}(A_2\#\|A_1\#)} = \Delta t_{\text{push}(A_2\#)} + \Delta t_{\text{push}(A_1\#)}$ when these two actions are executed serially. The second step can be optimized as follows: pop up A_1 from A_2's stack, and then push A_1 into A_1's stack and pop up $\#$ from A_2's stack in parallel, finally push $\#$ into A_1's stack. So the time needed for the second step is $\Delta t_{\text{pop}(A_1)} + \Delta t_{\text{push}(A_1)\|\text{pop}(\#)} + \Delta t_{\text{push}(\#)} = \Delta t_{\text{pop}(A_1)} + \max\{\Delta t_{\text{push}(A_1)}, \Delta t_{\text{pop}(\#)}\} + \Delta t_{\text{push}(\#)}$. The maximum time needed for the second step is $\Delta t_{\text{pop}(A_1)} + \Delta t_{\text{push}(A_1)} + \Delta t_{\text{pop}(\#)} + \Delta t_{\text{push}(\#)}$ when pop$(\#)$ happens after push(A_1). The time needed for the third step is $\Delta t_{\text{pop}(\#A_1A_2\#)}$. Apart from the self-replication, CA can be extended to model other malicious behavior and the corresponding time for this optional extension is expressed by $[+\Delta t_{\text{mal}}]$. The pure time needed for the self-replication via fusion Δt_{fusion} is: $\Delta t_{\text{push}(A_2\#\|A_1\#)} + \Delta t_{\text{pop}(A_1)} + \max\{\Delta t_{\text{push}(A_1)}, \Delta t_{\text{pop}(\#)}\} + \Delta t_{\text{push}(\#)} + \Delta t_{\text{pop}(\#A_1A_2\#)}[+\Delta t_{\text{mal}}]$. It is assumed that Δt_{apu} is the average time allocated to CA by the operating system per unit time. One cycle T in such a system is Δt_{fusion} for $\Delta t_{\text{fusion}} \leq \Delta t_{\text{apu}}$ or $\frac{\Delta t_{\text{fusion}}}{\Delta t_{\text{apu}}}$ for $\Delta t_{\text{fusion}} > \Delta t_{\text{apu}}$.

Obviously, $f = \frac{1}{T}$ and it means the number of clones generated by one father CA per unit time. The self-replicating velocity of one father CA is defined as the size of cloning automata generated by this father CA per unit time: $v_{\text{one}} = s \cdot f$, which shows the ability of a father CA to generate direct descendants. If time starts at 0, there are $\frac{t}{T}$ or $(t \cdot f)$ cycles at the time t. Volume is defined as the total size of all cloning automata at the time t: $vol(t) = s \cdot 2^{\frac{t}{T}} = s \cdot 2^{t \cdot f}$. The instantaneous self-replicating velocity of the CA system is the total size of cloning automata generated directly and indirectly by the ancestor per unit time with respect to the time t: $v_i(t) = (vol(t) - vol(t-1))/(t - (t-1)) = vol(t) - vol(t-1) = s \cdot (2^{t \cdot f} - 2^{(t-1) \cdot f}) = s \cdot (2^{t \cdot f} - 2^{t \cdot f} \cdot 2^{-f}) = s \cdot 2^{t \cdot f}(1 - (1/2)^f)$, which represents the instantaneous fusion ability of the whole CA system at the time t. The self-replicating acceleration of the CA system with respect to the time t is: $a(t) = \Delta v_i(t)/\Delta t = (v_i(t) - v_i(t-1))/(t - (t-1)) = (s \cdot 2^{t \cdot f}(1 - (1/2)^f) - s \cdot 2^{(t-1) \cdot f}(1 - (1/2)^f))/1 = s \cdot 2^{t \cdot f}(1 - (1/2)^f)(1 - 2^{-f}) = s \cdot 2^{t \cdot f}(1 - (1/2)^f)^2$, which is the accelerating reproduction ability of the whole CA system. The average self-replicating velocity of the CA system with respect to the time t: $v_a(t) = \frac{vol(t)}{t} = \frac{s \cdot 2^{\frac{t}{T}}}{t} = \frac{s \cdot 2^{t \cdot f}}{t}$, which represents the average ability of the whole CA system to generate direct and indirect descendants.

The analysis above relates to DoS attacks launched by computer bacteria. With the self-replication of computer bacteria, available resources become less and less and computer systems become busier and busier. At a time, it is too hard for the system to respond to users' request before a deadline and DoS

occurs: $\frac{vol(t_{\mathrm{DoS}})}{max} \geq r \Rightarrow \frac{s \cdot 2^{t_{\mathrm{DoS}} \cdot f}}{max} \geq r \Rightarrow 2^{t_{\mathrm{DoS}} \cdot f} \geq \frac{max \cdot r}{s} \Rightarrow t_{\mathrm{DoS}} \cdot f \geq \log \frac{max \cdot r}{s} \Rightarrow$
$t_{\mathrm{DoS}} \geq T \cdot \log \frac{max \cdot r}{s}$. That is, DoS occurs after $T \cdot \log \frac{max \cdot r}{s}$ if no countermeasures are adopted. In the case that there is a limit to the maximum processes of the process table and there is no limit to the number of processes created by one user, the operating system's process table will be saturated by processes of bacteria after $\frac{max_{\mathrm{pt}}}{2^{t \cdot f}} = 1 \Rightarrow t = T \cdot \log max_{\mathrm{pt}}$, where max_{pt} is the maximum processes of the process table.

3.2   Analysis of Fission

As defined in Sect. 2.2 and depicted in Fig. 2, one cycle of fission consists of two steps: the first step is to push a copy of A_2 into A_1's stack and to push a copy of A_1 into A_2's stack; the second step is that A_1 and A_2 break away from the original CA and that A_1 pops up a new A_2 from A_1's own stack and A_2 pops up a new A_1 from A_2's own stack in parallel. The time needed for the first step is $\Delta t_{\mathrm{push}(A_2\|A_1)}$ and $\Delta t_{\mathrm{push}(A_2\|A_1)} = \Delta t_{\mathrm{push}(A_2)} + \Delta t_{\mathrm{push}(A_1)}$ when the two push actions are executed serially. The time needed for the second step is $\Delta t_{\mathrm{pop}(A_2\|A_1)}$ and $\Delta t_{\mathrm{pop}(A_2\|A_1)} = \Delta t_{\mathrm{pop}(A_2)} + \Delta t_{\mathrm{pop}(A_1)}$ in the serial case. Except for fission, the CA model can be extended to simulate optional malicious behavior and the corresponding time is $[+\Delta t_{\mathrm{mal}}]$. So, the pure time needed for one clone via fission is: $\Delta t_{\mathrm{fission}} = \Delta t_{\mathrm{push}(A_2\|A_1)} + \Delta t_{\mathrm{pop}(A_2\|A_1)}[+\Delta t_{\mathrm{mal}}]$. If Δt_{apu} is the average time allocated to CA every unit time, one cycle of fission can be obtained: $T = \Delta t_{\mathrm{fission}}$ when $\Delta t_{\mathrm{fission}} \leq \Delta t_{\mathrm{apu}}$, otherwise $T = \frac{\Delta t_{\mathrm{fission}}}{\Delta t_{\mathrm{apu}}}$.

The frequency of fission is $f = \frac{1}{T}$, which is defined as the number of break-ups of an ancestor CA per unit time. The self-replicating velocity of one ancestor CA is defined as the size of new cloning automata split off from this ancestor CA per unit time: $v_{\mathrm{one}} = s \cdot 2^f$. If the starting time is 0, there exist $\frac{t}{T}$ or $(t \cdot f)$ cycles at the time t. Volume is defined as the total size of all existing cloning automata at the time t: $vol(t) = s \cdot 2^{\frac{t}{T}} = s \cdot 2^{t \cdot f}$. The instantaneous self-replicating velocity of the CA system is the total size of new cloning automata per unit time with respect to the time t: $v_{\mathrm{i}}(t) = s \cdot 2^{t \cdot f}(1 - (1/2)^f)$, which represents the instantaneous fission ability of the CA system. The self-replicating acceleration of the CA system with respect to the time t is: $a(t) = s \cdot 2^{t \cdot f}(1 - 2^{-f})^2$. $a(t)$ is the accelerating fission ability of the whole CA system. The average self-replicating velocity of the CA system with respect to the time t: $v_{\mathrm{a}}(t) = \frac{s \cdot 2^{t \cdot f}}{t}$.

Properties analyzed above are relevant to the analysis of DoS caused by the fission. The final result of the analysis of DoS caused by the fission is similar in form to that caused by the fusion, but may be different in content for different T's: $t_{\mathrm{DoS}} \geq T \cdot \log \frac{max \cdot r}{s}$. Similarly, the operating system's process table will be saturated by processes of bacteria after $t = T \cdot \log max_{\mathrm{pt}}$, where max_{pt} is the maximum processes of the process table.

4   Interruption of the Self-replication

By the analysis in Sect. 3, it is clear that cloning automata deplete resources quickly and launch DoS attacks. So it is necessary to interrupt the self-replicating processes and then to clean them. Before cleaning, interruption methods should be found out to stop the self-replicating processes which deplete resources. Once all self-replicating processes are stopped, resources occupied by cloning automata would not increase anymore and this is the foundation of cleaning.

According to the CA model, two of the three steps in the fusion process can be interrupted. When both stacks of A_{2i-1} and A_{2i} are empty, conditions for the first step to be executed are satisfied. In order to hinder the execution of the first step, symbols like \$ should be pushed into empty stacks and $\delta(p_0, \epsilon, \text{nil}; q_0, \epsilon, \text{nil}) = \{(p_1, A_{2i}\#; q_1, A_{2i-1}\#)\}$ will not be executed since conditions are not satisfied. Without the execution of the first step, A_{2i} will not be on the top of A_{2i-1}'s stack and A_{2i-1} will not be on the top of A_{2i}'s stack either. So the second step $\delta(p_1, \epsilon, A_{2i}; q_1, \epsilon, A_{2i-1}) = \{(p_2, \#A_{2i-1}A_{2i}\#; q_2, \text{nil})\}$ cannot be executed. Without the execution of the second step, $\#$ will not be on A_{2i-1}'s stack and the third step $\delta(p_2, \epsilon, \#; q_2, \epsilon, \text{nil}) = \{(p_0, \text{nil}; q_0, \text{nil})\}$ will not be executed either. All cloning automata will be stopped immediately in the case that they are synchronous. In the case that they are asynchronous for some reason, that is, some cloning automata will execute the first step, some will execute the second step and the remainder will execute the third step. In the asynchronous case, all of them will be stopped after one or two steps. Thus after two steps at most, resources will not be depleted anymore. The other way to stop the fusion processes is to replace every $\#$ with \$ and this hinders the third step $\delta(p_2, \epsilon, \#; q_2, \epsilon, \text{nil}) = \{(p_0, \text{nil}; q_0, \text{nil})\}$. In the synchronous case, all cloning automata will be stopped at once. In the asynchronous case, after two steps at most, all cloning automata will be stopped by this interruption method.

According to the CA model, the first of the two steps in the fission process can be obstructed by pushing $\#$ or \$ into stacks whenever empty. $\delta(p_0, \epsilon, \text{nil}; q_0, \epsilon, \text{nil}) = \{(p_1, A_{2i}; q_1, A_{2i-1})\}$ will not be executed. In the synchronous case, all cloning automata will be stopped immediately. In the asynchronous case, some cloning automata will execute the first step while others will execute the second step. So after one step at most, all cloning automata will be stopped and resources will not be depleted anymore.

All methods mentioned above may be changed with situations, but it is fixed that the basic principle is to wreck conditions of the self-replicating process.

5 Simulation and Analysis of Fork Bomb

Unlike worms, wabbits [1] are a form of self-replicating malware which do not spread across networks but merely generate numerous copies of themselves on the local system. As an instance of wabbits, processes of fork bomb continually replicate themselves exponentially to deplete system resources available. The self-replication causes resource starvation and in some cases brings a denial of service [2,3,6,10]. As a computer bacterium, fork bomb can be modeled by CA. As defined in Sect. 2, CA can also model other malicious behavior as well as the self-replication. At this point, fork bomb does not have other malicious behavior except for the self-replication. Thus CA can model it in a brief way.

5.1   Operational Principles of Fork Bomb

There are many variants of fork bomb in different operating systems such as ":(){ :|:& };:" in Unix-like shells, "%0|%0" in Microsoft Windows and so on. Nevertheless, a basic implementation of fork bomb is an infinite loop that repeatedly launches the same process. In traditional Unix-like operating systems, fork bomb is usually written to use the fork system call. The fork system call creates a new process which is in essence a copy of the parent process. Details of the fork system call are connected with the process structure. In Unix, a process consists of a process control block (PCB), a shared-text segment, a data segment (or data area), and a working segment (or working area). PCB has the proc structure which contains basic control information and the user structure which contains information not related to the run of a process. The working segment consists of a kernel stack whose working space is at the kernel state and a user stack whose working space is at the user state. In Linux, a process has the following segments: code, data, heap and stack. The code segment, data segment, heap and stack segments in Linux are similar to the shared-text segment, data segment and working segment in Unix, respectively. For convenience, the structure of a process in Unix-like operating systems is considered as follows: a shared-text segment, a data segment and a working segment. In traditional Unix-like operating systems, the fork system call copies the virtual address space of the shared-text segment of the current process to the new process called a forked process and the forked process has a pointer to the physical space of the current process. The forked process has no physical space of its own unless executing a different image. Different from the operation on the shared-text segment, the fork system call copies the virtual address space of the data segment to the forked process and creates a physical space for the forked process. Thus, the forked process has a pointer from its virtual space to its own physical space. Similar to the operation on the data segment, the fork system call copies the virtual address space of the working segment to the forked process and creates a physical space for the forked process. The forked process has a pointer from its virtual space to its own physical space. The detail of the fork system call in traditional Unix-like operating systems can be depicted in Fig. 3 ("$\twoheadrightarrow$" stands for "copy to" and "$\rightarrow$" stands for "point to"). As forked processes are copies of the original process, they also seek to create copies of themselves upon execution. This cycle has the effect of causing an exponential growth in processes. Fork bombs work both by consuming CPU time in the process of forking and by saturating the operating system's process table.

5.2   Simulation of Fork Bomb with CA

Based on the operational principles, fork bomb can be modeled by CA via fusion as follows. ForkBombs = $(\bigcup_{i=2^0}^{2^n} Q_{2i-1} \cup Q_{2i}, \bigcup_{i=2^0}^{2^n} \Sigma_{2i-1} \cup \Sigma_{2i}, \bigcup_{i=2^0}^{2^n} \Gamma_{2i-1} \cup$ $\Gamma_{2i}, \bigcup_{i=2^0}^{2^n} \delta_{2i-1,2i}, S)$, where for $j = 2i - 1$ and $k = 2i$, (1) $Q_j = \{p_m^j | 0 \leq m \leq 6\}$ and $Q_k = \{q_m^k | 0 \leq m \leq 6\}$; (2) $\Sigma_j = \{\epsilon, w, d, s\}$ and $\Sigma_k = \{\epsilon, a_w, a_d, a_s\}$; (3) $\Gamma_j = \{\#, \$\} \cup \Sigma_j \cup \Sigma_k$ and $\Gamma_k = \{\#, \$\} \cup \Sigma_k$; (4)

P_1 virtual space   physical space   physical space   virtual space P_2

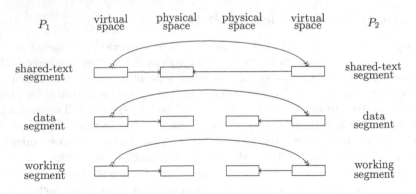

shared-text segment   shared-text segment

data segment   data segment

working segment   working segment

Fig. 3. The fork system call in traditional Unix-like OSs

$\delta_{2i-1,2i}$ is the transition function, which will be explained in detail later; (5) S is the set of initial states. Thus, a forkbomb consists of internal automata with stacks: $A_{2i-1}=(Q_{2i-1}, \Sigma_{2i-1}, \Gamma_{2i-1}, \delta_{2i-1}, p_0^{2i-1})$ and $A_{2i} = (Q_{2i}, \Sigma_{2i}, \Gamma_{2i}, \delta_{2i}, q_0^{2i})$. Superscripts and subscripts are omitted where no ambiguity exits.

The simulation via fusion is an infinite loop based on Algorithm 1 and consists of the following seven steps: (1) A_{2i-1} is at its initial state p_0 and A_{2i} is at its initial state q_0. When the process table is not saturated and virtual spaces for the shared-text segment, the data segment and the working segment and physical spaces for the latter two segments are ready to be allocated for the new process, all virtual and physical spaces are allocated. Virtual spaces of the data segment and the working segment of the new process have pointers to corresponding physical spaces. The virtual space of the shared-text segment has a pointer to the physical space of the current process. Contents of virtual spaces of the shared-text segment, the data segment and the working segment are denoted by s, d and w, respectively. The corresponding addresses of virtual spaces of the three segments above are denoted by a_s, a_d and a_w, respectively. A_{2i-1} reads w and pushes #, w and \$ into its stack, with A_{2i-1} going to the state p_1. A_{2i} reads a_w and pushes \$, a_w and \$ into its stack, with A_{2i} going to the state q_1. The corresponding δ is $\delta(p_0, w, \text{nil}; q_0, a_w, \text{nil}) = \{(p_1, \$w\#; q_1, \$a_w\$)\}$. (2) A_{2i} pops up \$, a_w and \$ from its stack and pushes a_w and \$ into A_{2i-1}'s stack. A_{2i-1} and A_{2i} go to p_2 and q_2, respectively. The corresponding δ is: $\delta(p_1, \epsilon, \$; q_1, \epsilon, \$) = \{(p_2, \$a_w\$w\#; q_2, \text{nil})\}$. (3) When A_{2i-1}'s state is p_2 and A_{2i}'s state is q_2, A_{2i-1} reads d and then pushes d and \$ into its stack, with its state changed to p_3. A_{2i} reads a_d and then pushes \$, a_d and \$ into its stack, with its state changed to q_3. The corresponding δ is $\delta(p_2, d, \$; q_2, a_d, \text{nil}) = \{(p_3, \$d\$a_w\$w\#; q_3, \$a_d\$)\}$. (4) A_{2i} pops up \$, a_d and \$ from its stack and pushes a_d and \$ into A_{2i-1}'s stack. A_{2i-1} and A_{2i} go to p_4 and q_4, respectively. δ is defined by $\delta(p_3, \epsilon, \$; q_3, \epsilon, \$) = \{(p_4, \$a_d\$d\$a_w\$w\#; q_4, \text{nil})\}$. (5) A_{2i-1} reads s and then pushes s and \$ into its stack, with its state changed to p_5. A_{2i} reads a_s and pushes #, a_s and \$ into its stack, with its state changed to q_5.

The corresponding δ is $\delta(p_4, s, \$; q_4, a_s, \text{nil}) = \{(p_5, \$s\$a_d\$d\$a_w\$w\#; q_5, \$a_s\#)\}$. (6) A_{2i} pops up \$, a_s and \# from its stack and pushes a_s and \# into A_{2i-1}'s stack in order. States of A_{2i-1} and A_{2i} are changed to p_6 and q_6, respectively. The corresponding δ is $\delta(p_5, \epsilon, \$; q_5, \epsilon, \$) = \{(p_6, \#a_s\$s\$a_d\$d\$a_w\$w\#; q_6, \text{nil})\}$. (7) When \# is on the top of A_{2i-1}'s stack, A_{2i-1} pops up "$\#a_s\$s\$a_d\$d\$a_w\$w\#$" from its stack and writes the contents to virtual spaces of the shared-text segment, the data segment and the working segment of the new process according to the corresponding addresses, respectively. A_{2i-1} goes to the state p_0 and A_{2i} goes to the state q_0. Information about the new process is written to the process table. The corresponding δ is $\delta(p_6, \epsilon, \#; q_6, \epsilon, \text{nil}) = \{(p_0, \text{nil}; q_0, \text{nil})\}$. This simulation of fork bomb using CA via fusion can be depicted briefly in Fig. 4.

5.3   Analysis of Fork Bomb

Based on the simulation, let Δt_{vp} denote the time needed for the allocation of virtual and physical spaces, Δt_{cs}, Δt_{cd} and Δt_{cw} for copying the virtual spaces of the shared-text segment, the data segment and the working segment respectively, and Δt_{ca} for copying addresses of virtual spaces. Δt_{w} denotes the time needed for writing the virtual spaces of the data, working and shared-text segments and the time to write the process table. Thus, the pure time for creating a new process is: $\Delta t_{\text{pure}} = \Delta t_{\text{vp}} + \Delta t_{\text{cs}} + \Delta t_{\text{cd}} + \Delta t_{\text{cw}} + \Delta t_{\text{ca}} + \Delta t_{\text{w}}$.

It is assumed that the target machine has m $(m \geq 1)$ CPUs (or equal executors) and that one process can only be executed on one CPU at a moment. For convenience, it is also assumed that no other process will be executed except for processes pid = 0, pid = 1 and fork bomb, and that the ratio of time used for process to switch to the time slice is r_s. Without limitation of maximum processes, there will be $n = m \cdot \frac{t \cdot (1-r_s)}{\Delta t_{\text{pure}}}$ processes of fork bomb at the time t. Fork bomb will eat up all memory available after $\frac{max_{\text{m}} \cdot \Delta t_{\text{pure}}}{s \cdot m \cdot (1-r_s)}$, where max_{m} denotes the maximum memory available and s denotes the size of memory allocated to one process (e.g. 8 KB in Linux). If there is one upper bound on the number of processes, fork bomb will saturate the operating system's process table after $\frac{(max_{\text{pt}}-2) \cdot \Delta t_{\text{pure}}}{m \cdot (1-r_s)}$, where max_{pt} denotes the maximum number of processes.

5.4   Interruption of Fork Bomb

According to Fig. 3, the key to the self-replication is that the new process share the malicious shared-text segment with the current process which is reflected by the fifth step of Fig. 4. In order to interrupt the self-replication of new processes, a safe virtual space of the shared-text segment s' which points to a safe physical space of the shared-text segment instead of s will be pushed into the stack. For example, the simplest s' does nothing. Thus, new processes cannot replicate themselves. In order to interrupt the self-replication of current processes, all shared-text segments including infinite loops or recursions depicted by the last two steps of Fig. 4 should be found out and then should be replaced with a safe shared-text segment. By the two ways, fork bomb will be interrupted entirely.

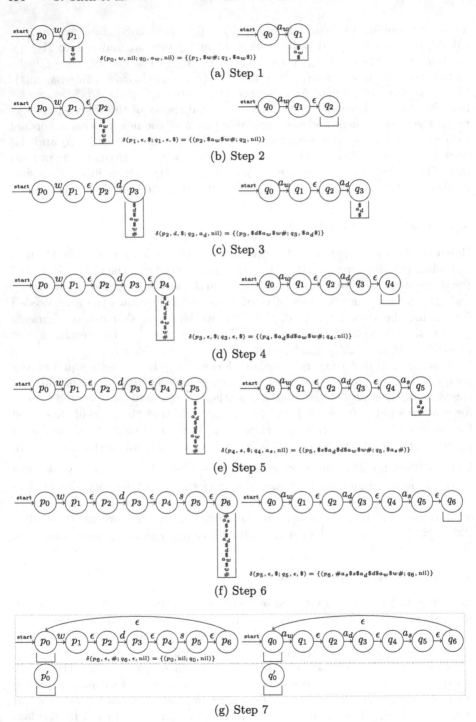

Fig. 4. Simulation of fork bomb using CA via fusion

6   Conclusion

A new model named cloning automata is constructed to model computer bacteria in two ways: fusion and fission. The analysis of important properties related to fusion and fission is made for a better understanding of computer bacteria. Methods for interrupting the self-replicating process of the CA model are given. Simulation of fork bomb and corresponding analysis show the effectiveness of this model. In the future, it will be investigated that how to judge any software whether it has features of computer bacteria or not and how to clean the CA model. Further more, it will be considered how to extend cloning automata to model hybrid malicious behavior to solve more complex problems.

Acknowledgments. This research was supported by NSFC with Grant Nos. 61732013 and 61420106004.

References

1. https://www.virusbtn.com/resources/glossary/wabbit.xml
2. https://www.virusbtn.com/resources/glossary/fork_bomb.xml
3. https://en.wikipedia.org/wiki/Fork_bomb
4. Mcafee labs threats report, May 2015. http://www.mcafee.com/us/resources/reports/rp-quarterly-threat-q1-2015.pdf
5. Symantec: 2015 internet security threat report, vol. 20. http://www.symantec.com/security_response/publications/threatreport.jsp
6. Bohra, A., Neamtiu, I., Gallard, P., Sultan, F., Iftode, L.: Remote repair of operating system state using backdoors. In: Proceedings, International Conference on Autonomic Computing, pp. 256–263, May 2004
7. Cohen, F.: Computer viruses. Comput. Secur. **6**(1), 22–35 (1987)
8. Gazet, A.: Comparative analysis of various ransomware virii. J. Comput. Virol. **6**(1), 77–90 (2010)
9. Joy, J., John, A., Joy, J.: Rootkit detection mechanism: a survey. In: Nagamalai, D., Renault, E., Dhanuskodi, M. (eds.) PDCTA 2011. CCIS, vol. 203, pp. 366–374. Springer, Heidelberg (2011). https://doi.org/10.1007/978-3-642-24037-9_36
10. Matthews, J.N., Hu, W., Hapuarachchi, M., Deshane, T., Dimatos, D., Hamilton, G., McCabe, M., Owens, J.: Quantifying the performance isolation properties of virtualization systems. In: Proceedings of the 2007 Workshop on Experimental Computer Science, ExpCS 2007. ACM, New York (2007)
11. Neumann, J.V., Burks, A.W.: Theory of Self-reproducing Automata. University of Illinois Press, London (1966)
12. Pelaez, C., Bowles, J.: Computer viruses. In: Twenty-Third Southeastern Symposium on System Theory, Proceedings, pp. 513–517, March 1991
13. Pickover, C.A.: The Math Book: From Pythagoras to the 57th Dimension, 250 Milestones in the History of Mathematics. Sterling Publishing Company, Inc. (2012)
14. Qing, S., Wen, W.: A survey and trends on internet worms. Comput. Secur. **24**(4), 334–346 (2005)
15. Rossow, C., Dietrich, C., Bos, H.: Large-scale analysis of malware downloaders. In: Flegel, U., Markatos, E., Robertson, W. (eds.) DIMVA 2012. LNCS, vol. 7591, pp. 42–61. Springer, Heidelberg (2013). https://doi.org/10.1007/978-3-642-37300-8_3

16. Mohd Saudi, M., Abuzaid, A.M., Taib, B.M., Abdullah, Z.H.: Designing a new model for trojan horse detection using sequential minimal optimization. In: Sulaiman, H.A., Othman, M.A., Othman, M.F.I., Rahim, Y.A., Pee, N.C. (eds.) Advanced Computer and Communication Engineering Technology. LNEE, vol. 315, pp. 739–746. Springer, Cham (2015). https://doi.org/10.1007/978-3-319-07674-4_69
17. Spafford, E.H.: Computer viruses as artificial life. Artif. Life **1**(3), 249–265 (1994)
18. Wolfram, S.: Statistical mechanics of cellular automata. Rev. Modern Phys. **55**(3), 601 (1983)
19. Wu, M.W., Wang, Y.M., Kuo, S.Y., Huang, Y.: Self-healing spyware: detection, and remediation. IEEE Trans. Reliab. **56**(4), 588–596 (2007)

Research on Arrival Integration Method for Point Merge System in Tactical Operation

Yannan Qi[1,2(✉)], Xinglong Wang[1], and Chen Chen[3]

[1] Civil Aviation University of China,
Tianjin, China
yannan.qi@yahoo.com
[2] International Civil Aviation Organization,
Montreal, Canada
[3] East China Normal University, Shanghai, China

Abstract. In this paper, a Point Merge arrival integration method of tactical operation is introduced under current Communication, Navigation and Surveillance technologies. In present research, the multi-agent theory is used to build and stimulate the arrival integration system. Agents involved in point merge operation and action modules are designed to realize automatic trajectory generating, adjustment, sequencing and conflict detection. The architecture of point merge operation is obtained as well. In order to verify the method, historical ADS-B data is analyzed and the point merge procedures are designed for single runway and two runway arrival separately for different verification. The outcome proves the correctness and efficiency of the method and demonstrates the advantage of Point merge procedure on reducing flight time, fuel consumption, delay time and ATC workload.

Keywords: Arrival integration · Point merge · Continuous descent approach · Multi-agent system

1 Introduction

Point Merge System (PMS) provides systematic method of sequencing arrival flows instead using heading vector. Both efficient sequencing and optimum descent are achieved simultaneously with a built-in continuous descent (CDA) [1]. Along with the capacity and efficiency taken from point merge system, poor trajectory prediction ability is a vital problem in density TMA implementation.

A PMS is defined as an RNAV STAR, transition or initial approach procedure, or a portion thereof, and is characterized by the following feature [5]:

Q. Yannan—This project was Supported by National Natural Science Foundation of China (Grant No. 61571441); National Key Basic Research Program of China (2016YFB0502405); Fundamental Research Funds for the Central Universities (ZXH2012M002, 3122014D036, 3122015C024); State Key Laboratory of Air Traffic Management System and Technology, NO: SKLATM201705 for this project are gratefully acknowledged.

X. Gao et al. (Eds.): COCOA 2017, Part I, LNCS 10627, pp. 417–425, 2017.
https://doi.org/10.1007/978-3-319-71150-8_34

- A single point – denoted 'merge point' (MP), is used for traffic integration;
- Pre-defined legs – denoted 'sequencing legs', equidistant from the merge point, are dedicated to path stretching/shortening or each inbound flow (Fig. 1).

Using PMS and CDA in Arrival integration refers to sequencing, level/vertical trajectory preplanning and spacing maintenance. Current research is primarily concerned with 3 aspects: performance assessment, trajectory optimization and air flow integration in TMA. [1] reports on flight simulation of Point merge procedure, and the outcome shows good performance on environmental adaptability and ATC convenience. [2, 3] present the time based arrival control concept. As for the trajectory optimization, [6–8] introduce vertical trajectory optimization method based on optimal control theory; [3, 9] use controlled time and geometry method to detect and solve the conflict; [10] researches the trajectory prediction uncertainty modeling for CDA. In 2014, LiangMan introduced an agent-based approach to automated merge 4D arrival, but only outline is presented [4].

The main purpose of this paper is to introduce PMS in TMA to improve the arrival efficiency and capacity. Multi-agent technique is adopted in solving arrival integration for PMS in Tactical Operation, and detailed algorithm is presented.

2   Construction of Multi-agent System for Merge Point Arrival

In this chapter, the overall architecture and agents are introduced to give a description of agents and information transfer process. The realization algorithms of each agent are put forward.

Arrival aircraft control implicates arrival time management, sequencing, conflict detection and space maintenance as well as information exchange between aircraft and air traffic controller. Five agents are designed: aircraft agent (AA), vertical trajectory agent (VTA), runway control agent (RCA), arrival trajectory agent (ATA), space maintain agent (SMA).

The information interaction process is showed in Fig. 2.

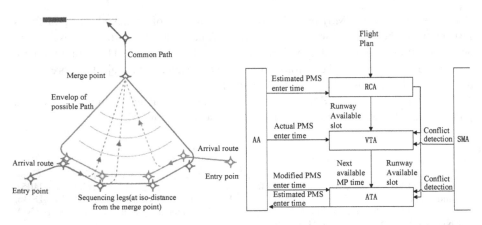

Fig. 1. Point merge system   **Fig. 2.** Information interaction process

2.1 Aircraft Agent (AA)

Aircraft agent manages arrival flights in PMS. Each flight contains: flight no., estimated/actual enter PMS time, pre-planned runway, enter PMS speed, aircraft speed, aircraft position, Direct-to time, Direct-to speed, pass MP time, sequencing leg No., PM No., wake turbulence category (Cat) and predefined trajectory. AA stores information on all arrival aircraft and transit information with other agents.

2.2 Runway Control Agent (RCA)

Runway Control agent is designed to control the runway usage through limiting the merge point passing time according to specific runway operation strategy. In RCA, merge point available slot module is responsible for calculating the available slot of a runway according to flight plan and runway operation strategy.

2.3 Vertical Trajectory Agent (VTA)

The objective of VTA is to adjust the vertical trajectory, and calculate the estimated MP passing time and earliest Direct-to time for succeeding aircraft. VTA incorporate aircraft performance database.

In VTA vertical trajectory adjustment module is to generate the optimized descend trajectory for each aircraft. Vertical trajectories optimization is commonly formulated as optimal control problems with a fixed range [6–8]. In this paper, we use a multiple phase optimal control problem with respect to two performance indicators, minimum arrival time and minimum fuel consumption. The state $x = [V_T \ x_s \ h]^T$, and the control input $u = [T \ \mu]$, where V_T is true airspeed, x_s is along track distance, h is altitude, T is the thrust force, μ is aerodynamic roll angle. A pseudospectral method is used to obtain the optimal trajectory for a CDA operation in this paper.

2.4 Arrival Trajectory Agent (ATA)

Arrival trajectory agent is designed to determine the sequence and adjust trajectory in sequencing leg. With regard to multi-merge system, each point merge system is a separate agent, and they are parallel and in charge of aircraft operation in PM system respectively. In the process of operation, deviation caused by environment and operation will change the actual Direct-to time and passing MP time of leading aircraft. Then, new earliest Direct-to time will be compared with the estimated Direct-to time. There are two situations:

The difference does not exceed a time threshold. If aircraft should be delayed, ATA will assess if the delay time could be mitigated with the help of speed adjustment. If not, next waypoint along the sequencing leg is chosen as the Direct-to point.

The difference exceeds a time threshold. If there is available time slot from other runway, the pre-planned runway of the aircraft will be changed in AA, and this aircraft will be delivered to available runway. There are two modules in Arrival trajectory agent: sequence generation module and trajectory adjustment module.

The detailed model is illustrated in Sect. 3.

2.5   Space Maintain Agent (SMA)

Space maintain agent is designed to detect conflicts in sequencing and descending phase, which receives the estimated trajectories from ATA and VTA to assess the conflict and transmit the results to ATA and VTA as the basis for trajectory adjustment.

3   Module Design

3.1   Trajectory Generation Module

We apply single-objective optimization problem to achieve the goal of minimum overall complete time for trajectory generation problem. The trajectory for an aircraft is expressed as a waypoint sequence. With each waypoint, the speed limitation is attached. We assume speed adjustment only happen on waypoints.

Suppose there are n waypoints on leg l, waypoints set is denoted as $wp_l = \{wp_1^l, wp_2^l, \ldots, wp_m^l, \ldots\}$. A trajectory of aircraft is expressed as a string, and the length of a trajectory string is decided by the number of waypoints on leg. $wp_l = \{(D_{p1}^i, v_{p1}^i), (D_{p2}^i, v_{p2}^i), \ldots, (D_{pm}^i, v_{pm}^i), \ldots\}$.

D_{pm}^i designates the action when aircraft i passing waypoint m, $m = (1, 2, \ldots, M)$, if turning at m, $D_{pm}^i = 1$, else $D_{pm}^i = 0$.

The speed of aircraft i at waypoint m is designated as v_{pm}^i, $m = (1, 2, \ldots, M)$.

Suppose t_i^{ed}, t_j^{ed} are direct-to time on a sequencing leg, $m \leq M$, t_i^{ee} is the estimated PMS enter time, S_i is the distance between two neighboring waypoints, if $D_{pm}^i \neq 0$, N is the number of Direct-to waypoint. For aircraft i, the Direct-to time is t_i^{ed}:

$$t_i^{ed} = \sum_{n=1}^{N} \frac{S_i}{v_{p_i}} + t_i^{ee} \tag{1}$$

A Polar coordinate system is set up with the merge point as the origin, and the first link is the polar axis. Because the waypoints are scattered on sequence leg evenly, so the coordinate of m's waypoint could be calculated easily:

$$x^{pm} = r \cdot \cos \theta \tag{2}$$

$$y^{pm} = r \cdot \sin \theta \tag{3}$$

At given time t, the coordinates of aircraft i could be expressed as:

$$x_i^c(t) = r \cdot \cos \left(m \cdot \theta + \frac{\left(t - \sum_{j=1}^{m-1} \frac{S_i}{v_{p_j}} \right) v_{p_m}}{\pi r_n} \right) \tag{4}$$

$$y_i^c(t) = r \cdot \sin \left(m \cdot \theta + \frac{\left(t - \sum_{j=1}^{m-1} \frac{S_i}{v_{p_j}} \right) v_{p_m}}{\pi r_n} \right) \tag{5}$$

r_l is the radius of sequencing leg l, S_l is the distance between neighboring way-points on sequencing leg l, x^{pm} is the closest waypoint before current position of aircraft i. The single-objective optimization problem is build as:

Minimum overall complete time T_o.

$$\min T_o = \min \sum_{i=1}^{N} \left| t_i^{ed} - t_i^{ee} \right| \tag{6}$$

Constraints:

$$\left| v_{pm}^i - v_{pm-1}^i \right| \leq \Delta v \tag{7}$$

$$v_{\min}^i \leq v_{pm}^i \leq v_{\max}^i \tag{8}$$

$$\sum_{m=2}^{N} \left\lceil \left| \frac{v_{pm}^i - v_{pm-1}^i}{\Delta v} \right| \right\rceil \leq n_v \tag{9}$$

$v_{\min}^i$ and $v_{\max}^i$ are the minimum and maximum aircraft performance limited speed of aircraft i. Δv is the increment or decrement for each speed adjustment, n_v is the limited time for speed change for each trajectory.

3.2 Trajectory Adjustment Module

In our research, maximum arrival flow speed is considered as the main objective in the premise of conflict free. Due to closed path and level flight on sequencing leg, only speed adjustment and changing Direct-to waypoint are used in trajectory adjustment process. If the delay or advanced time cannot be consumed by both speed (exceed the performance of aircraft) or Direct-to point adjustment (overflow sequencing leg), flow management is activated through adjusting the enter PMS time of succeeding aircraft.

- Direct-to point changing
  Suppose N is the maximum number of waypoints on sequencing leg j, original Direct-to waypoint is wp_m^j with speed vp_m^j, t_{rd}.

Direct-to point changing	
1 **initialization**: $p=0$;	5 End Direct-to point adjustment, change and save trajectory;
2 $p=p+1$, $vp=vp_m^j$;	6 *Else*
3 If $\sum_{i=0}^{p}(\dfrac{S}{v_p}) \geq t_{rd}$ is satisfied	7 Flow management module is activated
	8 *Else*
4 *If* $p \leq N$	9 n=n+1, and go to step 2.

- Speed adjustment
  When leading aircraft need delay, speed adjustment module is activated to conduct speed adjustment for succeeding aircraft i. Speed adjustment is processed step by step, assume Δv is the amount of speed change for each adjustment, t_{rd} is the required delay time, the process of speed adjustment is as below.
- Flow management module

$$t_i^{ee} = t_i^{ee} + t_{rd}$$

speed adjustment	
1 **initialization:** k=0, vk=0, Reverse check from Direct-to waypoint, and obtain the waypoint (wp_m^j) with maximum speed (vp_m^j), vk=vp_m^j	6 *Else*
	7 **End** speed adjustment failed
	8 *Else*
	9 *If* current way point is the entry waypoint:
2 k=k+1;	10 **End** speed adjustmentfailed;
3 *If* $\sum_{i=0}^{k}(\dfrac{S}{v_k - \Delta v} - \dfrac{S}{v_k}) \geq t_{rd}$ is satisfied	11 *Else*
4 *If* $v_{min} \leq (v_k - \Delta v) \leq v_{max}$ is satisfied	12 check from current waypoint, and obtain the waypoint wp_m^j with maximum speed vp_m^j, vk=vp_m^j, and go to step 2.
5 **End** speed adjustment, change and save trajectory;	

3.3   Conflict Detection Module

Two types of conflicts can happen through analyzing the structure of PMS: catch-up (direct-to from same waypoint) and merging (direct-to from different waypoint) conflict could be occur.

With regard to catch-up conflict in sequencing leg, distance based radar separation is used. For aircraft i and j, (i, j $\in$ F) in same sequencing leg. Assume D_{ij} as the distance between aircraft i and j, Sepr is the radar separation minima, so on the sequencing leg:

$$D_{ij} = \sqrt{(x_i(t) - x_j(t))^2 + (y_i(t) - y_j(t))^2} \geq Sep^r \qquad (10)$$

For catch-up conflict, time based separation is used to control the aircraft separation passing by the merge point. For aircraft i, j, assume t_i^v, t_j^v are the flight duration on descend link.

t_i^{ep}, t_j^{ep} are the estimated passing time on merging point, t_i^{ed}, t_j^{ed} are the

$$t_i^{ep} - t_j^{ep} \geq Sep^w \qquad (11)$$

$$t_i^{ep} = t_i^{ed} + t_i^v \qquad (12)$$

$$t_j^{ep} = t_j^{ed} + t_j^v \qquad (13)$$

4 Verification

Simulation is made in Tianjin airport, and 3 scenarios on RWY 16 are designed to verify different aspect of the new method: current operation scenario, PM procedure for single runway and PM procedure for parallel runways (Independent operation mode). Historical flight data is used to extract current operation scenario (Fig. 3) and traffic distribution (Table 1). The PM procedures are designed according to traffic distribution.

Simulation of multi-agent system is conducted on NetLogo 5.3, the vertical trajectory generation and adjustment is achieved with the help of GPOSP toolbox in Matlab. Genetic Algorithm is adopted to solve the objective problem in trajectory generation module.

Table 1. Arrival distribution

Enter point	Flow distribution
VYK	41%
NIRON	32%
KALBA	17%
HCX	10%

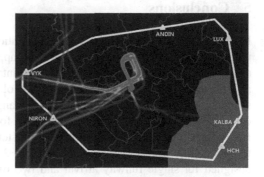

Fig. 3. Heat-map of arrival trajectories

In experiment, the fuel consumption, complete time, and ATC workload are compared to prove the performance of point merge procedure. The instruction number is calculated to denote the ATC workload simply. In scenario 1, the instruction number

Table 2. Experimental results

Experimental results	Scenario 1	Scenario 2	Scenario 3
Average fuel consumption per flight (kg)	958.43	718.4	758.4
Average flight time per flight (min)	22	16.8	18.9
Average flight distance in TMA per flight (km)	165.32	151.45	169.87
Runway change	–	–	5
Enter time change	–	–	8
Average ATC instruction number per flight	12	7	8

Initial Parameters: Enter PMS speed = 210kt; Lateral Separation = 3NM; Initial weight (A330) = 160T; Initial weight (A320/B737/A319) = 60T; Speed adjustment increment = ±20kt; Maximum speed change for one trajectory = ±20kt

is obtained through historical data statistics. And in scenario 2 and 3, the instruction contains the speed adjustment, direct to instruction, runway change instruction and link to other PMS instruction. So, the ATC workload is obtained by counting the instructions.

According to outcome of experiment (Table 2), conflict free trajectories are obtained both in scenario 2 and scenario 3. Compared with scenario 1, the fuel consumption is reduced by 25% and 21%, and the flight time and distance reduced 24%, 14% and 8.3%, –2.8% separately. Compare to flight fuel consumption, the flight distance does not reduce too much. The reason is that the PMS contains arcs with relative large radius for better maneuvering ATC ability, and the fuel consumption rate is pretty low in descend flight in terminal airspace. Consequently, PMS shows good performance to reduce the level flight in TMA. In scenario 3, there are 5 aircraft change landing runway from runway 16R to 16L and 8 aircraft from west change enter time because the main arrival flow is from west and exceed the capacity of PMS.

5 Conclusions

In this paper, point merge arrival integration method has been studied, and a multi-agent based modeling and simulation approach is proposed to improve the efficiency and environmental performance of point merge operation. 5 agents are designed and modules that lead to the achievement of trajectory generation and adjustment, sequencing, conflict detection and data exchange are presented. In order to verify the correctness and efficiency of the method, we focus on Tianjin airport. Three scenarios are designed to display different situations. Historical flight data is analyzed in scenario 1 to illustrate the current situation. In scenario 2 and 3, the point merge procedures are designed for single runway arrival and two runway arrival situation. Comparison is made on aspect of flight time, fuel consumption and ATC workload. The results show the good performance of Point merge procedure on arrival management efficiency, reduced environment impact and less air-ground communication requirement.

References

1. Favennec, B., Hoffman, E., Trzmiel, A., Vergne, F., Zeghal, K.: The point merge arrival flow integration technique: towards more complex environments and advanced continuous descent. In: 9th AIAA-6921, South Carolina (2009)
2. Klooster, J.K., de Smedt, D.: Controlled time of arrival spacing analysis. In: Proceedings of the Ninth USA/Europe Air Traffic Management Research and Development Seminar, Berlin (2011)
3. Man, L.: An agent-based approach to automated merge 4D arrival trajectories in busy terminal maneuvering area. In: 2014 Asia-Pacific International Symposium on Aerospace Technology (2014)
4. Eurocontrol: Point merge integration of arrival flows enabling extensive RNAV application and continuous descent operational services and environment definition, version 2.0 (2010)
5. Park, S.G., Clarke, J.-P.: Vertical trajectory optimization for continuous descent arrival procedure. In: AIAA Guidance, Navigation, and Control Conference (AIAA 2012-4757), Minneapolis, Minnesota (2012)
6. Zhao, Y., Tsiotras, P.: Analysis of energy-optimal aircraft landing operation trajectories. J. Guid. Control Dyn. **36**, 833–845 (2013)
7. Park, G., Clarke, J.-P.: Trajectory generation for optimized profile descent using hybrid optimal control. In: AIAA Guidance, Navigation, and Control Conference (2013)
8. Delahaye, D., Puechmorel, S., Tsiotras, P., Feron, E.: Mathematical models for aircraft trajectory design: a survey. In: Electronic Navigation Research Institute (ed.) Air Traffic Management and Systems. LNEE, vol. 290, pp. 205–247. Springer, Tokyo (2014). https://doi.org/10.1007/978-4-431-54475-3_12
9. Patel, R.B., Goulart, P.J.: Trajectory generation for aircraft avoidance maneuvers using online optimization. J. Guid. Control Dyn. **34**(1), 218–230 (2011)
10. Enea, G., Vivona, R., Karr, D., Cate, K.: Trajectory prediction uncertainty modeling for continuous descent. In: 27th Congress of International Council of the Aeronautical Sciences, Nice, Paper ICAS 2010-11.11.1 (2010)

Repair Position Selection for Inconsistent Data

Xianmin Liu[1]([⊠]), Yingshu Li[2], and Jianzhong Li[1]

[1] Harbin Institute of Technology, Harbin, China
{liuxianmin,lijzh}@hit.edu.cn
[2] Georgia State University, Atlanta, USA
yili@gsu.edu

Abstract. Inconsistent data indicates that there is conflicted information in the data, which can be formalized as the violations of given semantic constraints. To improve data quality, repair means to make the data
consistent by modifying the original data. Using the feedbacks of users to
direct the repair operations is a popular solution. Under the setting of big
data, it is unrealistic to let users give their feedbacks on the whole data
set. In this paper, the repair position selection problem (RPS for short)
is formally defined and studied. Intuitively, the RPS problem tries to find
an optimal set of repair positions under the limitation of repairing cost
such that we can obtain consistent data as many as possible. First, the
RPS problem is formalized. Then, by considering three different repair
strategies, the complexities and approximabilities of the corresponding
RPS problems are studied.

Keywords: Inconsistent data · Repair · Position selection · RPS

1 Introduction

Managing inconsistent data is a key problem in the area of data quality and
database management. Since dirty data has been widely viewed in practical
applications and caused many research interests [16], many works focus on the
data quality problems [3,4,7,21] and a central problem in this area is how to
make data *consistent*. In this paper, we focus on the inconsistencies caused by
FD and consider the corresponding data repair problem.

The following example can be used to explain the inconsistent data problem
caused by FD.

Example 1. An FD rule $\varphi = AB \rightarrow C$ defined over relation $R = \{A, B, C\}$ has
the following semantics. For any two tuples in the instance I_R over relation R, if

This work was supported in part by the General Program of the National Natural
Science Foundation of China under grants 61502121, 61402130, 61772157, U1509216,
the China Postdoctoral Science Foundation under grant 2016M590284, the Fundamental Research Funds for the Central Universities (Grant No. HIT.NSRIF.201649),
and Heilongjiang Postdoctoral Foundation (Grant No. LBH-Z15094).

X. Gao et al. (Eds.): COCOA 2017, Part I, LNCS 10627, pp. 426–438, 2017.
https://doi.org/10.1007/978-3-319-71150-8_35

their have equal values on attributes A and B, they must also have equal values on attribute C. For consistent data, it is required that the given FD rules are valid on each pair of data tuples. Consider an instance I_R, assume that there are two tuples $t_1 = \{A : x, B : a, C : m\}$ and $t_2 = \{A : x, B : a, C : n\}$ in I_R, also, we assume that $m \neq n$. Then, we have I_R does not satisfy the rule φ. That is, there are inconsistencies caused by t_1 and t_2 under limitation of the FD rule φ.

The inconsistencies shown in Example 1 indicate that there are errors in the data which violate the semantic defined by the FD rule φ. To eliminate such inconsistencies, a common method is to repair data by operations according to the users' feedback [19]. Therefore, under the constraints of limited computing resources, how to select the set inconsistent tuples for users to review such that maximum benefits will be obtained becomes a challenge and key problem.

In this paper, to meet the new challenges introduced above, we study the Repair Position Selection Problem (RPS for short) under the constraints defined by FD rules. Intuitively, given a database instance I_R and a set of FD rules Σ, RPS aims to find a subset S of I_R with limited size such that we can repair the data as many as possible after receiving the feedbacks of users. For special I_R, Σ and S, which data can be repaired also depends on the repair strategies adopted. Three repair strategies are considered in this paper, *simple deletion* (SD for short), *full modify* (FM for short) and *half modify* (HM for short). We are not aware of any previous works on the RPS problem. Therefore, we firstly give the formal definition of the RPS problem and study its complexities and approximabilities under three different repair strategies. The main results obtained by us can be summarized as follows: RPS_{SD} is NP-*complete* even assuming $|\Sigma| = 2$, it can be solved in PTime when $|\Sigma| = 1$ and its general version can be approximated with ratio 2; RPS_{FM} is NP-*complete* even if $|\Sigma| = 3$, and it can be approximated within ratio $O(n^{\frac{1}{3}})$; RPS_{HM} is NP-*complete* and it can be approximated within ratio $(1 - 1/e)$.

2 Preliminary

2.1 Basic Notations

In this paper, we use R to represent a relation schema, and use $attri(R)$ to represent the attribute set of R, which is also denoted as R with clear context. The symbol I_R is usually used to represent an instance of given relation schema R. For any attribute $A \in R$, $dom(A)$ denotes the domain of attribute A in R. For given relation $R = \{A_1, A_2, ..., A_n\}$, instance I_R is composed of one set of *tuples*, where each tuple t belongs to $dom(A_1) \times dom(A_2) \times ... \times dom(A_n)$. Assuming that t is a tuple of I_R and $S \subseteq R$, $t[S]$ is the restriction of t over S.

A FD rule φ can be represented by $(R : X \to Y)$ (or $X \to Y$ when the context is clear), where (1) both X and Y are subsets of R and (2) $X \cap Y = \emptyset$. Given I_R and a FD rule $\varphi = X \to Y$, consider two tuples t_1 and t_2 in I_R satisfying $t_1[X] = t_2[Y]$. If $t_1[Y] = t_2[Y]$, they *satisfy* the rule φ, otherwise, they are called to be a violation of φ, denoted by $(t_1, t_2) \nvDash \varphi$.

Given I_R and a dependency rule set Σ, we can use the set $C = \{(t_1, t_2)|(t_1, t_2) \not\models \varphi, \text{s.t.}\ \varphi \in \Sigma\}$ to represent the inconsistencies of I_R under the constraints defined by Σ. If C is not empty, we say that I_R is *inconsistent* and there are *inconsistencies* in I_R.

Some useful notations are also needed here. Given a dependency rule φ, $L(\varphi)$ represents the set of attributes involved in the left side of φ and $R(\varphi)$ represents the set of attributes on the right side of φ. A dependency rule is called to be *standard* if and only if the right side only involves one attribute. For example, the FD rule $A \to B$ is standard but the rule $A \to BC$ is not.

2.2 Data Repair Strategies and the Repair Position Selection Problem

In this paper, to solve the problem of inconsistent data, *data repair* means eliminating the inconsistencies by special repairing strategies. Using feedback of users on inconsistent data is a popular method for data repair. Generally speaking, a repair position is represented by one tuple or one attribute of some tuple which representation is used depends on special repairing strategies. To collect the feedbacks of users, a set S of positions will be offered to the users and they will return us the correct decisions (such as deleting some tuples, modifying some values and so on) to fix inconsistencies. The following three repairing strategies will be considered by this paper.

- **Simple Deletion** (SD for short). Given the repairing position S where each position is one tuple in I_R, the tuples in S will be determined by users whether or not to be removed from the whole data set I_R. Here, we assume that the feedbacks on different repairing positions made by users are independent from each others. Intuitively, removing all tuples in S from I_R will eliminate the most inconsistencies, and all inconsistencies involving positions in S can be eliminated trivially by removing all tuples from S. However, usually, we do not need to remove the whole S but only some of them. Therefore, a good *simple deletion* strategy will eliminate inconsistencies as many as possible while remove tuples from I_R as few as possible.
- **Full Modify** (FM for short). In this strategy, a repairing position in S is explained to be one tuple $t \in I_R$. For special FD rule $X \to Y$ where $|Y| = 1$, the users will eliminate the inconsistencies by modifying the value of $t[Y]$ or $t[X]$. The main idea of FM strategy is to try to fix the inconsistent data in worst cases, intuitively, for two tuples s and t which are inconsistent, the inconsistency may be caused by either side of the FD rule, therefore, the inconsistency can be removed only after $s[X \cup Y]$ and $t[X \cup Y]$ are both repaired by the feedbacks of users. Of course, there is an implicit assumption behind the FM strategy that the users know the true values of $t[X \cup Y]$ and $s[X \cup Y]$. Obviously, when limiting the size of S, an optimal *full modify* strategy will eliminate inconsistencies as many as possible.
- **Half Modify** (HM for short). This strategy is similar with the *full modify* strategy, but one position in HM strategy can be treated to be an attribute

on some tuple. The main difference is that there is an additional assumption that the inconsistencies caused by $\varphi : X \to Y$ can be always repaired by modifying the values of Y. This assumption can be satisfied in many real applications and is utilized in most previous research works. In the *half modify* strategy, after receiving all repairs from users, further repairs can be made by refining the dependency rules based on the repairs provided by users. For example, let φ to be $A \to B$, two tuples s and t satisfying $s[A, B] = \{1, x\}$ and $t[A, B] = \{1, y\}$ are inconsistent obviously. Suppose the users repair s to be $s[A, B] = \{1, z\}$, then another rule φ' with more details can be generated to be $A = 1 \to B = z$. Directed by the new rule φ', more tuples such as t can be repaired. Finally, an optimal *half modify* strategy will choose the position set S which eliminate the most inconsistencies.

According to the details of the three repairing strategies above, obviously, selecting different S for users will obtain different repairing results, and the choice of S which can produce the most consistent data is preferred. Therefore, the optimization goals of the RPS considered by this paper is to choose a perfect S such that the inconsistencies eliminated by repairing strategies are maximized. Suppose we have defined a function f such that $f(S)$ can measure how many inconsistencies are repaired for given S, we can give the following general formal definition of the RPS problem.

Definition 1 (Repair Position Selection Problem, RPS for short).
Given database I_R and dependency rule set Σ and integer k, find a subset S of I_R with k repairing positions such that the repair gain $f(S)$ is maximized.

In the following parts, for different data repairing strategies, the complexities and approximabilities of the corresponding RPS problems are studied.

3 RPS Problem for Simple Deletion Strategy

We first consider the repair position selection problem for simple deletion strategy. Given a tuple set S, the experts will determine whether to delete the tuples in S from the whole data I_R. In this followings, it is assumed that all tuples will be removed by experts with the same probability p independently. Let $V(S)$ be the set of inconsistent tuples pairs which involve at least one tuple in S. Then, for SD strategy, we can define the function value of $f_{SD}(S)$ to be the expected number of inconsistencies reduced by randomly removing S from I_R, that is $f_{SD}(S) = |V(S)| \cdot p$. In this part, when the context is clear we will use f to represent f_{SD}, and the definition $f_{SD}(S) = |V(S)|$ will be adopted for the simplicity since p is a constant value.

It is easy to verify that the RPS_{SD} problem is NP-*hard* by making a direct reduction from *vertex cover* problem. Here, we give a stronger complexity result of RPS problem by limiting the size of dependency rules.

Theorem 1. *The RPS_{SD} problem is NP-*complete *even for the case $|\Sigma| = 2$.*

Proof (sketch). The proof can be finished by making a reduction from the classic 3SAT problem to the decision version of RPS_{SD} problem.   □

Since the general RPS_{SD} is NP-*complete*, efficient approximation algorithms are needed in the practical applications. Here, we give the approximation algorithm by greedy idea as shown in Fig. 1.

Algorithm GreedySD
Input: Database I_R, rule set Σ and an integer m
Output: The repair position set S
1. Let $V = \{(t,i)|t \in I_R, 1 \leq i \leq |R|\}$;
2. Let graph $G = (V, E = \emptyset)$;
3. **for** each pair $\{(t,i),(s,j)\} \in V \times V$ **do**
4.   **if** (t,i) and (s,j) are inconsistent according to $\varphi \in \Sigma$ **then**
5.   add the edge $((t,i),(s,j))$ to E;
6. Remove all nodes with degree 0 from V;
7. **if** $|V| \leq m$ **then return** V;
8. Let $S = \emptyset$;
9. **while** $m > 0$ **do**
10.   Let $(t,i) \in V$ be the node with maximum degree in G;
11.   insert (t,i) to S;
12.   remove (t,i) from V and update G;
13. $m = m - 1$;
14.**return** S;

Fig. 1. Greedy Algorithm for RPS_{SD}

Theorem 2. *The Algorithm GreedySD is a 2-approximation algorithm for RPS problem under SD strategy.*

Proof. It can be proved simply by using the similar analysis with approximation algorithms for the minimum vertex cover problem. The details are omitted in this paper.   □

4   RPS Problem for Full Modify Strategy

In this part, we consider the repair position selection problem for full modify strategy. Under the full modify strategy, each item v in the repair position set S indicates a tuple t and an attribute index i. Given the set S, the experts will repair the data by modifying the values on repair positions in S, then the inconsistency caused by (t_1, i_1) and (t_2, i_2) is repaired if both of them belong to S. Let $I(S)$ be the set of inconsistent pairs of repair positions in S. Then, for FM strategy, we can define the function value of $f_{FM}(S)$ to be the size of $I(S)$, that is $f_{FM}(S) = |I(S)|$. In this part, when the context is clear we will use f to represent f_{FM}.

In this part, we utilize the dense k-subgraph (DkS) maximization problem [11], of computing the dense k-vertex subgraph of a given graph. Given a graph G and a parameter k, the DkS problem is to find a set of k vertices with maximum average degree in the subgraph induced by this set. It is still NP-*hard* even when restricted to bipartite graphs of maximum degree 3 [12].

Theorem 3. *The problem* RPS_{FM} *is NP-complete even if there are only 3 rules in* Σ.

Proof (sketch). We can show the RPS_{FM} problem is NP-*complete* by a reduction from the DkS problem whose definition is given as follows. DkS problem: given a bipartite graph $G = (V, E)$ of maximum degree 3 and two integer parameters k and m, the DkS problem is to determine whether there is a set S of k vertices such that the edge size of the induced graph G_S is not smaller than m.   □

The following lemma will show the correctness of Property PI used in the proof of Theorem 3.

Lemma 1 (Property PI). *Given a bipartite graph G with maximum degree 3, the MatchE3 Algorithm will output a maximum matching M such that all nodes with degree 3 will be included in M.*

Proof. As shown in Fig. 2, the main idea of MatchE3 Algorithm can be described as follows. First, a *maximum matching* $\hat{M}$ of G is built by using the Hungarian Algorithm [15] whose running time can be bounded by $O(|V_G| \cdot |E_G|)$ (line 1). Then, let V_1 be the set of nodes with degree 3 which are not covered by $\hat{M}$ (line 2–3). Then, to cover V_1, $\hat{M}$ is modified to M by using an idea similar with searching augmenting paths used in Hungarian Algorithm (line 4–24). The node in V_1 is processed iteratively. For each node $s \in V_1$, we will search M-*extending* path starting at s (line 6–18). An M-*extending* path P is a path with even edges whose edges alternate between E_M and $E_G \setminus E_M$. Also, if we label all nodes on P sequentially as $\{n_1 = s, n_2, \ldots, n_{2k+1}\}$, all nodes in $\{n_{2i+1} | 0 \leq i < k\}$ are required to have degree 3 and the degree of the node n_{2k+1} is required to be less than 3. Intuitively, after finding such a path P, we can use $(E_P \setminus E_M) \cup (E_M \setminus E_P)$ to construct a new matching M' such that (1) $|M'| = |M|$, (2) M' covers node s, and (3) the nodes with degree 3 covered by M can be still covered by M' (line 19–23). Therefore, after all nodes in V_1 are processed, the final matching M will satisfied the conditions required and be outputted (line 25).

The process of searching M-*extending* paths starting from s can be explained as follows. Since G is a bipartite graph, we can assume that $V_G = X \cup Y$ such that $E_G \subseteq X \times Y$. Also, without loss of generality, we assume that $s \in X$. During the procedure of searching M-*extending* paths, we use S to represent the set of nodes visited in X, vR to represent the set of nodes visited in Y. The set pre includes all edges visited when searching M-*extending* paths. Finally, if an M-*extending* path is found, we use t to represent the ending node of that path. The set $S \setminus vS$

```
Algorithm MatchE3
Input: Graph G
Output: a maximum matching M of G
1. Let M̂ be a maximum matching of G;
2. Let V₁ = V_G \ V_M̂;
3. Remove nodes with degree less than 3 from V₁;
4. M = M̂;
5. while V₁ is not ∅ do
6.       Let s be some node in V₁;
7.       S = {s}; vS = ∅; vR = ∅; pre = ∅; t = null;
8.       while |S| ≠ |vS| do
9.              if there is a node v ∈ S \ vS satisfying deg(v) < 3 then
10.                   t = v;
11.                   break ;
12.             for each node v ∈ S \ vS do
13.                   insert v to vS;
14.                   for each node r ∈ neighbor(v) \ vR do
15.                         Let u be the node satisfying (r, u) ∈ M;
16.                         insert r to vR;
17.                         insert u to S;
18.                         insert (v, r) and (r, u) to pre;
19.      while t! = s do
20.             Let x be the node satisfying (x, t) ∈ pre;
21.             Let y be the node satisfying (y, x) ∈ pre;
22.             M = (M/(x, t)) ∪ (y, x);
23.             t = y;
24.      Remove s from V₁;
25.return M;
```

Fig. 2. MatchE3 Algorithm for graph G

represent the nodes visited in the last iteration step. MatchE3 tries to find the *M-extending* path by extending the paths through adding edges starting from nodes in $S \setminus vS$. In details, for each node $v \in S \setminus vS$ and each unvisited neighbor node r of v, we can find an edge $(r, u) \in M$ and use (v, r) and (r, u) to extend the current path. If the new added node u satisfies $deg(u) < 3$, an *M-extending* path is found and let $t = u$.

The correctness of MatchE3 can be obtained by following results.

(1) When MatchE3 is trying to extend the paths by adding edges, for node $r \in neighbor(v) \setminus vR$, there must be a node u satisfying $(r, u) \in M$. We prove this by contradiction. Suppose there is a node r such that there does not exist a node u satisfying $(r, u) \in M$. That is we find a path P connecting s and r. P is composed of edges alternating between E_M and $E_G \setminus E_M$, and both the first edge and the last edge on P belong to $E_G \setminus E_M$. Intuitively, P is an augmenting path in the Hungarian Algorithm, and we can construct a maximum matching M' by letting $M' = (E_P \setminus E_M) \cup (E_M \setminus E_P)$ satisfying $|M'| = |M| + 1$.

That is a contradiction, since M is a maximum matching found by the Hungarian Algorithm.

(2) After MatchE3 has finished the steps of searching M-*extending* paths (line 8–18), t must be a node in G such that the path between s and t is an M-*extending* path. There are two possible ways for MatchE3 to quit the loop control defined between line 8 and 18. The first one is that t is assigned to be a node v satisfying $deg(v) < 3$, and the second one is that $|S| = |vS|$. In the followings, we will show that the second way is impossible by contradiction. Suppose that $|S| = |vS|$ and $t = null$. Obviously, we have $S = vS$. Let consider the induced graph G_{SR} of G on vertex set $vS \cup vR$. According to the definition of vS and vR, obviously, all nodes in vS have degree 3 in G_{SR}. In addition, the maximum matching M can be divided into two parts M_1 and M_2, where M_1 includes all edges of M also belonging to G_{SR} and M_2 includes all other edges. Consider a maximum matching M_{SR} of G_{SR}. According to the König-Egerváry theorem [22], $|M_{SR}| = |VC(G_{SR})|$, where $VC(G_{SR})$ is the minimum vertex cover of G_{SR}. Since the maximum node degree in G_{SR} is 3 and G_{SR} is a bipartite graph, we have $3VC(G_{SR}) \geq |E_{G_{SR}}|$. Then, we have $|M_{SR}| \geq |E_{G_{SR}}|/3$. According to the definition of G_{SR}, $|E_{G_{SR}}| = 3|vS|$. Thus, $|M_{SR}| \geq |E_{G_{SR}}|/3 = |vS|$. Since G_{SR} is connected and bipartite, we have $|M_{SR}| = |vS|$. According to the definition of vS, expect s, every node in vS has an incoming edge in M_1. Therefore, $|M_1| = |vS| - 1$. Consider $M' = M_{SR} \cup M_2$, we will show that M' is a matching of G which is a contradiction since M is assumed to be the maximum matching and $|M'| = |M| + 1$. To show that M' is a matching of G, we only need to verify that all edges in M_{SR} and M_2 are disjoint. Consider some node x satisfying there is an edge $(x, y) \in M_2$. First, x can not belong to vR, otherwise, y will belong to S and (x, y) will belong to G_{SR} which is a contradiction with the assumption. Additionally, x can not belong to S, otherwise, according to the Algorithm MatchE3, y and (x, y) will belong to vR and G_{SR} respectively which is a contradiction also.

(3) After using the M-*extending* path to modify the maximum matching M (line 19–23), the new matching M' will cover one more vertex with degree 3 in G. Consider the M-*extending* path between s and t. All nodes except t on path P have degree 3. Let $N(M)$ and $N(M')$ be the nodes covered by matching M and M' respectively. Obviously, we can obtain $N(M')$ by replacing t with s in $N(M)$. According to the Algorithm MatchE3, $deg(t) = 2$ and $deg(s) = 3$. Therefore, M' covers one more node with degree 3 in G.

Finally, after showing the previous 3 properties, the correctness of Algorithm MatchE3 can be proved.   □

We can build an approximation algorithm for the RPS_{FM} problem by giving a reduction from RPS_{FM} to DkS problem.

Theorem 4. *The RPS_{FM} problem can be approximated within ratio $O(n^{\frac{1}{3}})$.*

Proof. Given an RPS_{FM} instance $I = \{I_R, \Sigma, m, k\}$, a linear reduction to the DkS instance $I' = \{G, m, k\}$ can be built by following steps. First, for each tuple $t \in I_R$ and each attribute $A \in R$, build one node v_{tA} in V. Then, for tuples t

and t', if there is a rule $\varphi \in \Sigma$ such that t and t' are inconsistent on attribute X, add one edge $(v_{tX}, v_{t'X})$ into E. It is easy to verify that the reduction shown above is a linear reduction from RPS_{FM} to DkS problem. Finally, since DkS problem can be approximated within ratio $O(n^{\frac{1}{3}})$ [11], the RPS_{FM} problem can be approximated within ratio $O(n^{\frac{1}{3}})$ also.   □

5   RPS Problem for Half Modify Strategy

In this part, we consider the RPS problem for HM (Half Modify) strategy. For HM strategy, we can define the function value of $f_{HM}(S)$ to be size of tuples which can be repaired by S in I_R. In this part, when the context is clear, we also use f to represent f_{HM}.

Before analyzing the RPS_{HM} problem, one property about the Half Modify strategy is introduced which actually shows how the HM strategy is utilized to repair the inconsistent data.

Proposition 1. *Given relational instance I_R and a standard dependency rule $\varphi : X \rightarrow Y$, for a special tuple t, the set S_t is defined to be $\{t'|t' \in I_R \land t'[X] = t[X]\}$. Using the HM strategy, if the tuple t is repaired, the tuples in S_t can also be repaired automatically.*

Proof. After t has been repaired, no matter what value the expert give to the Y attribute of t, it essentially refines the rule φ to the form $R : X = t[X] \rightarrow Y = t[Y]$. The new refined rule has only constant values in the both sides of the rule, therefore, the tuples in S_t can also be repaired.   □

Theorem 5. *The problem RPS_{HM} is NP-complete.*

Proof (sketch). It can be proved by a reduction from VC (vertex cover for short) problem.   □

In the followings, a greedy algorithm is introduced to approximate the RPS_{FM} problem.

Theorem 6. *The Algorithm GreedyHM is a $(1 - 1/e)$-approximation algorithm for RPS problem under HM strategy.*

Proof. Using the similar ideas in [10], we can simply obtain the approximation ratio $(1 - 1/e)$. As shown in Fig. 3, the first step of Algorithm GreedyHM is to construct the set rel (line 1–12) which is composed of items with structure (key, val). After that, for each item $e \in rel$, $e.key$ is composed of *equivalent* repair positions where "equivalent" means that the data sets finally repaired by repairing any one of them are same, and $e.val$ is the data set that can be repaired by $e.key$. Let the set of all possible repairing positions be Ω. By rel, we can define $F_p \subseteq \Omega$ for each repair position p such that $F_p = e.val$ where $p \in e.key$. The second step in Algorithm GreedyHM is a greedy iteration during which the position p with maximal uncovered F_p is selected (line 13–19). Suppose

---

Algorithm GreedyHM
Input: Database I_R, rule set Σ and an integer k
Output: The repair position set S
1. Initialize rel to be $\emptyset$;
2. Initialize res and S to be $\emptyset$;
3. **for** each rule $r : X \rightarrow y$ in Σ **do**
4.   **for** each inconsistent pair of positions $(t[y], s[y])$ of r **do**
5.   Let $T = \{[t', y] | t' \in I_R \wedge t'[X] = t[X]\}$;
6.   Let $tmp.key = T$, $tmp.val = T$;
7.   **for** each item $e \in rel$ **do**
8.   **if** $e.key \cap tmp.key \neq \emptyset$ **then**
9.   Insert $(key = e.key \cap tmp.key, val = e.val \cup tmp.val)$ into rel;
10. $e.key = e.key \setminus tmp.key$;
11. $tmp.key = tmp.key \setminus e.key$;
12.   Insert tmp to rel;
13.**for** $i \in [1, k]$ **do**
14.   Let $e \in rel$ such that $|e.val|$ is maximized;
15.   Add $e.key$ to res;
16.   Remove e from rel;
17.   **for** each $e' \in rel$ **do**
18. $e'.val = e'.val \setminus e.val$;
19.   Select arbitrary $p \in e.key$ and add p to S;
20.**return** S;

Fig. 3. Greedy Algorithm for RPS_{HM}

S_{opt} be the optimal solution and let C_{opt} be the set $\cup_{p \in S_{opt}} F_p$. Let C be the set $\cup_{p \in S} F_p$, and let C_i be the set $\cup_{p \in S_i} F_p$ where S_i is the set of positions collected after the ith iteration. We have that

$$|C_i| - |C_{i-1}| \geq \frac{|C_{opt}| - |C_{i-1}|}{k},$$

since Algorithm GreedyHM is greedy based and the optimal solution can be repaired by selecting k positions only. Finally, we have $|C_k| \geq |C_{opt}|(1 - (1 - 1/k)^k)$, that is $|C_k|/|C_{opt}| \geq 1 - 1/e$ (Fig. 3). $\square$

6   Related Work

Inconsistency is a common data quality problem, which is also the main focus of data quality research area. There are several kinds of methods to solve the inconsistency problem. An intuitive idea is to fix the inconsistencies using sophisticated tools. Data repair aims to find a repair with a minimum modifications on the given database, where usually a repair can be defined to be a minimal set of repair operations (such as insert, delete and so on) which can make the database consistent. In [2], a database repair is defined to be a set of value modifications, based on the cost model defined over database repair, the minimum-cost repair problem is shown to be NP-*complete* and heuristic algorithms are designed to solve this problem in practical applications. [6] studies the problem of minimal-change integrity maintenance using tuple deletions. [18] considers the problem of repairing a database that is inconsistent with respect to a set of integrity constraints by updating numerical values. Based on an unified cost model, heuristic algorithms are designed to repair data under the unified cost model in [5]. The limitation of those methods is that there may be many different optimal repairs which is hard to explain when using the intuitive idea that optimal repair is the correct repair. In [14], to help decide which repair is preferred by users or systems, a repair is defined to be a set of insertion and deletion, and a function f is used to measure the *score* of each repair (for example, f just computes the size of insertions). Then, based on the f a logical method is proposed to compute all preferred repairs and answers over repairs. In [8], without the usual assumption that database is consistent before updates, the researchers give an extended definition of *sound* and *complete* inconsistency check and show whether previous methods involving integrity check satisfy those conditions. The second kind of method is consistent query answering. The main idea of consistent query answering (CQA) is to answer the queries on inconsistent databases without repairing them first. It is similar with the certain answers over incomplete database. [1] studies the problem of the logical characterization of the notion of consistent answer in a relational database. A method for computing consistent answers is given and its soundness and completeness (for some classes of constraints and queries) are proved. In [17], the algorithmic and complexity theoretic issues of CQA under database repairs that minimally depart from the original database are investigated. To solve CQA problem, one important strategy is to rewrite the queries according to the inconsistencies and constraints [9,13,20]. Since automatic repair methods can not guarantee to find the true values when repairing errors in database, using the information of experts is a possible way to solve this problem. [19] considers to use the feedback on query results from users to guide the data repairing so that a more reasonable repairing solutions can be obtained. Under different query classes, both the data and combined complexities of the data repairing problem are studied. Some close related works are in the name of view update or delete propagation. The view update problem is to translate given updates on a fixed view into a series corresponding updates over the original data.

7   Conclusion

In this paper, to develop the idea of using user feedbacks to guide the data repair, the RPS problem is studied under three different data repairing strategies, SD, FM and HM. To our best knowledge, this is the first work on the RPS problem. We firstly give the formal definition of the RPS problem and study its complexities and approximabilities. All corresponding RPS problems under the three repairing strategies are shown to be NP-*complete*, and efficient approximation algorithms for them are also designed.

References

1. Arenas, M., Bertossi, L., Chomicki, J.: Consistent query answers in inconsistent databases. In: Proceedings of the Eighteenth ACM SIGMOD-SIGACT-SIGART Symposium on Principles of Database Systems (PODS 1999), New York, pp. 68–79. ACM (1999)
2. Bohannon, P., Fan, W., Flaster, M., Rastogi, R.: A cost-based model and effective heuristic for repairing constraints by value modification. In: Proceedings of the 2005 ACM SIGMOD International Conference on Management of Data (SIGMOD 2005), New York, pp. 143–154. ACM (2005)
3. Bohannon, P., Fan, W., Geerts, F., Jia, X., Kementsietsidis, A.: Conditional functional dependencies for data cleaning. In: 2007 IEEE 23rd International Conference on Data Engineering, pp. 746–755, April 2007
4. Cai, Z., Heydari, M., Lin, G.: Iterated local least squares microarray missing value imputation. J. Bioinform. Computat. Biol. **4**, 935–958 (2006)
5. Chiang, F., Miller, R.J.: A unified model for data and constraint repair. In: Proceedings of the 2011 IEEE 27th International Conference on Data Engineering (ICDE 2011), Washington, DC, pp. 446–457. IEEE Computer Society (2011)
6. Chomicki, J., Marcinkowski, J.: Minimal-change integrity maintenance using tuple deletions. Inf. Comput. **197**, 90–121 (2005)
7. Cong, G., Fan, W., Geerts, F., Jia, X., Ma, S.: Improving data quality: consistency and accuracy. In: Proceedings of the 33rd International Conference on Very Large Data Bases (VLDB 2007), pp. 315–326. VLDB Endowment (2007)
8. Decker, H., Martinenghi, D.: Inconsistency-tolerant integrity checking. IEEE Trans. Knowl. Data Eng. **23**, 218–234 (2011)
9. Eiter, T., Fink, M., Greco, G., Lembo, D.: Repair localization for query answering from inconsistent databases. ACM Trans. Database Syst. **33**, 10:1–10:51 (2008)
10. Feige, U.: A threshold of ln n for approximating set cover. J. ACM **45**, 634–652 (1998)
11. Feige, U., Peleg, D., Kortsarz, G.: The dense k-subgraph problem. Algorithmica **29**, 410–421 (2001)
12. Feige, U., Seltser, M.: On the densest k-subgraph problems, technical report, The Weizmann Institute, Jerusalem, Israel (1997)
13. Fuxman, A., Miller, R.J.: First-order query rewriting for inconsistent databases. J. Comput. Syst. Sci. **73**, 610–635 (2007)
14. Greco, S., Sirangelo, C., Trubitsyna, I., Zumpano, E.: Preferred repairs for inconsistent databases. In: Proceedings of the Seventh International Database Engineering and Applications Symposium, pp. 202–211, July 2003
15. Kuhn, H.: The Hungarian method for the assignment problem. Nav. Res. Logist. Q. **2**, 83–97 (1955)
16. Li, J., Liu, X.: An important aspect of big data: data usability. J. Comput. Res. Dev. **50**, 1147–1162 (2013)
17. Lopatenko, A., Bertossi, L.: Complexity of consistent query answering in databases under cardinality-based and incremental repair semantics. In: Schwentick, T., Suciu, D. (eds.) ICDT 2007. LNCS, vol. 4353, pp. 179–193. Springer, Heidelberg (2006). https://doi.org/10.1007/11965893_13
18. Lopatenko, A., Bravo, L.: Efficient approximation algorithms for repairing inconsistent databases. In: 2007 IEEE 23rd International Conference on Data Engineering, pp. 216–225, April 2007

19. Miao, D., Liu, X., Li, J.: On the complexity of sampling query feedback restricted database repair of functional dependency violations. Theor. Comput. Sci. **609**, 594–605 (2016)
20. Staworko, S., Chomicki, J.: Consistent query answers in the presence of universal constraints. Inf. Syst. **35**, 1–22 (2010)
21. Wang, Y., Cai, Z., Stothard, P., Moore, S., Goebel, R., Wang, L., Lin, G.: Fast accurate missing SNP genotype local imputation. BMC Res. Notes **5**, 404 (2012)
22. West, D.B.: Introduction to Graph Theory. Prentice Hall, New York (2001)

Unbounded One-Way Trading on Distributions with Monotone Hazard Rate

Francis Y.L. Chin[1], Francis C.M. Lau[2], Haisheng Tan[3], Hing-Fung Ting[2], and Yong Zhang[4(✉)]

[1] Hang Seng Management College, Shatin, Hong Kong
francischin@hsmc.edu.hk
[2] Department of Computer Science, The University of Hong Kong,
Pokfulam, Hong Kong
{fcmlau,hfting}@cs.hku.hk
[3] School of Computer Science and Technology,
University of Science and Technology of China, Hefei, China
hstan@ustc.edu.cn
[4] Shenzhen Institutes of Advanced Technology, Chinese Academy of Sciences,
Shenzhen, China
zhangyong@siat.ac.cn

Abstract. One-way trading is a fundamental problem in the online algorithms. A seller has some product to be sold to a sequence of buyers $\{u_1, u_2, \ldots\}$ in an online fashion and each buyer u_i is associated with his accepted unit price p_i, which is known to the seller on the arrival of u_i. The seller needs to decide the amount of products to be sold to u_i at the then-prevailing price p_i. The objective is to maximize the total revenue of the seller. In this paper, we study the unbounded one-way trading, i.e., the highest unit price among all buyers is unbounded. We also assume that the highest prices of buyers follow some distribution with monotone hazard rate, which is well-adopted in Economics. We investigate two variants, (1) the distribution is on the highest price among all buyers, and (2) a general variant that the prices of buyers is independent and identically distributed. To measure the performance of the algorithms, the expected competitive ratios, $E[OPT]/E[ALG]$ and $E[OPT/ALG]$, are considered and constant-competitive algorithms are given if the distributions satisfy the monotone hazard rate.

1 Introduction

Revenue maximization is an important problem studied by researchers in the fields of economics, mathematics and computer science. This problem has many variations but generally involves the question of how to sell or assign products (goods or services) to various buyers. The assignment of products includes determining both the price and the amount of products sold to each buyer, which is a fundamental problem related to markets and market mechanisms in economics. Accordingly, there are two ways for a seller to maximize revenue: controlling the selling price and controlling the amount sold.

© Springer International Publishing AG 2017
X. Gao et al. (Eds.): COCOA 2017, Part I, LNCS 10627, pp. 439–449, 2017.
https://doi.org/10.1007/978-3-319-71150-8_36

In this paper, we focus on the design of an online strategy to determine how much should be sold at the prevailing market price (which cannot be controlled by the seller) at different times. This problem was first studied by El-Yaniv et al. [12,13], which was called named *one-way trading*. In the one-way trading problem, a player has some initial asset (e.g., dollar) to be changed to a target asset (e.g., yen). The exchange rate fluctuates over time. To maximize the revenue, the player must decide the amount of the initial asset to be changed when the exchange rate on each day is known. The offline version of this problem is straightforward as the seller can know all the future information: the seller can simply exchange all initial assets to the target asset on the day with the highest exchange rate. However, in the online version where the player has no knowledge of the future, at no point will the player be sure that the prevailing exchange rate is the highest one. The key features of the one-way trading problem are: (1) the player has no control of the exchange rate which fluctuates over time; (2) the player has no knowledge, or incomplete knowledge, of the future; and (3) the player can decide the amount to be changed only upon the arrival of each rate.

The one-way trading problem studied in [13] is the bounded version, i.e., the range of the exchange rate is in $[m, M]$, where m and M are fixed values. Based on the relationship between m and M, El-Yaniv et al. presented an optimal online algorithm by using a threat-based policy, of which the competitive ratio is $\Theta(\log(M/m))$. If the highest possible rate is unbounded, even for a fixed number of transactions, the threat-based policy cannot be implemented since the ratio between any two rates can be arbitrary large. In the bounded one-way trading problem, the remaining amount of the initial asset after the last transaction will be changed to the target asset with the minimum rate m. However, if the highest possible rate is unbounded, in the worst case, the total revenue is dominated by the revenue from high rates and the revenue from the remaining asset using the minimum rate is very tiny and ignoring this part will hardly affect the performance. For the one-way trading with unbounded value, Chin et al. [10] gave a near optimal algorithm with competitive ratio $O(\log r^* (\log^{(2)} r^*) \dots (\log^{(h-1)} r^*)(\log^{(h)} r^*)^{1+\epsilon})$ if the value of $r^* = p^*/p_1$, the ratio between the highest market price $p^* = \max_i p_i$ and the first price p_1, is large and satisfies $\log^{(h)} r^* > 1$, where $\log^{(i)} x$ denotes the application of the logarithm function i times to x; otherwise, the algorithm has a constant competitive ratio. A lower bound was also proved in [10]. Given any positive integer h and any one-way trading algorithm A, a sequence of buyers σ with $\log^{(h)} r^* > 1$ exist such that the ratio between the optimal revenue and the revenue obtained by A is at least $\Omega(\log r^* (\log^{(2)} r^*) \dots (\log^{(h-1)} r^*)(\log^{(h)} r^*))$.

In some sense, the one-way trading problem can be regarded as a *time series search problem*, the objective of which is to find the maximum (or the minimum) value among a sequence of values in an online fashion. For the 1-max-search variant, i.e., determining the highest value among the whole sequence in an online fashion, El-Yaniv et al. [13] presented a randomized $O(\log M/m)$-competitive algorithm if the values fluctuate between m and M; when M/m is unknown in advance, a randomized online algorithm with competitive ratio $O(\log(M/m) \cdot$

$\log^{1+\epsilon}(\log(M/m))$ can be achieved. In [16], Lorenz et al. gave an optimal online algorithm for the k-search problem, in which the player's target is to find the k highest (or lowest) values among all values in a sequence and the values are chosen from $[m, M]$.

In this paper, we assume that items can be sold fractionally, thus, the amount of items can be normalized to be 1. A sequence of buyers come one after one and each buyer i is associated with a price p_i, which is his accepted unit price for buying the items. Only upon the arrival of a buyer i will his accepted price p_i be known to the seller, who will immediately determine the amount of items to be sold to the buyer with unit price p_i. The objective is to maximize the total revenue of the seller. In the unbounded one-way trading, the range of the accepted prices is in $(0, +\infty)$.

In all previous studies, if there is no information about the future prices, no algorithm achieved a competitive ratio better than a logarithm factor. However, given some partial information about the prices, the performance could be improved greatly. In this paper, we assume that the distribution of the highest accepted price is the partial information that is known. Firstly, assume that the distribution is on the highest price among all buyers, i.e., $\max_i p_i$. We then consider a general variant where the sequence of prices of buyers is *independent and identically distributed* (i.i.d.).

To measure the performance of the online algorithm, the competitive ratio is often used, which denotes the ratio between the result form the online algorithm and the optimal offline algorithm. For the online algorithm with distributions, we use the expected competitive ratio for evaluation. There are mainly two kinds of expectation of competitive ratio, i.e., $\frac{E[OPT]}{E[ALG]}$ and $E[\frac{OPT}{ALG}]$. Both of them are considered with respect to different situations and the values of them may vary a lot. For the former measure, since the expected value of the optimal solution is independent of the algorithm solution, the target is to maximize the expected output of the algorithm.

The paper is organized as follows: Sect. 2 describes the one-way trading with distributions and the measurement of the algorithm; in Sect. 3, constant competitive algorithms are given if the distribution is on the highest price among all buyers; in Sect. 4, we prove that the variant with i.i.d. distribution on each buyer can be reduced to the variant in Sect. 3, and thus constant-competitive algorithms can be obtained too.

2   One-Way Trading with Distribution

In the one-way trading problem, we may regard the first price as a unit value. This assumption is reasonable since in the remaining part of the price sequence, values lower than the first one could be ignored and will not affect the performance. Let f be the density function and F be the accumulated distribution with respect to the highest price among all buyers. We assume that f and F are continuous in $[1, +\infty)$. Given F, the expected revenue received from the optimal algorithm is

$$E[OPT] = \int_1^{+\infty} x dF(x) = \int_1^{+\infty} x f(x) dx.$$

El-Yaniv et al. showed that for the bounded one-way trading problem, the adversary can choose the worst distribution on the highest selling price and force the online algorithm to achieve the competitive ratio no less than $\Omega(\log M/m)$ (Theorem 7 in [13]), where the highest price $p \in [m, M]$. This result can be extended to the unbounded one-way trading problem.

Fact 1. *There exists the worst distribution F such that no online algorithm can solve the unbounded one-way trading problem with the competitive ratio better than a logarithm factor if the highest price is drawn from F.*

Proof. The distribution on the bounded one-way trading can be also used as the distribution on the unbounded version such that the probabilities on the highest price higher than M and lower than m are both zero. Thus, setting the distribution F to be the worst distribution w.r.t. the bounded one-way trading implies the competitive ratio of any online algorithm cannot be better than a logarithm factor.   □

This negative result is unimportant in reality since most frequently used distributions in economics are far from the worst distribution. If the highest price among all buyers is uniformly distributed, Fujiwara et al. [14] considered the selling strategy according to several measures, e.g., $E[ALG/OPT]$, $E[OPT/ALG]$, $E[ALG]/E[OPT]$, $E[OPT]/E[ALG]$. The algorithms for the average case analysis of the bounded one-way trading are based on the threat-based policy. However, such a strategy does not work for the unbounded variant since the lowest price m and highest price M may not be known in advance.

The hazard rate, a.k.a. the failure rate, is the probability of observing an outcome within a neighborhood of some value x, conditional on the outcome being no less than x. The concept of the hazard rate is well-adopted in economics. For example, in English auctions, the hazard rate on x denotes the probability of the auction ending at x, conditional on the bidders' prices reach x. In this paper, we consider *the monotone hazard rate*, which is reasonable and also has been considered in theoretical computer science [8,17]. Formally speaking,

Definition 1 *(Monotone Hazard Rate). A distribution F with density f is said to satisfy the monotone hazard rate (MHR) if $\frac{1-F(x)}{f(x)}$ is monotonically non-increasing for all $x > 0$.*

3   Distribution on the Highest Price Among All Buyers

In this part, we consider the variant that the distribution on the highest price among all buyers is known in advance and satisfies the monotone hazard rate.

3.1 Measure of $\frac{E[OPT]}{E[ALG]}$

The following two lemmas from Chawla et al. [8] and Dhangwatnotai et al. [11] respectively can be regarded as the consequences of Myerson's optimal strategy [18]. They also gave the idea to maximize the algorithm's expected revenue.

Lemma 1 ([8]). *If the distribution F with density f satisfies MHR, then there exists x_0 such that (1) $x_0(1 - F(x_0))$ is maximized, (2) for any $x_0 < x_1 < x_2$, $x_0(1 - F(x_0)) > x_1(1 - F(x_1)) > x_2(1 - F(x_2))$ and, (3) for any $x_0 > x_1 > x_2$, $x_0(1 - F(x_0)) > x_1(1 - F(x_1)) > x_2(1 - F(x_2))$.*

From Lemma 1, it is natural to assign all products to any buyer with value no less than x_0. With probability $1 - F(x_0)$, all products are assigned with price no less than x_0, which means the expected revenue from the algorithm is at least $x_0 \cdot (1 - F(x_0))$.

Lemma 2 ([11]). $E[OPT] = O(x_0 \cdot (1 - F(x_0)))$

According to the above two lemmas, the algorithm can be simply described as follows.

Algorithm 1. Online Selling for the measure of $E[OPT]/E[ALG]$

1: Let $x_0 = \arg\max_x x \cdot (1 - F(x))$
2: Sell the whole product to the first buyer with price no less than x_0.

Thus, we have the following conclusion.

Theorem 1. *When considering the measure of $\frac{E[OPT]}{E[ALG]}$, the expected competitive ratio of Algorithm 1 is a constant.*

3.2 Measure of $E[\frac{OPT}{ALG}]$

For the measure of $E[\frac{OPT}{ALG}]$, the competitive ratio of Algorithm 1 is unbounded since the seller does not assign any product to the buyer with price less than $\arg\max_x x(1 - F(x))$ and the ratio in such case is unbounded. Thus, we have to investigate the intrinsic property and find other way to achieve good performance for this measurement.

Lemma 3. *Given a distribution F satisfying MHR, $h(x) = \frac{1-F(x)}{1-F(2x)}$ is monotone non-decreasing.*

Proof.

$$h'(x) = \frac{-(1 - F(2x))f(x) + 2(1 - F(x))f(2x)}{(1 - F(2x))^2}$$

$$= \frac{2f(2x)(1 - F(x)) - f(x)(1 - F(2x))}{(1 - F(2x))^2}$$

$$= \frac{2f(2x)}{1 - F(2x)} \cdot \frac{1 - F(x)}{1 - F(2x)} - \frac{f(x)}{1 - F(2x)}$$

Since F satisfies MHR, i.e., $\frac{1-F(x)}{f(x)} \geq \frac{1-F(2x)}{f(2x)}$, we have $h'(x) \geq 0$, which means that $h(x)$ is monotone non-decreasing.   □

From Lemma 1, we know that if the distribution of the highest price satisfies MHR, there exists p such that $p \cdot (1 - F(p))$ is maximized. W.l.o.g., assume that $2^k \leq p < 2^{k+1}$. As mentioned before, if the coming price is no less than p, selling the whole item to this buyer is a good idea. But for the remaining case that the highest price is strictly less than p, the assignment is also critical. In our algorithm, the item is partitioned with respect to the range of the price. Upon the arrival of a buyer, if his price is the first in some range, the corresponding amount of item will be assigned to him. The description of the algorithm is shown in Algorithm 2.

Algorithm 2. Online Selling for the measure of $E[OPT/ALG]$

1: **if** v is the first value no less than p **then**
2:    Assign $1/2$ product to this buyer.
3: **else**
4:    **if** v is the first value within $[2^{-i} \cdot p, 2^{1-i} \cdot p)$ **then**
5:       Assign 2^{-i-1} product to this buyer.
6:    **end if**
7: **end if**

Theorem 2. *When considering the measure of* $E[\frac{OPT}{ALG}]$, *the expected competitive ratio of the above algorithm is a constant.*

Proof. For a sequence of buyers, suppose that the highest price among all buyers is x. The maximal revenue for this sequence is x by assigning the whole product to the buyer with the highest price. Let $ALG(x)$ be the revenue received by the online algorithm on a buyer sequence with the highest price x.

According to the online algorithm, if $x \geq p$, the algorithm assigns $1/2$ of a product to a buyer with price no less than p; if $x \in [2^{-i} \cdot p, 2^{1-i} \cdot p)$, the algorithm assigns 2^{-i-1} products to a buyer with price no less than $2^{-i} \cdot p$. For any sequence of buyers, the total amount of products assign to buyers is at most $1/2 + 1/4 + \cdots < 1$. The whole product is sufficient to be assigned to all buyers according to the algorithm.

The expected competitive ratio is

$$E[\frac{OPT}{ALG}] = \int_1^{+\infty} \frac{x}{ALG(x)} dF(x)$$

$$= (\int_1^{2^{-k} \cdot p} + \sum_{-k}^{-1} \int_{2^i \cdot p}^{2^{i+1} \cdot p} + \sum_0^{+\infty} \int_{2^i \cdot p}^{2^{i+1} \cdot p}) \frac{x}{ALG(x)} dF(x)$$

$$\leq (\sum_{-k-1}^{-1} \int_{2^i \cdot p}^{2^{i+1} \cdot p} + \sum_0^{+\infty} \int_{2^i \cdot p}^{2^{i+1} \cdot p}) \frac{x}{ALG(x)} dF(x)$$

The above formula has two parts and we analyze them as follows.

(i) $-k - 1 \leq i \leq -1$.

In this case, $ALG(x) \geq 2^{i-1} \cdot 2^i \cdot p$ while $x \leq 2^{i+1} \cdot p$. Thus,

$$\int_{2^i \cdot p}^{2^{i+1} \cdot p} \frac{x}{ALG(x)} dF(x) \leq 2^{2-i} \int_{2^i \cdot p}^{2^{i+1} \cdot p} dF(x) = 2^{2-i}(F(2^{i+1} \cdot p) - F(2^i \cdot p))$$

(ii) $i \geq 0$.

In this case, $ALG(x) \geq p/2$ while $x \leq 2^{i+1} \cdot p$. Thus,

$$\int_{2^i \cdot p}^{2^{i+1} \cdot p} \frac{x}{ALG(x)} dF(x) \leq 2^{i+2} \int_{2^i \cdot p}^{2^{i+1} \cdot p} dF(x) = 2^{i+2}(F(2^{i+1} \cdot p) - F(2^i \cdot p))$$

From Lemma 1, if $i \geq 0$, we have $2^i \cdot p(1 - F(2^i \cdot p)) > 2^{i+1} \cdot p(1 - F(2^{i+1} \cdot p))$. Thus, $1 - F(2^{i+1} \cdot p) < (1 - F(2^i \cdot p))/2$ and $F(2^{i+1} \cdot p) - F(2^i \cdot p) > (1 - F(2^i \cdot p))/2$. Let $1 - F(2^{i+1} \cdot p) = (1 - F(2^i \cdot p)) \cdot \delta_i$ and $F(2^{i+1} \cdot p) - F(2^i \cdot p) = (1 - F(2^i \cdot p)) \cdot \gamma_i$, where $\delta_i < 1/2$, $\gamma_i > 1/2$ and $\delta_i + \gamma_i = 1$.

From Lemma 3, $\frac{1 - F(2x)}{1 - F(x)}$ is monotone non-increasing when $x > p$, thus, δ_i is monotone non-increasing and γ_i is monotone non-decreasing when i increasing.

Thus, if $i \geq 0$,

$$\int_{2^i \cdot p}^{2^{i+1} \cdot p} \frac{x}{ALG(x)} dF(x) \leq 2^{i+2}(F(2^{i+1} \cdot p) - F(2^i \cdot p))$$

$$= 2^{i+2} \cdot (1 - F(2^i \cdot p)) \cdot \gamma_i$$

$$= 2^{i+2} \cdot (1 - F(p)) \cdot \prod_{k=0}^{i-1} \delta_k \cdot \gamma_i$$

$$\leq 2^{i+2} \cdot (1 - F(p)) \cdot \delta_0^i$$

$$= 4 \cdot (1 - F(p)) \cdot (2\delta_0)^i$$

$$\int_p^{+\infty} \frac{x}{ALG(x)} dF(x) \leq 4 \cdot (1 - F(p)) \cdot \sum_i (2\delta_0)^i \tag{1}$$

$$= \frac{4 \cdot (1 - F(p))}{1 - 2\delta_0}$$

From Lemma 1, if $i \leq -1$, we have $2^i \cdot p(1 - F(2^i \cdot p)) < 2^{i+1} \cdot p(1 - F(2^{i+1} \cdot p))$. Thus, $1 - F(2^{i+1} \cdot p) > (1 - F(2^i \cdot p))/2$ and $F(2^{i+1} \cdot p) - F(2^i \cdot p) < (1 - F(2^i \cdot p))/2$. Let $1 - F(2^{i+1} \cdot p) = (1 - F(2^i \cdot p)) \cdot \mu_i$ and $F(2^{i+1} \cdot p) - F(2^i \cdot p) = (1 - F(2^i \cdot p)) \cdot \nu_i$, where $\mu_i > 1/2$, $\nu_i < 1/2$ and $\mu_i + \nu_i = 1$.

From Lemma 3, $\frac{1-F(x)}{1-F(2x)}$ is monotone non-decreasing when $2x < p$, and thus, μ_i is monotone non-decreasing and ν_i is monotone non-increasing when i increases.

Since

$$F(2^{i+2} \cdot p) - F(2^{i+1} \cdot p) = (1 - F(2^{i+1} \cdot p)) \cdot \nu_{i+1}$$
$$= (1 - F(2^i \cdot p)) \cdot \mu_i \cdot \nu_{i+1}$$
$$= (F(2^{i+1} \cdot p) - F(2^i \cdot p)) \cdot \mu_i \cdot \nu_{i+1}/\nu_i.$$

We have

$$F(2^{i+1} \cdot p) - F(2^i \cdot p) = (F(2^{i+2} \cdot p) - F(2^{i+1} \cdot p)) \cdot \frac{\nu_i}{\mu_i \cdot \nu_{i+1}}.$$

Thus, if $i \leq -1$,

$$\int_{2^i \cdot p}^{2^{i+1} \cdot p} \frac{x}{ALG(x)} dF(x) \leq 2^{2-i}(F(2^{i+1} \cdot p) - F(2^i \cdot p))$$
$$= 2^{2-i} \cdot (F(2^{i+2} \cdot p) - F(2^{i+1} \cdot p)) \cdot \frac{\nu_i}{\mu_i \cdot \nu_{i+1}}$$
$$= 2^{2-i} \cdot (F(p) - F(p/2)) \cdot \frac{\nu_i}{\nu_0} \cdot \frac{1}{\prod_{k=i}^{0} \mu_k}$$
$$\leq 8 \cdot (F(p) - F(p/2)) \cdot \frac{1}{\prod_{k=i}^{0} 2\mu_k}$$
$$\leq 8 \cdot (F(p) - F(p/2)) \cdot (\frac{1}{2\mu_0})^{i+1}$$

Therefore,

$$\int_1^p \frac{x}{ALG(x)} dF(x) \leq 8 \cdot (F(p) - F(p/2)) \cdot \sum_i (\frac{1}{2\mu_0})^i \qquad (2)$$
$$= \frac{8 \cdot (F(p) - F(p/2))}{1 - 1/(2\mu_0)}$$

Combining the inequalities (1) and (2), we can say that the excepted competitive ratio of the algorithm is

$$E[\frac{OPT}{ALG}] \leq \frac{4 \cdot (1 - F(p))}{1 - 2\delta_0} + \frac{8 \cdot (F(p) - F(p/2))}{1 - 1/(2\mu_0)} = O(1).$$

□

4   Distribution on the Highest Price of Each Buyer

In the previous part, we study the case that the distribution is on the highest price among all buyers. Now we assume that the distribution on the price of each buyer is known in advance, and the distribution on the buyers is under the i.i.d. assumption. We also assume that the number of buyers is bounded by n. Otherwise, even for a distribution with a very tiny value in some high price, the adversary can force the probability of the high price to be close to 1 by sending sufficiently large number of buyers.

Formally speaking, there are at most n buyers who will come to the seller to buy products; the price of each buyer is drawn from the accumulated distribution $F(x)$ with the density function $f(x)$, where $f(x)$ is derivable.

For this variant, if the distribution of the highest price among all buyers also satisfies the MHR property, the algorithms in Sect. 3 can be implemented. This gives us a heuristic to reduce this variant to the previous one. Let $\tilde{F}(\mathrm{x})$ and $\tilde{f}(x)$ be the accumulated distribution and density function on the highest price among all buyers, thus, $\tilde{F}(x) = F^n(x)$ and $\tilde{f}(x) = nF^{n-1}(x)f(x)$, respectively.

Lemma 4. *If $f(x)$ satisfies the monotone hazard rate, then $\tilde{f}(x)$ also satisfies the monotone hazard rate.*

Proof. If $f(x)$ satisfies the monotone hazard rate (MHR), i.e., $\frac{1-F(x)}{f(x)}$ is non-increasing, we have $(\frac{1-F(x)}{f(x)})' \leq 0$, thus, $f'(x) \leq \frac{f^2(x)}{F(x)-1}$. Now we consider $\frac{1-\tilde{F}(x)}{\tilde{f}(x)}$. If $(\frac{1-\tilde{F}(x)}{\tilde{f}(x)})' \leq 0$, this lemma is true.

$$
\begin{aligned}
&(\frac{1-\tilde{F}(x)}{\tilde{f}(x)})' \\
=&(\frac{1-F^n(x)}{nF^{n-1}(x)f(x)})' \\
=&\frac{-(nF^{n-1}(x)f(x))^2 - (1-F^n(x))[n(n-1)F^{n-2}(x)f^2(x) + nF^{n-1}(x)f'(x)]}{(nF^{n-1}(x)f(x))^2} \\
=&\frac{-nF^n(x)f^2(x) - (n-1)f^2(x) - F(x)f'(x) + (n-1)F^n(x)f^2(x) + F^{n+1}(x)f'(x)}{nF^n(x)f^2(x)} \\
=&\frac{-F^n(x)f^2(x) - (n-1)f^2(x) + (F^{n+1}(x) - F(x))f'(x)}{nF^n(x)f^2(x)} \\
\leq&\frac{-F^n(x)f^2(x) - (n-1)f^2(x) + (F^{n+1}(x) - F(x))\frac{f^2(x)}{F(x)-1}}{nF^n(x)f^2(x)} \\
=&\frac{-F^{n+1}(x) + F^n(x) - (n-1)F(x) + (n-1) + F^{n+1}(x) - F(x)}{nF^n(x)(F(x)-1)} \\
=&\frac{(F(x)-1)(F^{n-1}(x)-1) - (n-1)(F(x)-1) + F^{n-1}(x) - 1}{nF^n(x)(F(x)-1)} \\
=&\frac{F^{n-1}(x) - 1 - (n-1) + F^{n-2}(x) + F^{n-3}(x) + \cdots + 1}{nF^n(x)} \\
=&\frac{F^{n-1}(x) + F^{n-2}(x) + \cdots + F(x) - (n-1)}{nF^n(x)} \\
\leq& 0
\end{aligned}
$$

Therefore, $\tilde{f}(x)$ also satisfies the monotone hazard rate.   □

Since $\tilde{F}(x)$ and $\tilde{f}(x)$ satisfy the monotone hazard rate, Algorithms 1 and 2 can be used to handle this variant. Thus, we have the following conclusion.

Theorem 3. *In the unbounded one-way trading problem, if the number of buyers is bounded, the distribution on price of each buyer is i.i.d. and satisfies the monotone hazard rate, online algorithms with constant competitive ratios can be obtained under the measures of* $E[OPT/ALG]$ *and* $E[OPT]/E[ALG]$.

5   Concluding Remark

Design selling mechanisms to maximize the seller's revenue is well-studied in the field of economy whereas related research in theoretical computer science is relatively more recent and ongoing. Many variants of the problem have been found to be computationally difficult when cast in a realistic setting. The challenge has been to identify special cases for which a solution can be efficiently computed while keeping their relevance to real-life situations. Traditional worst case analyses in which the algorithm designer usually knows nothing about the future may not match the reality well. Average case analysis of the expected ratio is a direct measure of performance. This paper is an attempt to model the real case where the seller has some partial information about the buyers. For future research, it may be worthwhile to determine which information is critical and how to fully utilize the partial information to design selling strategies.

Acknowledgements. This research is supported by National Key Research and Development Program of China under Grant 2016YFB0201401, China's NSFC grants (No. 61433012, U1435215, 61402461, 61772489), Hong Kong GRF grant (17210017, HKU 7114/13E), and Shenzhen basic research grant JCYJ20160229195940462.

References

1. Babaioff, M., Dughmi, S., Kleinberg, R., Slivkins, A.: Dynamic pricing with limited supply. In: Proceedings of the 13th ACM Conference on Electronic Commerce (EC 2012), pp. 74–91 (2012)
2. Badanidiyuru, A., Kleinberg, R., Singer, Y.: Learning on a budget: posted price mechanisms for online procurement. In: Proceedings of the 13th ACM Conference on Electronic Commerce (EC 2012), pp. 128–145 (2012)
3. Balcan, M.-F., Blum, A., Mansour, Y.: Item pricing for revenue maximization. In: Proceedings of the 9th ACM Conference on Electronic Commerce (EC 2008), pp. 50–59 (2008)
4. Blum, A., Hartline, J.D.: Near-optimal online auctions. In: Proceedings of the 16th Annual ACM-SIAM Symposium on Discrete Algorithms (SODA 2005), pp. 1156–1163 (2005)
5. Blum, A., Gupta, A., Mansour, Y., Sharma, A.: Welfare and profit maximization with production costs. In: Proceedings of 52th Annual IEEE Symposium on Foundations of Computer Science (FOCS 2011), pp. 77–86 (2011)

6. Chakraborty, T., Even-Dar, E., Guha, S., Mansour, Y., Muthukrishnan, S.: Approximation schemes for sequential posted pricing in multi-unit auctions. In: Saberi, A. (ed.) WINE 2010. LNCS, vol. 6484, pp. 158–169. Springer, Heidelberg (2010). https://doi.org/10.1007/978-3-642-17572-5_13

7. Chakraborty, T., Huang, Z., Khanna, S.: Dynamic and non-uniform pricing strategies for revenue maximization. In: Proceedings of 50th Annual IEEE Symposium on Foundations of Computer Science (FOCS 2009), pp. 495–504 (2009)

8. Chawla, S., Hartline, J.D., Kleinberg, R.: Algorithmic pricing via virtual valuations. In: Proceedings of the 8th ACM Conference on Electronic Commerce (EC 2007), pp. 243–251 (2007)

9. Chen, G.-H., Kao, M.-Y., Lyuu, Y.-D., Wong, H.-K.: Optimal buy-and-hold strategies for financial markets with bounded daily returns. SIAM J. Compt. **31**(2), 447–459 (2001)

10. Chin, F.Y.L., Fu, B., Guo, J., Han, S., Hu, J., Jiang, M., Lin, G., Ting, H.-F., Zhang, L., Zhang, Y., Zhou, D.: Competitive algorithms for unbounded one-way trading. Theor. Comput. Sci. **607**(1), 35–48 (2015)

11. Dhangwatnotai, P., Roughgarden, T., Yan, Q.: Revenue maximization with a single sample. Games Econ. Behav. **91**, 318–333 (2015)

12. El-Yaniv, R., Fiat, A., Karp, R.M., Turpin, G.: Competitive analysis of financial games. In: Proceedings of 33th Annual IEEE Symposium on Foundations of Computer Science (FOCS 1992), pp. 372–333 (1992)

13. El-Yaniv, R., Fiat, A., Karp, R.M., Turpin, G.: Optimal search and one-way trading online algorithms. Algorithmica **30**(1), 101–139 (2001)

14. Fujiwara, H., Iwama, K., Sekiguchi, Y.: Average-case competitive analyses for one-way trading. J. Comb. Optim. **21**(1), 83–107 (2011)

15. Koutsoupias, E., Pierrakos, G.: On the competitive ratio of online sampling auctions. In: Saberi, A. (ed.) WINE 2010. LNCS, vol. 6484, pp. 327–338. Springer, Heidelberg (2010). https://doi.org/10.1007/978-3-642-17572-5_27

16. Lorenz, J., Panagiotou, K., Steger, A.: Optimal algorithms for k-search with application in option pricing. Algorithmica **55**(2), 311–328 (2009)

17. Mahdian, M., McAfee, R.P., Pennock, D.: The secretary problem with a hazard rate condition. In: Papadimitriou, C., Zhang, S. (eds.) WINE 2008. LNCS, vol. 5385, pp. 708–715. Springer, Heidelberg (2008). https://doi.org/10.1007/978-3-540-92185-1_76

18. Myerson, R.B.: Optimal auction design. Math. Oper. Res. **6**(1), 58–73 (1981)

19. Zhang, Y., Chin, F.Y.L., Ting, H.-F.: Competitive algorithms for online pricing. In: Fu, B., Du, D.-Z. (eds.) COCOON 2011. LNCS, vol. 6842, pp. 391–401. Springer, Heidelberg (2011). https://doi.org/10.1007/978-3-642-22685-4_35

20. Zhang, Y., Chin, F.Y.L., Ting, H.-F.: Online pricing for bundles of multiple items. J. Glob. Optim. **58**(2), 377–387 (2014)

Generalized Bidirectional Limited Magnitude Error Correcting Code for MLC Flash Memories

Akram Hussain[✉], Xinchun Yu, and Yuan Luo

Department of Computer Science and Engineering, Shanghai Jiao Tong University,
Shanghai, China
hakram940@gmail.com, {moonyuyu,yuanluo}@sjtu.edu.cn

Abstract. The flash memories have gained considerable attention to replace hard-disk drives in modern storage applications because of the following excellent features such as the low cost, low power consumption, and high storage densities as compared to other non-volatile technologies. However, some error types are associated with flash memories such as charge leakage and inter-cell interference errors. It leads to the bidirectional limited magnitude channel model if both the error types are considered together. It has been observed that these error types are data value dependent for 2-bit MLC flash; they have different probabilities to become erroneous. In this paper, we consider the bidirectional limited magnitude errors by considering the data value dependencies of these error sources. A code construction to correct bidirectional limited magnitude errors is provided as well. The proposed code construction is the generalized case of asymmetric, symmetric, and bidirectional limited magnitude error correcting codes.

Keywords: Bidirectional error correcting code · Data value dependent errors · Limited magnitude errors · Flash memories

1 Introduction

Non-volatile memories (NVMs) hold the stored data even after the power to memories is cut off. They provide significant advantages such as faster data access, low power consumption, and improved physical resilience. These significant features of NVMs made them considerable as the primary replacement of the hard disk drives for modern storage applications, like mobile computing, enterprise storage, and data warehouses. Although there are many NVM based memories such as phase change memory (PCM), magneto-resistive random access memory (MRAM), and spin transfer torque random access memory (STT-RAM), unique characteristics of flash devices made them most popular among others [1,2].

(NAND) flash memory is comprised of floating gate transistors (cells) which are organized in blocks. The amount of charge, present at the gate of a transistor/cell, represents the data. The single level cell (SLC) flash memories store the only single bit in each cell.

© Springer International Publishing AG 2017
X. Gao et al. (Eds.): COCOA 2017, Part I, LNCS 10627, pp. 450–461, 2017.
https://doi.org/10.1007/978-3-319-71150-8_37

The two main issues are associated with flash devices, which are: data retention errors (charge leakage) and the inter-cell interference errors. In the first problem, the device leaks out the stored data over time, and in the second the cells gain some amount of charge while neighbor cells are being programmed.

The demand of large storage density has yielded the concept of multi-level cell (MLC) flash memories in which the device can store more than single information bit in each cell. However, the device capacity is increased by storing more than one bit in a cell which leads to low reliability (less margin between adjacent levels) and worsens the two main issues as discussed previously [3]. Cai *et al.* in [4] have studied the different types of errors in 2-bit MLC flash memory. One of their main observations is that the data retention and inter-cell interference errors are data value dependent. Moreover, they also observe that these errors are asymmetric errors of small magnitudes. This observation is exploited recently for placing more thresholds between different levels where more errors are expected, which reduces the decoding latency of an LDPC code [5].

In [6], they give code a construction and encoding/decoding algorithms for t asymmetric l–limited magnitude error correcting code to handle programming overshoot/inter-cell interference errors. The given code is constructed from the base code over the small size alphabet. The significant advantage of the code construction is that the encoding/decoding complexities are reduced. In the study of [7], they observe that a large number of cells experience asymmetric errors of smaller magnitudes, and they design a code to correct t_1 and t_2 number of errors of magnitude l_1 and l_2, respectively, where $(l_1 < l_2)$. Furthermore, Gabrys *et al.* study a more refined model for three-level-cell (TLC) [1,8]. In this model (Graded bit errors), most of the time only single bit becomes erroneous when data/symbol is stored as a triplet in TLC. They provide code constructions based on Tensor product to correct graded bit errors.

The data retention errors have been considered in [9,10] for all asymmetric l–limited magnitude errors detecting and correcting codes, respectively. They derive a lower bound on the number of check digits. Their encoding and decoding algorithms are very efficient. In general, errors do not happen to all the symbols in a codeword.

An intrinsic/neutral distribution appears when flash memory is used in high radiation environments such as satellite or probes. This intrinsic distribution is typically located between the erased state and first programmed state. In high radiation environment this distribution attracts all the distributions/levels towards itself, which causes asymmetric errors [11]. They derive an upper bound on the maximum code size, and provide code constructions to handle the radiation errors.

1.1 Bidirectional Limited Magnitude Channel Model

In general, charge leakage and inter-cell interference errors are considered individually which leads to asymmetric limited magnitude channel models to simplify the system. However, if both error types are considered together, then it is possible that the practical systems can perform better in terms of bit error rate.

The consideration of both error types together induces the bidirectional limited magnitude channel model; the magnitude of one direction may be larger than the other. The bidirectional (l_u, l_d)–limited magnitude channel model is shown in Fig. 1, where we set the upward magnitude $l_u = 2$ and downward magnitude $l_d = 1$. In [12], they design code to correct asymmetric and bidirectional errors by assuming that both sources of errors are equally significant. The designed code performs better sometimes over large alphabet size.

Inter-cell interference causes the threshold voltage shift of the neighbor cells when the charge is stored in an adjacent cell. The number of errors due to inter-cell interference can be minimized by adjusting the V_{read} (reading voltage) but this adjustment makes the other errors appear due to different error sources in the downward direction. These downward errors have small magnitude, however, they are significant. In [13,14], code constructions and encoding/decoding algorithms are given for bidirectional limited magnitude errors which also happen due to adjustment of V_{read}. The given code constructions are based on the construction in [6].

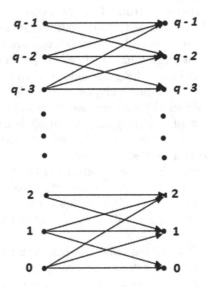

Fig. 1. q–ary bidirectional $(l_u = 2, l_d = 1)$–limited magnitude channel.

Problem Statement: As discussed in [4], the data retention and intercell interference errors are data value dependent errors. These errors have different probabilities for each symbol to be confused for upper symbol $(a+1)$ and lower $(a-1)$ symbols when a was the original symbol. For instance, if we store a codeword $x = (1, 0, 1, 3, 1, 2)$ which contains different components over $4 - ary$ (alphabet size is 4), then each component of this codeword has different probabilities to transit to lower and upper values (levels). Such as probabilities of data retention and intercell interference errors for level $(L_2 = 2)$ are $Pr(L_1/L_2) = 0.44$

and $Pr(L_3/L_2) = 0.004$, respectively, similarly probabilities for level $(L_1 = 1)$ are $Pr(L_0/L_1) = 0.02$ and $Pr(L_2/L_1) = 0.24$, respectively. This codeword contains many lower symbols which implies that many upward errors will happen to this codeword and few downward errors are possible. This situation implies that the given code constructions in [12–14] will penalize in term of code size if those constructions are used whenever both error sources are considered equally significant for the data value dependent errors (data retention and intercell interference). **In general, we consider the (t_+, t_-) bidirectional (l_u, l_d)–limited magnitude errors whenever both error types are equally significant. This consideration will lead to the higher code rate.**

The paper is organized as follows: some definitions are reviewed in Sect. 2, and some important propositions are proved too. In Sect. 3, the code is constructed, moreover, encoding, decoding, size of code, and some special cases of the code are discussed there as well. The constructed code is the generalization form of asymmetric, symmetric, and bidirectional limited magnitude error correcting codes. In last, we conclude the paper.

We use following notations in this paper:

t_+: the number of upward errors, for instance $(0 \rightarrow 1)$,
t_-: the number of downward errors, for instance $(1 \rightarrow 0)$,
l_u and l_d represent the error magnitude in upward and downward directions, respectively,
Q is the alphabet whose size is q, and defined as $Q = \{0, 1, ..., q - 1\}$, and
$N(x, y) = |\{i \mid x_i > y_i\}|$ which denotes the number of indices where $x > y$.

2   Preliminaries

The following Definition 1 is the description of a code in a more general way [15–17].

Definition 1. *We say a code Σ (t_+, t_-) asymmetric error correcting code if and only if Σ is capable of correcting (t_+) number of errors in upward direction and (t_-) number of errors in downward direction.*

If $t = (t_+ = t_-)$, then the code Σ is called t symmetric error correcting code.

The code, defined in Definition 1, was designed in [15, 16] over binary and non-binary alphabets, respectively. They develop key equations using elementary symmetric functions, which are used to decode the designed codes. For the binary field, they use isomorphism between sets and binary strings, and for non-binary they use isomorphism between non-binary strings and multi-sets.

Definition 2 [18]. *For two vectors $\mathbf{x} = (x_1, ..., x_n) \in Q^n$ and $\mathbf{y} = (y_1, ..., y_n) \in Q^n$, the asymmetric distance is defined as*

$$d_A(x, y) = max\{N(x, y), N(y, x)\}.$$

The asymmetric ball $\mathcal{B}(x)$, centered at x, is the set of all vectors obtained by at most t_+ and t_- errors, and is defined as

$$\mathcal{B}(x) = \{x' \in Q^n \mid N(x, x') \leq t_- \text{ and } N(x', x) \leq t_+\}.$$

Let $x' \in \mathcal{B}(x)$ and $y' \in \mathcal{B}(y)$ be the received vectors such that $N(x, x') \leq t_-$, $N(x', x) \leq t_+$, $N(y, y') \leq t_-$, and $N(y', y) \leq t_+$.

A similar theorem is proved in [15,16] but the following proposition is for asymmetric distance in Definition 2.

Proposition 1. *A code Σ corrects at most (t_+, t_-) asymmetric errors if and only if it has minimum distance $d_A(x, y) > t_+ + t_-$ for all distinct x, $y \in \Sigma$.*

Proof. We prove this proposition by showing that $\mathcal{B}(x) \cap \mathcal{B}(y) = \phi$ for all distinct codewords $x, y \in \Sigma$.

($\Leftarrow$) Without loss of generality (w.l.o.g), assume that $d_A(x, y) = N(x, y) > t_+ + t_-$ for any two codewords $x, y \in \Sigma$. We have $t_+ + t_- < N(x, y) \leq N(x, x') + N(x', y') + N(y', y) \leq t_- + N(x', y') + t_+$. So $1 \leq N(x', y')$, which implies that $x' \neq y'$, and we have $\mathcal{B}(x) \cap \mathcal{B}(y) = \phi$. Hence, the code can correct at most t_+ and t_- errors simultaneously.

($\Rightarrow$) Suppose that for some distinct $x, y \in \Sigma$, $d_A(x, y) \leq t_+ + t_-$. To prove in this direction, we consider the following two cases. (C1) We must assume that $d_A(x, y) = N(x, y) \leq t_+ + t_-$. In that case, the inequality is $N(x, y) \leq N(x, x') + N(x', y') + N(y', y) \leq t_- + N(x', y') + t_+$, so $N(x', y') = 0$. (C2) Let $d_A(x, y) = N(y, x) \leq t_+ + t_-$. Using inequality $N(y, x) \leq N(y, y') + N(y', x') + N(x', x) \leq t_- + N(y', x') + t_+$, so we have $N(y', x') = 0$.

Both of the cases imply that $x' = y'$, and we have $x' \in \mathcal{B}(x) \cap \mathcal{B}(y) \neq \phi$. Therefore, the code is not capable of correcting at most t_+ and t_- errors simultaneously. □

The flash devices experience limited magnitude errors. When both sources of errors (data retention and intercell interference errors) are considered together, then it will induce the bidirectional error which is defined as

Definition 3. *We say a vector of integers $\mathbf{e} = (e_1, ..., e_n)$, a (t_+, t_-) bidirectional (l_u, l_d)–limited magnitude error if*

1. $|\{i : e_i \neq 0\}| \leq t_+ + t_-$.
2. $-l_d \leq e_i \leq l_u$.

The distance to capture correction capability for the channel error in the Definition 3, is defined as follows:

Definition 4. *The distance between two codewords $\mathbf{x}$ and $\mathbf{y}$ over Q^n, is defined as*

$$\tilde{d}_{(l_u, l_d)}(\mathbf{x}, \mathbf{y}) = \begin{cases} n + 1, & \text{if } \exists\, i : \{|x_i - y_i| \geq l_u + l_d + 1\} \\ max\{N(x, y), N(y, x)\}, & \text{otherwise.} \end{cases}$$

The asymmetric ball $\mathcal{B}_{l_u,l_d}(x)$, centered at x, is the set of all vectors obtained by at most (t_+, t_-) bidirectional (l_u, l_d)–limited magnitude errors, and is defined as

$$\mathcal{B}_{l_u,l_d}(x) = \{x' \in Q^n \mid \forall i \; |x_i - x'_i| \leq l_u \text{ and } l_d,$$
$$N(x, x') \leq t_- \text{ and } N(x', x) \leq t_+\}.$$

Let $x' \in \mathcal{B}_{l_u,l_d}(x)$ and $y' \in \mathcal{B}_{l_u,l_d}(y)$ be the received vectors such that $\forall i$ $|x_i - x'_i| \leq l_u$ and l_d and $\forall i \; |y_i - y'_i| \leq l_u$ and l_d, and $N(x, x') \leq t_-$, $N(x', x) \leq t_+$, $N(y, y') \leq t_-$, and $N(y', y) \leq t_+$.

Proposition 2. *A code C is capable of correcting (t_+, t_-) bidirectional (l_u, l_d)–limited magnitude errors if and only if the minimum distance is $\tilde{d}_{(l_u,l_d)}(x, y) > t_+ + t_-$ for all distinct codewords $x, y \in C$.*

Proof. A code cannot correct (t_+, t_-) bidirectional (l_u, l_d)–limited magnitude errors if and only if there exist (t_+, t_-) bidirectional (l_u, l_d)–limited magnitude error words e and f such that $x' = y'$ for some distinct codewords $x, y \in C$, where $x' = x + e$ and $y' = y + f$. We prove this proposition by showing that $\mathcal{B}_{l_u,l_d}(x) \bigcap \mathcal{B}_{l_u,l_d}(y) = \phi$.

($\Leftarrow$) We assume that $\tilde{d}_{(l_u,l_d)}(x, y) > t_+ + t_-$ for any pair of codewords x, y. Then, we can show that $x' \neq y'$ by considering one of the following cases:

1. $N(x, y) > t_+ + t_-$ or $N(y, x) > t_+ + t_-$.
2. $|x_i - y_i| \geq l_u + l_d + 1$ for any $i \in \{1, ..., n\}$.

For case 1, w.l.o.g we assume that $N(x, y) > t_+ + t_-$ for any pair of codewords $x, y \in C$. We have $t_+ + t_- < N(x, y) \leq N(x, x') + N(x', y') + N(y', y) \leq t_- + N(x', y') + t_+$. So $1 \leq N(x', y')$, which implies that $x' \neq y'$, and we have $\mathcal{B}_{l_u,l_d}(x) \bigcap \mathcal{B}_{l_u,l_d}(y) = \phi$.

In case 2, it implies that either e_i or f_i has magnitude larger than l_u or l_d for at least one i although both cannot be possible as defined in the Definition 3. To clarify, we let the magnitude of errors, happen to x and y, be $-l_d = e$ and $f = l_u$, respectively, so it is not possible that $|f_i - e_i| \geq l_u + l_d + 1$ for any i.

Thus, both the cases imply that the code can correct (t_+, t_-) bidirectional (l_u, l_d)–limited magnitude errors successfully.

($\Rightarrow$) Suppose that for some distinct codewords $x, y \in C$, the distance is $\tilde{d}_{(l_u,l_d)}(x, y) \leq t_+ + t_-$. Then, we have $N(x, y) \leq t_+ + t_-$ and $N(y, x) \leq t_+ + t_-$, and for all $i \; |x_i - y_i| \leq l_u$ and l_d. Similarly we did before, let $x' \in \mathcal{B}_{l_u,l_d}(x)$ and $y' \in \mathcal{B}_{l_u,l_d}(y)$ be the received vectors. To prove in this direction, we consider the following two cases. (C1) We must assume that $\tilde{d}_{(l_u,l_d)}(x, y) = N(x, y) \leq t_+ + t_-$. In that case, the inequality is $N(x, y) \leq N(x, x') + N(x', y') + N(y', y) \leq t_- + N(x', y') + t_+$, so $N(x', y') = 0$. (C2) Let $\tilde{d}_{(l_u,l_d)}(x, y) = N(y, x) \leq t_+ + t_-$ such that $N(x', x) \leq t_+$ and $N(y, y') \leq t_-$. Using inequality $N(y, x) \leq N(y, y') + N(y', x') + N(x', x) \leq t_- + N(y', x) + t_+$, so we have $N(y', x') = 0$.

Both of the cases imply that $x' = y'$, and we have $x' \in \mathcal{B}_l(x) \bigcap \mathcal{B}_l(y) \neq \phi$. Therefore, the code is not capable of correcting (t_+, t_-) bidirectional (l_u, l_d)–limited magnitude errors. $\qquad\square$

3 Code Construction for (t_+, t_-) Bidirectional (l_u, l_d)–Limited Magnitude Error Correcting Codes

The code construction is based on the construction given in [6].

Construction 1. Let Σ be a (t_+, t_-) asymmetric error correcting code over alphabet size $q' = l_u + l_d + 1$. Then, the code C over alphabet size q $(q > q')$ is defined as

$$C = \{\mathbf{c} = (c_1, ..., c_n) \mid c \bmod q' \in \Sigma\}. \tag{1}$$

The error correction capability of the code in (1) is summarized in the following theorem. The proof of the theorem has some similarity with the Theorem 5 in [6].

Theorem 1. *The code C corrects (t_+, t_-) bidirectional (l_u, l_d)–limited magnitude errors if and only if Σ corrects (t_+, t_-) asymmetric errors.*

Proof. For all distinct pair of codewords $x, y \in C$, the distance between them is at least $\tilde{d}_{(l_u, l_d)}(x, y) \geq t_+ + t_- + 1$. We have shown in the Proposition 2 that the code C is capable of correcting all (t_+, t_-) bidirectional (l_u, l_d)–limited magnitude errors.

There are two cases to be considered:

1. If $x \bmod q' = y \bmod q'$ for all distinct pair of codewords $x, y \in C$, then $|x_i - y_i| \geq l_u + l_d + 1$ for at least one position $i \in \{1, ..., n\}$. This makes the $\tilde{d}_{(l_u, l_d)}(x, y)$ to be $n + 1$.
2. For $x \bmod q' \neq y \bmod q'$, we know that the code Σ has minimum distance $d_A \geq t_+ + t_- + 1$, which implies that bidirectional distance should be $\tilde{d}_{(l_u, l_d)}(x, y) = max\{N(x, y), N(y, x)\} \geq t_+ + t_- + 1$. The Proposition 1 shows that Σ corrects all the (t_+, t_-) asymmetric errors, and so code C corrects (t_+, t_-) bidirectional (l_u, l_d)–limited magnitude errors.

The converse part of this theorem is easy to follow, and can be referred in the Proposition 2. □

3.1 Encoding

There may be many procedures to encode the $q' - ary$ codeword from base code, such as (Σ), to the $q - ary$ codeword in C. In this paper, we just describe two encoding methods from [6,13], respectively.

1. For simplicity, we assume that $q = A.q'$ where both A and q' are integers. We have $(\chi_i, ..., \chi_n)$ as the codeword of the base code Σ over the alphabet size q', and we set the n symbols $(b_1, ..., b_n)$ over alphabet size A as pure information symbols. Then we get each symbol x_i of the codeword of C over the alphabet size q by $x_i = b_i.q' + \chi_i$.

2. In the second encoding method, assume $x = (x_1, ..., x_k)$ to be the $q - ary$ message. The $q' - ary$ remainder is obtained by modular operation on the message vector x $(x \bmod q')$. The check symbols are calculated by the base code encoder using $(x \bmod q')$ as the input of the encoder. These check symbols are then converted to $q - ary$ symbols, and finally, we have $c = [x\ p]$ as the $q - ary$ codeword, where $p = (p_1, ..., p_r)$ is the $q - ary$ check symbols vector.

Although the first encoding method is very simple, it is non-systematic; message symbols are mixed with parity symbols in a codeword. In the second method, encoding is systematic; message symbols and parity symbols are separated in a codeword but this procedure has message correction and parity code writing problems [13]. To avoid these problems, we need to use base code over large alphabet size p $(p > q > q')$.

3.2 Decoding

One of the main advantages of the Construction 1 is the reduction in encoding and decoding complexities which are performed by the base code and after that codewords from the base code are mapped to the codewords in the code C [6]. In this paper, we only describe the decoding method for non-systematic case. The decoding steps are similar for the systematic case.

Let $y = (y_1, ..., y_n)$ be the received vector and the transmitted codeword was $\mathbf{x} = (x_1, ..., x_n) \in C$. We denote the corresponding Σ codeword by $\chi = \mathbf{x} \bmod q'$. To decode the received vector by the base code, first modular operation by q' is performed on the received vector $(\psi = y \bmod q')$. This ψ is decoded using the base code Σ where the base decoder is invoked just to find out the estimated error word $\hat{\epsilon}$. Finally, the estimated codeword of C is $\hat{\mathbf{x}} = \mathbf{y} - \hat{\epsilon}$.

We consider upward and downward errors separately to avoid confusion in modular operation with negative integer [13].

Let $\epsilon = (\epsilon_1, ..., \epsilon_n)$ be the (t_+, t_-) bidirectional (l_u, l_d)–limited magnitude error vector, so

$$\psi = \mathbf{y} \bmod q' = (\mathbf{x} + \epsilon) \bmod q'$$
$$= (\mathbf{x} \bmod q' + \epsilon \bmod q') \bmod q'$$
$$= (\chi + \epsilon) \bmod q'.$$

We need to consider positive and negative errors separately to bound the estimated error word in the proper range of $-l_d \leq \hat{\epsilon} \leq l_u$, as:

1. Downward error: $\epsilon_i = \epsilon_- (-l_d \leq \epsilon_- \leq -1)$, then
$\psi_i = (\chi_i + \epsilon_i) \bmod q' = (\chi_i + \epsilon_-) \bmod q'$

$$\psi_i = \begin{cases} \chi_i + \epsilon_- + q', & if\ 0 < \epsilon_- + q' < q' \\ \chi_i + \epsilon_-, & if\ q' \leq \epsilon_- + q' < 2q'. \end{cases}$$

2. Upward error: $\epsilon_i = \epsilon_+ (0 \leq \epsilon_+ \leq l_u)$, then
$$\psi_i = (\chi_i + \epsilon_i) \; mod \; q' = (\chi_i + \epsilon_+) \; mod \; q'$$

$$\psi_i = \begin{cases} \chi_i + \epsilon_+, & if \; 0 < \epsilon_+ + q' < q' \\ \chi_i + \epsilon_+ - q', & if \; q' \leq \epsilon_+ + q' < 2q'. \end{cases}$$

Therefore, by considering upward and downward errors separately we are able to distinguish and recover the estimated error vector from the distinct error ranges as shown in the Fig. 2 [13].

Fig. 2. Estimated error vector adjustment for ($l_u = l_d$).

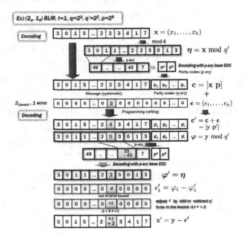

Fig. 3. Encoding/decoding for $t = 1$, $l_u = 2$ and $l_d = 1$.

3.3   Size of the Code

The upper and lower bounds on the size of the code in (1) can be easily found. As defined in the encoding method 1, we get each symbol x_i in a valid codeword of the code C by replacing every χ_i by any element of the set $\Theta = \{y \in Q : y = \chi_i \; mod \; q'\}$. The size of this set is $\lceil q/q' \rceil$ if $\chi_i < q \; mod \; q'$ and $\lfloor q/q' \rfloor$ otherwise. Then, the size of the code in (1) is bounded as

$$\lfloor \frac{q}{l_u + l_d + 1} \rfloor^n . |\Sigma| \leq |C| \leq \lceil \frac{q}{l_u + l_d + 1} \rceil^n . |\Sigma|. \tag{2}$$

We used Σ as the base code in Eq. (2) for the (t_+, t_-) bidirectional (l_u, l_d)–limited magnitude error correcting code C. In general, Σ corrects (t_+, t_-) asymmetric errors where number of upward errors (t_+) and downward errors (t_-) are different. The detailed code design for Σ can be referred in [15] over the binary field and in [16] over F_q.

3.4 Decoding Error Probability

Let p and q be the crossover probabilities of the upward and downward errors, respectively as shown in Fig. 1, assuming that the magnitudes of the errors are bounded. We have t_+ and t_- numbers of errors in upward and downward directions, respectively. The correct (block) decoding probability is given by

$$P_t(x) = \sum_{i=0}^{t_+} \binom{n}{i} p^i \cdot \sum_{j=0}^{t_-} \binom{n-i}{j} q^j (1-p-q)^{n-i-j}. \tag{3}$$

Then the decoding (block) error probability is

$$P_B(x) \le 1 - P_t(x). \tag{4}$$

We have proved the error correction capability of the code construction (1), and Σ is used as base code there. It is very difficult to find some encoding and decoding procedures generally for Σ and the size of the code as well. Therefore, some special cases of the construction (1) are provided in the next subsection and some examples are given there too.

3.5 Special Cases of the Code Construction 1

In this subsection, we consider the different cases of the main construction in (1) as follows:

1. For $t = t_+ = t_-$: If numbers of errors in both directions are equal, then the construction in (1) becomes the code construction as given in [13]. In addition, the base code Σ can correct symmetric errors using the Hamming distance instead of defined asymmetric distance. Therefore, the encoding and decoding complexities reduce, and we can easily find out the size of resultant code using Eq. (2); t bidirectional (l_u, l_d)–limited magnitude error correcting code.

Example 1. Let C be the code over alphabet size $(q = 8)$ as defined in construction (1). We consider these parameters $l_u = 2, l_d = 1$ and $t = 1$, and we use Hamming code $[n = 5, k = 3]$ as the base code over alphabet size $q' = 4$ $(q' = l_u + l_d + 1)$.

$$H = \begin{bmatrix} 0 & 1 & 1 & 1 & 1 \\ 1 & 1 & 0 & x & x+1 \end{bmatrix}$$

Let $(1, x, x+1, 0, 1)$ be the Hamming codeword. Using the encoding method 1 in *Encoding*, one of the codewords of C is $\mathbf{x} = (5, 6, 7, 4, 5)$. Let $\epsilon = (0, -1, 0, 0, 0)$ be the error word. Then the received vector is $\mathbf{y} = (5, 5, 7, 4, 5)$. To decode this received vector, we invoke Hamming decoder after q'–modular and find out that the error word is $\hat{\epsilon} = (0, 3, 0, 0, 0)$. This error word is out of the bound of (l_u, l_d)–limited magnitudes, so as discussed in the *Decoding*, we need to subtract $(q' = 4)$ from that component of the error word $\hat{\epsilon}_2 - q' = (3 - 4) = -1$ which makes the error word $\hat{\epsilon} = (0, -1, 0, 0, 0)$. Hence, the estimated codeword is $\hat{\mathbf{x}} = \mathbf{y} - \hat{\epsilon} = (5, 5, 7, 4, 5) - (0, -1, 0, 0, 0) = (5, 6, 7, 4, 5)$. The code rate of this code is 0.7333, and the decoding error probability is upper bounded by 0.0226 whenever crossover probability is $p = 0.05$.

The supercode is defined on the integer ring. For instance, the Hamming code is used as a subcode in Example 1, and it is mapped to the supercode, which is defined over integer ring, using *Encoding* method 1. Furthermore, the code rate in that example is same as the code rate in [13] if *Encoding* method 1 is used. The bound (2) is achieved because $q' \mid q$ and due to Hamming code as well.

Example 2. Figure 3 [13] shows another example of encoding/decoding of t bidirectional (l_u, l_d)–limited magnitude error correcting code. In this example, message is encoded with encoding method 2 as described in the *Encoding* section. Furthermore, the RS code is used as the base code over finite field F_{2^6}.

2. For $(t = t_+ = t_-)$ and $(l = l_u = l_d)$: If the numbers and magnitudes of errors in both directions are equal, then the construction in (1) can correct t symmetric l–limited magnitude errors.

3. For $t_- = 0$, $l_d = 0$ and $t = t_+$, $l = l_u$: The code construction in (1) will converge to the code construction 1 in [6], and the base code Σ uses the Hamming distance instead. Then, the code will correct t asymmetric l–limited magnitude errors.

4   Conclusion

In this paper, we have studied the bidirectional limited magnitude channel model by considering data retention and inter-cell interference errors together with their data value dependencies. We proposed the code construction for bidirectional limited magnitude channel model. The proposed construction is the generalized case of the bidirectional, asymmetric, and symmetric limited magnitude error correcting codes. Some special cases of the construction were presented as well. In future, we will try to find some encoding and decoding algorithms for the base code on which the proposed construction 1 is based.

Acknowledgment. This work was supported by China Program of International S&T Cooperation 2016YFE0100300 and National Natural Science Foundation of China under Grant 61571293.

References

1. Dolecek, L., Sala, F.: Channel coding methods for non-volatile memories. Found. Trends Commun. Inf. Theory **13**(1), 1–28 (2016). Boston-Delft
2. Sala, F., Immink Schouhamer, K.A., Delecek, L.: Error control schemes for modern flash memories. IEEE Consum. Electron. Magz. **4**(1), 66–73 (2015)
3. Huang, X., Kavcic, A., Ma, X., Dong, G., Zhang, T.: Multilevel flash memories: channel modeling, capacities and optimal coding rates. Int. J. Adv. Syst. Measur. **6**(3 and 4), 364–373 (2013)
4. Cai, Y., Haratsch, E. F., Mutlu, O., Mai, K.: Error patterns in MLC NAND flash memory: measurement, characterization, and analysis. In: IEEE Design, Automation and Test in Europe Conference and Exhibition (DATE), pp. 521–526. IEEE, Dresden (2012)
5. Li, Q., et al.: Improving LDPC performance via asymmetric sensing level placement on flash memory. In: 22nd Asia and South Pacific Design Automation (ASP-DAC), pp. 560–565. IEEE, Tokyo (2017)
6. Cassuto, Y., Schwartz, M., Bohossian, V., Bruck, J.: Codes for asymmetric limited magnitude errors with application to multi-level flash memories. IEEE Trans. Inf. Theory **56**(4), 1582–1595 (2010)
7. Yaakobi, E., Siegal, P.H., Vardy, A., Wolf, J.K.: On codes that correct asymmetric errors with graded magnitude distribution. In: Proceeding of IEEE International Symposium on Information Theory, pp. 1056–1060. IEEE, Saint-Petersburg (2011)
8. Gabrys, R., Yaakobi, E., Dolecek, L.: Graded bit-error-correcting codes with applications to flash memory. IEEE Trans. Inf. Theory **59**(4), 2315–2327 (2013)
9. Elarief, N., Bose, B., Elmougy, S.: Limited magnitude error detecting codes over Z_q. IEEE Trans. Comput. **62**(5), 984–989 (2013)
10. Elarief, N., Bose, B.: Optimal, systematic, q-ary codes correcting all asymmetric and symmetric error of limited magnitude. IEEE Trans. Inf. Theory **56**(3), 979–983 (2010)
11. Sala, F., et al.: Asymmetric error correcting codes for flash memories in high radiation environments. In: Proceeding of IEEE International Symposium on Information Theory, pp. 2096–2100. IEEE, Hong Kong (2015)
12. Kotaki, S., Kitakami, M.: Codes correcting asymmetric/unidirectional errors along with bidirectional errors of small magnitude. In: IEEE 20th Pacific Rim International Symposium on Dependable Computing, pp. 159–160. IEEE, Singapore (2014)
13. Jeon, M., Lee, J.: On codes correcting bidirectional limited-magnitude errors for Flash memories. In: International Symposium on Information Theory and its Applications, pp. 96–100. IEEE, Hawaii (2012)
14. Jeon, M., Lee, J.: Bidirectional limited-magnitude error correction codes for flash memories. IEICE Trans. Fundam. **E96-A**(7), 1602–1608 (2013)
15. Tallini, L.G., Bose, B.: On a new class of error control codes and symmetric functions. In: Proceeding of IEEE International Symposium on Information Theory, pp. 980–984. IEEE, Ontario (2008)
16. Tallini, L.G., Bose, B.: On L_1-distance error control codes. In: Proceeding of IEEE International Symposium on Information Theory, pp. 1061–1065. IEEE, St. Petersburg (2011)
17. Zhou, H., Jiang, A., Bruck, J.: Nonuniform codes for correcting asymmetric errors in data storage. IEEE Trans. Inf. Theory **59**(5), 2988–3002 (2013)
18. Kløve, T.: Error correcting codes for the asymmetric channel. University of Bergen, Department of Informatics, Bergen, Norway (1995)

Optimal Topology Design of High Altitude Platform Based Maritime Broadband Communication Networks

Jianli Duan[1,2(✉)], Tiange Zhao[1], and Bin Lin[1]

[1] Dalian Maritime University, Dalian 116026, China
15140412232@163.com, binlin@dlmu.edu.cn
[2] Qingdao University of Technology, Qingdao 266033, China
duanjianli@qut.edu.cn

Abstract. To satisfy the increasing demand for various types of coastal wireless communication, the pursuit of fully functional and low cost means of maritime broadband communication is imminent. High Altitude Platform (HAP) based maritime communication is a promising solution to improve the quality of coverage on the sea. In this paper, we present Integrated HAP-Sea-Land Network (IHSL) architecture for maritime broadband communication. And then we focus on the problem of Optimal Topology Design (OTD) which is an important issue for the IHSL network deployment in practice. We formulate the OTD problem as a generic integer linear programming (ILP) model with the objective of minimizing the total deployment cost subject to coverage, reliability, and topology constraints of the network. The linear programming solver Gurobi is applied to solve the ILP model. A series of case studies are conducted to validate the optimization framework and demonstrated the solvability and scalability of the ILP model. The simulation results show the significant performance benefits of IHSL in terms of cost and reliability under 1-coverage and 2-coverage. The proposed optimization framework is expected to provide a guideline for the IHSL network deployment in practice.

Keywords: High Altitude Platforms (HAP) · Maritime broadband communication · Topology optimization design (OTD) · Integer linear programming (ILP)

1 Introduction

With the rapid development of shipping industry and ocean trade, the number of vessels shoot up, thus the demands for maritime broadband communications increase explosively [1, 2]. At present, the most common way of maritime communication is the global satellite communication, which has obvious shortcomings such as high cost, low data rate, large transmission delay, and specific shipboard equipment terminals are required to be installed on board [3]. In contrast, High Altitude Platform (HAP) based maritime communication is a promising solution to improve the quality of communications on the sea [4] due to lower cost, wider coverage, higher loading and reusable capabilities, easy deployment and maintenance [5]. The HAP and the 4G communication technology can

© Springer International Publishing AG 2017
X. Gao et al. (Eds.): COCOA 2017, Part I, LNCS 10627, pp. 462–470, 2017.
https://doi.org/10.1007/978-3-319-71150-8_38

be integrated and applied to the maritime communication such that the cellular network can extend to the ocean, and constitute the sea-land integrated communication system. The maritime broadband communication technology should embark on a new phase.

In this paper, by exploiting the HAP-centered near-space communication technology, we propose an Integrated HAP-Sea-Land (IHSL) Network for maritime broadband communication. IHSL can extend the terrestrial broadband network to the ocean through cellular networks, HAP and the ship-borne relay station (RS). As shown in the Fig. 1, IHSL comprises four parts, i.e., maritime user terminals, shipborne relays, HAPs which carry base stations (BSs) and ground BSs (coast control center). Then we investigate the Optimal Topology Design (OTD) problem which is important for the IHSL net-work architecture in practical deployment.

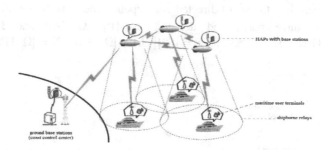

Fig. 1. Integrated HAP-Sea-Land (IHSL) network

Driven by the needs of coverage extension to ocean, a tree based structure is applied in the IHSL network by taking advantages of essential scalability and robustness of tree topology. To capture the complex relations among the OTD problem and the tree topology structure, we formulate the OTD problem mathematically. Given the information such as candidate points, the cost of deployment, constricts of network entities and network coverage requirements within the Regions of Interest (RoI) on the sea, the OTD problem can optimally and simultaneously: (i) minimize the total cost; (ii) identify the locations of HAPs; (iii) take reliability of network constraints into consideration; (iv) satisfy the coverage requirements so as to meet the demands of broadband users at sea. The ILP model is solved by Gurobi, which is the state-of-the-art mathematical programming solver [6]. Note that, considering the importance of reliability for maritime communications, especially for rescue and emergency communication scenarios, this paper studies k-coverage (k = 1, 2) as different minimum coverage requirements in the OTD formulations.

2   Network Model and Problem Formulation

2.1   Network Model

As shown in Fig. 2, in the network model of HICS it mainly consists the following parts: HAP0, HAP1, HAP2, HAP3 … and test points on the sea (TP). The zero-order

HAP is deployed near hot ports and is primarily responsible for connecting ground BSs and other HAPs, while the first-order HAP, the second-order HAP, and so on are responsible for connecting upper-order HAP/next-order HAP and TPs. The first-order HAP, the second-order HAP, and so on use a tree structure to be connected with the zero-order HAP which is root node, i.e., the first-order HAP is directly connected with the zero-order HAP, the second-order HAP is connected with the zero-order HAP by the first-order HAP and so on. TP is used to test whether the coverage rate, connectivity and QoS indicators of the network meet the requirement. In Fig. 2, TP4 and TP5 are covered by the first-order HAPi, TP6 and TP7 are covered by the first-order HAPj, TP1, TP2 and TP3 are covered by the second-order HAPk, and TP8, TP9 and TP10 are covered by the second-order HAPm. According to the character of tree based integrated network topology as shown in Fig. 2, we model the OTD as a directed graphs: $\vec{G} = (\Omega, \vec{E})$, where $\vec{E}$ is the set of directed edges/paths, and Ω is the set of nodes. Ω is partitioned into three parts, and denoted as $\{P\}$, Ω_HAP and Ω_TP, i.e., $\Omega = \{P\} \cup \Omega\_HAP \cup \Omega\_TP$, and let $S = |\{P\}|$, $M = |\Omega\_HAP|$, $N = |\Omega\_TP|$.

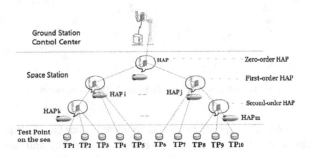

Fig. 2. Tree based IHSL network mode

2.2   Problem Statement

The OTD problem for the IHSL network architecture can be stated as follows.

Given:
As shown in Fig. 2, the zero-order HAP is taken as the root node, and HAPs are taken as the middle node/leaf nodes to cover the ship on the sea. The following are given: the position of candidate points (CPs) for the zero-order HAP in the region; the locations of CPs for HAP in the region; the location of TP on the sea; the cost of HAP and its carrying load; related parameters of HAP, such as coverage radius of HAP, capacity, the maximum communication distance between HAP and the minimum anti-collision distance, etc.

Variables:
Whether the location of nodes and links between nodes are usable can be represented by 0–1 variables (1: the node is selected/link between nodes exists; 0: otherwise). Specifically, we define:

(1) HAP and the location incidence vector $A = (a_m)_{1 \times M}$, $m \in \Omega_{HAP}$, if CP_m is selected to place a HAP, a_m is set to 1, otherwise a_m is 0;

(2) The location incidence vector of the zero-order HAP $B = (b_s)_{1 \times S}$, $S \in \{P\}$, if the CP is selected to place a zero-order HAP, b_s is set to 1, otherwise b_s is 0;

(3) The allocation matrix between HAP and TP $Z = (z_{mn})_{M \times N}$, $m \in \Omega_{HAP}$, $n \in \Omega_{TP}$, if HAP at CP_m exists and covers TP_n, z_{mn} is set to 1, otherwise z_{mn} is 0;

(4) The coverage matrix of TP $Q = (q_n)_{1 \times N}$, $n \in \Omega_{TP}$, if TP_n is covered, q_n is set to 1, otherwise q_n is 0;

(5) The Connection matrix between nodes $E = (e_{ij})_{(S+M) \times (M+N)}$, $i \in \{P\} \cup \Omega_{HAP}$, $j \in \Omega_{HAP} \cup \Omega_{TP}$, if there is a connection between node i to node j, e_{ij} is set to 1, otherwise e_{ij} is 0;

(6) To control the flow direction and the number of hops effectively, the inflow control matrix is defined as $F = (f_{ij}^k)_{(S+M) \times (M) \times (M)}$, $i \in \{P\} \cup \Omega_{HAP}$, $j \in \Omega_{HAP}$, $k \in \Omega_{HAP}$, and if the flow from zero-order HAP to HAP_k passes through e_{ij}, f_{ij}^k, is set to 1, otherwise f_{ij}^k is 0.

Constraints:

(1) Determine the parent-child relationship between nodes to ensure that the network satisfies the tree topology;

(2) Each candidate point for HAP can only place one HAP;

(3) Distance restriction between HAP;

(4) I/O links number limitation of Network element devices;

(5) The k-coverage requirement: all TPs in the model must meet the minimum number of coverage requirements in the network, that is, each TP is covered by at least k HAP carried BSs, k = 1, 2...;

(6) HAP capacity limitation.

Objective:

The optimization objective is to minimize the total HAP deployment cost for the IHSL network over RoI on the sea.

The important symbols for the problem formulation are defined as following:

(1) Parameters of the OTD problem: $\{P\}$-The set of Zero-order HAP nodes; N-The number of TPs in network; S-The number of Zero-order HAP nodes in network; M-The number of candidate HAP nodes in network; Ω_{HAP}:-The set of candidate HAP nodes $\Omega_{HAP} = \{CP_m | m = 0, 1, \cdots, M - 1\}$; k-The minimum coverage requirement of TPs; R-The maximum coverage radius of HAPs to sea surface; Ω_{TP}-The set of TPs $\Omega_{TAP} = \{TP_n | n = 0, 1, \cdots, N - 1\}$; $L_{max}^{HAP-HAP}$-The maximum communication distance between HAPs; ρ_0-The minimum coverage rate of target region $L_{min}^{HAP-HAP}$-The minimum anti-collision distance between HAPs; C^{HAP}-The cost of HAP; C^L-The loading cost of airship; d_{ij}-The distance between node i and node j.

(2) Variables of the OTD problem: A-HAP and the location incidence vector $A = (a_m)_{1 \times M}$; B-The location incidence vector of the zero-order HAP $B = (b_s)_{1 \times S}$; Z-The relation matrix between HAP and TP $Z = (z_{mn})_{M \times N}$; Q:-The coverage matrix of TP $Q = (q_n)_{1 \times N}$; E-The Connection matrix between nodes $E = (e_{ij})_{(S+M) \times (M+N)}$; F-The inflow control matrix $F = (f_{ij}^k)_{(S+M) \times (M) \times (M)}$.

2.3 Problem Formulation

Objective:

$$\text{minimize } C = (C^{HAP} + C^L) \sum\nolimits_{s \in S} b_s + (C^{HAP} + C^L) \sum\nolimits_{m \in M} a_m \qquad (1)$$

Equation (1) is to minimize the total deployment cost of OTD.

Subject to:

$$e_{ij} + e_{ji} \leq 1, \forall i, j \in \{p\} \cup \Omega_{HAP}, i \neq j \qquad (2)$$

$$\sum\nolimits_{M \in \Omega_{HAP}} z_{mn} \geq k, \forall n \in \Omega_{TP} \qquad (3)$$

$$z_{mn} \leq a_m, \forall m \in \Omega_{HAP}, \forall n \in \Omega_{TP} \qquad (4)$$

Equation (2) ensures unidirectional connection between network element devices to avoid forming loops; Eq. (3) guarantees each TP meets the k-coverage requirements of the network; Eq. (4) defines the coverage relationship between the HAP node and the TP point.

$$e_{ij} d_{ij} \leq L_{max}^{HAP-HAP}, \forall i \in \{p\} \cup \Omega_{HAP}, \forall j \in \Omega_{HAP} \qquad (5)$$

$$e_{ij} d_{ij} \geq L_{min}^{HAP-HAP}, \forall i \in \{p\} \cup \Omega_{HAP}, \forall j \in \Omega_{HAP} \qquad (6)$$

Equations (5, 6) constrain the distance between nodes. If the distance between e_{ij} and d_{ij} meets the distance requirement, the connection $e_{ij} d_{ij}$ is feasible.

$$\sum\nolimits_{i \in \{P\}} b_i \geq 1 \qquad (7)$$

$$\sum\nolimits_{j \in \Omega_{HAP}} e_{ij} \geq b_i, \forall i \in \{P\} \qquad (8)$$

Equations (7, 8) constrain zero-order HAP nodes. Constrain (7) specifies that there is at least one zero-order HAP node as the output source point of the network; Constrain (8) indicates that if there is a zero-order HAP node in the network, there is at least one output from the zero-order HAP node to the rest of the HAP nodes.

$$\sum\nolimits_{i \in \{p\} \cup \Omega_{HAP}} e_{ij} = a_j, \forall j \in \Omega_{HAP}, i \neq j \tag{9}$$

$$\sum\nolimits_{j \in \Omega_{HAP} \cup \Omega_{TP}} e_{ij} \geq a_i, \forall i \in \Omega_{HAP}, i \neq j \tag{10}$$

Equations (9, 10) restrict the input and output of the HAP node. Equation (9) indicates that if the HAP node is selected, it has only one input from the zero-order HAP node or the rest of the HAP node. Equation (10) constrains that the selected HAP node have at least one output.

$$f_{ij}^k \leq a_k, \forall i \in \{p\} \cup \Omega_{HAP}, \forall j \in \Omega_{HAP}, \forall k \in \Omega_{HAP}, i \neq k \tag{11}$$

$$f_{ij}^k \leq e_{ij}, \forall i \in \{p\} \cup \Omega_{HAP}, \forall j \in \Omega_{HAP}, \forall k \in \Omega_{HAP}, i \neq k, i \neq j \tag{12}$$

Equations (11, 12) constrains inflows of a HAP node. Equation (11) associates the HAP node with the flow, and only when the HAP node is selected can there be a inflow from other nodes; Eq. (12) associates edges with flows, and an inflow possibly passes the edge only when there is a direct connection between nodes (i.e., $e_{ij} = 1$).

$$\sum\nolimits_{j \in \Omega_{HAP}} f_{ij}^k = a_k, \forall k \in \Omega_{HAP}, \forall i \in \{P\} \tag{13}$$

$$\sum\nolimits_{j \in \Omega_{HAP}} \sum\nolimits_{k \in \Omega_{HAP}} f_{ij}^k = \sum\nolimits_{k \in \Omega_{HAP}} a_k, \forall k \in \Omega_{HAP}, \forall i \in \{P\} \tag{14}$$

$$\sum\nolimits_{i \in \{P\} \cup \Omega_{HAP}} f_{ij}^k = a_k, \forall k \in \Omega_{HAP}, i \neq k \tag{15}$$

Equation (13) shows that a unique flow from the zero-order HAP node to the node exists only when the HAP node is selected. Equation (14) indicates the node will be linked up to the tree network topology only when the HAP node is selected. If the HAP node is selected, there is at least one inflow ending with this node (15).

$$\sum\nolimits_{i \in \{P\} \cup \Omega_{HAP}} \sum\nolimits_{j \in \Omega_{HAP}} f_{ij}^k \leq Hop\_max, \forall k \in \Omega_{HAP}, i \neq j, i \neq k \tag{16}$$

$$\sum\nolimits_{i \in \{P\} \cup \Omega_{HAP}} f_{ij}^k = \sum\nolimits_{m \in \Omega_{HAP}} f_{jm}^k, \forall j, k \in \Omega_{HAP}, i \neq j \neq m, i \neq j \neq k \tag{17}$$

Equation (16) constrains hop limit in the network. Equation (17) shows that if HAP_j exists in the inflow path of zero-order HAP to HAP_k, it must have an input path starting with zero-order HAP and an output path ending with HAP_k.

$$e_{ij} d_{ij} \leq R, \forall i \in \Omega_{HAP}, \forall j \in \Omega_{TP} \tag{18}$$

$$\frac{1}{N} \sum\nolimits_{n \in \Omega_{TP}} q_n \geq \rho_0 \times 100\% \tag{19}$$

Equation (18) defines the coverage relationship between HAP and TP; Eq. (19) ensures that the network meets coverage requirements.

$$a_m \in \{0, 1\}, \forall m \in \Omega_{HAP} \tag{20}$$

$$b_s \in \{0, 1\}, \forall s \in \{P\} \tag{21}$$

$$z_{mn} \in \{0, 1\}, \forall m \in \Omega_{HAP}, \forall n \in \Omega_{TP} \tag{22}$$

$$q_n \in \{0, 1\}, \forall n \in \Omega_{TP} \tag{23}$$

$$e_{ij} \in \{0, 1\}, \forall i \in \{P\} \cup \Omega_{HAP}, j \in \Omega_{HAP} \cup \Omega_{TP} \tag{24}$$

$$f_{ij}^k \in \{0, 1\}, \forall i \in \{P\} \cup \Omega_{HAP}, \forall j \in \Omega_{HAP}, \forall k \in \cup \Omega_{HAP}, i \neq k \tag{25}$$

Equations (20–25) states that A, B, Z, Q, E, and F are binary variables.

3 Numerical Analysis

A. Validation of the OTD formulation

In the numerical studies, we hope to validate the mathematical formulation of OTD problem firstly. We conduct the simulations in small-scale scenarios where the global optimal solutions can be obtained by using exhaustive search. We compare the results obtained by Gurobi, a state-of-the-art ILP solver, with the optimal solutions and then examine whether the OTD formulation is correct or not. A serious of simulations in small-scale scenarios are carried out. Scenario 0 is only one of them and illustrated in Fig. 3, which includes 10 candidate nodes for HAP and 10 TPs We define a generic cost unit (gcu) to evaluate the network costs [7].

For the convenience of research, the RoI on the sea is divided into a set of rectangular grids with a uniform size according to the desired accuracy. Figure 3 also shows the zero-order HAP, the candidate points for HAP deployment and the TPs.

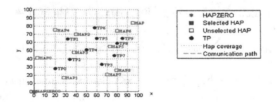

Fig. 3. Layout graph of scenario 0 before optimization

To improve the reliability of the network, the minimum requirement of network coverage is 2-coverage based on 1-coverage. We analyze and compare the deployment solutions under the requirement of 1-coverage and 2-coverage for scenario 0 as shown in Table 1.

Figure 4 shows the deployment solutions under the requirement of 1-coverage and 2-coverage for Scenario 0. The minimum cost deployment scheme is obtained and the

Table 1. The comparison of the solutions under 1-coverage and 2-coverage for Scenario 0

Solution	ID of selected HAPs	Cost (gcu)	Running Time(S)
1-coverage	HAP1, HAP2, HAP3, HAP5	200	0.05
2-coverage	HAP0, HAP1, HAP2, HAP3, HAP5, HAP6, HAP8	350	0.03

requirements of network coverage and coverage are satisfied. As shown in left part of Fig. 4, each HAP node can communicate with the zero-order HAP node, which ensure the network connectivity and the tree topology. Compare with the scheme in left part, as the number of coverage weights is raised, the scheme in right part of Fig. 4 selects 3 more HAP candidate nodes: HAP0, HAP6 and HAP8 to achieve 2-coverage of TPs in the network. The above OTD solution obtained by Gurobi are the same as the global optimal one obtained by exhaustive search. So we validate our proposed OTD formulation.

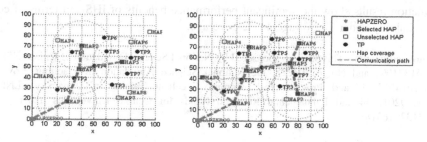

Fig. 4. The OTD solution for Scenario 0 (1-coverage and 2-coverage)

B. Feasibility and Scalability of OTD

Considering the practical network deployment, the network scale and the location of CPs and TPs, the values of key parameters of HAPs may affect the results of network deployment, so we conduct a series of simulations. Here we only select representative scenarios with an increasing network scale as listed in Table 2 to test the feasibility and scalability of the OTD model.

The simulation results illustrate that the OTD model can solve medium-scale and large-scale network deployment problem with both 1-coverage and 2-coverage. As shown in Table 2, the data values of 2-coverage deployment solutions are roughly double those of 1-coverage, and running time rises rapidly with the increase of net-work scale. In practical applications, if an island or offshore drilling platform which need stable communications exists in the target area, the 2-coverage scheme should be adopted to provide high-quality maritime wireless communication services. So we validate the feasibility and scalability of the OTD model.

Table 2. The comparison of deployment solutions in different scenarios

Scenario	The # of CPs for HAP, TP	The # of selected HAPs		Cost (gcu)		Running time (S)	
		1-coverage	2-coverage	1-coverage	2-coverage	1-coverage	2-coverage
1	20,40	7	12	350	600	0.75	0.53
2	30,60	7	15	350	750	1.59	1.93
3	40,80	9	17	450	850	41.39	17.64
4	50,100	10	21	500	1050	6753.61	3422.41
5	60,120	12	24	600	1200	31957.50	143913.89

4 Conclusion

In this paper, we present the IHSL Network architecture to meet the increasing demands of broad-band communication users at sea. And then we study the OTD problem which is important for the IHSL network architecture from theory to practical deployment. A series of case studies are conducted to validate the optimization framework and demonstrated the solvability and scalability of the ILP model. The simulation results show the significant performance benefits of IHSL in terms of cost and reliability under 1-coverage and 2-coverage.

Acknowledgments. This study is sponsored by National Science Foundation of China (NSFC) No. 61371091 and No. 61301228, the National Science Foundation of Liaoning Province No. 2014025001, and Program for Liaoning Excellent Talents in University (LNET) No. LJQ2013054 and Fundamental Research Funds for Central Universities under grant No. 3132016318.

References

1. Ejaz, W., Manzoor, K., Kim, H.J., et al.: Two-state routing protocol for maritime multi-hop wireless networks. Comput. Electr. Eng. **39**(6), 1854–1866 (2013)
2. Roste T, Yang K, Bekkadal F.: Coastal coverage for maritime broadband communications. In: Oceans 2013, pp. 1–8
3. Liu, F., Kong, X.: Research and implementation of a novel maritime wireless integrated network. Ship. Sci. Technol. **36**(12), 141–144 (2014)
4. Mohammed, A., Mehmood, A., Pavlidou, F.N., Mohorcic, M.: The Role of High-Altitude Platforms (HAPs) in the global wireless connectivity. Proc. IEEE **99**(11), 1939–1953 (2011)
5. Yang, Z., Mohammed, A.: Wireless communications from high altitude platforms: applications, deployment and development. In: 2010 12th IEEE International Conference on Communication Technology (ICCT), pp. 1476–1479. IEEE Xplore, Nanjing (2010)
6. Gurobi Optimizer 4.6, Gurobi Optimization Inc. (2012)
7. Alkandari, A., Alnasheet, M., Alabduljader, Y., Moein, S.M.: Water monitoring system using wireless sensor network (WSN): case study of Kuwait beaches. In: IEEE 2nd International Conference on Digital Information Processing and Communications, pp. 10–15. IEEE (2012)

On Adaptive Bitprobe Schemes for Storing Two Elements

Deepanjan Kesh[(⊠)]

Indian Institute of Technology Guwahati, Guwahati, India
deepkesh@iitg.ernet.in

Abstract. In this paper, we look into the problem of storing a subset $\mathcal{S}$ containing at most two elements of the universe $\mathcal{U}$ in the adaptive bitprobe model. Due to the work of Radhakrishnan *et al.* [3], and more recently of Lewenstein *et al.* [2], we have excellent schemes for storing such an $\mathcal{S}$, and answering membership queries using two or more bitprobes. Yet, Nicholson *et al.* [4] in their survey of the area noted that the space lower bound of even the first non-trivial scenario, namely that of answering membership of $\mathcal{S}$ using two bitprobes, is still open. Towards that end, we propose an unified geometric approach to designing schemes in this domain. If t is the number of bitprobes allowed, we arrange the universe $\mathcal{U}$ in a $(2t - 1)$-dimensional hypercube, and look at its two-dimensional faces. This approach matches the space bound of the best known schemes for certain cases, and gives results that are close to the best known schemes for others.

1 Introduction

Let $\mathcal{U}$ denote an universe of m elements. Consider a set $\mathcal{S}$ containing n elements of $\mathcal{U}$. The *bitprobe model* studies the following question – how efficiently can the subset $\mathcal{S}$ be stored such that the membership queries of the elements from the universe can be answered correctly, the constraint being that the number of bits of our datastructure that we are allowed to access is a fixed constant, say t. Each bit access is called a *bitprobe*, hence the name of this model. The bitprobe schemes in which the location of the current bitprobe is independent of the previous bitprobes are called *nonadaptive schemes*; otherwise, they are called *adaptive schemes*. We borrow the notation introduced by Radhakrishnan *et al.* [1], and denote the space used by adaptive and nonadaptive schemes by $s_A(n, m, t)$ and $s_N(n, m, t)$, respectively.

For a more detailed discussion about the bitprobe model, and about several other related models which are collectively referred to as the *membership problem*, the reader is adviced to read the excellent exposition by Nicholson *et al.* [4] where they survey the area and its current state of the art.

In this paper, we address the problem of designing adaptive schemes in the bitprobe model when the size of the subset $\mathcal{S}$ is two, i.e. $n = 2$. The number of bitprobes, denoted by t, though could be arbitrary, is a constant. The problem when $n = 1$ is more or less well understood – the space lower bound

© Springer International Publishing AG 2017
X. Gao et al. (Eds.): COCOA 2017, Part I, LNCS 10627, pp. 471–479, 2017.
https://doi.org/10.1007/978-3-319-71150-8_39

is $s_A(1, m, t) = \Omega(m^{1/t})$ (Alon and Feige [6]), and there is a folklore explicit scheme that matches the lower bound. So, as it stands, $n = 2$ is the first non-trivial scenario. For the case of $t = 1$, Buhrman et al. [5] showed that no space saving is possible, i.e. $s_A(2, m, 1) = \Omega(m)$. We study the scenario of $t \geq 2$.

For two bitprobes, Radhakrishnan et al. [3] gave an explicit scheme that uses $\mathcal{O}(m^{2/3})$ space. It uses a clever way of arranging the elements of $\mathcal{U}$ into, what the authors referred to as, blocks and superblocks. The first major result for $t = 3$ was due to Radhakrishnan et al. [1] where they proposed a non-explicit scheme taking up $\Theta(m^{2/5})$ amount of space. They further conjectured the existence of an explicit scheme matching their space bound. Lewenstein et al. [2], recently, resolved this long-standing conjecture by proposing such an explicit deterministic scheme for the problem. In the same paper, the authors also proposed a generalised scheme for $t > 3$ and showed that $s_A(2, m, t) = \mathcal{O}(m^{1/(t-2^{2-t})})$.

We present a novel approach for designing schemes in the adaptive bitprobe model. The idea is to choose a hypercube of suitable dimensions, and arrange the elements of the universe $\mathcal{U}$ within and on the integral points of the hypercube. The dimension of the hypercube depends on the number of bitprobes t. We then look at the projection of those integral points on certain faces of the hypercube, the dimension of the face will depend on n, the size of our subset S. To take a concrete example, in our scenario where $n = 2$ we look at two-dimensional faces of our hypercube. The dimension of the hypercube is 3 when $t = 2$, 5 when $t = 3$, and more generally $2t - 1$ for any given t.

Though the results we present in itself might not be interesting, our contribution is an unified approach to designing explicit deterministic schemes in this domain. Our approach gives the following bounds – $s_A(2, m, 2) = \mathcal{O}(m^{2/3})$; $s_A(2, m, 3) = \mathcal{O}(m^{2/5})$; and $s_A(2, m, t) = \mathcal{O}(m^{1/(t-2^{-1})})$, for $t > 3$. As we can see, it matches the best known schemes for $t = 2$ and 3. More importantly, it gives an alternate scheme for $t = 3$ by answering the long standing conjecture of Radhakrishnan et al. [1] about the existence of an $\mathcal{O}(m^{2/5})$ scheme. However, it falls short for higher values of t. We believe that this approach is interesting because of the geometric interpretation it lends to the various schemes for different values of t and a discovers a nice interdependence between t and n, the dimension of our hypercube and the dimension of the face, respectively.

For lack of space, we only present the schemes for $t \geq 3$. For the much simpler case of $t = 2$ and for the proof of correctness of all the schemes, the reader is referred to [7].

2   The Three Probe Adaptive Scheme

In this section, we describe a new adaptive scheme that stores two elements of the universe and answers membership queries in three bitprobes. In this scheme, we consider projections of points of a five dimensional hypercube onto two-dimensional faces.

Consider the five-dimensional space with axes $v, w, x, y,$ and z. We emphasize the relationship between the coordinate system and the coordinates (a, b, c, d, e) of a point – a as its v-coordinate, b as its w-coordinate, c as its x-coordinate, and so on. In that space, consider the five-dimensional hypercube in the positive orthant with one of its vertices at the origin and each of its sides having magnitude $m^{1/5}$. There are m points with integer coordinates if we consider all the points within and on the hypercube. We arrange the m elements of $\mathcal{U}$ on those points. We would refer to any element by the coordinates of the point on which it lies.

We introduce the sets $(V, W)_{(a,b)}, (X, Y)_{(a,b)}, (X, Z)_{(a,b)}, (V, Z)_{(a,b)},$ $(W, Z)_{(a,b)}, (V, Y)_{(a,b)},$ and $(Y, Z)_{(a,b)},$ and their corresponding families $(\mathcal{V}, \mathcal{W}), (\mathcal{X}, \mathcal{Y}), (\mathcal{X}, \mathcal{Z}), (\mathcal{V}, \mathcal{Z}), (\mathcal{W}, \mathcal{Z}), (\mathcal{V}, \mathcal{Y}),$ and $(\mathcal{Y}, \mathcal{Z})$. For the sake of brevity and to avoid needless repetition, we provide the definition for only one pair of coordinates.

$$(V, W)_{(a,b)} = \Big\{ (p, q, r, s, t) \mid p = a \text{ and } q = b,\ 0 \le r, s, t < m^{1/5} \Big\},$$

where $0 \le a, b < m^{1/5}$. Moreover,

$$(\mathcal{V}, \mathcal{W}) = \Big\{ (V, W)_{(a,b)} \mid 0 \le a, b < m^{1/5} \Big\}.$$

Observation 1. $|(V, W)_{(a,b)}| = m^{3/5}$ and $|(\mathcal{V}, \mathcal{W})| = m^{2/5}$.

2.1   Our Scheme

For this three probe scheme, we will have seven tables corresponding to the seven families mentioned in the previous section. They also form the internal nodes of the decision tree of the scheme (Fig. 1). The tables will have one bit for each set in the corresponding families. We will abuse the notations and use $(\mathcal{V}, \mathcal{W})$ for the family name as well as for its corresponding table in the datastructure. Moreover, we will use $(V, W)_{(a,b)}$ to denote a member of the family $(\mathcal{V}, \mathcal{W})$ as well as one bit in the table $(\mathcal{V}, \mathcal{W})$.

Lemma 2. *The size of our datastructure is* $7m^{2/5}$.

The design of the query scheme is depicted in the decision tree T_3 in Fig. 1. Each internal node of the tree is a table in our datastructure, as indicated by the node labels. The children of a node tell us which table to query next, or in the case of the second last level, what is the output of the query scheme. We start the query scheme by making our first query into the table at the root of the decision tree. If the bit returned upon the query is 0, we move to the left child, otherwise we move to the right child. We repeat the process at the next node, and so on until we reach a leaf of the tree where we get our Yes or No answer for our query scheme.

The query locations for any element will depend on the coordinates of the point on which the element is placed. To take an example, the query location for

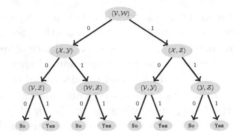

Fig. 1. T_3 : The decision tree of an element for three probe adaptive scheme.

the element (a, b, c, d, e) in table $(\mathcal{V}, \mathcal{W})$ will be $(V, W)_{(a,b)}$, and in table $(\mathcal{X}, \mathcal{Y})$ it will be $(X, Y)_{(c,d)}$, which are nothing but the projections of the point on the vw- and xy-planes, respectively.

Given a subset $\mathcal{S} = \{(a_1, b_1, c_1, d_1, e_1), (a_2, b_2, c_2, d_2, e_2)\}$ of $\mathcal{U}$, the storage scheme tells us how to set the bits of our datastructure such that membership queries can be answered correctly. As before, the scheme varies according to the elements we want to store, i.e. how the members of $\mathcal{S}$ are chosen from $\mathcal{U}$. We discuss each case separately.

Case I – Let us assume that $a_1 \neq a_2$. In table $(\mathcal{V}, \mathcal{W})$, we set the bit $(V, W)_{(a_1,b_1)}$ to 0 and the bit $(V, W)_{(a_2,b_2)}$ to 1. For the rest of the bits, we set all those bits $(V, W)_{(a,b)}$ such that $a = a_2$ to 0. The remaining bits are set to 1. In the second level of the tree, the table $(\mathcal{X}, \mathcal{Y})$ has $(X, Y)_{(c_1,d_1)}$ set to 0 and the rest of the bits set to 1. In table $(\mathcal{X}, \mathcal{Z})$, we set the bit $(X, Z)_{(c_2,e_2)}$ to 0 and the rest of the bits to 1. Now, we set the bits at third level. We set the bits $(V, Z)_{(a_1,e_1)}$ in table $(\mathcal{V}, \mathcal{Z})$ and $(V, Y)_{(a_2,d_2)}$ in table $(\mathcal{V}, \mathcal{Y})$ to 1. The rest of the bits in all of the tables are set to 0.

Case II – Let us consider the case when $a_1 = a_2 = a'$ (say), but $b_1 \neq b_2$. The members of $\mathcal{S}$ now are (a', b_1, c_1, d_1, e_1) and (a', b_2, c_2, d_2, e_2). We set the bit $(V, W)_{(a',b_1)}$ to 0 and $(V, W)_{(a',b_2)}$ to 1 in table $(\mathcal{V}, \mathcal{W})$. The rest of the bits of the table are assigned as follows. For the bits $(V, W)_{(a,b)}$ such that $a = a'$, we set them to 0. We set the rest of the bits to 1. Coming to the second level, the bit $(X, Y)_{(c_1,d_1)}$ in table $(\mathcal{X}, \mathcal{Y})$ is set to 1 and the others to 0. In table $(\mathcal{X}, \mathcal{Z})$, the bit $(X, Z)_{(c_2,e_2)}$ is set to 0 and the rest to 1. In the third level, we set the bits $(W, Z)_{(b_1,e_1)}$ in $(\mathcal{W}, \mathcal{Z})$ and $(V, Y)_{(a',d_2)}$ in $(\mathcal{V}, \mathcal{Y})$ to 1, and the rest of the bits in all the tables to 0.

Case III – We now consider the scenario where $a_1 = a_2 = a'$ (say), $b_1 = b_2 = b'$ (say), and $e_1 \neq e_2$. The members of $\mathcal{S}$ now look like (a', b', c_1, d_1, e_1) and (a', b', c_2, d_2, e_2). We set the bit $(V, W)_{(a',b')}$ in table $(\mathcal{V}, \mathcal{W})$ to 1, and the rest of its bits to 0. In the second level, the bits $(X, Z)_{(c_1,e_1)}$ and $(X, Z)_{(c_2,e_2)}$ in table $(\mathcal{X}, \mathcal{Z})$ are set to 1, and the rest of bits of all the tables in the level to 0. In the third level, all the bits in tables $(\mathcal{V}, \mathcal{Z}), (\mathcal{W}, \mathcal{Z})$, and $(\mathcal{V}, \mathcal{Y})$ are set to 0. In table $(\mathcal{Y}, \mathcal{Z})$, the bits $(Y, Z)_{(d_1,e_1)}$ and $(Y, Z)_{(d_2,e_2)}$ are set to 1, and the rest to 0.

Case IV – The final case to consider is $a_1 = a_2 = a'$ (say), $b_1 = b_2 = b'$ (say), and $e_1 = e_2 = e'$ (say), and hence the members of $\mathcal{S}$ are (a', b', c_1, d_1, e') and (a', b', c_2, d_2, e'). In table $(\mathcal{V}, \mathcal{W})$, the bit $(V, W)_{(a', b')}$ is set to 0, and the rest of the bits are set to 1. In the second level, the bits $(X, Y)_{(c_1, d_1)}$ and $(X, Y)_{(c_2, d_2)}$ in the table $(\mathcal{X}, \mathcal{Y})$ are set to 0, and every other bit to 1. In the last level, only the bit $(V, Z)_{(a', e')}$ in table $(\mathcal{V}, \mathcal{Z})$ is set to 1.

3 Adaptive Scheme for Four Probes or More

In this section, we present an adaptive scheme that stores subsets $\mathcal{S}$ of size at most two and answers membership queries using t bitprobes, where $t \geq 4$. Our scheme is a generalisation of the approaches for the two and three bitprobe schemes. For a scheme involving t bitprobes, we look at the two-dimensional faces of a $(2t - 1)$-dimensional hypercube. It is a simple recursive scheme that uses the schemes for $t - 2$ bitprobes as a subroutine. The base cases of $t = 2$ and $t = 3$ is already known, and we now present schemes for higher values of t.

Consider the $(2t - 1)$-dimensional space with coordinates $x_1, x_2, \ldots, x_{2t-1}$. In the first orthant of that space, consider a $(2t-1)$-dimensional hypercube with one of its vertices at the origin and each of its sides having magnitude $m^{1/(2t-1)}$. The total number of points within and on the cube with integral coordinates is m. We arrange the elements of our universe $\mathcal{U}$ on those points. As before, the identity of an element will be the coordinates of the point on which it is placed, e.g. $(a_1, a_2, \ldots, a_{2t-1})$.

We will stick to the nomenclature introduced in the previous section. For any two coordinates x_i and x_j, we define the set $(X_i, X_j)_{(c,d)}$ as the collection of all points whose projection on the $x_i x_j$-plane is (c, d).

$$(X_i, X_j)_{(c,d)} = \left\{ (a_1, a_2, \ldots, a_{2t-1}) \mid a_i = c, a_j = d; \ \forall k, 0 \leq a_k < m^{1/(2t-1)} \right\},$$

where $0 \leq c, d < m^{1/(2t-1)}$. We also define the corresponding family of sets $(\mathcal{X}_i, \mathcal{X}_j)$.

$$(\mathcal{X}_i, \mathcal{X}_j) = \left\{ (X_i, X_j)_{(c,d)} \mid 0 \leq c, d < m^{1/(2t-1)} \right\}$$

Observation 3. $|(X_i, X_j)_{(c,d)}| = m^{(2t-3)/(2t-1)}$ and $|(\mathcal{X}_i, \mathcal{X}_j)| = m^{2/(2t-1)}$.

3.1 Our Scheme

The decision tree for our scheme is shown in Fig. 2. It is a complete binary tree with $t + 1$ levels, of which the first t levels are shown. Similar to the earlier decision tree T_3, the last level has Yes and No nodes. For want of space, only those nodes of the tree which are necessary for our discussion have been drawn. The big triangle marked T_{t-2} denotes the decision tree for $t - 2$ bitprobes with the root as $(\mathcal{X}_5, \mathcal{X}_6)$ and the coordinates involved being $x_5, x_6, \ldots, x_{2t-1}$.

As before, there is one table in our datastructure for each of the nodes in the decision tree which, of course, are the various set families. We have one bit in the

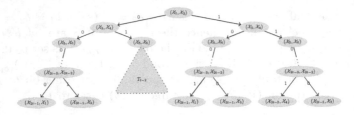

Fig. 2. T_t : The decision tree of an element for three probe adaptive scheme. The **Yes** and the **No** nodes are not shown to save space.

table for every member of the corresponding family, and as there are $m^{2/(2t-1)}$ sets in each family (Observation 3), the size of each table is $m^{2/(2t-1)}$. Combined with the fact that our decision tree has $2^t - 1$ nodes and we need one table for each node, the following lemma follows.

Lemma 4. *The size of our datastructure is* $(2^t - 1)m^{2/(2t-1)}$.

To verify whether an element $(a_1, a_2, \ldots, a_{2t-1})$ is in $\mathcal{S}$, all we need to do is to follow down a path in the tree T_t (Fig. 2). The label of a node in the tree tells us two things – on which two-dimensional plane we need to project the element, and in which table we need query in the current step. To take an example, T_t tells us that for the first query we need to project the element on the $x_1 x_2$-plane. The projection of $(a_1, a_2, \ldots, a_{2t-1})$ on the plane is (a_1, a_2). It also tells us to look into the table $(\mathcal{X}_1, \mathcal{X}_2)$ at the location $(X_1, X_2)_{(a_1, a_2)}$. The next query will be in the left child or the right child depending on whether the bit stored at that location is 0 or 1, respectively. If and only if in the last node our query gives us a 1, we deduce that the element is a member of $\mathcal{S}$.

As before, how the bits are set in our datastructure depends on how the members of $\mathcal{S}$ are chosen from $\mathcal{U}$. Let the elements we want to store be $(a_1, a_2, \ldots, a_{2t-1})$ and $(b_1, b_2, \ldots, b_{2t-1})$. The intuition behind the scheme is the following – if the first four coordinates of the two elements are equal to each other, we use the recursive part of the decision tree to store the elements, the part of the tree marked T_{t-2} which, as we have mentioned earlier, denotes the decision tree for the scheme with $t - 2$ probes; otherwise, we would use the ideas from the first two cases of the storage scheme for three probes (Sect. 2.1).

Case I – Let us assume that $a_1 \neq b_1$. The tables where we are going to store the two elements and the path one has to take to get a **Yes** are shown in Fig. 3.

Case II – Now, let us assume that $a_1 = b_1$ and $a_2 \neq b_2$. The tables where we are going to store the two elements and the path one has to take to get a **Yes** are shown in Fig. 4.

Cases III and IV – We now consider the two scenarios where $a_1 = b_1 = c_1(\text{say}), a_2 = b_2 = c_2(\text{say}), a_3 \neq b_3$ and where $a_1 = b_1, a_2 = b_2, a_3 = b_3, a_4 \neq b_4$. The solution in these cases is similar to the Cases I and II discussed above.

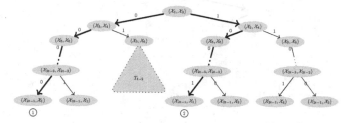

Fig. 3. $a_1 \neq b_1$.

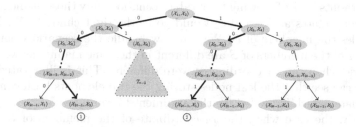

Fig. 4. $a_1 = b_1$ and $a_2 \neq b_2$.

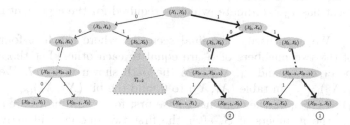

Fig. 5. $a_1 = b_1, a_2 = b_2$ and $a_3 \neq b_3$.

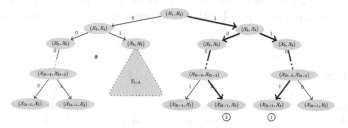

Fig. 6. $a_1 = b_1, a_2 = b_2, a_3 = b_3$ and $a_4 \neq b_4$.

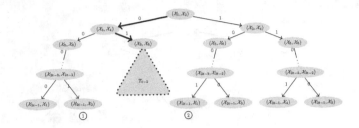

Fig. 7. $a_1 = b_1, a_2 = b_2, a_3 = b_3$ and $a_4 = b_4$.

We set the bit $(X_1, X_2)_{(c_1, c_2)}$ in table $(\mathcal{X}_1, \mathcal{X}_2)$ to 1 and the rest of the bits to 0. This results in the following favourable scenario – only those elements whose first two coordinates are c_1 and c_2 would go to the right child of $(\mathcal{X}_1, \mathcal{X}_2)$, and this includes the members of $\mathcal{S}$. We can now handle the scenario when the x_3 coordinates of the members of $\mathcal{S}$ are different in the same manner as we handled the scenario when the x_1 coordinates were different. This is illustrated by the Fig. 5. We can see that the leaf nodes storing the two elements both contain the x_3 coordinate, which is needed for the argument to work.

Similarly, the case when the x_3 coordinate of the members of $\mathcal{S}$ are also equal but the x_4 coordinates are different can be handled using the same logic which we used to handle the scenario when x_1 coordinates were equal and the x_2 coordinates were different. Figure 6 illustrates the situation. The leaf node storing the first element has x_4 coordinate, whereas the leaf node storing the second element has x_3 coordinate, which is required for the argument to work.

Case V – We now discuss the final scenario, where the first four coordinates of the two members of $\mathcal{S}$ are equal to each other. Let those coordinates be $c_1, c_2, c_3,$ and c_4. We take the path shown in Fig. 7. We set the bit $(X_1, X_2)_{(c_1, c_2)}$ in table $(\mathcal{X}_1, \mathcal{X}_2)$ to 0, and the bit $(X_3, X_4)_{(c_3, c_4)}$ in table $(\mathcal{X}_3, \mathcal{X}_4)$ to 1. So, all the elements whose first four coordinates are as above, including the members of $\mathcal{S}$, after the first two queries will arrive at the subtree T_{t-2} with root $(\mathcal{X}_5, \mathcal{X}_6)$. The elements that will be mapped to this subtree will all lie in a $(2t - 5)$-dimensional hypercube, and among these elements we have to store a subset of size two. The number of bitprobes left is $t - 2$. So, we can safely use the scheme for $t - 2$ probes, and the decision tree for that scheme should be able to store and answer membership queries correctly.

References

1. Radhakrishnan, J., Shah, S., Shannigrahi, S.: Data structures for storing small sets in the bitprobe model. In: de Berg, M., Meyer, U. (eds.) ESA 2010. LNCS, vol. 6347, pp. 159–170. Springer, Heidelberg (2010). https://doi.org/10.1007/978-3-642-15781-3_14

2. Lewenstein, M., Ian Munro, J., Nicholson, P.K., Raman, V.: Improved explicit data structures in the bitprobe model. In: Schulz, A.S., Wagner, D. (eds.) ESA 2014. LNCS, vol. 8737, pp. 630–641. Springer, Heidelberg (2014). https://doi.org/10.1007/978-3-662-44777-2_52

3. Radhakrishnan, J., Raman, V., Srinivasa Rao, S.: Explicit deterministic constructions for membership in the bitprobe model. In: Heide, F.M. (ed.) ESA 2001. LNCS, vol. 2161, pp. 290–299. Springer, Heidelberg (2001). https://doi.org/10.1007/3-540-44676-1_24

4. Nicholson, P.K., Raman, V., Rao, S.S.: A survey of data structures in the bitprobe model. In: Brodnik, A., López-Ortiz, A., Raman, V., Viola, A. (eds.) Space-Efficient Data Structures, Streams, and Algorithms. LNCS, vol. 8066, pp. 303–318. Springer, Heidelberg (2013). https://doi.org/10.1007/978-3-642-40273-9_19

5. Buhrman, H., Miltersen, P.B., Radhakrishnan, J., Venkatesh, S.: Are bitvectors optimal. In: Proceedings of the Thirty-Second Annual ACM Symposium on Theory of Computing, Portland, pp. 449–458, 21–23 May 2000

6. Alon, N., Feige, U.: On the power of two, three and four probes. In: Proceedings of the Twentieth Annual ACM-SIAM Symposium on Discrete Algorithms, SODA 2009, New York, pp. 346–354, 4–6 January 2009

7. https://www.iitg.ernet.in/deepkesh/main.pdf

Author Index

Printed in the United States
By Bookmasters